ALLE·ZEIT·WACH 1842

Friedrich Liebau

Structural Chemistry of Silicates

Structure, Bonding, and Classification

With 136 Figures

Springer-Verlag
Berlin Heidelberg New York Tokyo

Professor Dr. FRIEDRICH LIEBAU
Mineralogisch-Petrographisches Institut
Universität Kiel
Olshausenstraße 40–60
2300 Kiel

Cover illustration:
Mutual adjustment between tetrahedral single chains (*black*) of different periodicities and slabs of edge-shared cation – oxygen octahedra in pyroxenes and pyroxenoids.
See Fig. 10.3, page 173.

ISBN 3-540-13747-5 Springer-Verlag Berlin Heidelberg New York Tokyo
ISBN 0-387-13747-5 Springer-Verlag New York Heidelberg Berlin Tokyo

Library of Congress Cataloging in Publication Data. Liebau, F. (Friedrich), 1926– . Structural chemistry of silicates. Bibliography: p. . Includes index. 1. Silicates. I. Title. QD181.S6L614 1985 546′.68324 84-23532

Typesetting, Printing and Bookbinding: Graphischer Betrieb Konrad Triltsch, Würzburg
2132/3130-543210

Preface

As natural minerals, silica and silicates constitute by far the largest part of the earth's crust and mantle. They are equally important as raw materials and as mass produced items. For this reason they have been the subject of scientific research by geoscientists as well as by applied scientists in cement, ceramic, glass, and other industries. Moreover, intensive fundamental research on silicates has been carried out for many years because silicates are, due to their enormous variability, ideally suited for the study of general chemical and crystallographic principles.

Several excellent books on mineralogy and cement, ceramics, glass, etc. give brief, usually descriptive synopses of the structure of silicates, but do not contain detailed discussions of their structural chemistry. A number of monographs on special groups of silicates, such as the micas and clay minerals, amphiboles, feldspars, and zeolites have been published which contain more crystal chemical information. However, no modern text has been published which is devoted to the structural chemistry of silicates as a whole.

Within the last 2 decades experimental and theoretical methods have been so much improved to the extent that not only have a large number of silicate structures been accurately determined, but also a better understanding has been obtained of the correlation between the chemical composition of a silicate and its structure. Therefore, the time has been reached when a modern review of the structural chemistry of silicates has become necessary.

The primary purpose of this book is to demonstrate the exceptional variability of silicate structures and to describe the correlations between chemical composition and structure. This is facilitated by applying a classification scheme which is an extension of the classical Bragg–Náray-Szabó classification and which is entirely established on crystallochemical parameters.

To aspire to this goal a considerable number of silicate structures must be discussed together with the chemical formulae of an even larger number of silicates, natural as well as synthetic. However, no attempt has been made to give a full description of the structures or to aim at completeness. Although this book has not been intended as a reference work, many references to the original literature are included to assist in more detailed studies. In general, reference has been made to the most recent work only in which references to earlier work can be found. To keep the book to a reasonable size much material is presented in tabular form.

Most of structure diagrams, such as Figs. 6.1, 7.9, and 10.16, are drawn to a scale of approx. 1 Å = 1.90 mm, several idealized structure diagrams, such as Figs. 4.4 and 4.8, are drawn to a scale of approx. 1 Å = 2.02 mm. Other structure diagrams have an arbitrary scale. Figures 9.3, 5, 10.34, and 36 have been prepared with the program SCHAKAL which has kindly been provided by Dr. E. Keller, Freiburg, Germany.

Although I have taken great care to avoid errors and inconsistencies, some errors will have remained undiscovered during the treatment of such a large amount of material. I ask for and will be grateful to be informed of any omissions or errors of fact.

I am deeply grateful to my sister Elisabeth who, by her manual labor during the hard times after World War II, made it possible for me to study science. I also owe particular debt of gratitude to the late Professor Erich Thilo for having introduced me to the field of silicate science.

The book developed over more than a decade, in particular while I was a visiting professor in the Department of Geological Science at Virginia Polytechnic Institute and State University in Blacksburg, Virginia, U.S.A., at the Earth and Space Science Department of the State University of New York in Stony Brook, New York, U.S.A., and at the Chemistry Department of the University of Aberdeen, Old Aberdeen, U.K. I thank Professors F. D. Bloss, C. T. Prewitt, and H. F. W. Taylor for their kind invitations and for the assistance they and many of their colleagues, in particular Professors G. V. Gibbs, P. H. Ribbe, R. C. Liebermann, L. S. Dent Glasser, and F. P. Glasser have given and for their warm hospitality. Particular thanks are due to my colleagues in Kiel, Professors H. Küppers, F. Seifert, P. K. Hörmann, and H. Kern and to Professors W. Eysel, Heidelberg, and O. Jarchow, Hamburg, who took over part of my duties during my absence. I would not have been able to finish the manuscript without the opportunity to spend 5 months at the Max-Planck-Institut für Festkörperforschung in Stuttgart away from a desk full of everyday work and distraction. I am grateful to Professor A. Rabenau and his colleagues for their hospitality and help.

By the time I had finished this book, the list of individuals who provided reprints, preprints, and data had become so long that it is not possible to name them all. The same is true for many colleagues as well as students who gave generously of their time for invaluable discussions. Their contributions are gratefully acknowledged.

I particularly thank Professor A. Pabst, Berkeley, California, U.S.A. and Dr. R. J. Hill, Port Melbourne, Victoria, Australia for critically reading an early draft and the last version of the manuscript, respectively. Their comments and suggestions considerably improved the text.

I also thank Dr. M. Czank, Kiel, Germany and Professor D. R. Veblen, Baltimore, Maryland, U.S.A., and K. Yada, Sendai, Japan, for supplying electronmicrographs for the Figs. 7.14, 7.17, and 10.26.

I am deeply grateful to Mrs. E. Richert, who skillfully and with great care and patience, prepared most of the figures. In particular I wish to express my sincere gratitude to my wife Waltrude who typed most of the several versions of the manuscript and who saved me from error and inconsistency by checking and cross-checking formulae and references. For all remaining errors and deficiencies I alone am responsible. Finally, I thank my wife and my children for their patience, understanding, and encouragement during the preparation of the book.

The scientific work of the author, which eventually led him to write this book, has been supported substantially by the Deutsche Forschungsgemeinschaft. Without a special grant from the Bundesministerium für Forschung und Technologie for the preparation of the book, I would never have found the time and rest to write.

Kiel, Spring 1985 FRIEDRICH LIEBAU

Contents

1 Introduction

1.1 Occurrence and Abundance of Silicon 1
1.2 Technical Importance of Silicon 5
1.3 Diversity of Silicon Compounds 5

2 Methods to Describe the Atomic Structure of Silicates

2.1 Model Representations of Silicate Structures 7
2.2 Treatment of Tetrahedrally Coordinated Cations in Silicates 12

3 Chemical Bonds in Silicates

3.1 The Silicon – Oxygen Bond 14

3.1.1 Coordination Numbers of Silicon 14
3.1.2 Bond Lengths and Bond Angles in Silicates Containing $[SiO_4]$ Tetrahedra (Tetraoxosilicates) 15

3.1.2.1 Correlations Between Si–O Distance, Oxygen Coordination Number, and Bond Strength 16
3.1.2.2 Correlations Between Si–O Distance and Bond Angles . . 20
3.1.2.3 Correlations Between Si–O Distance, Oxygen Coordination Number, Bond Strength, and Si–O–Si Angle 21
3.1.2.4 Influence of Structural Disorder on Si–O Distances . . . 22
3.1.2.5 Si–O–Si Angles 24

3.1.3 Bond Lengths and Bond Angles in Silicates Containing $[SiO_6]$ Octahedra (Hexaoxosilicates) 30
3.1.4 The Nature of the Si–O Bond 32

3.1.4.1 The Ionic Model 33
3.1.4.2 The Covalent Model 34
3.1.4.3 Nonbonded Interactions 44
3.1.4.4 The Ionicity of the Si–O Bond 46

3.2 The Cation – Oxygen Bond, M–O, and Its Influence on the Si–O Bond 48

3.2.1 Electropositive Cations 48
3.2.2 Electronegative Cations 51

4 **Crystal Chemical Classification of Silicate Anions**

4.1 General Principles of the Classification 52
4.2 Parameters Used in the Crystal Chemical Classification of Silicate Anions 52
4.3 The Crystal Chemical System of Silicate Anions 63
4.4 Periodic Character of the Crystal Chemical Classification of Silicate Anions 67

5 **Nomenclature and Structural Formulae of Silicate Anions and Silicates**

5.1 Nomenclature 69
5.1.1 Chemical Nomenclature 69
5.1.2 Mineralogical Nomenclature 71
5.2 Structural Formulae 72

6 **Crystal Chemical Classification of Silicates: General Part**

6.1 Mixed-Anion Silicates 76
6.2 The Crystal Chemical System of Silicates 76
6.3 Further Subdivision of Silicates 78
6.3.1 Atomic Ratio Si : O of Silicate Anions 78
6.3.2 Degree of Chain Stretching 80
6.3.3 Silicate Anion Symmetry 82
6.3.4 Cation – Oxygen Polyhedra 82
6.4 Silicate Classification Procedure 82
6.5 Shortcomings of the Crystal Chemical Classification . . . 85

7 **Crystal Chemical Classification of Silicates: Special Part**

7.1 Silicon Compounds with $[SiO_6]$ Octahedra (Hexaoxosilicates) 90
7.2 Silicates Containing $[SiO_4]$ Tetrahedra (Tetraoxosilicates) 93
7.2.1 Silicates with Edge-Sharing Tetrahedra 93
7.2.2 Silicates with Corner-Sharing Tetrahedra 93
7.2.2.1 Oligosilicates (Sorosilicates) 93
7.2.2.2 Ring Silicates (Cyclosilicates) 96
7.2.2.3 Single Chain Silicates (Monopolysilicates) 102
7.2.2.4 Multiple Chain Silicates (Oligopolysilicates) 108
7.2.2.5 Single Layer Silicates (Monophyllosilicates) 113
7.2.2.6 Double Layer Silicates (Diphyllosilicates) 121
7.2.2.7 Tectosilicates 126
7.2.2.8 Silicates with Interpenetrating Anions 128
7.3 Mixed-Anion Silicates 129
7.4 Estimated Frequency Distribution of Silicate Species . . 130

8 **Other Classifications of Silicates**

8.1 Early Classifications of Silicates 136
8.2 Kostov's Classification of Silicates 137
8.3 Zoltai's Classification of Silicates 138
8.4 Geometrical Classification of Tectosilicates 143

8.4.1 Connectedness 144
8.4.2 Secondary Building Units 145
8.4.3 Chain-Like Building Units 146
8.4.4 Layer-Like Building Units 151
8.4.5 Polyhedral Building Units; Framework Density 155
8.4.6 Concluding Remarks 159

8.5 Silicate Classification Based on Non-Silicon Cation – Oxygen Polyhedra 159

9 **General Rules for Silicate Anion Topology** 161

10 **Influence of Non-Tetrahedral Cation Properties on the Structure of Silicate Anions**

10.1 Influence of Cation Properties on the Conformation of Unbranched Single Chain Anions 170

10.1.1 Conformation of Unbranched Silicate Single Chains . . . 170
10.1.2 Qualitative Correlations Between Chain Conformation and Cation Properties 171
10.1.3 Semiquantitative Correlations Between Chain Conformation and Cation Properties 176
10.1.4 Crystallochemical Interpretation of the Correlations . . . 178
10.1.4.1 Correlation Between Stretching Factor and Average Cation Electronegativity 179
10.1.4.2 Correlation Between Stretching Factor and Average Cation Valence 182
10.1.4.3 Correlation Between Stretching Factor and Average Cation Radius 182
10.1.4.4 Correlation Between Stretching Factor and Chain Periodicity 183

10.2 Influence of Cation Properties on the Shape of Unbranched Multiple Chain Anions 184
10.3 Influence of Cation Properties on the Shape of Branched Silicate Anions 187
10.4 Influence of Cation Properties on the Formation of Cyclic Silicate Anions 191
10.5 Influence of Cation Properties on the Shape of Single Layer Silicate Anions 195

10.5.1 Anhydrous Single Layer Silicates 198
10.5.2 The Shape of Silicate Anions in Anhydrous Silica-Rich Barium Silicates 206
10.5.3 Hydrous Single Layer Silicates 212
10.5.3.1 Kaolinite-Like and Mica-Like Arrangements 213
10.5.3.2 Strain Reduction Mechanisms in Cation-Poor Phyllosilicates 214
10.5.3.3 Strain Reduction Mechanisms in Cation-Rich Phyllosilicates 223
10.5.4 Comparison of Anhydrous and Hydrous Single Layer Silicates 229

10.6 Influence of Cation Properties on the Shape of Double Layer Silicate Anions 231
10.7 Influence of Cation Properties on the Shape of Silicate Frameworks 239

10.7.1 Silica and Clathrasils 240
10.7.2 Zeolites 244
10.7.3 Aluminosilicates $M[(Al, Si)_4O_8]$: Feldspars and Their Polymorphs 246
10.7.3.1 $M[AlSi_3O_8]$ Phases with M = Li, Na, K, Rb, Cs 246
10.7.3.2 $M[Al_2Si_2O_8]$ Phases with M = Mg, Ca, Sr, Ba 252
10.7.4 Aluminosilicates $M[AlSi_2O_6]$ 253
10.7.5 Aluminosilicates $M[AlSiO_4]$ 255
10.7.6 Comparison of Framework Silicates 260

11 Conclusion

11.1 Correlation Between Classification Parameters and Cation Properties 266
11.2 Influence of Cation Radius on the Structure of Silicates . 274
11.3 Characteristics of the Crystal Chemical Classification of Silicates 277

References 279

Appendices I–III 304

Subject Index 319

Substance Index 325

Formula Index 330

List of Symbols

B	Temperature factor determined from X-ray diffraction data
B_{iso}	Isotropic temperature factor
br	Branched
B	Branchedness
r	Cyclic
CN	Coordination number of silicon
CN (X)	Coordination number of atom X
d	Distance between atoms
d_{ind}	Distance between individual atoms
$d^{rö}$	Distance determined from X-ray diffraction data
d^{real}	Real bond length
d_f	Framework density
D	Dimensionality
$\mathbb{D}$	Directedness "down"
Δ	Directedness
f_s	Stretching factor
f_{sh}	sharing coefficient
F_{tetr}	Area per tetrahedron in silicate layers
hB	Hybrid
L	Linkedness
lB	Loop-branched
M	Multiplicity
N_{an}	Number of anion types
oB	Open-branched
O_{br}	Bridging oxygen atom
olB	Mixed-branched
O_{term}	Terminal oxygen atom
P	Pressure
P	Chain periodicity
$\boldsymbol{P^r}$	Ring periodicity
p_X	Sum of electrostatic bond strengths received by an anion X
Q^s	Tetrahedron of connectedness ***s***
r	Radius
r_{nb}	Nonbonded radius
s	Electrostatic bond strength
s	Connectedness
t	Terminated
T	Temperature
T	Tetrahedrally coordinated cation
$\mathbb{U}$	Directedness "up"
uB	Unbranched
v	Valence
V_{ox}	Volume per oxygen atom
χ	Electronegativity
z	Formal charge

1 Introduction

This book deals with inorganic compounds of silicon in which Si atoms are coordinated by oxygen atoms, i.e., silicon dioxide (silica) is considered, along with the compounds in which silicon–oxygen groups constitute the anionic part of the phase (silicates) and compounds, such as SiP_2O_7 in which silicon is considered to be the cationic part. Since all these compounds follow the same structural chemical rules they are summarized under the general term silicate in this book.

1.1 Occurrence and Abundance of Silicon

Silicon is one of the most abundant and widespread elements in terms of its total mass, number of phases, frequency of occurrence, and range of its distribution in the universe.

Simple molecular compounds of silicon, such as SiO and SiS, have been detected by radiowave spectroscopic studies as interstellar material (Zuckerman 1977; Cosmovici 1980). Although in most regions of space the density of these gaseous compounds is extremely low (less than about 10^5 molecules per cm^3) the total amount of silicon in the interstellar gas may reach a level equivalent to several percent of that concentrated in the stars.

In addition to these gaseous silicon compounds solid silicates have been found in interstellar space by spectroscopy in and near the visible range (Winnewisser et al. 1974). They form particles ranging from fractions of a millimeter to more than tens of centimeters. Silicate phases found in cosmic dust collected by jet planes (Wada et al. 1983) are listed in Table 1.1. With the exception of roedderite, all the silicates detected in cosmic dust have also been found to be constituents of meteorites (Kwasha 1976).

From a recent table of the primordial abundances of the elements in the solar system (Holweger 1979), which are assumed to be silimar to those in the universe, it can be seen that the abundance of silicon is considerably exceeded only by those of the elements hydrogen (31 000 times), helium (3100 times), carbon (15 times), nitrogen (3 times), and oxygen (26 times) (Table 1.2).

As yet we have no reliable knowledge of the bulk chemical composition of the earth since we have no direct access to the material in the earth's mantle and core. However, from geochemical studies we have obtained a fairly good estimate of the chemical composition of the crust. In terms of number of atoms as well as weight, silicon is exceeded only by oxygen in the earth's crust (Table 1.2) (Ronov and Yaroshevsky 1972; Mielke 1979).

Table 1.1 Silicon compounds identified in extraterrestrial material

Chemical formula	Name	Occurrence
Gaseous		
SiO	Silicon monoxide	IG
SiS	Silicon monosulfide	IG
Solid		
Si_2N_2O	Sinoite	CD, M
SiO_2	Quartz, cristobalite, tridymite	M
$(Mg, Fe)_2[SiO_4]$	Forsterite, olivine, hortonolite, fayalite	M
β-$(Mg, Fe)_2SiO_4$	Wadsleyite	M
$M_3^{+2}M_2^{+3}[SiO_4]_3$	Garnets (almandine, andradite, grossular)	M
$Mg_3(Mg, Si)_2^{[6]}[SiO_4]_3$	Majorite	CD, M
$(Fe, Mg)^{[4]}Fe^{[6]}Si^{[6]}O_4$	Ringwoodite	CD, M
$Ca(Mg, Fe)[SiO_4]$	Monticellite	M
$Zr[SiO_4]$	Zircon	M
$Ca_2Al[AlSiO_7]$	Gehlenite	M
$Ca_2(Mg, Al)[(Al, Si)_2O_7]$	Melilite	M
$NaMg_2CrSi_3O_{10}$	Krinovite	CD, M
$M_2[Si_2O_6]$	Enstatite, clinoenstatite, bronzite, clinobronzite, hypersthene, clinohypersthene, diopside, pigeonite	CD, M
$NaCr[Si_2O_6]$	Ureyite, kosmochlor	CD, M
$(M^{+2}, M^{+3})_2[(Si, Al)_2O_6]$	Augite, fassaite	M
$Ca_3[Si_3O_9]$	Wollastonite	M
$Mg_2Al_3[AlSi_5O_{18}]$	Cordierite	M
$\sim CaMg_2Ti[Al_2SiO_{10}]$	Rhönite	M
$Mg_3[Si_2O_5](OH)_4$	Clinochrysotile	M
$Mg_{48}[Si_{34}O_{85}](OH)_{62}$	Antigorite	M
$(Fe^{+2}, Fe^{+3})_{<6}[Si_4O_{10}](OH)_8$	Greenalite	M
$(Fe, Mg)_6[AlSi_3O_{10}](O, OH)_8$	Chamosite	M
$Mg_5[Si_2O_5]_4(OH)_2 \cdot 8\,H_2O$	Palygorskite	M
$Mg_4[Si_2O_5]_3(OH)_2 \cdot 4\,H_2O$	Sepiolite	M
$Na_x(Mg, Al)_2[Si_4O_{10}](OH)_2 \cdot 4\,H_2O$	Montmorillonite	M
$(Mg, Fe, Al)_6[(Al, Si)_4O_{10}](OH)_8$	Rhipidolite	M
$M_2^{+}(M^{+2}, M^{+3})_5[(Si, Al)_{12}O_{30}]$	Merrihueite, roedderite, yagiite	CD, M
$(M^{+}, M^{+2})[(Si, Al)_4O_8]$	Feldspars (albite, orthoclase, (K, Na)-feldspar, plagioclase, anorthite)	M
$Na[AlSiO_4]$	Nepheline	M
$Na_4[Al_3Si_3O_{12}]Cl$	Sodalite	M

IG = interstellar gas (Zuckerman 1977); CD = cosmic dust (Bradley et al. 1983; Wada et al. 1983); M = meteorites (Kwasha 1976; Price et al. 1983)

Table 1.2 Elemental abundances in the primordial solar system (Holweger 1979) and the earth's crust (adapted from Mielke 1979)

Element	Solar abundance [atoms/10^6 Si]	Earth's crust abundance [atoms/10^6 Si]	Earth's crust abundance weight [ppm]
1 H	$31\,000 \times 10^6$	155 140	1 520
2 He	$3\,100 \times 10^6$	–	–
3 Li	60	267	18
4 Be	0.81	23	2
5 B	44	86	9
6 C	15×10^6	1 542	180
7 N	3.1×10^6	140	19
8 O	26×10^6	2.9×10^6	456 000
9 F	1 000	2 946	544
10 Ne	1.6×10^6	–	–
11 Na	60 000	101 582	22 700
12 Mg	1.06×10^6	116 996	27 640
13 Al	85 000	318 763	83 600
14 Si	1×10^6	1×10^6	273 000
15 P	7 000	3 720	1 120
16 S	502 000	1 091	340
17 Cl	5 700	366	126
18 Ar	210 000	–	–
19 K	3 500	48 411	18 400
20 Ca	72 000	119 615	46 600
21 Sc	31	57	25
22 Ti	2 400	13 574	6 320
23 V	254	275	136
24 Cr	12 700	241	122
25 Mn	9 300	1 985	1 060
26 Fe	901 000	114 582	62 200
27 Co	2 200	51	29
28 Ni	47 800	173	99
29 Cu	540	110	68
30 Zn	1 260	120	76
31 Ga	14	28	19
32 Ge	117	2.1	1.5
33 As	6.2	2.5	1.8
34 Se	67	0.07	0.05
35 Br	14	3.2	2.5
36 Kr	40	–	–
37 Rb	6.0	94	78
38 Sr	27	451	384
39 Y	4.8	36	31
40 Zr	9.1	183	162
41 Nb	0.9	22	20
42 Mo	4.0	1.3	1.2
44 Ru	1.9	–	–
45 Rh	0.40	–	–
46 Pd	1.3	0.015	0.015
47 Ag	0.46	0.08	0.08
48 Cd	1.55	0.15	0.16
49 In	0.19	0.22	0.24
50 Sn	3.7	1.82	2.1
51 Sb	0.31	0.17	0.2
52 Te	6.5	–	–
53 I	1.16	0.37	0.46
54 Xe	6.3	–	–
55 Cs	0.39	2.01	2.6
56 Ba	4.8	292	390
57 La	0.37	26	34.6
58 Ce	1.2	49	66.4
59 Pr	0.18	6.6	9.1
60 Nd	0.79	28	39.6
61 Pm		–	–
62 Sm	0.24	4.8	7.02
63 Eu	0.094	1.45	2.14
64 Gd	0.42	4.02	6.14
65 Tb	0.076	0.76	1.18
66 Dy	0.37	–	–
67 Ho	0.092	0.79	1.26
68 Er	0.23	2.13	3.46
69 Tm	0.035	0.30	0.5
70 Yb	0.20	1.84	3.1
71 Lu	0.035	–	–
72 Hf	0.17	1.61	2.8
73 Ta	0.020	0.97	1.7
74 W	0.30	0.67	1.2
75 Re	0.051	0.00039	0.0007
76 Os	0.69	–	–
77 Ir	0.72	–	–
78 Pt	1.4	–	–
79 Au	0.21	0.002	0.004
80 Hg	~1.4	0.044	0.086
81 Tb	0.19	0.36	0.72
82 Pb	2.6	6.46	13
83 Bi	0.14	0.004	0.0082
90 Th	0.057	3.6	8.1
92 U	0.027	0.99	2.3

Table 1.3 Crustal abundances of the most important minerals (after Wedepohl 1971)

Mineral	Vol%
Plagioclase	42
Potash feldspar	22
Quartz	18
Amphibole	5
Pyroxene	4
Biotite	4
Magnetite, ilmenite	2
Olivine	1.5
Apatite	0.5

From the high cosmic and terrestrial abundance of silicon it is evident that detailed knowledge and understanding of the physical and chemical properties of silicates are essential in attempts to deduce the history of the universe in general and, in particular, the history of the solar system, of the planets, and especially of the geologic history of the earth. For example, while the sequence of events in the history of a particular meteorite can be inferred from isotope analyses, information about the specific nature of these events can be derived from crystallographic information obtained from high resolution transmission electron microscopic studies of the silicate constituents (Grove 1982; Müller and Wlotzka 1982; Czank and Liebau 1983; Ashworth et al. 1984).

The same is of course true in the case of the determination of the geologic history of terrestrial rocks inferred from studies of their minerals. More than 95% of the volume of the earth's crust is composed of quartz and a small number of other rock-forming silicates (Table 1.3) (Wedepohl 1971). Therefore, in regard to the geological history and present state of the solid earth, the prominence of silica and silicates can hardly be overestimated. However, compounds of silicon are not only constituents of the lithosphere: they are also present in the hydrosphere, mainly as dissolved silica.

The importance of silicon for life on earth primarily results from the fact that the fertility of soil relies in large part on the ability of clay minerals to absorb and release water and several cations that are indispensible for plant nutrition. This process is fundamental to the life of higher plants and, based on these, to the life of those animals which, in turn, feed upon the plants.

In addition, silicon compounds play a substantial role in the cells of living organisms. Large amounts of silica are found in horse-tail, rice, feather grass, reed, and bamboo, where it contributes to the strength of the leaves and stems, and in the skeletons of diatoms which consist of very pure SiO_2 (Werner 1966). Silicon compounds are vital components in the metabolism of many bacteria, especially those living in hot springs. The bacterium proteus mirabilis even substitutes silicon for phosphorus in phospholipids (Heinen 1967a, b).

Silicon is also present as a trace element in higher animals and in man where it constitutes about 10 mg per 1 kg live weight. It is essential in the cells of connective tissues and is involved in the biosynthesis of collagen, the substance that forms hair and nails, and in the formation of bony tissue (Carlisle 1970).

Considering the widespread occurrence of silicon compounds in living organisms it is not surprising that research on the biological activity of organosilicon compounds is a rapidly expanding field (Voronkov 1973, 1979; Tacke and Wannagat 1979; Tacke et al. 1982).

1.2 Technical Importance of Silicon

Several of the silicates and silica itself play an important role as raw materials as well as products in technical processes. In the form of granite and sandstone they are used as natural building materials. A mixture of clay and limestone is used to produce about 880 million tons of Portland cement each year (Bucchi and Pesenti 1982), and the world production of silicate glass amounts to approx. 62 million tons per year (Ullmanns Encyklop. techn. Chemie 1976, p. 363). The list of ceramic materials based on silicates includes stoneware, whiteware, porcelain, vitroceram, and refractories used to line furnaces in the manufacture of steel, cement, glass, etc. One can scarcely imagine present day life without these materials.

1.3 Diversity of Silicon Compounds

The immensely widespread application of silicates is due to the large variability of their properties and this, in turn, is due to the very large number of different silicate phases and the diversity of their structures. Second to carbon, silicon forms the largest number of compounds with other elements.

Since silicon follows carbon in the fourth main group of the periodic table, it could be expected that the large variety of silicon compounds may be a result of the same factors that produce so large a variety of carbon compounds. However, this is not the case.

The large number of carbon compounds is due to the fact that the bond energies of the C–C, C–O, and C–H bonds (Table 1.4) are of about equal magnitude and they will, therefore, be formed with about the same probability. In contrast, the bond energy of the Si–O bond is considerably higher than that of the

Table 1.4 Mean bond energies in kJ/mol of several bonds of carbon and silicon (Cottrell 1958)

Bond	C	Si
X–X	346	222
X–O	358	452
X–H	413	318

Si–H bond and more than twice that of the Si–Si bond. As a consequence, instead of the common –C–C–C– chains of carbon chemistry, chains of the type –Si–O–Si–O–Si– are the skeletons of silicon chemistry, and only a rather small number of silicon compounds are known which are the analogues of the organic carbon compounds. Moreover, since only one of the three kinds of bonds, X–X, X–O, and X–H, is energetically strongly favored in silicon chemistry, while all three are almost equally favorable in carbon chemistry, the number of silicon compounds is smaller than the number of carbon compounds.

This does not, however, explain why the number of silicates is much larger than the number of phosphates, sulfates, and the like since the P–O and S–O bonds, etc. are also considerably stronger than the corresponding X–X and X–H bonds. Instead, this observation can be explained by considering that cations A form $[AO_n]$ polyhedra with oxygen and that the formation of –A–O–A–O–A– chains links such $[AO_n]$ polyhedra via common oxygen atoms. Since linking of polyhedra is accompanied by a closer approach of the cations to each other, the repulsive forces between the cations increase and this increase is in the order Na–Mg–Al–Si–P–S–Cl due to increasing charge of the cation A. The tendency to link $[AO_n]$ polyhedra decreases in the same sequence, and as a result, the number of compounds decreases from silicates to phosphates, sulfates, and perchlorates. Sodium and magnesium do form more compounds than silicon, but they are different from the other cations in the series in that their bonds to oxygen are weaker. Therefore, compounds such as $Na_4P_2O_7$, Na_2SO_4, and Na_2CrO_4 are usually not described as sodium compounds, but rather as phosphates, sulfates, chromates, etc.

2 Methods to Describe the Atomic Structure of Silicates

2.1 Model Representations of Silicate Structures

When the atomic structure of any material has been determined, models are needed to visualize it. Whether these models are two-dimensional drawings or three-dimensional structures made from balls glued together or linked by sticks, they give only a static picture of the atomic arrangement at a specific moment or the time average of the geometric arrangement of the atoms in space. It is not easily possible to construct models which illustrate the dynamic behavior of the atoms.

Like other substances silicates can exist as solids with crystalline or amorphous form, as liquids in melts or in solution, and sometimes as a vapor. We are able to obtain an accurate knowledge of the structure on an atomic scale only in the case of crystalline compounds because in these materials the structure is defined by a basic building unit that seldom exceeds a volume of 1000 $Å^3$, and which usually contains less than 100 atoms. This basic unit, called the unit cell, is periodically repeated in three dimensions. It is, therefore, sufficient to have a model of one unit cell to visualize the whole structure.

In the liquid and vapor states there is no periodic structure and it is necessary to describe the atomic structure of essentially the whole sample in order to get a complete description of the material in space. For silicate melts, solutions, and vapors, in which there is continuous breaking and rebuilding of bonds between atoms, even such a large model would not be able to describe those structural details that are responsible for the low viscosity, reactivity, and other properties which are typical of silicates in these states.

Because of these difficulties only those methods which are used to visualize crystalline structures are described here. They can be applied to noncrystalline structures if their shortcomings due to disorder in space and time are kept in mind.

Taking into consideration that matter consists of atoms bonded together, and that covalent, ionic, metallic, and other bond types are only models used to describe the bonding between atoms, we can in similar ways describe silicates as composed either of ions or of neutral atoms or of groups of ions or atoms packed together. In fact, there are at least six different methods used to describe silicate structures.

(1) Dense Packing of Ions. Starting with the ionic approach silicates may be regarded as composed of oxygen ions O^{-2} with an approximate radius of 1.40 Å (Appendix III), very often packed as densely as possible. This oxygen ion packing is usually either cubic (or almost cubic) or hexagonal (or almost hexagonal). The smaller cations as well as the very small Si^{+4} ions (radius 0.26 Å) then fill sites between the oxygen ions. Hydrogen is usually regarded as part of hydroxyl groups which have approximately the same size as the O^{-2} ions. If the cations are small enough to fit easily into the tetrahedral and octahedral holes between the oxygen ions, then there is very little distortion of the structure. With increasing size of cations the oxygen ions are displaced from their positions of closest packing, thereby reducing the density of the structure and decreasing the symmetry of the oxygen packing.

Such considerations can give a reasonable estimate of the density of a silicate structure and of the influence of cations of various sizes upon the structure. Unfortunately, however, it is often hard to visualize the complete structure from three-dimensional models or two-dimensional drawings of this kind due to severe overlap of the ions in projection. As a result only very thin slices of a structure can be depicted in this way as shown for the simple case of forsterite in Fig. 2.1 a.

(2) Dense Packing of Atoms. Slater (1972) has pointed out that the structures of solids might be described as composed of neutral atoms without considering the character of the bonds between the atoms. This procedure makes the metal atoms and the silicon atoms larger than the oxygen atoms [$r_{Si} = 1.10$ Å, $r_O = 0.60$ Å] (Appendix II). Although models and drawings of structures with these radii are uncommon and, therefore, appear rather strange to us, more recent calculations of ionic charges of the atoms in silicates and other complex inorganic compounds are more consistent with this atomic model than with the ionic one.

When the olivine structure is drawn with the sizes of the atoms proportional to the Slater atomic radii (Fig. 2.1 b), such a model is slightly more transparent than the ionic case shown in Fig. 2.1 a. Nevertheless, it is still rather difficult to visualize slices of the structure which are more than 3 or 4 Å thick.

(3) Ball and Stick Models. To improve this situation different kinds of abstraction from the "reality" of close packing are used to describe silicate structures. One of these uses balls (for three-dimensional models) or circles (for two-dimensional drawings) to represent the ions (atoms) with sizes such that the sum of the radii of neighboring balls (circles) is less than the distance between them. Chemical bonds between ions (atoms) are then represented by sticks or lines, respectively. Sometimes, in two-dimensional drawings the thickness of the lines is made proportional to the height above the plane of projection in order to give some impression of the third dimension of the structure (Fig. 2.1 c). By this method thicker sections of a structure can be presented with little loss of transparency, especially when care is taken in choosing the direction in which to project the structure.

(4) Models Using Anion Polyhedra. In another method of structure modelling, groups of atoms connected by particularly strong bonds are represented by polyhedra with boundary faces formed by constructing planes through the centers of the outer atoms of these groups.

In the case of silicates, in which each silicon atom is strongly bonded to four oxygen atoms, the $[SiO_4]$ group is represented by a solid tetrahedron. The silicon atom at the center of the tetrahedron is usually not shown. The metal atoms M

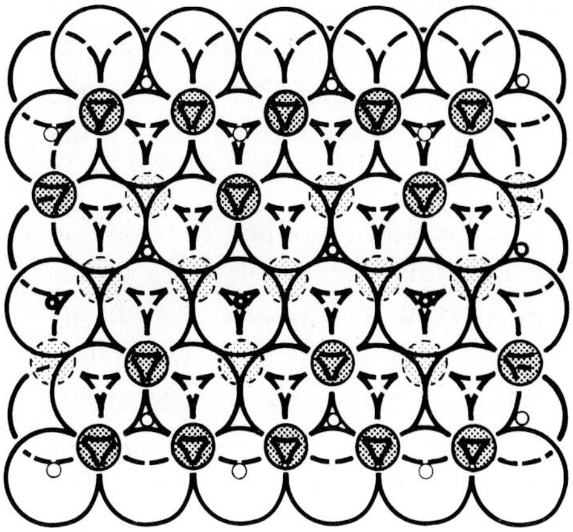

a

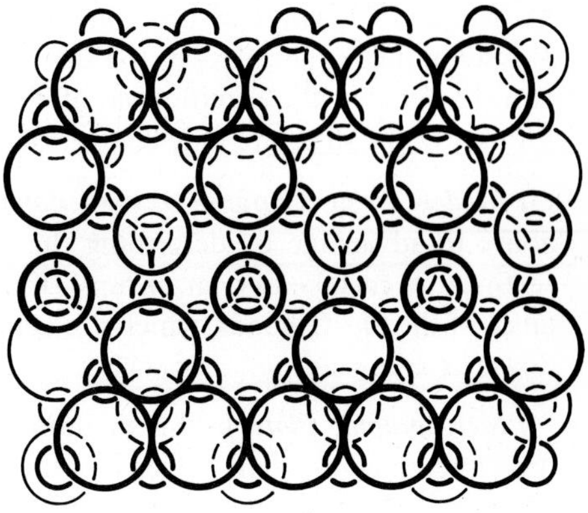

b

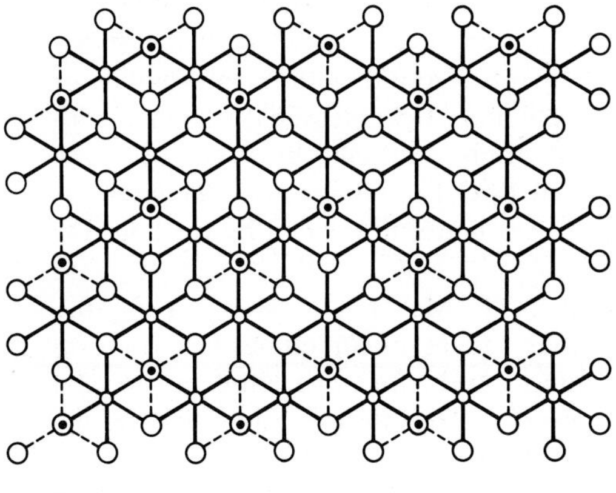

c

d

Fig. 2.1 a – e. Different ways to present the structure of forsterite, $Mg_2[SiO_4]$.

a Packing of spheres with radii proportional to their ionic radii (*small spheres* silicon, $r_{Si^{+4}} = 0.26$ Å; *medium spheres* magnesium, $r_{Mg^{+2}} = 0.72$ Å; *large spheres* oxygen, $r_{O^{-2}} = 1.40$ Å). **b** Packing of spheres with radii proportional to their atomic radii (*small spheres* oxygen, $r_O = 0.60$ Å; *medium spheres* silicon, $r_{Si} = 1.10$ Å; *large spheres* magnesium, $r_{Mg} = 1.50$ Å). **c** Ball and stick model (*black dots* silicon, *small spheres* magnesium, *larger spheres* oxygen, *solid lines* Mg – O bonds, *broken lines* Si – O bonds). **d** Packing of cations and $[SiO_4]$ tetrahedra. **e** Packing of silicon ions and $[MgO_6]$ octahedra

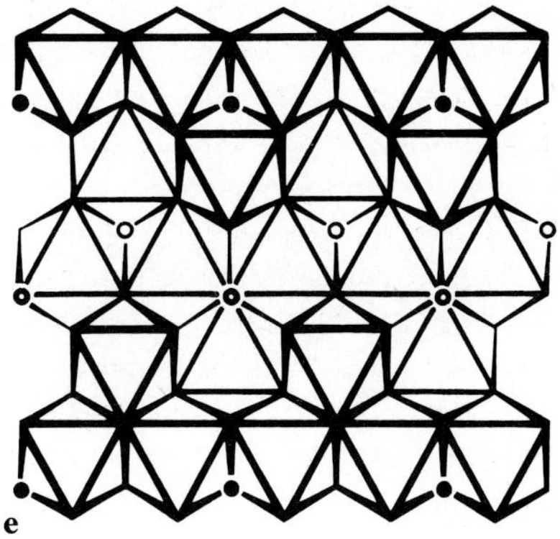

e

(cations) are represented by balls or circles between these tetrahedra (Fig. 2.1 d). Models and drawings of this kind have an advantage over those described previously in that larger parts of the structure can be easily visualized. However, there is a danger that the significance of the metal–oxygen bonds in determining the stability of the structure may be overlooked.

(5) Models Using Cation–Oxygen Polyhedra. This method attempts to avoid the above disadvantage by describing the structure in terms of the polyhedra representing the metal–oxygen groups (Fig. 2.1 e). However, in this case the importance of the Si–O bonds, which are actually stronger than the M–O bonds, is now downplayed.

(6) Stereographic Plots. In this method of describing structures, the atoms and ions are represented as small circles and the chemical bonds between them as lines, but the model is supplemented by making pairs of stereographic diagrams (Fig. 2.2). Viewing them with a stereo viewer gives a three-dimensional image of slabs of considerable thickness.

The last four methods are the most commonly used. Since the "ball and stick" methods, numbers (3) and (6), require no statement of the relative importance of the silicon–oxygen bonds and the metal–oxygen bonds, they avoid the shortcomings of methods (4) and (5) which overemphasize either the Si–O bonds or the cation–oxygen bonds. However, the ball and stick methods are unable to demonstrate the presence of larger groups of atoms connected by strong bonds in the form of chains or layers of corner-linked $[SiO_4]$ tetrahedra or of corner- or edge-linked $[MO_n]$ octahedra. Such structures are better illustrated by methods (4) and (5).

In silicates with atomic ratios $M:Si \geq 2$ ($M = Me^+$, $Me^{+2}_{0.5}$, $Me^{+3}_{0.33}$, etc.), the $[MO_n]$ polyhedra are linked via common corners, edges, or sometimes faces to form clusters of infinite extension, while the $[SiO_4]$ tetrahedra have either no oxygen atoms in common or are linked together in small clusters only. In these silicates the metal polyhedral part is the strong skeleton of the structure to which the small silicate ions must conform. In this case method (5) is the appropriate one to describe the structure.

In most silicates with $M:Si \leq 2$, the $[SiO_4]$ tetrahedra form anions of infinite extension. Since the M–O bond is weaker than the Si–O bond we have to regard

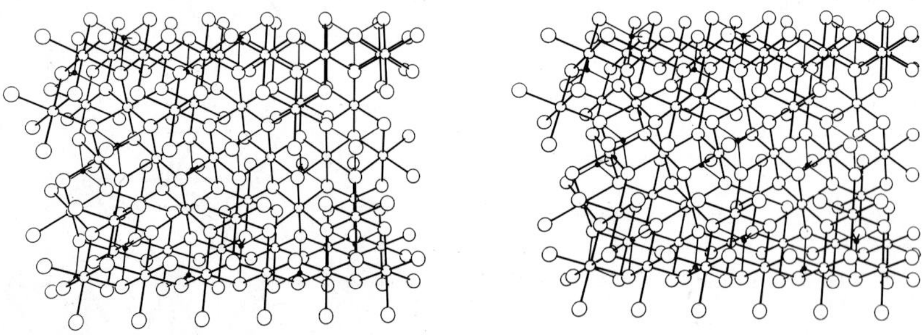

Fig. 2.2. Stereographic plot of the ball and stick model of forsterite

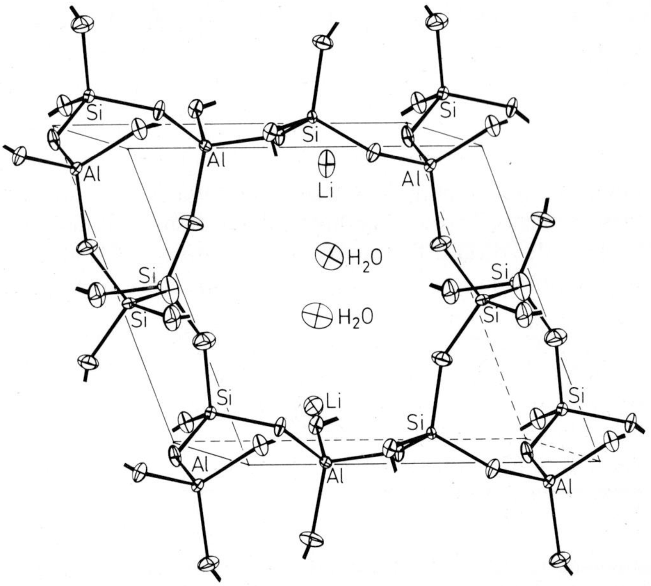

Fig. 2.3. Projection of the structure of bikitaite, $Li[AlSi_2O_6] \cdot H_2O$. The atoms are represented by their thermal ellipsoids scaled to 70% probability (Bissert 1985)

the silicate anions as the skeleton of the structure even when the $[MO_n]$ polyhedra also form infinite clusters. For these silicates the structure is best described by method (4), the suitability increasing with decreasing M : Si ratio.

In this book we will mainly use the $[SiO_4]$ tetrahedra and their clusters to illustrate the silicate structures, but we must always be aware that such models, like others, are unable to adequately describe every aspect of a structure.

X-rays interact with electrons. Consequently, X-ray diffraction methods give information about the electron density distribution within the material studied. Modern methods of crystal structure analysis are accurate enough not only to reveal the locations of the atomic centers, but can also measure deviations of the electron density distribution from spherical symmetry. The electron density distribution is then represented by a three-axial ellipsoid which is drawn at a certain probability level. For example, in Fig. 2.3 the so-called thermal ellipsoids of the atoms of bikitaite, $Li[AlSi_2O_6] \cdot H_2O$, are scaled to 70% probability, i.e., the ellipsoid encloses 70% of the electron density, averaged over space and time, of each atom.

While phases containing $[SiO_4]$ tetrahedra as the anionic part are called tetraoxosilicates or, in general, simply silicates, phases with $[SiO_6]$ octahedra are named hexaoxosilicon compounds because these octahedra are considered to be the cationic part of the structure. As yet, the crystal structures of only a few hexaoxosilicon compounds have been described so that at present no general rules for these can be given.

For convenience, unless a distinction between silica, tetraoxosilicates, and hexaoxosilicon compounds is given, all three types are summed up under the term silicates in this book.

2.2 Treatment of Tetrahedrally Coordinated Cations in Silicates

In silicates there is extensive replacement of silicon by aluminum and, usually, minor replacement by other elements, such as germanium, titanium, iron, beryllium, and phosphorus. We must, therefore, decide on some way of dealing with these elements. Throughout this book the commonly used procedure is as follows.

If, under certain thermodynamic conditions, an element replaces some of the silicon atoms in a given silicate crystal structure so that the crystallographic positions are statistically occupied by silicon and the other cation, then this cation is regarded as a part of the silicate ion. This is explained with some examples.

In high-sanidine, the high temperature phase of potassium feldspar, $KAlSi_3O_8$, there is a completely random distribution of silicon and aluminum atoms in the two eightfold crystallographic so-called T[1] positions, and the structural formula of sanidine is, therefore, written as $K_2[(Al_2Si_6)O_{16}]$ or, in short, $K[(AlSi_3)O_8]$. According to the definition given above, aluminum is then part of the silicate anion.

On the other hand, in the low temperature form of potassium feldspar, ideal low-microcline, there is complete ordering of silicon and aluminum among the tetrahedral sites in the structure so that the formula might be written as $KAl[Si_3O_8]$. However, even though there is no longer a statistical distribution of Si and Al in the T positions, the aluminum is still regarded as part of the silicate anion since the Al/Si distribution becomes increasingly disordered with increasing temperature without going through a first-order phase transition. Therefore, the structural formula of fully ordered low-microcline is also written as $K[AlSi_3O_8]$. This means that both modifications of potassium feldspar are regarded as so-called framework silicates, with the $[AlSi_3O_8]$ unit forming the framework.

In petalite, $LiAlSi_4O_{10}$, lithium, aluminum, and silicon are all tetrahedrally coordinated by oxygen atoms. This mineral is sometimes regarded as a layer silicate, $LiAl[Si_2O_5]_2$, and sometimes as a framework silicate, $Li[AlSi_4O_{10}]$. The crystal structure determinations indicate that there is complete cation ordering. When heated the structure breaks up into $LiAl[Si_2O_6]$ and SiO_2 by a reconstructive phase transformation before Al/Si disorder sets in. Therefore, in accordance with the definition given above, petalite should be regarded as a layer silicate even though aluminum is tetrahedrally coordinated.

Using the symbol T for the tetrahedral cations of the silicate anion, the general formula of the feldspars can be written as $M[T_4O_8]$ with $M = Na^+$, K^+, Ca^{+2}, Ba^{+2}, NH_4^+ and $T = Si^{+4}$, Al^{+3} in the natural feldspars, and, in addition to these, $M = Rb^+$, Cs^+, Sr^{+2}, Mn^{+2}, Pb^{+2}, Eu^{+2}, La^{+3} and $T = Ge^{+4}$, Ga^{+3}, Fe^{+3} in the synthetic varieties. For comparison, the general formula of petalite would have to be written as $LiAl[T_2O_5]_2$.

---

1 In ordinary silicates, in which silicon is tetrahedrally coordinated (in contrast to silicon compounds with octahedrally coordinated silicon), the position usually occupied by silicon is more generally called a T position in order to allow for partial or complete substitution of silicon by other atoms. The atoms themselves are then often called T atoms regardless of their chemical nature.

In general, silicates which contain tetrahedrally coordinated aluminum are called aluminosilicates in contrast to silicates containing octahedrally coordinated aluminum for which the term aluminum silicates is used. However, it appears to be more accurate to describe only those silicates as aluminosilicates in which Al is regarded as part of the silicate anions, designating the rest as aluminum silicates irrespective of the coordination number of aluminum.

If there is no misinterpretation possible we will use the terms Si, $[SiO_4]$, silicon, silicate anion, etc. instead of T, $[TO_4]$, etc., respectively, regardless of some replacement of silicon or not.

It should be pointed out that there are other definitions of what constitutes a part of the anion cluster.

Zoltai (1960), for example, considers that every cation that is tetrahedrally coordinated by anions is part of the tetrahedral complex (see Sec. 8.3). This more general and purely geometric approach is appropriate for an unequivocal formalistic classification of coordination compounds with tetrahedral complexes of any kind, but it is less suitable in giving an insight into the specific structural chemistry of the silicates considered in this book.

3 Chemical Bonds in Silicates

3.1 The Silicon – Oxygen Bond

The character of the chemical bonds in silicates has been studied extensively for a long time and with particular intensity during the last two decades. Among the various methods that have been used, the following may be included:

1. those which are predominantly experimental, such as infrared spectroscopy (IR), X-ray fluorescence spectroscopy (XFS and EXAFS), and photoelectron spectroscopy (ESCA, PES, XPS, UPS);
2. those which compare structural, chemical, and physical properties of silicates and related substances in order to deduce the bond character indirectly; and
3. predominantly theoretical methods, such as molecular orbital calculations.

However, in spite of all these studies, much controversy remains and still no completely satisfactory picture of the chemical bonds in silicates has yet emerged. In view of this continued controversy and since this book is not intended to evaluate these different methods and their results, experimental observations will be reported in detail, but only short descriptions will be given of their import on theory.

3.1.1 Coordination Numbers of Silicon

It is common practice to give the chemical formula of a complex silicate either in the general form $M'_{r'}M''_{r''}M'''_{r'''} \ldots Si_sO_t$ or in the oxide form $q'\,M'_{r'}O \cdot q''\,M''_{r''}O \cdot q'''\,M'''_{r'''}O \cdots s\,SiO_2$. Silicates then contain Si–O–M bonds where the M atoms can range from monovalent to heptavalent, from very small to rather large, and from strongly metallic to strongly nonmetallic.

Silicon is an amphoteric element. Consequently, SiO_2 acts as an acid anhydride with metal oxides M_rO, but as a basic oxide if M_rO is an acid anhydride.

In general, metal ions are larger and have a lower valence than silicon so that their M–O bonds are weaker than the Si–O bonds. As a result silicon attracts oxygen ions more strongly than the metal ions do, forming $[SiO_4]$ tetrahedra with a mean bond length $\langle d(Si^{[4]}-O)\rangle = 1.62$ Å[1] (Fig. 3.1). This situation occurs for all tetraoxosilicates in which silica acts as an acid anhydride, i.e., as the anionic component of the compound.

1 Throughout this text the coordination number of an atom is indicated in square brackets and given as a right superscript to the element symbol, e.g., $Si^{[4]}$ denotes four-coordinated silicon, and $O^{[2]}$ denotes two-coordinated oxygen.

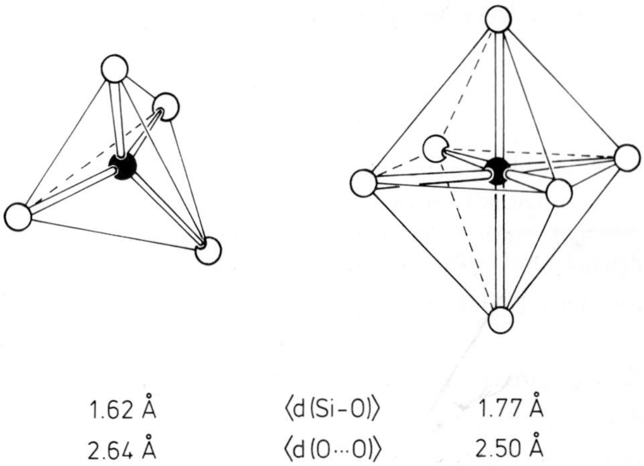

Fig. 3.1. [SiO_4] tetrahedra and [SiO_6] octahedra and their average dimensions

On the other hand, if M is a metalloid the strength of the M–O bond is roughly comparable to that of the Si–O bond. The small metalloid atoms which have a high formal charge can then compete successfully with the silicon atoms for the oxygen ions. As a result, the silicon atoms of such compounds can attract oxygen ions only to a mean distance $\langle d(\mathrm{Si}^{[6]}-\mathrm{O})\rangle = 1.77$ Å leaving space for six oxygen ions around each silicon atom in octahedral coordination of these hexaoxosilicates.

This argument and the fact that Pauling's bond strength sums are only 4/6 for each bond in the octahedron, rather than 4/4 in the tetrahedron clearly demonstrate that the higher coordination number of silicon in phases containing [SiO_6] octahedra is an indication that the Si–O bonds are weaker than the ones existing in phases containing [SiO_4] tetrahedra. Therefore, under ordinary conditions for a given composition the phase containing silicon in octahedral coordination should be energetically less favorable than the one with tetrahedrally coordinated silicon. This may also be deduced from the shorter oxygen–oxygen distances of 2.50 Å – and, hence, higher repulsive forces between the oxygen ions – in an [SiO_6] octahedron relative to the longer O ··· O distance of 2.64 Å in the [SiO_4] tetrahedron (Fig. 3.1).

3.1.2 Bond Lengths and Bond Angles in Silicates Containing [SiO_4] Tetrahedra (Tetraoxosilicates)

Within the last two decades a sufficiently large number of silicate structures have been accurately determined to reveal the ranges within which individual atomic distances d(Si–O) and bond angles ∢ O–Si–O and ∢ Si–O–Si vary. For the silicate structures published in *Structure Reports* between 1970 and 1980 with R values[2] below 0.08 using X-ray or neutron diffraction intensities collected near or

2 In crystal structure determinations with diffraction methods, the reliability index R is an estimate of the accuracy of the structure obtained. For details see textbooks on X-ray crystallography, for example, Stout and Jensen (1968).

below room temperature, these ranges are

$$1.57\,\text{Å} < d(\text{Si}-\text{O}) < 1.72\,\text{Å},$$
$$98° < \sphericalangle\,\text{O}-\text{Si}-\text{O} < 122°, \text{ and}$$
$$120° < \sphericalangle\,\text{Si}-\text{O}-\text{Si} \le 180°.$$

When interatomic distances and bond angles determined at room temperature are reported near or outside the limits of these ranges, the deviation is probably a result of inaccuracies in the structure determination or of considerable static or dynamic disorder in the structure (see Sec. 3.1.2.4). Only in very rare cases will they be caused by unusual chemical or sterical effects.

3.1.2.1 Correlations Between Si – O Distance, Oxygen Coordination Number, and Bond Strength

Irrespective of the character of the Si – O bond, i.e., whether this bond is predominantly ionic or covalent, there is an equilibrium distance, $d_{eq}(\text{Si}-\text{O})$, between a silicon and an oxygen atom. At this distance the attractive and repulsive forces between the two atoms are evenly balanced. The grand mean value of all Si – O atomic distances reported for accurately determined silicate structures, $\langle d(\text{Si}^{[4]}-\text{O})\rangle \simeq 1.62$ Å, together with the frequency distribution of the individual observed distances, $d_{ind}(\text{Si}-\text{O})$, suggests that the equilibrium distance, $d_{eq}(\text{Si}-\text{O})$, is somewhere near 1.60 Å.

Energy is required to expand as well as to compress the Si – O bond. Since the shape of the energy curve of the diatomic system as function of the Si – O distance (Fig. 3.2) is not symmetric about the energy minimum, the same amount of energy that is required to compress the Si – O bond by, for example, 0.02 Å from the equilibrium distance, will produce an expansion of the Si – O bond by a considerably larger amount.

Other atoms, M, in the vicinity of the oxygen atom of an Si – O bond also attract the oxygen atom and enter into competition with the silicon atom, thereby weakening the bond. In general, this leads to a lengthening of the individual Si – O bond. The greater the number of M – O bonds involving a particular oxygen atom and the stronger these M – O bonds are, the weaker the Si – O bond becomes and the longer the bond length $d_{ind}(\text{Si}-\text{O})$ will be. As a consequence, the individual Si – O bond length should be positively correlated with the coordination number of the oxygen atom and with the sum of the bond strengths of all bonds between

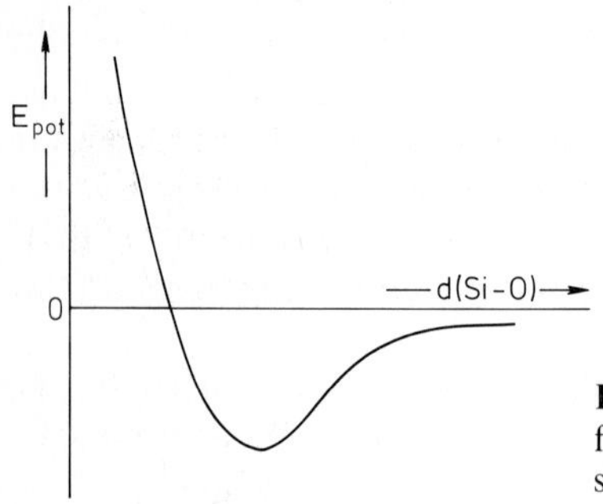

Fig. 3.2. The potential energy of an oxygen atom as function of its distance $d(\text{Si}-\text{O})$ from a neighboring silicon atom

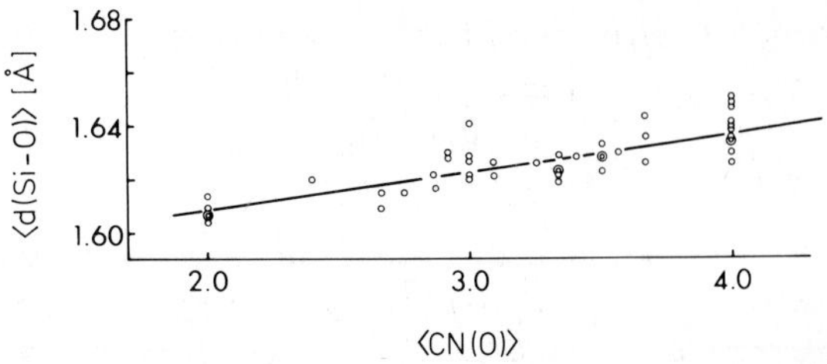

Fig. 3.3. Correlation between the mean values $\langle d(\mathrm{Si-O})\rangle$ of the Si – O distances and the mean coordination number $\langle \mathrm{CN(O)}\rangle$ of oxygen in 46 crystal structures with tetrahedrally coordinated silicon (after Brown and Gibbs 1969)

this oxygen atom and the surrounding atoms, Si as well as M. Such correlations have in fact been found and extensively studied.

The first correlation of this kind, although not immediately interpretable along these lines, was published by Smith and Bailey (1963). They found that the average of all of the Si – O distances in a silicate structure, $\langle d(\mathrm{Si-O})\rangle$, increases from 1.61 Å to 1.63 Å as the O : Si ratio (counting only those oxygen atoms that are linked to silicon atoms) increases from 2 for the three-dimensional tetrahedral frameworks of SiO_2 to 4 for monosilicates containing only $[SiO_4]$ tetrahedra not linked to other $[SiO_4]$ tetrahedra by common oxygen atoms.

With the aid of a regression analysis for 46 well-refined structures, Brown and Gibbs (1969) demonstrated that about 60% of the variation of the average of all Si – O distances in a structure can be explained in terms of a linear dependence of these $\langle d(\mathrm{Si-O})\rangle$ on the mean coordination number of the oxygen atoms bonded to silicon, $\langle \mathrm{CN(O)}\rangle$ (Fig. 3.3):

$$\langle d(\mathrm{Si}^{[4]}-\mathrm{O})\rangle = 1.579 + 0.015\,\langle \mathrm{CN(O)}\rangle\ [\text{Å}]. \tag{1}$$

In good agreement with the values obtained by Smith and Bailey (1963), Brown and Gibbs obtained a grand mean value $\langle d(\mathrm{Si-O})\rangle = 1.608$ Å for $\langle \mathrm{CN(O)}\rangle = 2$ (i.e., silica) and 1.638 Å for $\langle \mathrm{CN(O)}\rangle = 4$ (i.e., typical monosilicates). The remaining 40% of the variation of the $\langle d(\mathrm{Si-O})\rangle$ values must be due to factors other than the mean coordination number, including the strength of the M – O bonds.

The high valence of silicon implies that a second silicon atom linked to the oxygen atom of an Si – O bond weakens this Si – O bond more than an M atom (Na, K, Ca, etc.). It was, in fact, observed long ago that the $\mathrm{Si-O_{br}}$ bonds between silicon and an oxygen atom O_{br} linking two Si atoms are, on average, longer by approx. 0.025 Å than the average $\mathrm{Si-O_{term}}$ bonds between silicon and a terminal oxygen atom O_{term} linked to only one silicon atom. This is clearly visible from the data given in Table 3.1 for accurately determined silicate structures which contain $[\mathrm{O_3Si-O_{br}-SiO_3}]$ groups. Similar differences between $\langle d(\mathrm{Si-O_{br}})\rangle$ and $\langle d(\mathrm{Si-O_{term}})\rangle$ exist for silicates in which the $[SiO_4]$ tetrahedra are linked into chains, rings, or layers.

The influence of both the oxygen coordination number and the strength of the M – O bonds on Si – O distances has been demonstrated by Baur (1970) who applied and extended Pauling's (1967, p. 547 f.) second rule for complex ionic compounds. If the electrostatic bond strength received by an anion X from a cation is defined as

$$s = \frac{z}{\mathrm{CN}} \tag{2}$$

Table 3.1 Bond lengths and bond angles in the $[Si_2O_7]$ groups of silicates for which the crystal structures have been accurately determined

$^6\langle d(Si-O_{term})\rangle$: mean value of the bond lengths of the six $Si-O_{term}$ bonds of the $[Si_2O_7]$ group
$^2\langle d(Si-O_{br})\rangle$: mean value of the bond lengths of the two $Si-O_{br}$ bonds of the $[Si_2O_7]$ group
Δd $= {}^2\langle d(Si-O_{br})\rangle - {}^6\langle d(Si-O_{term})\rangle$
$^6\langle \measuredangle O_{term}-Si-O_{term}\rangle$: mean value of the six $O_{term}-Si-O_{term}$ angles of the $[Si_2O_7]$ group
$^6\langle \measuredangle O_{br}-Si-O_{term}\rangle$: mean value of the six $O_{br}-Si-O_{term}$ angles of the $[Si_2O_7]$ group

Phase (Mineral name)	$^6\langle d(Si-O_{term})\rangle$	$^2\langle d(Si-O_{br})\rangle$	Δd	$^6\langle \measuredangle O_{term}-Si-O_{term}\rangle$	$^6\langle \measuredangle O_{br}-Si-O_{term}\rangle$	$\measuredangle Si-O-Si$	Ref.
$La_2[Si_2O_7]$	1.621	1.666	0.045	110.1	108.7	128.2	[1]
	1.632	1.664	0.032	110.3	108.4	130.9	
$Sm_4[Si_2O_7]S_3$	1.628	1.623	−0.005	109.3	109.7	129	[2]
$CaFe_2(Fe, Mn)[Si_2O_7]O(OH)$ (ilvaite)	1.625	1.653	0.028	112.5	106.2	129.6	[3]
$Sm_2[Si_2O_7]$ (IT)	1.624	1.649	0.025	110.7	108.1	129.7	[4]
	1.612	1.644	0.032	111.3	107.5	136.0	
$Li_2Cu_5[Si_2O_7]_2$	1.618	1.645	0.027	111.6	107.2	130.2	[5]
$Nd_2[Si_2O_7]$	1.631	1.613	−0.016	107.0	111.6	132.6	[6]
$Na_2(Sr, Ba)_2Ti_3[Si_2O_7]_2(OH, F)_2O_2$ (lamprophyllite)	1.612	1.654	0.042	112.7	106.0	135.6	[7]
$Ca_2Be[Si_2O_7]$ (gugiaite)	1.609	1.663	0.054	113.5	104.9	135.9	[8]
$CaNaAl[Si_2O_7]$ (melilite)	1.613	1.648	0.035	112.2	106.3	136.2	[9]
$Ca_3[Si_2O_7]$ (rankinite)	1.605	1.668	0.063	112.2	107.7	136.2	[10]
$CaB_2[Si_2O_7]O$ (danburite)	1.617	1.614	−0.003	109.0	109.9	136.8	[11]
$CaAl_2[Si_2O_7](OH)_2 \cdot H_2O$ (lawsonite)	1.626	1.654	0.028	112.4	106.4	136.9	[12]
$Ag_6[Si_2O_7]$	1.626	1.653	0.027	110.2	108.7	137.9	[13]
$Ca_2Mg[Si_2O_7]$ (åkermanite)	1.609	1.649	0.040	114.3	105.0	139.4	[14]
$(Ba, Sr)(Mn, Fe)_2(Fe, Ti)[Si_2O_7](O, OH)_2$ (orthoericssonite)	1.619	1.639	0.020	113.1	105.5	148.8	[15]
$Zn_4[Si_2O_7](OH)_2 \cdot H_2O$ (hemimorphite)	1.621	1.629	0.008	110.8	108.1	149.5	[16]

$Na_2Ca_4Zr(Nb, Ti)[Si_2O_7]_2$	1.613	1.646	0.033	113.3	105.2	149.5	[17]
$O_2F(O, F)$ (wöhlerite)	1.613	1.640	0.027	112.8	105.9	159.8	
$Ca_4[Si_2O_7](OH, F)_2$ (cuspidine)	1.606	1.659	0.053	114.2	104.2	155.4	[18]
$(Na, Ca)_3(Ca, Ce)_4(Ti, Nb)[Si_2O_7]_2(O, F)_4$ (rinkite)	1.615	1.645	0.030	112.9	105.9	155.5	[19]
$Ca_5[Si_2O_7][CO_3]_2$ (tilleyite)	1.610	1.667	0.057	114.4	103.9	157.3	[20]
$Gd_2[Si_2O_7]$	1.614	1.673	0.059	114.8	103.4	158.7	[6]
$(Na, Ca)_2(Ca, Fe, Mn, Ti)(Zr, Nb)[Si_2O_7]OF$ (låvenite)	1.617	1.619	0.002	112.3	106.4	171.3	[21]
$Sc_2[Si_2O_7]$ (thortveitite)	1.630	1.605	−0.025	112.3	106.5	180	[22]
$Yb_2[Si_2O_7]$	1.626	1.626	0.000	111.2	107.7	180	[6]
$Er_2[Si_2O_7]$	1.617	1.632	0.015	111.1	107.7	180	[6]
$K_6[Si_2O_7]$	1.621	1.675	0.054	111.8	107.0	180	[23]
$Ba_3Nb_6[Si_2O_7]_2O_{12}$	1.629	1.599	−0.030	105.7	113.0	180	[24]
Mean values	1.619	1.644	0.025	111.7	107.1	148.2	

References: [1] Dago et al. 1980; [2] Siegrist et al. 1982; [3] Beran and Bittner 1974; [4] Smolin et al. 1970; [5] Kawamura et al. 1978; [6] Smolin and Shepelev 1970; [7] Saf'yanov et al. 1983; [8] Kimata and Ohashi 1982; [9] Louisnathan 1970; [10] Saburi et al. 1976; [11] Phillips et al. 1974; [12] Baur 1978; [13] Jansen 1977; [14] Kimata and Ii 1981; [15] Matsubara 1980a; [16] Hill et al. 1977; [17] Mellini and Merlino 1979; [18] Saburi et al. 1977; [19] Galli and Alberti 1971; [20] Louisnathan and Smith 1970; [21] Mellini 1981; [22] Smolin et al. 1973; [23] Jansen 1982; [24] Shannon and Katz 1970

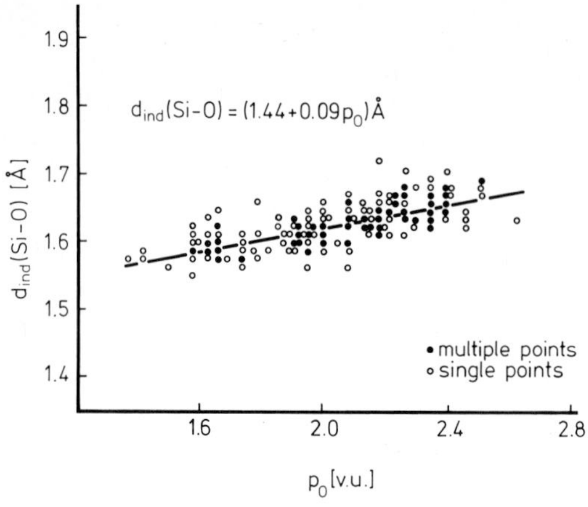

Fig. 3.4. Correlation between the individual bond length d_{ind} (Si–O) and the sum p_O of the electrostatic bond strengths received by each oxygen atom (after Baur 1970)

(z = formal charge of the cation, CN = coordination number of the cation), then the sum of electrostatic bond strengths

$$p_X = \sum_i s_i \tag{3}$$

"received by the anions in a crystal structure is on the average equal, with changed sign, to the valences of the anions. For any individual anion, however, p_X can deviate from this value up to about 40 percent." According to Baur there is a direct correlation between the individual distances d_{ind} (A–X) within a coordination polyhedron [AX_n] and the p_X values received by the individual X anions. The correlation equations for a number of the more commonly occurring [AX_n] polyhedra have been obtained by Baur from regression analyses. Figure 3.4 presents the relationship between d_{ind} (Si–O) and p_O for silicates containing [SiO_4] tetrahedra.

A similar empirical relationship

$$s = \left[\frac{1.605}{d(\text{Si}-\text{O})}\right]^4 \tag{4}$$

between the bond strength s and Si–O distance in silicates containing tetrahedrally coordinated silicon has been given by Brown and Shannon (1973).

3.1.2.2 Correlations Between Si–O Distance and Bond Angles

Although the above bond distance/bond strength correlations can be interpreted in a relatively simple way, several other empirical relationships have been reported for which interpretations are less obvious.

From the data presented in Table 3.1, it is not only obvious that the $Si-O_{term}$ distances in the [Si_2O_7] groups are on the average shorter than the $Si-O_{br}$ distances, but it is also evident that the mean value of the six angles $\measuredangle\, O_{term}-Si-O_{term}$ is in general larger than the mean value of the remaining six angles $\measuredangle\, O_{br}-Si-O_{term}$ in which the bridging oxygen atom O_{br} is involved. Taken together, these observations suggest that the silicon atoms are displaced from the centers of the [SiO_4]

tetrahedra, away from the bridging oxygen atoms, as a result of the repulsion between the two Si atoms.

Similar results apply in the cases of silicates with $[SiO_4]$ tetrahedra containing two or three bridging oxygen atoms.

In a series of papers Gibbs and others have shown that the distances $d(Si-O_{br})$ are correlated with the value of the $Si-O_{br}-Si$ angle. The mean value of the $Si-O_{br}$ distances within a structure increases with decreasing $Si-O_{br}-Si$ angle. Hill and Gibbs (1979) give a correlation equation

$$d(Si^{[4]}-O_{br}) = 1.530 - 0.080 \sec(\sphericalangle\, Si-O-Si)\ [\text{Å}] \tag{5}$$

for silicates in general and

$$d(Si^{[4]}-O_{br}) = 1.526 - 0.068 \sec(\sphericalangle\, Si-O-Si)\ [\text{Å}] \tag{6}$$

for silica in particular[3].

Table 3.2 Correlation between the mean bond length $d(Si-O_{br})$ and the $Si-O_{br}-Si$ bond angle for silicates and for silica, calculated from the regression equations given by Hill and Gibbs (1979)

$\sphericalangle\, Si-O_{br}-Si$ [°]	120	130	140	150	160	170	180
$d(Si-O_{br})$ [Å] for silicates	1.690	1.655	1.634	1.622	1.615	1.611	1.610
$d(Si-O_{br})$ [Å] for SiO_2	1.662	1.632	1.615	1.605	1.598	1.595	1.594

Table 3.2 illustrates the decrease in $d(Si-O_{br})$ with increasing $\sphericalangle\, Si-O-Si$ in silicates and SiO_2. In addition, since for a given $Si-O-Si$ angle the $Si-O_{br}$ distance is larger in silicates than in silica, the data shows that $d(Si-O_{br})$ decreases with increasing degree of condensation of the tetrahedra, i.e., with decreasing average coordination number of the oxygen atoms, as observed earlier by Smith and Bailey (1963), (see Sec. 3.1.2.1).

3.1.2.3 Correlations Between Si – O Distance, Oxygen Coordination Number, Bond Strength, and Si – O – Si Angle

The difficulty with regression analyses is that the presence of a strong correlation between the data used as the dependent variable and the property used as the independent variable does not necessarily mean that this dependence actually exists, i.e., does not necessarily imply a cause-and-effect relationship. Moreover, when a factor that has a strong influence on the property studied is not used as an independent variable in the regression analysis, part of the variation of the data due to the omitted variable may be falsely attributed to other included independent

3 $\sec\alpha = (\cos\alpha)^{-1}$.

variables, as demonstrated by Baur and Ohta (1982). These authors took account of the coordination number of the oxygen atom, the bond strengths of the Si–O bonds, and the value of the Si–O–Si angle and obtained the regression equation

$$d_{ind}(Si-O) = 1.560 + 0.131\,\Delta(p_O)_n + 0.014\,CN(O)_{mean} - 0.0108\,CN(O) - 0.019\,\sec(\measuredangle\,Si-O-Si) - 0.013\,[\sec(\measuredangle\,Si-O-Si)]_{mean}\ [Å] \quad (7)$$

for bonds to both bridging and nonbridging oxygen atoms. Here $\Delta(p_O)_n$ is the difference between the individual bond strength p_O and the mean p_O for the $[SiO_4]$ tetrahedron normalized by multiplying each p_O from a cation A by $[\langle d(A-O)\rangle / d_{ind}(A-O)]^2$; $CN(O)_{mean}$ is the mean coordination number of all oxygen atoms of one $[SiO_4]$ tetrahedron, CN(O) the coordination number of the oxygen atom involved in the Si–O bond considered, and $[\sec(\measuredangle\,Si-O-Si)]_{mean}$ is the mean value of sec ($\measuredangle$ Si–O–Si) of all O_{br} atoms in the tetrahedron. This regression equation explains 78% of the variation of $d_{ind}(Si-O_{br})$. The relative importance of the independent variables decreases dramatically from $\Delta(p_O)_n$, which explains 65.7%, to $CN(O)_{mean}$, CN(O), sec ($\measuredangle$ Si–O–Si), and $[\sec(\measuredangle\,Si-O-Si)]_{mean}$ which explain 4.6, 4.4, 2.7, and 0.7% of the variation, respectively. However, as Baur and Ohta point out, even the least important of these terms is statistically significant.

3.1.2.4 Influence of Structural Disorder on Si–O Distances

For the silica polymorphs, Δp_O and CN(O) have fixed values of 0 and 2, respectively, so that according to Eq. (7) the individual $Si-O_{br}$ distance should be controlled by the angle Si–O–Si only. However, the regression equation

$$d_{ind}(Si-O_{br}) = 1.575 - 0.026\,\sec(\measuredangle\,Si-O-Si)\ [Å] \quad (8)$$

obtained from the 100 $d_{ind}(Si-O_{br})$ values available in 1981 from exact structure analyses of the various SiO_2 polymorphs explains only 9% of the variation of the distances (Baur and Ohta 1982). This indicates that one or even several important factors were not considered in the analysis.

One obvious factor omitted is the thermal motion of the atoms. Atomic distances determined from diffraction data are distances between the mean positions of the moving atoms which do not necessarily correspond exactly to the true bond lengths. The atomic distances determined from diffraction methods differ from the actual bond lengths to an extent which depends on the amplitudes of the thermal motions and the strength of coupling of the individual motions of the atoms involved in the bond. The corrected distances are either shorter than the actual bond lengths or longer, depending on the model applied to the "experimental" distances (either correlated thermal motion, often called "riding motion", or noncorrelated thermal motion). Unfortunately, no suitable method is currently available which can accurately correct atomic distances determined from diffraction data for thermal motion errors [see Megaw (1973), Chap. 14, in general and Hazen (1976), in particular]. However, as a general rule, the uncorrected Si–O distances become increasingly shorter than the actual Si–O bond lengths as the amplitudes of thermal motion increase. At higher temperatures the difference may be as large as 0.02 Å.

With regard to bond length errors, static disorder of atoms has the same effect as thermal motion (dynamic disorder).

As a consequence, the regression Eqs. (1), (4) to (8), and many others in the literature, represent relationships between apparent distances and angles rather than between actual bond lengths and bond angles, although they are usually designated as such. Conclusions drawn from such correlations about the influence of crystal chemical factors on bond lengths and bond angles should, therefore, be considered with reservation; especially when the level of significance for the corresponding term in the regression equation is small.

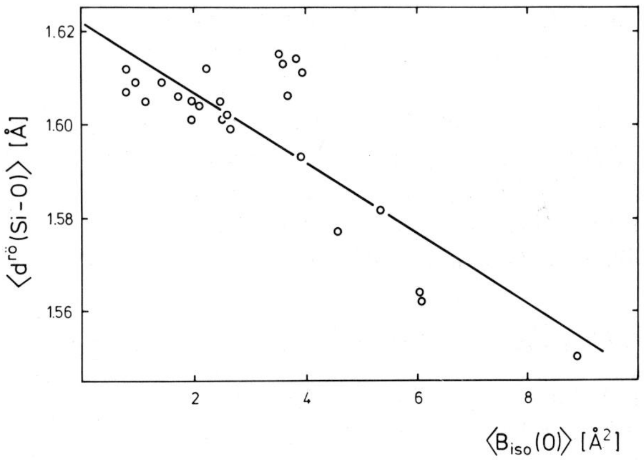

Fig. 3.5. Correlation between the mean values $\langle d^{rö}(Si-O)\rangle$ of the Si – O distances determined with X-rays and the mean isotropic temperature factors $\langle B_{iso}(O)\rangle$ of the oxygen atoms in 25 structure determinations of silica polymorphs and clathrasils (Liebau 1984a)

The influence of even a relatively inaccurate correction of $Si-O_{br}$ distances for thermal motion on the correlation between $d(Si-O_{br})$ and Si – O – Si angle has been studied by Taylor (1972) using a rather limited data set for the silica polymorphs.

Following the discovery of pure silica frameworks with exceptionally strong disorder of the oxygen atoms in three clathrasils (Gies 1983, 1984; Gerke and Gies 1984) (see Sec. 10.7.1), the study of the correlation between the degree of static and/or dynamic disorder and atomic distances $d(Si-O)$ and angles $\measuredangle$ Si – O – Si has been resumed (Liebau 1984a). Based on 25 structure determinations of silica polymorphs and clathrasils performed at various temperatures, the regression equation

$$\langle d^{rö}(Si-O)\rangle = 1.6220 - 0.0075\,(11)\,\langle B(O)\rangle \text{ [Å]} \qquad (9)$$

has been deduced (Fig. 3.5). Here $\langle d^{rö}(Si-O)\rangle$ is the mean value of all Si – O distances, and $\langle B(O)\rangle$ is the mean value of the isotropic temperature factors[4] of the oxygen atoms within a particular structure, determined from X-ray diffraction data. This equation explains 67% of the total variation of the 25 mean Si – O dis-

4 For thermal disorder the temperature factor, B, of an atom derived during a crystal structure determination is described by the equation $B = 8\pi^2 \overline{u^2}$, where $\overline{u^2}$ is the time-averaged mean square displacement of the atom in any direction from its mean position.

tances. The corresponding equation for the 85 individual Si−O bonds from these 25 structure determinations

$$d_{\text{ind}}^{\text{rö}}(\text{Si}-\text{O}) = 1.6157 - 0.0070\,(6)\ \text{B(O)}\ [\text{Å}] \qquad (10)$$

explains 63% of the variation in the individual Si−O distances.

The significance of these two equations lies not so much in the fact that they give a mean value for the Si−O distances in silica, but rather in the fact that they enable us to correct the Si−O distances obtained from X-ray structure analyses of silica frameworks for dynamic and static disorder and to transform them into actual Si−O bond lengths with the aid of

$$d_{\text{ind}}^{\text{real}}(\text{Si}-\text{O}) = d_{\text{ind}}^{\text{rö}}(\text{Si}-\text{O}) + 0.007\ \text{B(O)}\ [\text{Å}]. \qquad (11)$$

Since Eqs. (9) to (11) were obtained from data on SiO_2 frameworks they should, for the present, be applied only to such frameworks until it has been proven that the slope value of 0.007 holds for $Si-O_{br}$ bonds in other silicates as well. It might be expected that large deviations from this value will not occur for bridging oxygen atoms, but that a substantially different correction factor will be necessary for the lengths of $Si-O_{term}$ bonds.

3.1.2.5 Si−O−Si Angles

The Si−O−Si angles subtended at the oxygen atoms bridging between two tetrahedrally coordinated silicon atoms are of considerable interest since they seem to be clear indicators of the character of the Si−O bond.

A mean value of 140° was reported for the angle ∢ Si−O−Si (Liebau 1961a) in 17 structures determined with reasonable accuracy prior to 1961. From the much larger and more accurate data sets now available Tossell and Gibbs (1978) and Baur (1980) constructed very similar histograms of Si−O−Si angle frequencies observed in a large number of silicates. Based on 468 angles, the histogram of Fig. 3.6a shows a large maximum at 139° and small maxima at 157° and 180°.

For a smaller data set of 39 Si−O−Si angles obtained from accurate structure determinations of the silica polymorphs (Fig. 3.6b) (Baur 1980), the histogram has a maximum at about 147° with a shoulder at about 157° and an additional three angles at or very near 180°.

The maximum at 139° in the frequency distribution of Si−O−Si angles in silicates may be interpreted as the value which is energetically most favorable for a strain-free Si−O−Si bond. Larger deviations from 139° may be attributed either to steric effects, such as strain due to the forced accommodation of cations of unfavorable size, or to purely chemical effects, such as the influence of cations of extremely high or low electronegativity or high valence.

As the degree of condensation of $[SiO_4]$ tetrahedra increases, the freedom of a tetrahedron to assume a strain-free orientation relative to its neighbors decreases. This is particularly true when the $[SiO_4]$ tetrahedra are corner-linked to four others in a three-dimensional framework. Therefore, it is perhaps not surprising that the maximum in the SiO_2 histogram of Fig. 3.6b is at a different angle from that in the silicate histogram of Fig. 3.6a. Since less energy is necessary to widen an unstrained Si−O−Si bond by, for example, 10° than to narrow it by 10°, one would

Fig. 3.6. Histograms of Si–O–Si angles observed in **a** silicates (468 angles) and **b** silica polymorphs (80 angles). The histograms are prepared from the "uncorrected" values of Baur's (1980) table

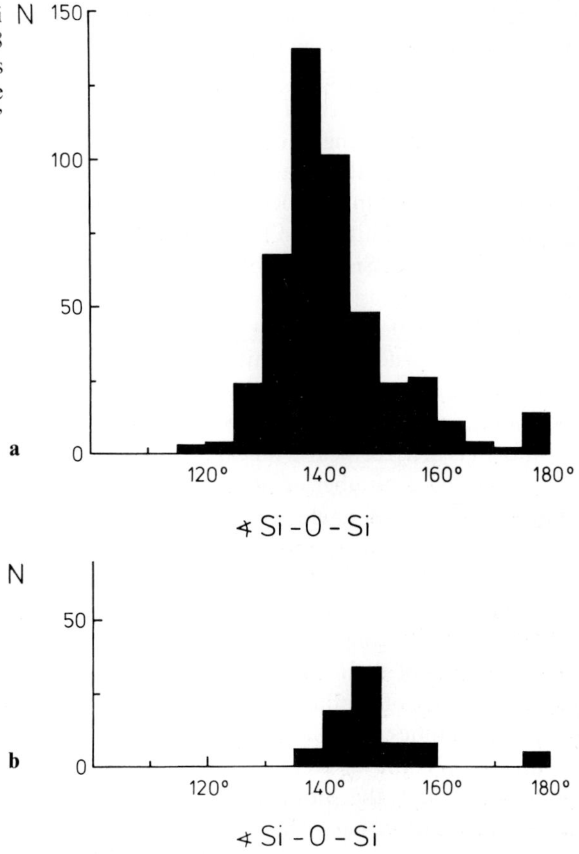

expect the maximum in the silica histogram to be at higher angles than in the silicate histogram, as observed.

The shift of the main maximum in the two histograms has been associated with changes in the coordination numbers of the bridging oxygen atoms (Gibbs 1982). A statistical survey of Si–O–Si angles observed in silicates containing $[Si_2O_7]$ groups (Table 3.1) reveals that the bond angle for three-coordinated oxygen ranges between about 124° and 137°, whereas the angle for two-coordinate oxygen is more variable, ranging between 130° and 180°. This increase in the average value and the range of the Si–O–Si angle as the coordination number of the bridging oxygen changes from three to two is in agreement with the results of recent molecular orbital calculations (Fig. 13 in Gibbs 1982). Since all oxygen atoms are two-coordinated in SiO_2, the mean bond angle should be larger than for silicates, again as observed.

The small maximum at 157° in the silicate histogram and the shoulder at the same value in the silica histogram (Fig. 3.6) has been associated (Baur 1980) with a hypothetic bimodal Si–O–Si distribution in vitreous silica (Vukcevich 1972). However, it is also possible that this peak in both histograms is merely due to non-representative data sets in which certain angles are overrepresented.

The maximum at 180° in both frequency distributions of Si–O–Si angles calls for a more detailed discussion.

In a number of silicate structures reported in the early literature, bridging oxygen atoms had been located on crystallographic symmetry elements in such a way that Si–O–Si bond angles of 180° were forced upon the structure model. In some instances a more careful analysis later showed that these particular O_{br} atoms were not located exactly on the special positions, obviously in order to avoid the straight Si–O–Si bonds. This led to the conclusion that, at least under normal energetical conditions, straight Si–O–Si bonds should not exist in crystalline silicates (Liebau 1961a).

This statement prompted a number of careful diffraction studies designed to determine the true upper limit of the ∢ Si–O–Si value. As a result of these studies, it is now clear that in a number of silicates bridging oxygen atoms do indeed reside on special positions and that straight Si–O–Si bonds do exist. This conclusion is in agreement with careful IR spectroscopic studies.

In addition, a number of structure determinations of high accuracy have shown that Si–O–Si bond angles can have values very close to 180° even in structures in which the corresponding O_{br} atoms are not associated with special positions. These cases need a more thorough discussion in terms of the thermal vibration and positional disorder of the atoms.

Let us start by considering an Si–O–Si group with its three atoms at rest (i.e., the amplitudes of their thermal vibrations are zero) and with the bridging oxygen atom at a center of symmetry (Fig. 3.7a). For this group the Si–O–Si bond angle is 180° at all times.

In Fig. 3.7b the Si atoms are still at rest, but the oxygen atom now vibrates isotropically, i.e., the atomic displacements from the equilibrium position are spherically symmetrical. The time-averaged position of the oxygen atom is then midway between the two silicon atoms and the angle Si–O–Si calculated using the time-averaged position of oxygen is again 180°.

Fig. 3.7c represents a "snapshot" of an Si–O–Si group with an isotropically vibrating oxygen atom as described in Fig. 3.7b. When the displacement vector u_t at a given time t has a non-zero component $u_t^{\perp}$ perpendicular to the Si ··· Si direction, the momentary Si–O–Si angle is then less than 180°. Indeed, for most of the time the three atoms will not be on a straight line and the Si–O–Si bond will not be straight.

When all three atoms of the Si–O–Si bond vibrate, either isotropically (Fig. 3.7d) or anisotropically, the situation becomes more complicated. However, for

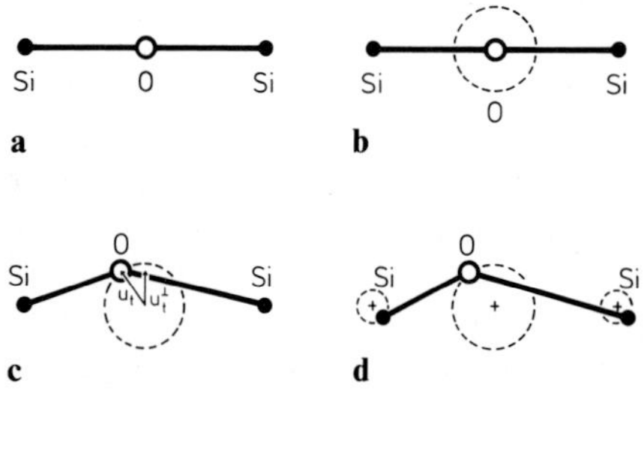

Fig. 3.7a–d. The influence of thermal vibrations on the Si–O–Si bond angle.
a Silicon and oxygen atoms at rest with the bridging oxygen atom midway between the two silicon atoms. **b** Silicon atoms at rest, oxygen atom vibrates isotropically (time average). **c** Snapshot of the vibration shown in **b**. **d** Silicon and oxygen atoms vibrate isotropically (snapshot)

most of the time, the Si–O–Si angle is once again smaller than 180°, even when the time-averaged centers of the electron densities of the three atoms are collinear.

Accurate structure analyses in which the anisotropy of the thermal vibrations is taken into account during the refinement have shown that, in general, the amplitude of vibration of the X atom of a T–X–T bond is higher in a direction within or near a plane perpendicular to the T···T vector. As a typical example, the $[Si_5O_{16}]$ group in zunyite is presented in Fig. 3.8 (Baur and Ohta 1982). In this group, one $[SiO_4]$ tetrahedron shares each of its four (symmetrically equivalent) corners with another $[SiO_4]$ tetrahedron. The ellipsoids shown enclose the volume in which a particular thermally vibrating atom can be found with 50% probability. The thermal ellipsoid of the bridging oxygen atom, shown in four different orientations (one for each of the symmetrically equivalent O_{br} atoms of the group), clearly demonstrates that this atom prefers to vibrate in directions at right angles to the Si ··· Si vector.

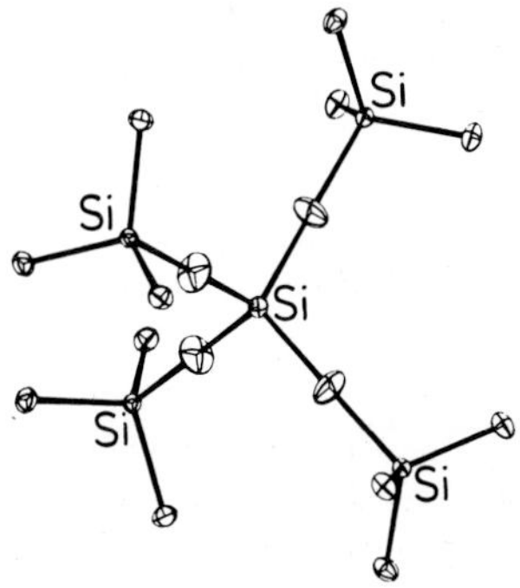

Fig. 3.8. The silicate anion $[Si_5O_{16}]$ of zunyite. The thermal ellipsoids are drawn at the 50% probability level of the electron density of each atom (after Baur and Ohta 1982)

From diffraction experiments it is not possible to distinguish between an atom thermally vibrating about a time-averaged mean position (dynamic disorder) and a nonvibrating atom statistically occupying two or more positions near a space-averaged mean position (static disorder). Consequently, the conclusions about the Si–O–Si angle drawn from dynamical models like those in Fig. 3.7 hold equally well for static models of positional disorder.

Table 3.3 is a list of accurately determined silicate structures in which the reported Si–O–Si angles are close or equal to 180°.

Examination of the isotropic temperature factors of the oxygen atoms shows that in most cases the amplitudes of thermal vibration are considerably higher for the oxygen atoms in straight or nearly straight Si–O–Si bonds than for the oxygen atoms in the same structure, which are bonded to one silicon atom only. This is also clearly demonstrated by the relative sizes of the 50% probability ellipsoids of the $[Si_5O_{16}]$ group in zunyite shown in Fig. 3.8. (At the present state of diffraction methods it is not reasonable to quantitatively compare temperature factors derived from different structure analyses.)

Regression analyses (Liebau 1984a) based on 25 X-ray diffraction structure determinations of silica polymorphs and clathrasils (see Sec. 3.1.2.4) containing a

Table 3.3 Bond angles Si–O–Si of straight and nearly straight Si–O–Si bonds and the corresponding isotropic temperature factors (B) of the oxygen atoms linked to silicon in accurately determined silicate structures

Phase (Mineral name)	$Si-O_{br}-Si$ [°][a]	Sym.[b]	B(O) [Å²][c]	B(O′) [Å²][d]	Ref.
$Sc_2[Si_2O_7]$ (thortveitite)	180	$\bar{1}$	0.98	0.55, 0.65	[1]
$Yb_2[Si_2O_7]$	180	$\bar{1}$	1.02	0.50, 0.54	[2]
$Er_2[Si_2O_7]$	180	$\bar{1}$	0.91	0.56–0.64	[2]
$Al_{13}[Si_5O_{16}](OH, F)_{18}O_4Cl$	180	3m	0.89	0.44[e]	[3]
(zunyite)	180	3m	1.07	0.49[f]	
$Ca_2Ba_4(Fe, Mg)_4[Si_{12}O_{34}]$ (pellyite)	180	$\bar{1}$	2.58	1.02–1.95	[4]
$Ca_6[Si_6O_{17}](OH)_2$ (xonotlite)	180	$\bar{1}$	1.65	0.49–1.21	[5]
$Na_6Be_2[Al_2Si_{16}O_{39}](OH)_2 \cdot 1.5\,H_2O$ (leifite)	180	2/m	1.18	0.95–2.58	[6]
SiO_2 (coesite)	180	$\bar{1}$	0.71	0.70–0.85[g]	[7]
	180	$\bar{1}$	0.94	0.55–0.80[h]	
$LiAl[Si_2O_5]_2$ (petalite)	180	$\bar{1}$	1.83	1.03–1.21	[8]
	163	2	1.62		
SiO_2 (low-tridymite, Cc)	179.1	1	2.52	1.02–2.16	[9]
SiO_2 (high-tridymite, $C222_1$)	179	2	8.32		[10]
	171	2	8.54		
	165	1	8.27		
$BaFe[Si_4O_{10}]I$ (gillespite)	177.7	2	2.33	1.38, 1.62	[11]
$CaV[Si_4O_{10}]O \cdot 4\,H_2O$ (pentagonite)	176.3	1	1.8	1.1 –2.5	[12]
$Ba_3Nb_6[Si_2O_7]_2O_{12}$	180	$\bar{6}$	0.61	0.48–0.74	[13]
$(C_6H_5)_3Si-O-Si(C_6H_5)_3$	180	$\bar{1}$	3.97		[14]
$Si_5O(PO_4)_6$	180	$\bar{1}$	0.87	0.38–0.50	[15]
$K_2Ba_7[Si_4O_{10}]_4$	180	$\bar{1}$	3.6	0.8 –1.8	[16]
	163	m	3.5		
$Rb_2Be_2[Si_2O_7]$	171	2	3.9	1.8 –2.8	[17]

[a] Only $Si-O_{br}-Si$ bonds with bond angles ≃ 180° have been considered
[b] Site symmetry of the bridging oxygen atoms of these $Si-O_{br}-Si$ groups
[c] Isotropic temperature factor of the bridging oxygen atoms of the straight and nearly straight $Si-O_{br}-Si$ groups
[d] Isotropic temperature factors of the other oxygen atoms linked to silicon
[e] Silica-rich zunyite with $[Si_5O_{16}]$ groups
[f] Silica-deficient zunyite with $[(Si_{0.76}Al_{0.24})Si_4O_{16}]$ groups
[g] At 10^5 Pa
[h] At $5.19 \cdot 10^9$ Pa

References: [1] Smolin et al. 1973; [2] Smolin and Shepelev 1970; [3] Baur and Ohta 1982; [4] Meagher 1976; [5] Kudoh and Takéuchi 1979; [6] Coda et al. 1974; [7] Levien and Prewitt 1981; [8] Effenberger 1980, Tagai et al. 1982; [9] Baur 1977b; [10] Dollase 1967; [11] Hazen and Finger 1983; [12] Evans Jr 1973; [13] Shannon and Katz 1970; [14] Glidewell and Liles 1978; [15] Mayer 1974; [16] Cervantes-Lee et al. 1982; [17] Howie and West 1977

total of 85 individual Si–O bonds led to relationships between the bond angles $\measuredangle^{rö}$ Si–O–Si and oxygen isotropic temperature factors B(O), of

$$\langle \sec(\measuredangle^{rö}\, \text{Si–O–Si}) \rangle = -1.243 + 0.031\,(4)\, \langle B(O) \rangle \qquad (12)$$

for the mean values (Fig. 3.9), and

$$\sec(\measuredangle^{rö}\, \text{Si–O–Si}) = -1.233 + 0.028\,(4)\; B(O) \qquad (13)$$

for the individual values. These equations explain 73 and 45%, respectively, of the total variation of the secant of the bridging angle.

Since $d^{rö}$(Si–O) is correlated with B(O) as well as with sec($\measuredangle^{rö}$ Si–O–Si), Eqs. (12) and (13) do not necessarily mean that such high percentages of the secant variations are caused by the disorder itself. The two regression functions extrapolate to Si–O–Si angles of 143.6° and 144.2° for ⟨B(O)⟩ = 0 and B(O) = 0, respectively, i.e., for no thermal vibration of the oxygen atoms. With a mean temperature factor ⟨B(O)⟩ = 1.0 Å², Eq. (12) leads to an apparent bond angle $\measuredangle^{rö}$ Si–O–Si of 145.6°, which is in good agreement with the value 147° for the maximum in the frequency distribution for silica frameworks shown in Fig. 3.6b.

The correlation functions (12) and (13) indicate that the amplitudes of the thermal vibrations of the oxygen atoms tend to increase as the Si–O–Si angle straightens. Although this conclusion is drawn from the crystal structures of silica frameworks, it is in agreement with the observation that, as a rule, also in silicates the B(O) values are high for those oxygen atoms for which the experimentally determined Si–O–Si angles are near or equal to 180°.

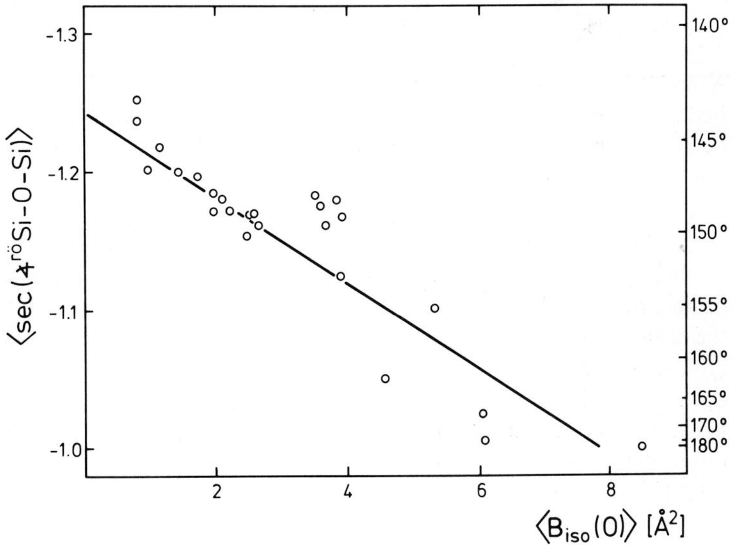

Fig. 3.9. Correlation between the mean values ⟨sec ($\measuredangle^{rö}$ Si–O–Si)⟩ of the secant of the Si–O–Si bond angles determined with X-rays and the mean isotropic temperature factors ⟨B_{iso}(O)⟩ of the oxygen atoms in 25 structure determinations of silica polymorphs and clathrasils (Liebau 1984a)

Inspection of the temperature factors in Table 3.3, however, shows that in a few structures the bridging oxygen atom does not deviate significantly from a straight Si−O−Si bond. This is particularly true for leifite and coesite, a polymorph of silica synthesized under high pressure, i.e., 6.5 GPa and 1375 K for the crystals studied by Levien and Prewitt (1981). In coesite at ambient pressure and temperature, the temperature factor of the oxygen atom at the center of symmetry between two silicon atoms is lower than the average of the temperature factors of the other oxygen atoms. As pressure increases, however, the amplitudes of thermal vibrations of the oxygen atom at the center of symmetry increase, while those of the other oxygen atoms decrease. This unusual behavior of the special oxygen atom has been explained by a destabilization of the symmetric Si−O−Si bond with increasing pressure (Levien and Prewitt 1981).

Conclusions. Information available to date about the Si−O−Si bond indicates that

(1) the bond angle of a strain-free Si−O−Si bond is near 140°;
(2) the space average of the Si−O−Si angle (calculated from the space average positions of the atoms) varies between ca. 120° and 180°;
(3) the time average of the Si−O−Si angle is always smaller than 180° due to thermal vibration;
(4) the contribution of the bridging angle to the energy of an Si−O−Si group is controlled by the time average of the bond angle rather than by the space average. It is, therefore, this time average of the Si−O−Si bond angle which is of crystal chemical significance.

3.1.3 Bond Lengths and Bond Angles in Silicates Containing $[SiO_6]$ Octahedra (Hexaoxosilicates)

X-ray diffraction crystal structure determinations of silicates containing octahedrally coordinated silicon are not common. Until 1984, 13 such structures had been published with 19 crystallographically independent $[SiO_6]$ octahedra. From the data presented in Table 3.4 the linear regression equation

$$^{6}\langle d(\mathrm{Si}^{[6]}-\mathrm{O})\rangle = 1.731 + 0.013(3)\,\langle \mathrm{CN(O)}\rangle\ [\text{Å}] \qquad (14)$$

is obtained between the mean value of the six Si−O distances within an $[SiO_6]$ octahedron and the mean value of the coordination numbers of all oxygen atoms of the octahedron (Fig. 3.10). It is in good agreement with the equation given by Baur (1977a) based on a smaller data set. The equation indicates that the mean Si−O distance of an $[SiO_6]$ octahedron increases with increasing mean coordination number of the oxygen atoms. It is analogous to the relationship obtained by Brown and Gibbs (1969) for the Si−O bond lengths in $[SiO_4]$ tetrahedra (see Eq. (1) in Sec. 3.1.2.1).

No silicate has yet been found in which the $[SiO_6]$ octahedron has ideal symmetry $4/m\,\bar{3}\,2/m$. Instead, the octahedra show angular and/or bond length distortions.

Individual $Si^{[6]}$−O distances vary between 1.70 and 1.84 Å, but seldom by more than 0.05 Å within a given $[SiO_6]$ octahedron. Deviations of the bond angles

Table 3.4 Dimensions of [SiO_6] octahedra in hexaoxosilicon compounds

Phase	Sym.[a]	⟨CN(O)⟩[b]	$^6\langle d(Si-O)\rangle$ [Å][c]	$d(Si-O)$ [Å]	$(\measuredangle O-Si-O)_{90}$ [°][d]	$(\measuredangle O-Si-O)_{180}$ [°][e]	$^{12}\langle d(O\cdots O)\rangle$ [Å][g]	$d(O\cdots O)$ [Å][f]	Ref.
$Si_3^{[6]}Si_2^{[4]}[PO_4]_6O$	$\bar{3}$	2	1.758	1.758	87.7, 92.3	180	2.486	2.436, 2.536	[1]
	3	2	1.768	1.744, 1.791	89.6–90.4	179.4, 180.6	2.500	2.458–2.542	
$Si[P_2O_7]AI$	1	2	1.750	1.704–1.786	87.5–91.4	178.0–182.0	2.475	2.427–2.542	[2]
	1	2	1.750	1.729–1.774	88.2–92.6	178.7–181.3	2.475	2.411–2.525	
	1	2	1.755	1.741–1.778	86.6–92.7	174.6–185.4	2.481	2.394–2.570	
	1	2	1.758	1.715–1.788	87.1–92.7	174.8–185.2	2.485	2.422–2.546	
	3	2	1.755	1.750–1.760	88.0–92.6	177.6, 182.4	2.482	2.445–2.536	
	$\bar{3}$	2	1.730	1.730	89.2–90.8	180	2.446	2.430–2.463	
$Si[P_2O_7]AIII$	1	2	1.763	1.732–1.793	88.1–92.0	173.4–186.6	2.493	2.454–2.559	[3]
$Si[P_2O_7]AIV$	1	2	1.766	1.736–1.786	88.6–91.1	178.5–181.5	2.498	2.466–2.528	[4]
$(NH_4)_2Si[P_4O_{13}]$	1	2	1.771	1.762–1.788	86.5–93.2	179.4–180.6	2.505	2.423–2.572	[5]
$[C_5H_5NH]_2[(C_6H_4O_2)_3Si]$	3	2.33	1.784	1.765–1.813	86.4–94.4	175.4–184.6	2.527	2.482–2.605	[6]
SiO_2 (hP) (stishovite)	mmm	3	1.774	1.757–1.809	81.3–98.7	180	2.379	2.290–2.666	[7]
$K_2Si[Si_3O_9]$ (hP) (wadeite-type)	$\bar{3}$	4	1.778	1.778	89.0–91.0	180	2.515	2.494–2.536	[8]
$Ca_3[Si(OH)_6][SO_4][CO_3] \cdot 12\,H_2O$ (thaumasite)	3	4	1.780	1.778, 1.781	86.5–93.4	169.1, 190.9	2.515	2.437–2.590	[9]
$Sc_2Si_2O_7$ (hP) (pyrochlore-type)	$\bar{3}$m	4	1.761	1.761	87.5–92.5	180	2.491	2.437–2.544	[10]
$In_2Si_2O_7$ (hP) (pyrochlore-type)	$\bar{3}$m	4	1.800	1.800	86.0–94.0	180	2.543	2.453–2.633	[11]
$MgSiO_3$ (hP) (ilmenite-type)	3	4	1.799	1.768, 1.830	80.8–97.2	166.4, 193.6	2.541	2.331–2.682	[12]
$MgSiO_3$ (hP) (perovskite-type]	$\bar{1}$	4.67	1.79	1.75, 1.79, 1.82	87.9–92.1	180	2.532	2.480–2.600	[13]
Ranges				1.704–1.835	80.8–98.7	166.4–193.6		2.290–2.682	
Mean values			1.768				2.497		

[a] Site symmetry of the silicon atom
[b] Mean coordination number of the oxygen atoms linked to silicon
[c] Mean value of the bond lengths of the six Si–O bonds
[d] Range of the 12 cis angles $\measuredangle$ O–Si–O
[e] Range of the six trans angles $\measuredangle$ O–Si–O
[f] Range of edge lengths
[g] Mean value of the 12 edge lengths

References: [1] Mayer 1974; [2] Tillmanns et al. 1973; [3] Bissert and Liebau 1970; [4] Hesse 1979a; [5] Durif et al. 1976; Sackerer and Nagorsen 1977; [6] Flynn and Boer 1969; [7] Sinclair and Ringwood 1978; Hill et al. 1983; [8] Swanson and Prewitt 1983; [9] Effenberger et al. 1983; [10, 11] Reid et al. 1977; [12] Horiuchi et al. 1982; [13] Yagi et al. 1978

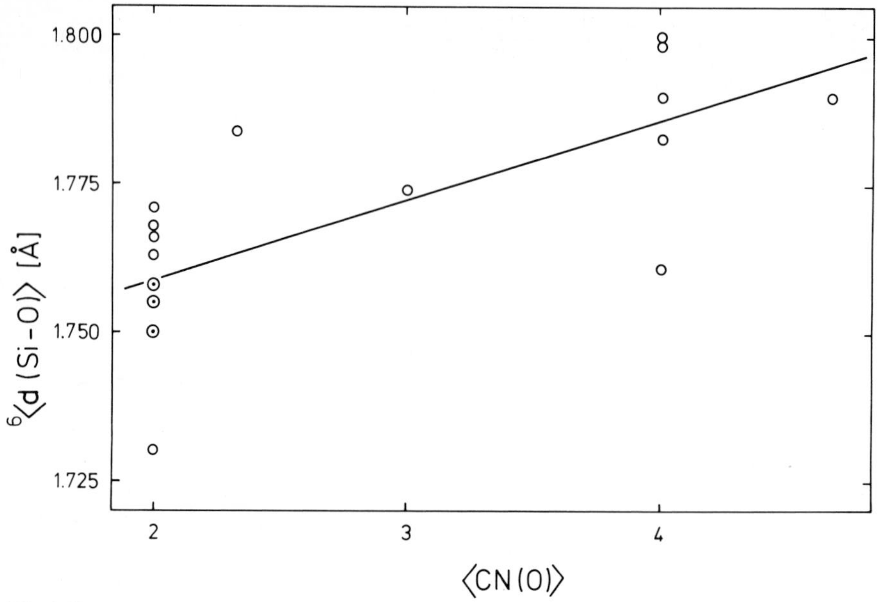

Fig. 3.10. Correlation between the mean values $^6\langle d(Si-O)\rangle$ of the six Si−O distances of an $[SiO_6]$ octahedron and the mean coordination number $\langle CN(O)\rangle$ of oxygen

∢O−Si−O by more than 5° from the ideal values of 90° and 180°, respectively, are rare (Table 3.4), indicating that the distortions of the $[SiO_6]$ octahedra are small.

The data in Table 3.4 gives overall mean values of 1.77 Å for the bond lengths $d(Si^{[6]}-O)$ and 2.50 Å for the lengths $d(O\cdots O)$ of the octahedral edges. A more detailed discussion of bond lengths and bond angles in $[SiO_6]$ octahedra is premature since only a rather small number of structures with six-coordinated silicon have been determined so far. Moreover, large deviations from the regression line for at least two of the data points in Fig. 3.10 suggest that not all these structure determinations are of very high accuracy.

3.1.4 The Nature of the Si−O Bond

In Sec. 1.3 it was demonstrated that the multitude of silicates, as compared to the smaller number of phosphates, sulfates, chromates, etc. can be explained by the relatively moderate repulsive forces between the silicon atoms, which allow extensive linking of $[SiO_4]$ tetrahedra by sharing common oxygen atoms. The large structural diversity of the silicates is further enhanced by peculiarities in their electronic structure, i.e., by special features of the Si−O bond. Although many studies have been made of the electronic structure of silicates, we are still far from a profound understanding of the detailed nature of the Si−O bond. Nevertheless, some progress has been made during the last decade by the comparison of X-ray fluorescence, photoelectron spectroscopic measurements, and accurate crystal structure refinements with the results of quantum mechanical calculations of various kinds. In this section an attempt is made to give only a brief and simple description of the subject. For a more detailed treatment it is recommended that the original literature be consulted. The list of references at the end of this chapter is a small selection, designed to provide only a starting point for access to the modern literature in this field.

3.1.4.1 The Ionic Model

The bond between silicon and oxygen is generally considered to be partly ionic and partly covalent. Therefore, up to a certain point the ionic model is about as successful in explaining structural and chemical properties of the silicates as the covalent model.

In a purely ionic model a silicate of general formula $M_r^{+q}Si_sO_t$ is described as composed of silicon ions Si^{+4}, oxygen ions O^{-2}, and ions M^{+q}, where $q=2(t-2s)/r$. These ions are held together by electrostatic forces. Since the Coulomb potential of an ion has spherical symmetry, the attractive and repulsive forces are also spherically symmetric, and the bonds between the ions are nondirectional. In the ionic model the structures of silicates are governed by their chemical composition, by the distances between the ions, and by their charges. In particular, for the most part, they obey Pauling's rules for ionic crystals (Pauling 1967, p. 543 ff.) (see Chap. 9). Applying Pauling's crystal radii (Pauling 1967, p. 514) for silicon, 0.41 Å, and O^{-2}, 1.40 Å, one discovers that the radius ratio $r_{Si^{+4}} : r_{O^{-2}} = 0.29$ lies well within the range $0.255 < r_{cat} : r_{an} \leq 0.414$ for tetrahedral coordination. This is in agreement with the fact that in the great majority of silicates silicon is tetrahedrally coordinated by four oxygen ions.

Shannon and Prewitt (1969, 1970; Shannon 1976) derived an extensive set of "effective" ionic radii for ions with different coordination numbers (see Appendix III). Their values for four- and six-coordinated silicon are $r_{Si^{[4]}} = 0.26$ Å and $r_{Si^{[6]}} = 0.400$ Å, respectively. The latter value has since been revised by Baur (1977a) to 0.407 Å. The Shannon/Prewitt radius of oxygen ranges from $r_{O^{[2]}} =$ 1.35 Å to $r_{O^{[8]}} = 1.42$ Å; for silicates, the value $r_{O^{-2}} = 1.37$ Å, which is intermediate between the radius values tabulated for three- and four-coordinated oxygen, seems to be adequate. With these ionic radii the ratio is $r_{cat} : r_{an} = 0.19$ for $Si^{[4]}$ and 0.29 for $Si^{[6]}$. These values lie below the ranges 0.225 to 0.414 and 0.414 to 0.732, corresponding to tetrahedral and octahedral coordination, respectively, indicating that the Si–O bond is not purely ionic.

According to the ionic model, the oxygen ions tend to be close-packed, the Si^{+4} ions filling tetrahedral or octahedral sites between the oxygen ions. The M cations also occupy tetrahedral and octahedral sites, provided that these cations are of the appropriate size. If not, they may still enter the vacant spaces between the oxygen ions, but they will distort the close-packed array of oxygen ions in so doing.

A close-packed array of oxygen ions of 1.37 Å radius has a specific volume per O^{-2} ion, V_{ox}, of 14.55 $Å^3$. The data in Table 3.5 show that with the exception of the very high pressure phase stishovite, only a few silicates have a density anywhere near that of a close-packed oxygen ion arrangement. The often considerable deviations from this ideal density are in some cases due to the presence of cations

Table 3.5 Specific volumes per oxygen atom, V_{ox}, for a representative selection of silicates compared with the theoretical value calculated using an oxygen ion radius of 1.37 Å

Silicate		V_{ox} [Å^3]	Radius ratio for CN		
Name	Formula		4	6	>6
Oxygen ion	O^{-2}	14.55			
1. Phases with six-coordinated silicon					
Stishovite	SiO_2 (hP)	11.6		0.30	
Synthetic	$Si[P_2O_7]$	14.9	0.12	0.30	
2. Phases with four-coordinated silicon					
a) Phases with four-coordinated cations only					
Phenakite	$Be_2[SiO_4]$	15.3	0.20		
Coesite	SiO_2 (hP)	17.7	0.19		
Quartz	SiO_2	18.8	0.19		
Synthetic	$Li_4[Si_2O_6]$	19.7	0.43		
Synthetic	$Li_2[Si_2O_5]$	20.4	0.43		
Synthetic	$Li_6[Si_2O_7]$	20.8	0.43		
b) Phases with six-coordinated M cations only					
Orthoenstatite	$Mg_2[Si_2O_6]$	17.4		0.53	
Forsterite	$Mg_2[SiO_4]$	18.4		0.53	
Thortveitite	$Sc_2[Si_2O_7]$	18.7		0.54	
Synthetic	$K_6[Si_2O_7]$	36.5		1.01	
c) Phases containing M cations with CN > 6					
Pyrope	$Al_2Mg_3[SiO_4]_3$	16.0		0.39	0.65
Zircon	$Zr[SiO_4]$	16.1			0.61
Andradite	$Fe_2Ca_3[SiO_4]_3$	18.2		0.47	0.82
Synthetic	$Ba_2[Si_2O_6]$ (hT)	19.7			1.04
Albite	$Na[AlSi_3O_8]$	20.8	0.28		0.82
Celsian	$Ba[Al_2Si_2O_8]$	23.0	0.28		1.07
Sanbornite	$Ba[Si_2O_5]$ (lT)	24.1			1.04

hT = high temperature phase; lT = low temperature phase; hP = high pressure phase; CN = coordination number of the cations

with sizes which are too large to fit into the tetrahedral and octahedral holes of the close-packed arrangement, e.g., synthetic $K_6[Si_2O_7]$ and sanbornite, $Ba[Si_2O_5]$ (lT). However, in phases like phenakite, forsterite, thortveitite, orthoenstatite, the silica polymorphs, and the lithium silicates, the cations have radius ratios well below the upper limit of 0.732 for octahedral coordination. In these phases the deviations of the V_{ox} values from the theoretical value for close-packing are due largely to the fact that the Si–O bond has a significant degree of covalent character rather than being purely ionic.

3.1.4.2 The Covalent Model

In this treatment the covalent nature of the Si–O bond is described in terms of the valence bond concept.

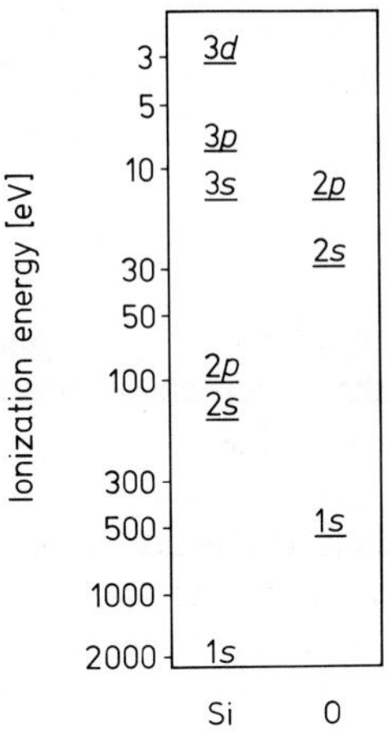

Fig. 3.11. Ionization energies of the electrons of silicon and oxygen

The electron configuration of a silicon atom in its ground state is

$$1s^2\,2s^2\,2p^6\,3s^2\,3p_x^1\,3p_y^1\,3p_z^0\,3d_{xy}^0\,3d_{yz}^0\,3d_{xz}^0\,3d_{x^2-y^2}^0\,3d_{z^2}^0.$$

An energy of only about 6 eV is required to promote one electron from the $3s$ level to a $3p$ level, leading to the valence state configuration

$$1s^2\,2s^2\,2p^6\,3s^1\,3p_x^1\,3p_y^1\,3p_z^1\,3d_{xy}^0\,3d_{yz}^0\,3d_{xz}^0\,3d_{x^2-y^2}^0\,3d_{z^2}^0.$$

The required promotion energy is readily gained by bond formation.

Since the energy difference between the $3s$ and $3d$ levels of silicon is small (~11 eV) and about the same size as that between Si $3d$ and the $2p$ level of oxygen (Fig. 3.11), all nine orbitals in the outer (valence) shell of silicon can participate in bonding. The degree of participation of the various Si $3d$ orbitals in the bonds formed depends on the kind of ligands, in particular on their electronegativity.

3.1.4.2.1 Four-Coordinated Silicon. Consider the tetrahedral Si–O bonds in a typical silicate such as Mg_2SiO_4. There is strong hybridization of the $3s$ and the three $3p$ orbitals of silicon to form an sp^3 hybrid with the four equivalent hybrid orbitals

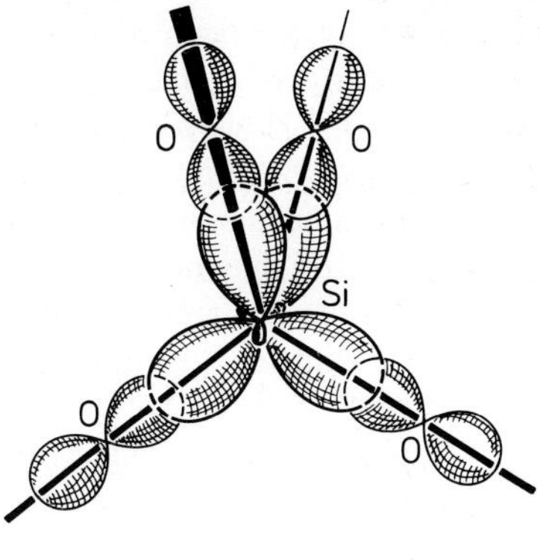

Fig. 3.12. Overlapping of Si $3s$ and O $2p$ orbitals in the sp^3 hybrid of an $[SiO_4]$ polyhedron

pointing towards the corners of a tetrahedron. Each lobe of an Si sp^3 hybrid can then overlap "head-on" with a $2p$ orbital of an oxygen atom to form a σ bond, i.e., a bond in which the maximum of the overlap electron density is on the Si–O line (Fig. 3.12).

In addition to the four σ bonds of the sp^3 hybrid there is also some overlap of the remaining $2p$ orbitals of oxygen with the d orbitals of silicon. Such overlap between d and p orbitals results in the formation of a π bond, i.e., a bond in which the maximum overlap of the orbitals lies off the straight line between the two atoms.

For typical silicates $M_rSi_sO_t$ containing electropositive cations M, the degree of participation of the $3d_{x^2-y^2}$ and $3d_{z^2}$ orbitals of silicon in the Si–O bonds is smaller than that of the other three Si $3d$ orbitals, $3d_{xy}$, $3d_{yz}$, and $3d_{xz}$. However, as the electronegativity of the M atoms increases the overlap of the O $2p$ orbitals (not participating in σ bonds) with the Si $3d_{x^2-y^2}$ and Si $3d_{z^2}$ orbitals becomes more favorable, possibly at the expense of the other three Si $3d$ orbitals. The weak bond formed in this way is also of the π-bond type.

While the σ bond between Si sp^3 and O $2p$ is a single covalent bond, the π-type Si $3d$–O $2p$ overlap adds some double-bond character to the resulting Si–O bond.

For more than two decades the question of the extent to which, if at all, Si $3d$ orbitals participate in bonding has been extensively debated in the literature. The difference between the sum of the covalent single-bond radii of silicon and oxygen $1.17\,Å + 0.66\,Å = 1.83\,Å$ and the observed bond lengths d_{ind}(Si–O), as well as their grand mean value 1.62 Å, has been ascribed to two effects: (i) partial ionic character of the bond (Schomaker and Stevenson 1941) and (ii) some double-bond character due to Si $3d$ – O $2p$ overlap (Cruickshank 1961).

In early research a considerable degree of d-orbital participation was assumed. Early molecular orbital calculations seemed to indicate that in fact substantial d-orbital participation was necessary to obtain good agreement between the calculations and spectroscopic measurements. However, with increasing advances in spectroscopy and computational quantum mechanical chemistry, the proposed degree of Si $3d$ participation has leveled off at a few percent for normal silicates. Indeed, the accuracy inherent in the various methods may now have reached a sufficiently high standard so that the errors are of a similar size to the differences in π-bond character in different silicon compounds.

The covalent bond model is in good agreement with the observation that the great majority of silicon compounds contain silicon in tetrahedral coordination and that the bond angles ∢ O–Si–O show only small deviations from the tetrahedral value 109.47°. However, the directional character of the covalent Si–O bond is even more convincingly displayed by the fact that the majority of Si–O–Si angles scatter in a rather small range near 140° (Fig. 3.6).

Whether, on the one hand, we assume that the four electron pairs in the valence shell of an oxygen atom avoid each other as much as possible or, on the other hand, we assume oxygen to form an sp^3 hybrid, the bond angle ∢ Si–O–Si should be near 109°.

The observed significant widening of this angle from 109° is often regarded as a reflection of the π-bond character of the Si–O bond resulting from participation

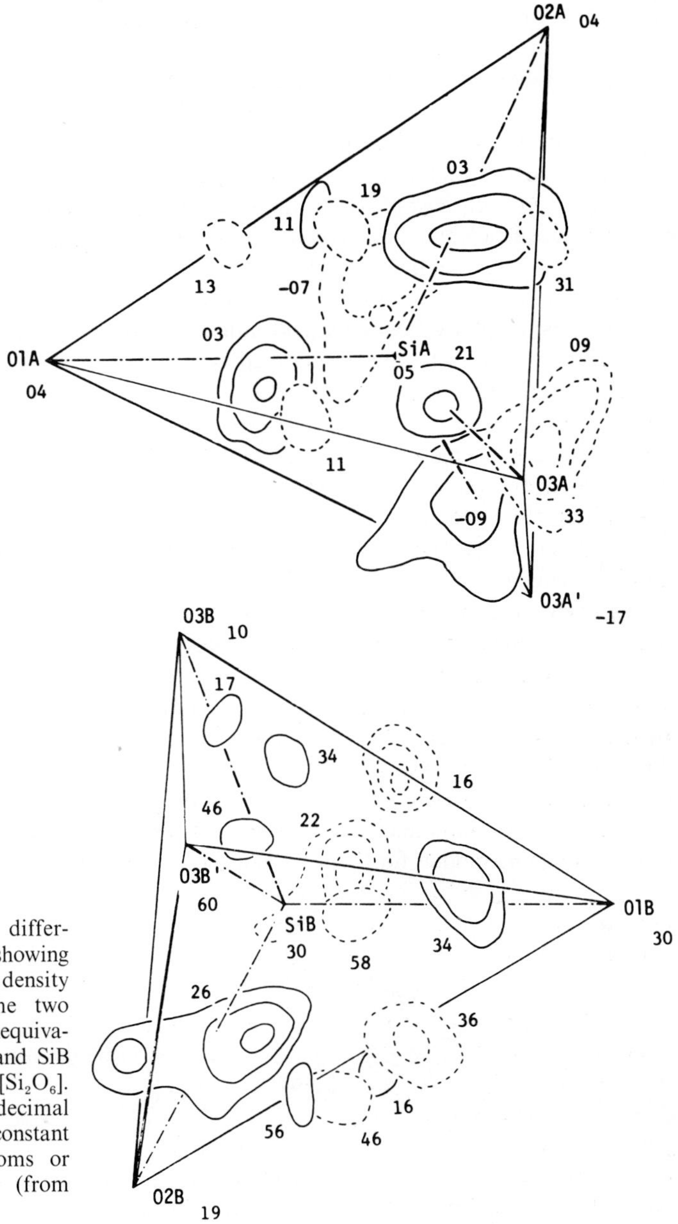

Fig. 3.13. Composite difference Fourier diagrams showing the residual electron density distribution around the two crystallographically nonequivalent silicon atoms SiA and SiB of orthoenstatite, $Mg_2[Si_2O_6]$. *Numbers* give, in decimal fractions of the lattice constant c_0, the heights of atoms or centers of the peak (from Sasaki et al. 1982)

of the Si 3*d* orbitals as described above. A different interpretation on the basis of nonbonded interaction between the silicon atoms is given in Sec. 3.1.4.3.

The observation that the bond lengths $d(Si-O_{term})$ are generally shorter than $d(Si-O_{br})$ (Table 3.1) is also often explained by a higher double-bond character of the (*d-p*) π-type component in the $Si-O_{term}$ bond. However, this difference in $d(Si-O)$ can also be explained by differences in the number and strength of bonds reaching the oxygen atoms, independent of the nature of the Si – O bond itself (see Sec. 3.1.2.1).

Considerable effort has recently been devoted to the determination of the distribution of the valence electrons from very accurately measured X-ray and neutron diffraction data. The instrumental and computational methods developed have now reached a stage where such "experimental" charge density distributions unambiguously demonstrate that there is an accumulation of electron density between the silicon and the oxygen atoms, suggesting that the Si – O bond has a significant covalent component. This is clearly visible in Fig. 3.13 which presents a composite map of the aspherical (deformation) electron density distribution around the two symmetrically nonequivalent silicon atoms in orthoenstatite, $Mg_2[Si_2O_6]$ (Sasaki et al. 1982). The density mapped is that part of the electron density which results when the spherical component of the electron density of each atom is subtracted. Similar residual electron density distributions have been obtained by other authors using a variety of slightly different methods. They are in reasonable agreement with corresponding "theoretical" charge density maps obtained from molecular orbital calculations.

At present, minor details relating to the location and magnitude of the residual peaks in many of the experimental electron density maps in the literature are still a matter of dispute. Deviations of the peaks from the lines connecting the silicon and oxygen atoms may, or may not, indicate some π-bond character. The fact that, for example, the straight Si – O(1) – Si bond in coesite, a high pressure polymorph of silica (see p. 30) shows considerable deviation of the residual electron density maxima from the Si···O line, whereas the maxima in the bent Si – O – Si bonds do less so (Fig. 3.14) (Ross 1980), seems to indicate that such minor details may be real in this case. Future improvement in the experimental and computational methods will no doubt enable more stringent tests to be made of the value of the different bonding models for the tetrahedral $[SiO_4]$ group.

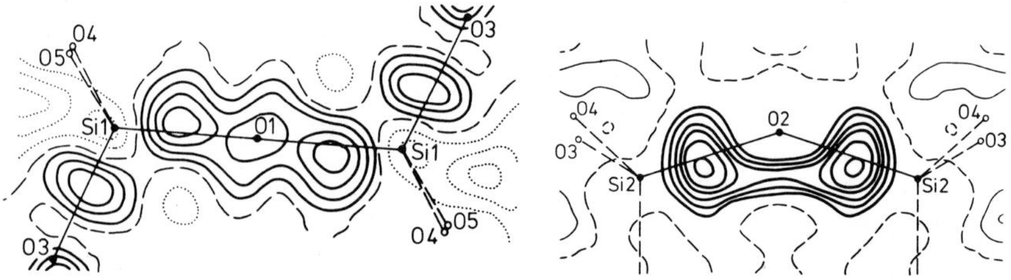

a **b**

Fig. 3.14. Sections through the difference Fourier maps of coesite, SiO_2, showing the residual electron density distribution in **a** a plane through the straight Si – O – Si group and **b** a corresponding plane through the bent Si – O – Si group (from Ross 1980)

Planar fourfold coordination of silicon by oxygen has recently been suggested for the crystalline orthosilicic ester bis(o-phenylenedioxy)silane,

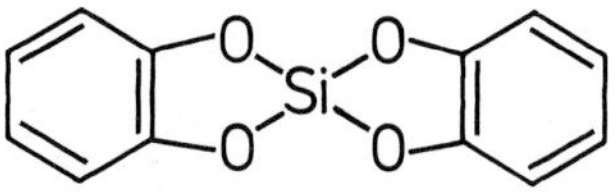

(Meyer and Nagorsen 1979; Würthwein and Schleyer 1979). Whether the silicon coordination in this compound is in reality square planar or is supplemented to octahedral coordination by two additional ligands, such as hydrate water molecules, or merely simulated by stacking disorder of the molecules, has yet to be proven (Dunitz 1980; Wojnowski et al. 1984). So far, no phase is known in which planar silicon coordination has definitely been observed (Bibber et al. 1983).

3.1.4.2.2 Five-Coordinated Silicon. In a number of quite stable phases silicon is five-coordinated. The structures of several such groups studied by single crystal X-ray diffraction methods are presented in Fig. 3.15. For a number of additional compounds the presence of five-coordinated silicon has been established by spectroscopic and purely chemical studies.

In compounds containing $[SiA_5]$ groups the first coordination sphere of the silicon atom contains at least one strongly electronegative atom other than oxygen, such as carbon ($\chi_C = 2.50$)[5], nitrogen ($\chi_N = 3.07$), chlorine ($\chi_{Cl} = 2.83$), or fluorine ($\chi_F = 4.10$). Although the atomic electronegativities of nitrogen, carbon and chlorine are lower than that of oxygen, the group electronegativity (Hinze 1967/1968) of all the ligands in these compounds is high due to the relatively high atomic electronegativity of carbon atoms in the second coordination sphere of the $[SiA_5]$ group. In other words, C and N increase the tendency of silicon to assume coordination numbers higher than four more than do the more electropositive cations Na^+, Mg^{+2}, Ca^{+2}, Fe^{+2}, H^+, etc. ($\chi_A = 1.01$, 1.23, 1.04, 1.64, 2.1, etc., respectively). This trend is illustrated by the data in Table 3.6.

In the phases containing five-coordinated silicon, the coordination polyhedron is usually a slightly distorted trigonal bipyramid. Such a bipyramid is commonly

Table 3.6 Correlation between the coordination number of silicon, *CN*, and the electronegativities of its nearest and next-nearest neighbors in normal pressure compounds (electronegativity values from Allred and Rochow 1958)

Si–A–M	χ_A	χ_M	$\lvert\chi_{Si} - \chi_A\rvert$	$\lvert\chi_A - \chi_M\rvert$	*CN*	Compounds
Si–O–Ca	3.50	1.00	1.76	2.50	4	Ca silicates
Si–O–Mg	3.50	1.23	1.76	2.27	4	Mg silicates
Si–O–Fe	3.50	1.64	1.76	1.86	4	Fe silicates
Si–O–Si	3.50	1.74	1.76	1.76	4	Silica
Si–O–P	3.50	2.06	1.76	1.44	4, 6	Silicon phosphates, e.g. SiP_2O_7
Si–O–H	3.50	2.1	1.76	1.4	4, 6	$Si(OH)_4$, thaumasite
Si–O–C	3.50	2.50	1.76	1.00	4, 6	Esters, chelates
Si–O–C[a]	3.50	2.50	1.76	1.00	4, 5	Organosilicon compounds
Si–N–Si	3.07	1.74	1.33	1.33	4	Si_3N_4
Si–N–C[a]	3.07	2.50	1.33	0.57	4, 5	Organosilicon compounds
Si–C–C[a]	2.50	2.50	0.76	0	4, 5	Organosilicon compounds
Si–C–N[a]	2.50	3.07	0.76	0.57	4, 5	Organosilicon compounds

[a] Compounds with $[SiA_n]$ having more than one kind of ligands A

5 Throughout this text atomic electronegativity values from the Allred and Rochow scale (1958) are used (Appendix II).

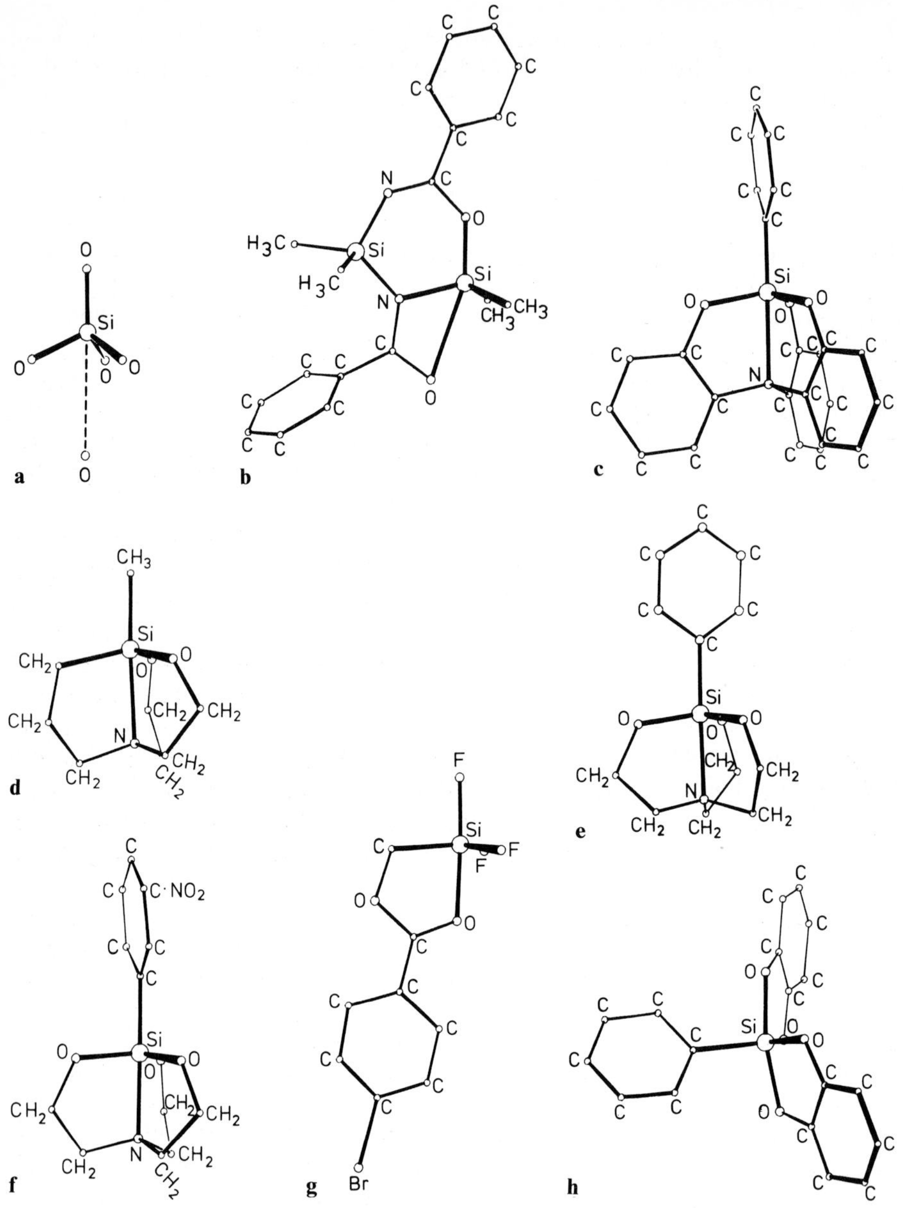

Fig. 3.15 a – h. Molecular geometry of compounds with pentacoordinate silicon. Numbers in parentheses refer to the data points in Fig. 3.16.

a Forsterite $Mg_2[SiO_4]$ (1); **b** cyclobis-(benzamidodimethylsilane) $C_{18}H_{22}N_2O_2Si_2$ (Boer and Remoortere 1970) (3); **c** phenyl-(2,2′,2″-nitrilotriphenoxy)silane $(C_6H_5)Si(OC_6H_4)_3N$ (Boer et al. 1968a) (5); **d** methyl-(2,2′,3-nitrilodiethoxypropyl)silane $C_8H_{17}NO_2Si$ (Boer and Turley 1969) (6); **e** phenyl-(2,2′,2″-nitrilotriethoxy)silane $(C_6H_5)Si(OCH_2CH_2)_3N$ (Turley and Boer 1968) (7); **f** m-nitrophenyl-(2,2′,2″-nitrilotriethoxy)silane $C_{12}H_{16}N_2O_5Si$ (Turley and Boer 1969) (10); **g** (4-bromobenzoyloxymethyl)-trifluorosilane $BrC_6H_4COOCH_2SiF_3$ (Voronkov et al. 1979b) (11); **h** tetramethylammonium-bis(o-phenylenedioxy)phenylsiliconate $(C_6H_5)Si(O_2C_6H_4)_2[N(CH_3)_4]$ (Boer et al. 1968b) (16) described in terms of the formation of an $sp^3 d_{z^2}$ hybrid on the central silicon atom. In the ideal $sp^3 d_{z^2}$ hybrid (and trigonal bipyramid), $[SiX_3Y_2]$, the bond angles between the equatorial ligands, ∢X−Si−X, are 120°, those between the apical ligands, ∢Y−Si−Y, are 180°, and those between equatorial and apical ligands, ∢X−Si−Y, are 90°.

In Fig. 3.16a the mean values of the three X−Si−X angles for the 15 accurately known structures containing pentacoordinate silicon are plotted versus the mean values of the angles ∢X−Si−Y and ∢X−Si−Y′, respectively, where Si−Y is the shorter and Si−Y′ the longer of the two apical bonds (Liebau 1984b). The data points fall on two curves which pass through the points for an ideal $sp^3 d_{z^2}$ hybrid, an ideal sp^3 hybrid in a hexagonal close-packed array of oxygen atoms and the data points for forsterite, Mg_2SiO_4, an olivine having a distorted tetrahedral silicon site. The plot indicates that there is a more or less continuous transition from the tetrahedral sp^3 hybrid to the trigonal bipyramidal $sp^3 d_{z^2}$ hybrid and that within this series there is probably variable participation of the $3d_{x^2-y^2}$ and the $3d_{z^2}$ orbitals of silicon.

This assumption is supported by a comparison of the observed bond lengths d(Si−A) plotted versus the mean angle ⟨∢X−Si−X⟩ as in Fig. 3.16b, c, d. Here, because atoms of different elements have different sizes, the lengths of the Si−O, Si−C, Si−N, etc. bonds must be plotted separately. The closer the mean X−Si−X angle approaches 120°, the shorter the bond length to the farther apical ligand Y′ becomes (Fig. 3.16d), and the longer the bond length to the nearer apical ligand Y (Fig. 3.16b) becomes. At the same time the bond lengths to the equatorial ligands, d(Si−X), become slightly longer (Fig. 3.16c) and the bond angles between equatorial and apical ligands, ∢X−Si−Y, approach 90° (Fig. 3.16a). As a result, the coordination number of silicon changes from 4 via (4+1) to 5. In phases with one rather long apical Si−A bond this bond is often called a dative bond.

All the $[SiA_n]$ conformations observed are stereochemical states of low energy. The plots in Fig. 3.16 then indicate that within "conformational space" (the space containing all imaginable conformations) of pentacoordinate silicon the regions of existing conformations mark a narrow trough of low energy.

Applying the "reaction path concept" of Dunitz (1979) and his co-workers, the existence of such a trough suggests that reactions of silicon compounds probably take place along this valley of low energy. It is, therefore, reasonable to assume that condensation and decondensation reactions of silicates and silicic acids in aqueous solution involve a short-lived transition state of pentacoordinate silicon rather than one of hexacoordinate octahedral silicon as suggested by Okkerse (1970) and others. Migration of silicon in diffusive phase transformations, solid state reactions, and flow processes in silicate melts may also involve pentacoordinate silicon intermediate states (Liebau 1984b).

3.1.4.2.3 Six-Coordinated Silicon. If the electronegativity χ_M in compounds of general formula $M_rSi_sO_t$ containing Si−O−M bonds reaches high values as, for example, in the cases of phosphorus ($\chi_P = 2.06$), hydrogen ($\chi_H = 2.1$), and carbon ($\chi_C = 2.50$), or if silicon is exclusively coordinated by fluorine, the most electronegative element ($\chi_F = 4.10$), then the energies of the valence orbitals of silicon are

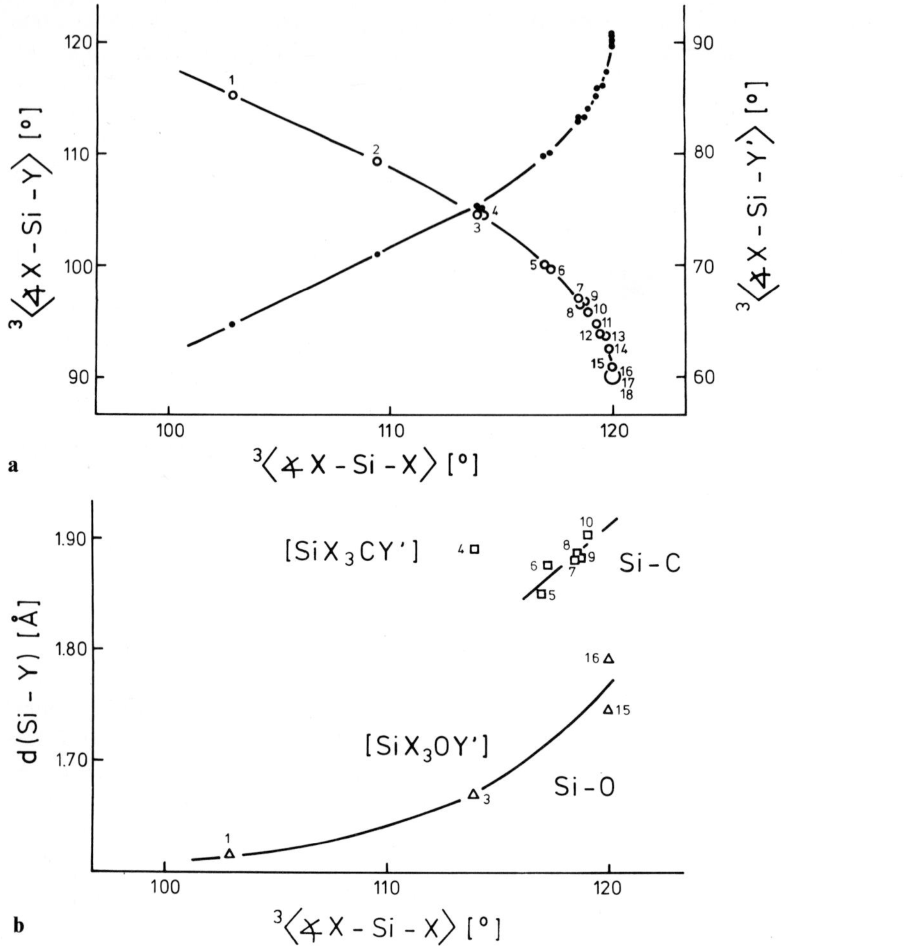

Fig. 3.16 a, b

changed in such a way that they form an sp^3d^2 hybrid. The six orbitals of this hybrid point to the corners of an octahedron. These hybrid orbitals overlap with the $2p$ orbitals of the coordinated oxygen or fluorine atoms, forming six covalent σ-type bonds. This is, for example, the case in the hexaoxosilicon compounds presented in Table 3.4 and in a number of phases containing the very stable complex $[SiF_6]^{-2}$.

Figure 3.17 is a purely qualitative picture of the correlation between the electronegativity of the ligands, the participation of the various valence shell orbitals of silicon to hybridization, and the coordination number of silicon. Since the state of hybridization of the valence orbitals of silicon manifests itself in the coordination of the ligands around the silicon atom, a combination of accurate structure determinations, detailed spectroscopic studies, and quantum mechanical calculations for a larger number of phases should be capable of placing this diagram on a more sound and quantitative basis.

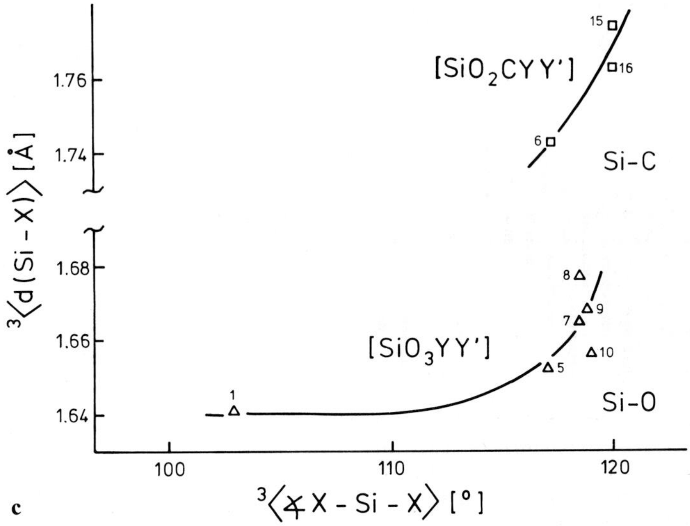

c

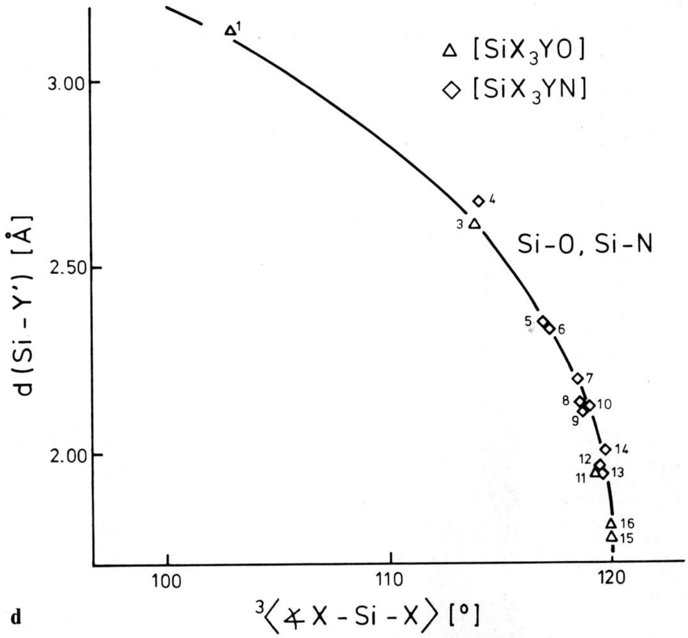

d

Fig. 3.16 a – d. Stereochemistry of compounds with pentacoordinate silicon (from Liebau 1984b). **a** Mean value of the three X–Si–X angles vs. the mean value of the three X–Si–Y and X–Si–Y′ angles. *Data points* No. 1 are for forsterite, $Mg_2[SiO_4]$, No. 2 for the ideal sp^3 hybrid in a hexagonal close-packed array, and No. 17 for the ideal $sp^3d_{z^2}$ hybrid. **b** Bond length between silicon and the nearer apical ligand vs. the mean value of X–Si–X angles. **c** Mean value of bond lengths between silicon and the three equatorial ligands vs. the mean value of X–Si–X angles. **d** Bond length between silicon and the more distant apical ligand vs. the mean value of X–Si–X angles

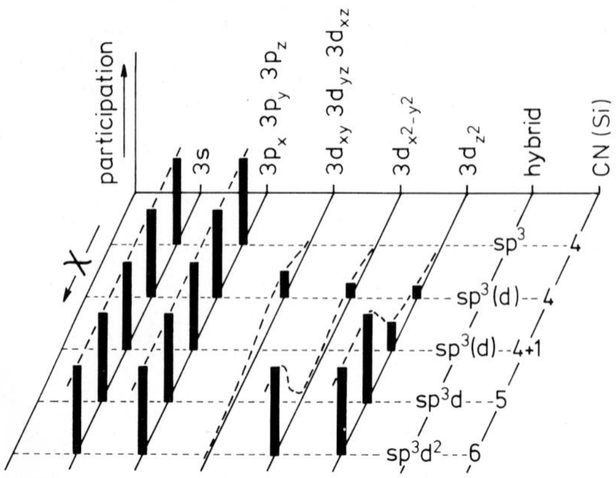

Fig. 3.17. Tentative diagram of the influence of electronegativity of the ligands to the degree of participation of the atomic orbitals of silicon

3.1.4.3 Nonbonded Interactions

As previously discussed (see Sec. 3.1.4.2.1) the overlap between Si $3d$ and O $2p$ orbitals is often used to explain the widening of the Si–O–Si angle from 109° for a covalent single bond to the observed values around 140°. However, a different explanation has recently been developed by Glidewell (1977, 1978) and O'Keeffe and Hyde (1978).

According to these authors it is less surprising – if repulsion between the silicon atoms is assumed – that the Si–O–Si angle has values much higher than the average of 140° (sometimes even 180°) than it is that the angles at the bridging oxygen atoms are very seldom narrower than 130°.

Evidence for Si···Si repulsion can be seen in the distribution of 141 nearest-neighbor Si···Si distances in Fig. 3.18: this histogram is skewed with a maximum

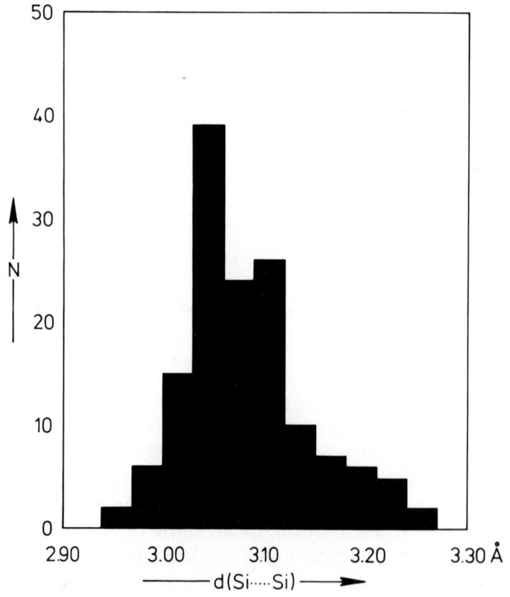

Fig. 3.18. Histogram of the distribution of 141 nearest-neighbor Si···Si distances in silicates and silica (after O'Keeffe and Hyde 1978)

at about 3.06 Å and a tail at larger distances. By analogy with the ionic and covalent radii, the authors derive a set of so-called *nonbonded atomic radii* (Appendix II) in such a way that "when in a fragment XMY of a noncyclic compound, the angle at M exceeds the tetrahedral value if M is 2-connected oxygen, 3-connected nitrogen or 4-connected carbon, or if the angle XMY exceeds 120° when M is 2-connected nitrogen or 3-connected carbon, then the minimum X···Y distance experimentally observed in such a fragment represents the close-contact limit and is the sum of the" nonbonded radii of the atoms X and Y. In practice, an X···Y distance larger than the minimum distance observed close to the peak of the distribution curve is chosen since the nonbonded repulsive forces can sometimes be partly offset by particular electronic or steric effects. According to O'Keeffe and Hyde (1978) the nonbonded radius of silicon deduced from 141 Si–O–Si bonds and a number of Si–N–Si and Si–C–Si bonds is $r_{nb}(\mathrm{Si}) = 1.53$ Å.

Following these authors the geometry of a T–X–T bond is as presented in Fig. 3.19. The bond length $d(\mathrm{T{-}X})$ is set equal to the sum of the ionic radii of T and X and the angle ∢T–X–T is as close as possible to the tetrahedral angle 109.5° subject to the limitation that, due to nonbonded repulsion, the distance $d(\mathrm{T\cdots T})$ cannot be smaller than twice the nonbonded radius of T. This means that the T–X–T angle is 109° if $d(\mathrm{T\cdots T}) > 2r_{nb}(\mathrm{T})$, and larger than 109° if $d(\mathrm{T\cdots T}) < 2r_{nb}(\mathrm{T})$.

Figure 3.19 indicates that for silicates

$$\begin{aligned}\sphericalangle\,\mathrm{Si{-}O{-}Si} &\leq 2\sin^{-1}\{r_{nb}(\mathrm{Si})/[r_{ion}(\mathrm{Si}^{[4]}) + r_{ion}(\mathrm{O}^{[2]})]\}\\ &= 2\sin^{-1}\{1.53/[0.26+1.35]\}\\ &= 144°,\end{aligned}$$

is in good agreement with the value 140° assumed to correspond to an unstrained Si–O–Si angle (see Sec. 3.1.2.5).

The decrease in the average T–O–T angles in the series Si, P, S, Cl is not necessarily caused by an increasing (Si $3d$–O $2p$) π-bonding effect, but can in-

a

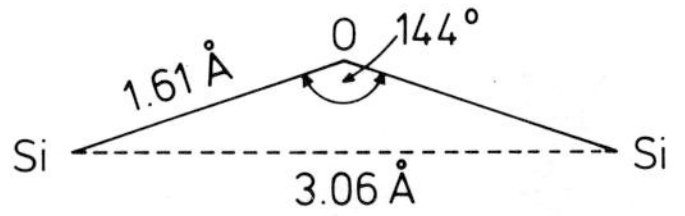

b

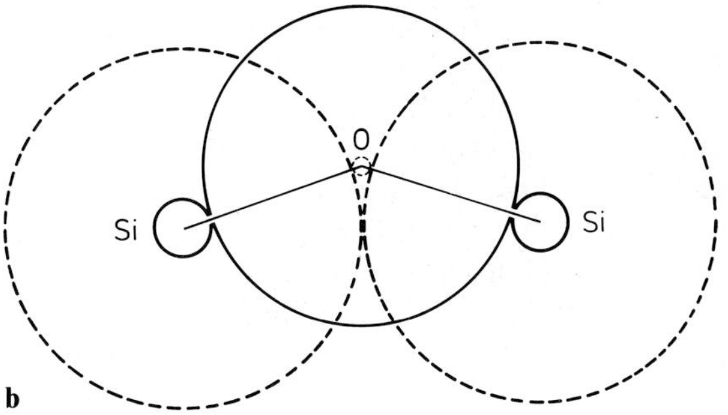

Fig. 3.19 a, b. Geometry of the Si–O–Si group.

a Geometry assumed by O'Keeffe and Hyde (1978) for an "unstrained" Si–O–Si group. **b** Hard-sphere model of the Si–O–Si group with ionic radii $r_{ion}(\mathrm{Si}^{[4]}) = 0.26$ Å and $r_{ion}(\mathrm{O}^{[2]}) = 1.35$ Å (*solid circles*) and nonbonded radii $r_{nb}(\mathrm{Si}^{[4]}) = 1.53$ Å and $r_{nb}(\mathrm{O}^{[2]}) = 0.08$ Å (*broken circles*)

stead be interpreted in terms of a nonbonding interaction within a series of compounds with T atoms of decreasing nonbonded radius (Glidewell 1978).

The good agreement between observed T–X–T angles and those calculated on the assumption of nonbonded repulsion does not favor either the ionic or the covalent model for a T–X bond, nor does it rule out the influence of π-bonding.

3.1.4.4 The Ionicity of the Si–O Bond

Not only is there a continuing dispute about the extent to which the various orbitals of silicon and oxygen participate in the bonding between Si and O (see Sec. 3.1.4.2), but there is also disagreement about the degree of ionicity and covalency of the Si–O bond.

For quite some time a value of 50% has been widely accepted as the ionicity of the Si–O bond. This value was deduced by Pauling (1939, p. 74) from an empirical correlation between atomic electronegativity differences and the observed values of the electric dipole moments of molecules. Different values for the ionicity are obtained if other electronegativity scales, such as the orbital electronegativities of Hinze (Hinze et al. 1963; Hinze 1967/68), are used and if the electronegativities are correlated with properties other than the electric dipole moments, such as optic dielectric constants or photoemission spectroscopic data (Hübner 1977).

Since the ionicity of a bond is reflected in the charge of an atom Stewart et al. (1980) defined ionicity as

$$i = 100 \times \text{residual charge/formal charge} \qquad (15)$$

expressed as a percentage. An ionicity of 50% for the Si–O bond would then mean that in SiO_2 the residual charges of Si and O are $+2$ and -1, respectively, since their formal charges are $+4$ and -2. This definition appears to be straightforward but its application is not simple since there is as yet no direct means of measuring the residual charges on atoms. The values obtained differ considerably, depending on the method used, such as X-ray diffraction, X-ray emission spectroscopy, or molecular orbital calculations.

Considering only the more recent publications, residual charges on the silicon and oxygen atoms have been obtained of $+0.95$ to $+3.29$ e and -0.3 to -1.5 e, respectively, with X-ray diffraction methods; $+1.0$ to $+1.5$ e and -0.8 to -1.2 e, respectively, from X-ray emission spectroscopy; and $+1.0$ to $+3.4$ e and -0.4 to -1.3 e, respectively, from molecular orbital calculations (Table 3.7).

These large variations in residual charge are, to a considerable extent, due to different definitions of the term ionicity and to inadequacies and inconsistencies in the methods applied, and will lead to incorrect estimates of ionicity. It is, therefore, wise not to take the absolute values of the residual charges reported too seriously and to make no comparisons between the residual charges obtained by different methods. However, in the more accurate studies, differences in the estimated residual charges of atoms of the same element within the same structure may be significant. For example, the residual negative charges on the bridging oxygen atoms were consistently slightly smaller than those on the terminal oxygen atoms of the same silicate, whether or not they were determined from X-ray dif-

Table 3.7 Residual charges of silicon and oxygen atoms obtained with different methods

Method	Silicate	Residual charge of			Reference
		Si	O_{br}	O_{term}	
X-ray diffraction	Mn_2SiO_4	+ 2.28	–	– 1.13 to – 1.29	Sasaki et al. 1980
	Mg_2SiO_4	+ 2.11	–	– 1.29 to – 1.52	Sasaki et al. 1982
	Co_2SiO_4	+ 2.21	–	– 1.24 to – 1.39	Sasaki et al. 1982
	Fe_2SiO_4	+ 2.43	–	– 1.13 to – 1.24	Sasaki et al. 1982
	$Mg_2Si_2O_6$	+ 2.20, + 2.36	– 1.22, – 1.31	– 1.36 to – 1.51	Sasaki et al. 1982
	$Co_2Si_2O_6$	+ 1.79, + 2.76	– 0.84, – 1.10	– 1.10 to – 1.16	Sasaki et al. 1982
	$Fe_2Si_2O_6$	+ 2.16, + 2.21	– 0.92, – 1.18	– 1.04 to – 1.20	Sasaki et al. 1982
	SiO_2 quartz	+ 1.0 (1)	– 0.5 (1)	–	Stewart et al. 1980
	SiO_2 coesite	+ 0.95 (19)	– 0.25 (15)	–	Ross 1980
	$K_2Si^{[6]}[Si_3^{[4]}O_9]$	+ 3.29 (15) for $Si^{[6]}$ + 2.52 (11) for $Si^{[4]}$	– 1.43 (8)	– 1.43 (4)	Swanson and Prewitt 1983
	SiO_2 stishovite	+ 1.71	– 0.86 (15)	–	Hill et al. 1983
X-ray emission spectra	SiO_2 quartz	+ 1.40	n.d.[a]	–	Urusov 1970
	Mg_2SiO_4	+ 1.33	–	n.d.	Urusov 1970
	Fe_2SiO_4	+ 1.33	–	n.d.	Urusov 1970
	Various other silicates	+ 1.40 to + 1.30			Urusov 1970
MO calculations	Various clusters	+ 2.8 to + 3.4	– 1.18 to – 1.28		Yip and Fowler 1974
	Various clusters	+ 1.0 to + 1.5		– 0.8 to – 1.2	Dikov et al. 1977
	H_4SiO_4	+ 1.36	–	– 0.52	Newton and Gibbs 1980
	$H_6Si_2O_7$	+ 1.4	– 0.7	– 0.50	Newton and Gibbs 1980
	$H_6Si_2O_7$	+ 1.04	– 0.49	– 0.41	Lasaga 1982
	SiO_2 cluster	+ 1.3	– 0.65	–	Newton and Gibbs 1980

[a] n.d. = not determined

fraction, X-ray emission spectra, or molecular orbital calculations (Table 3.7). There is even some indication from maps of the empirical aspherical electron density distribution (see Sec. 3.1.4.2.1) in stishovite, high pressure SiO_2 (Hill et al. 1983), in the high pressure phase $K_2Si^{[6]}Si_3^{[4]}O_9$ (Swanson and Prewitt 1983), and from molecular orbital calculations (Tossell 1975 b), that six-coordinated silicon has a slightly higher Si–O bond ionicity than tetrahedrally coordinated silicon.

Although the accuracy of the values reported for the residual charges of silicon and oxygen in silicates is rather low, it is clear that the ionicity of the Si–O bond, as defined by Eq. (15) is somewhere between 25 and 75% and that it is more likely to be below 50% than above.

3.2 The Cation–Oxygen Bond, M–O, and Its Influence on the Si–O Bond

In silicates of general formula $M_rSi_sO_t$ the electronegativity of M can vary considerably. This means that the number of electrons transferred from the M atoms to an oxygen atom bonded to silicon can vary substantially. Consequently, there may be a considerable variation in the ionicity of the Si–O bonds in Si–O–M groups.

3.2.1 Electropositive Cations

The transfer of electrons from the M cations to oxygen permits an explanation of the positions of the residual electron density maxima in the map obtained for orthoenstatite $Mg_2[Si_2O_6]$ (Fig. 3.13) (Sasaki et al. 1982). The average Si-to-peak distance is 0.71 Å for the $Si-O_{term}$ bonds and 1.14 Å for the $Si-O_{br}$ bonds, whereas the corresponding Si–O distances are 1.602 Å and 1.666 Å, respectively.

The electrons in the covalent component of the bonds are, therefore, nearer to the silicon atom than to the terminal oxygen atom, but are further from the silicon atom than from the bridging oxygen (Fig. 3.20).

If we consider a bond system $Si-O_{br}-Si-O_{term}-M$, where M is an electropositive cation, such as magnesium, then Mg transfers more electrons to its neighboring oxygen atom than does Si due to its lower electronegativity ($\chi_{Mg}=1.23$, $\chi_{Si}=1.74$). As a consequence, the slightly more negative O_{term} atom transfers part of its electron density to the adjacent silicon atom, thus shifting the electron den-

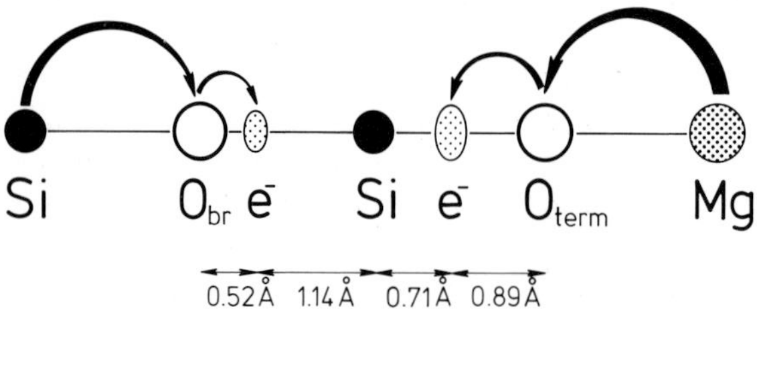

Fig. 3.20. Location of the residual electron density maxima e^- obtained from difference Fourier maps of orthoenstatite, $Mg_2[Si_2O_6]$, and suggested electron transfer in the bond system $Si-O_{br}-Si-O_{term}-Mg$

Table 3.8 Correlation between cation electronegativity, χ_M, and the charges averaged over the crystallographically nonequivalent atoms in orthopyroxenes as derived from X-ray diffraction data (Sasaki et al. 1982)

Atoms	Mean atomic charges		
	$Mg_2[Si_2O_6]$	$Fe_2[Si_2O_6]$	$Co_2[Si_2O_6]$
M	+ 1.82	+ 1.12	+ 0.95
Si	+ 2.28	+ 2.28	+ 2.19
O_{term}	− 1.42	− 1.13	− 1.13
O_{br}	− 1.27	− 0.97	− 1.05
O	− 1.37	− 1.10	− 1.08
χ_M	1.23	1.64	1.70

sity maximum further towards Si than does the less negative oxygen atom O_{br} (Fig. 3.20). In agreement with this model slightly higher charges have been found for the terminal than for the bridging oxygen atoms (Table 3.8). Moreover, as expected, the oxygen atom charges in orthopyroxenes of the more electronegative cations Fe^{+2} and Co^{+2} are lower than those in orthoenstatite. This explanation is also consistent with the observation that the SiK emission lines obtained from vitreous silica, crystalline $Mg_2[SiO_4]$, and sodium-silicate glasses of varying Na_2O content shift to lower energies when the M atoms of the Si−O−M system change from electronegative Si to electropositive Na (Wiech et al. 1976).

While the influence of cations of different electronegativity on the ionicity of the Si−O bond seems to be well-established, their influence on Si−O bond lengths and Si−O−Si angles is less obvious and is only revealed when the influence of the other parameters is kept as constant as possible. The only convincing reports of correlations between χ_M and Si−O distances have been made for silicates belonging to a family of isotypic structures, and even then the correlations were concealed by averaging over all structurally nonequivalent Si−O bonds.

For four amphiboles, $M_2M'_5[Si_4O_{11}]_2(OH)_2$, with different cations M and M′, Brown and Gibbs (1969, 1970) observed an increase of each of the individual crystallographically nonequivalent Si−O_{term} bond lengths as the mean electronegativity of the cations increased (Fig. 3.21 a). In contrast, reverse correlations between individual Si−O bond lengths and the electronegativity of the cations have been observed for olivines, $M_2[SiO_4]$, and garnets, $M_3^{[12]}M_2^{[6]}[SiO_4]_3$ (Novak and Gibbs 1971). This clearly indicates that the influence of χ_M on the Si−O distances is small and that it can be readily compensated and hidden by other parameters, such as cation size, which requires mutual adjustment between the cation−oxygen polyhedra and $[SiO_4]$ tetrahedra within the structure.

However, in some structures even the influence of certain atomic orbitals on the Si−O bond length can be detected. This has been demonstrated by Ohashi (1981) who correlated the mean values of the Si−O_{br} and Si−O_{term} distances of pyroxenes $NaM^{+3}[Si_2O_6]$ with the electronegativity of the trivalent cations. He concluded that of the six pyroxenes studied, the cations M = Cr, Fe, and In with all their 3*d* orbitals in the t_{2g} state filled show different correlations than the pyro-

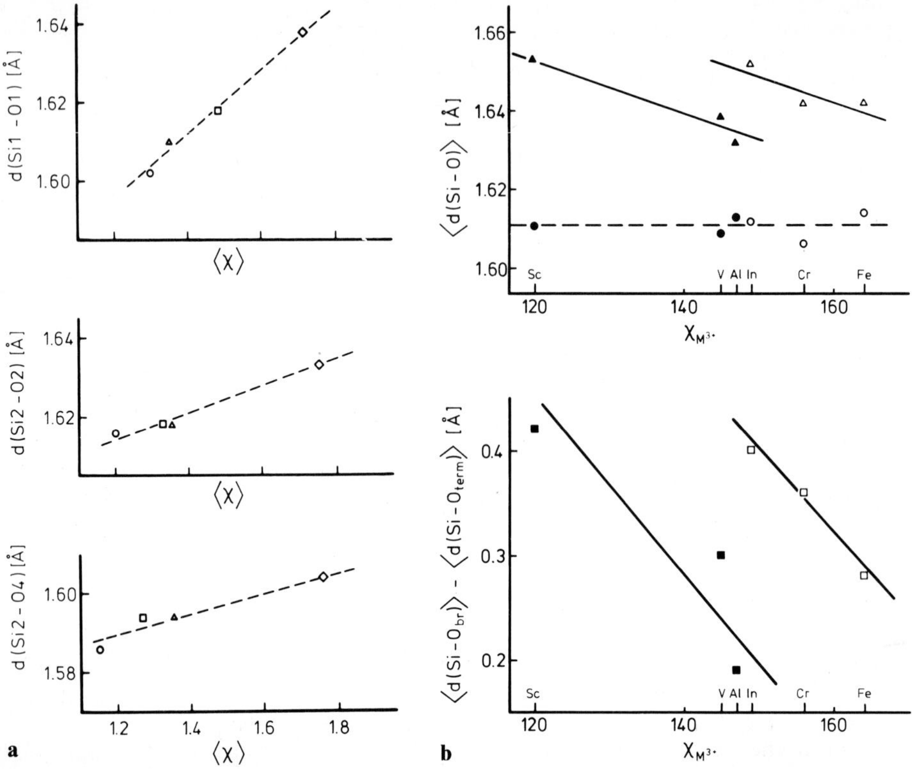

Fig. 3.21 a, b. Influence of the electronegativity χ of the cations on the Si–O bond length in several chain silicates.

a Three individual bond lengths $d(Si-O)$ in tremolite $Mg_5Ca_2[Si_4O_{11}]_2(OH)_2$ (○); manganoan cummingtonite (△); glaucophane $Na_2Mg_{2.4}Al_{1.6}Fe[Si_4O_{11}]_2(OH)_2$ (□); grunerite $Fe_2Fe_{4.2}Mg_{0.8}[Si_4O_{11}]_2(OH)_2$ (◇) vs. the average cation electronegativity $\langle\chi\rangle$ (after Brown and Gibbs 1969). **b** Average bond lengths $\langle d(Si-O_{br})\rangle$ (▲ and △) and $\langle d(Si-O_{term})\rangle$ (● and ○), and the differences between these average bond lengths, $\langle d(Si-O_{br})\rangle-\langle d(Si-O_{term})\rangle$ (■ and □), in synthetic pyroxenes $NaM^{+3}[Si_2O_6]$ with M = Al, V and Sc (*full symbols*) and M = Fe, Cr and In (*open symbols*) plotted vs. the electronegativity χ of the cations M^{+3}. (Adapted from Ohashi 1981)

Table 3.9 Correlation between the electronegativity of ligand X of molecules $X_3Si-O-SiX_3$ and the Si–O distance (for comparison, the grand mean of the $Si-O_{br}$ distances in the $[O_3Si-O-SiO_3]^{-6}$ ions of disilicates is also given)

X	χ_X	d(Si–O) [Å]
F	4.10	1.58
Cl	2.83	1.59
C_6H_5	2.50	1.616
H	2.1	1.634
O	3.50	1.64

xenes with M=Al, V, and Sc which have more than one empty $3d$ orbital in the t_{2g} state.

The difference between these two subfamilies is even more obvious if the differences $\langle d(Si-O_{br})\rangle-\langle d(Si-O_{term})\rangle$ are plotted against the electronegativity of the cation (Fig. 3.21 b). Similarly, aluminum silicate garnets $M_3^{[12]}Al_2^{[6]}[SiO_4]_3$ show different correlations between the mean Si–O bond lengths and cation electronegativities for garnets with $M^{[12]}=Mg^{+2}$ and Ca^{+2} which have empty $3d$ orbitals than those shown by garnets with $M^{[12]}=Mn^{+2}$, Fe^{+2}, Co^{+2} which have filled $3d$ orbitals (Ohashi et al. 1981).

3.2.2 Electronegative Cations

The influence of silicon as a cation of rather high electronegativity on the Si–O bond has already been demonstrated with the aid of Fig. 3.20. If the T atom in Si–O–T systems is varied from Si through Al, B, Be, to P no obvious correlation between the electronegativity of T and the Si–O bond lengths or Si–O–T angles has been found. This is not surprising since the nonbonded radii (Appendix II) of these tetrahedrally coordinated cations are not very different and the nonbonded interactions between Si and the particular T atom would conceal such correlations.

If, however, molecules of the formula $X_3Si-O-SiX_3$ containing strongly electronegative X atoms are compared, the Si–O bond length increases with decreasing electronegativity of X (Table 3.9). This trend cannot, however, be extrapolated to X = oxygen, i.e., to disilicates, since the resultant "molecules" are not neutral.

If in a phase of composition $M_rSi_sO_t$ the electronegativity of M is higher than that of Si, then the oxide of M is a stronger Lewis acid[6] than SiO_2. Consequently, in such a phase silicon has to be considered the cation in an oxysalt of M.

Typical examples are the various polymorphs of silicon diphosphate SiP_2O_7 (Liebau et al. 1968; Tillmanns et al. 1973; Hesse 1979a). Since $\chi_P=2.06>\chi_{Si}=1.74$, the P–O bond is more covalent than the Si–O bond. Furthermore, since the formal charge of phosphorus is + 5 and that of silicon is only + 4, phosphorus will be superior to silicon in its ability to attract the oxygen atoms. As a result, in SiP_2O_7 the P–O distances are shorter than the Si–O distances by about 15% ($\langle d(P-O)\rangle \simeq 1.52$ Å, $\langle d(Si-O)\rangle \simeq 1.76$ Å), and the P atom has a coordination number of only 4 compared with 6 for silicon. As explained above, other phases containing $[SiO_6]$ octahedra, so-called hexaoxosilicon compounds, which are formed at normal pressure are stabilized by the presence of Si–O–H ($\chi_H=2.1$) and Si–O–C ($\chi_C=2.50$) systems (Table 3.4).

Further Reading:

Cruickshank 1961; Robinson 1963; Urch 1969; Urusov 1970; Baur 1971; Gibbs et al. 1972; Louisnathan and Gibbs 1972; Collins et al. 1972; Tossell 1975a, b, 1977; Glidewell 1975, 1977, 1978; Tossell and Gibbs 1976, 1977, 1978; O'Keeffe and Hyde 1976, 1978; Wiech et al. 1976; Dikov et al. 1977; Griscom 1977; Meagher et al. 1979; Meagher 1980; Brückner et al. 1980; Meier and Ha 1980; Newton and Gibbs 1980; Oberhammer and Boggs 1980; Pauling 1980; Stewart et al. 1980; Binks and Duffy 1980; Newton 1981; Gibbs 1982; Lasaga 1982; Geisser and Hübner 1984.

6 A Lewis acid is a substance which accepts an unshared electron pair from a compound with which it forms a covalent bond. The compound donating electrons is the corresponding Lewis base.

4 Crystal Chemical Classification of Silicate Anions

4.1 General Principles of the Classification

The best classification which can be chosen is the one that is best able to serve the particular purpose under consideration. For the purpose of identifying silicate minerals in the field, a classification according to their color, density, hardness, and morphology will be appropriate. For the identification of natural and synthetic silicate materials in a well-equipped laboratory a classification according to their X-ray powder data is often employed.

In this book we are concerned with the structural chemistry of silicates, that is, the relationship between crystal structure, chemical composition, thermodynamic stability, and chemical reactivity of silicates. In this case a classification based either on chemical composition or on structure is appropriate. However, an even more suitable classification is one in which the structure is directly related to the chemical composition and is, therefore, based on both principles at the same time. The classification described in this section tries to come as near as possible to this aim.

4.2 Parameters Used in the Crystal Chemical Classification of Silicate Anions

We have seen in Chap. 3 that in almost every structure the Si–O bonds are stronger than the M–O bonds. Therefore, it seems reasonable to use the $[SiO_n]$ polyhedra and the way they are linked to each other to classify the silicate anions[1]. Indeed, this has been the procedure used by most authors from the early days of structural silicate chemistry.

In this classification, which has been described in detail recently (Liebau 1980a)[2], the following parameters have been applied.

(1) Coordination Number of Silicon. Although silicon has been assigned coordination numbers from one to six, the only well-established $[SiO_n]$ polyhedra encoun-

1 In the following discussion, the term "silicate anion" includes all the various kinds of $[SiO_n]$ polyhedra as well as the more complex aggregates formed by linking such polyhedra, irrespective of their formal charges and whether their silicon atoms have a chemically more acidic or more basic character.

2 A more recent description of the present state of silicate classification (Liebau 1982) is better not consulted since it contains a series of translation errors.

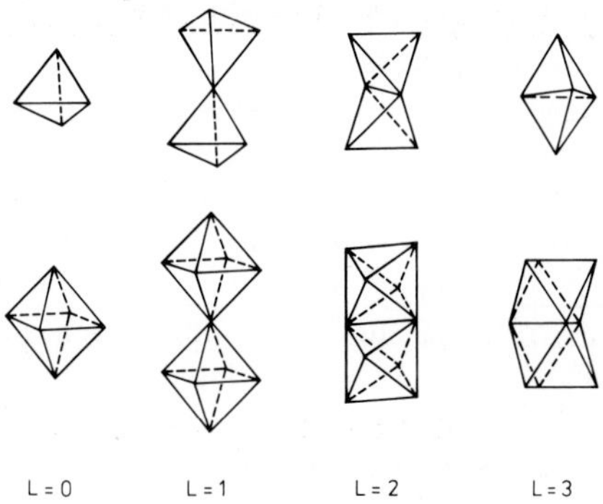

Fig. 4.1. Tetrahedra and octahedra with different linkedness ***L***.
L = 0 (isolated); ***L*** = 1 (corner-sharing); ***L*** = 2 (edge-sharing); ***L*** = 3 (face-sharing)

tered as yet in nature are $[SiO_4]$ tetrahedra and $[SiO_6]$ octahedra (Fig. 3.1). This parameter is expressed as

$$CN = \ldots, 4, 5, 6, \ldots, n.$$

(2) Linkedness of $[SiO_n]$ ***Polyhedra.*** An $[SiO_n]$ polyhedron may share zero, one, two, or three oxygen atoms with an adjacent $[SiO_n]$ polyhedron. In other words, $[SiO_n]$ polyhedra can either be isolated, or share corners, edges, or faces (Fig. 4.1). If the linkedness L is defined as the number of oxygen atoms shared between two $[SiO_n]$ polyhedra, it can have the values

$$L = 0, 1, 2, 3.$$

(3) Connectedness of $[SiO_n]$ ***Polyhedra.*** Irrespective of the specific value of its linkedness, an individual $[SiO_n]$ polyhedron may be distinguished by the number s of other $[SiO_n]$ polyhedra to which it is linked via common oxygen atoms. For

Table 4.1 Various types of tetrahedra with different values of connectedness, ***s***

Type	Name	Symbol
	Singular	Q^0
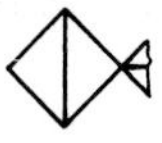	Primary	Q^1
	Secondary	Q^2
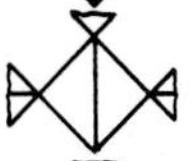	Tertiary	Q^3
	Quaternary	Q^4

example, an $[SiO_4]$ tetrahedron can share oxygen atoms with up to four other $[SiO_n]$ polyhedra, and an $[SiO_6]$ octahedron up to six. With the number s of elements (corners, edges, or faces) shared increasing from zero to six, a polyhedron is called *singular, primary, secondary, tertiary, quaternary, quinary,* or *senary.* The symbols $Q^s = Q^0$, Q^1, Q^2, Q^3, and Q^4 are used to denote the tetrahedra that share zero to four corners with others (Table 4.1).

The number s is the *connectedness* of the $[SiO_n]$ polyhedra, a term used to describe structural units independent of their size (atoms, polyhedra, etc.) or shape and extensively used to classify crystal structures of any kind (Wells 1975).

(4) Branchedness of Silicate Anions. Linear condensation of $[SiO_n]$ polyhedra, irrespective of their linkedness, leads to the formation of unbranched multiple polyhedra of finite length, to unbranched single chains, or to unbranched single rings. Such *unbranched anions* contain only primary and/or secondary $[SiO_n]$ polyhedra. Figures 4.2 and 4.3 give several examples of tetrahedral and octahedral unbranched anions, respectively.

A number of silicate anions are known which can be derived from an unbranched multiple polyhedron, single chain, or single ring by connecting additional $[SiO_n]$ polyhedra to it in a nonlinear way, i.e., in such a way that tertiary, quaternary, or in the case of $[SiO_6]$ octahedra, even quinary or senary polyhedra are formed. A small selection of such anions are presented in Figs. 4.4 and 4.5. Since in some of these anions, for example the one shown in Fig. 4.4d, the additional polyhedra stick out like branches on a stem, they are called *branched anions.*

If each branch is connected to the linear part of the anion by one corner, edge, or face only, we speak of *open-branched anions.* If the branch is attached to the linear part through more than one common element, a loop is formed and the term *loop-branched anion* is used. The anion of Fig. 4.4f would then be a *mixed-branched anion.*

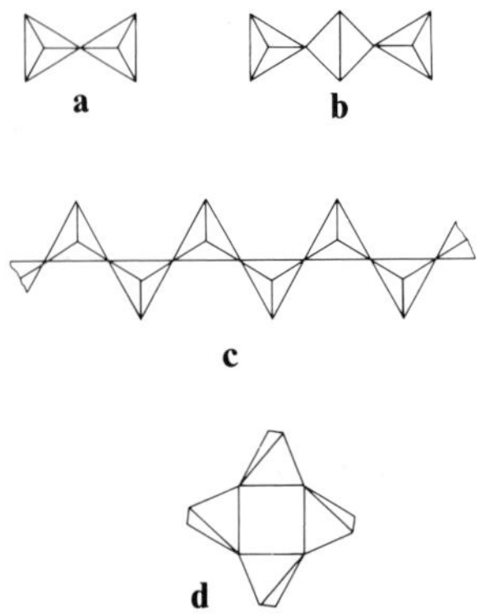

Fig. 4.2 a – d. Several unbranched fundamental anions formed by linear condensation of tetrahedra.

a Double tetrahedron; **b** unbranched triple tetrahedron; **c** unbranched single chain; **d** unbranched single ring

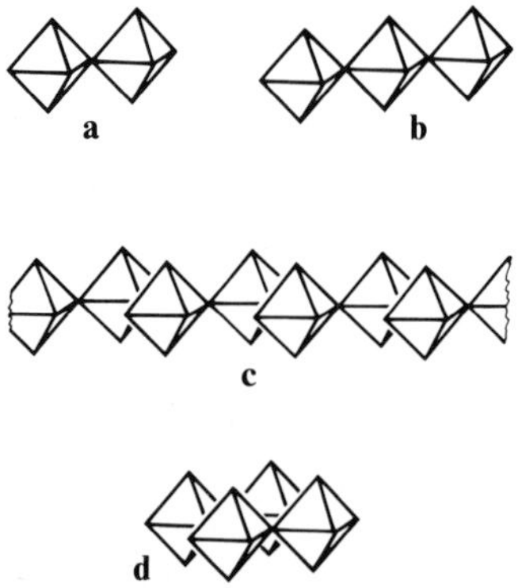

Fig. 4.3 a – d. Several unbranched fundamental anions formed by linear condensation of octahedra.

a Double octahedron; **b** unbranched triple octahedron; **c** unbranched single chain; **d** unbranched single ring

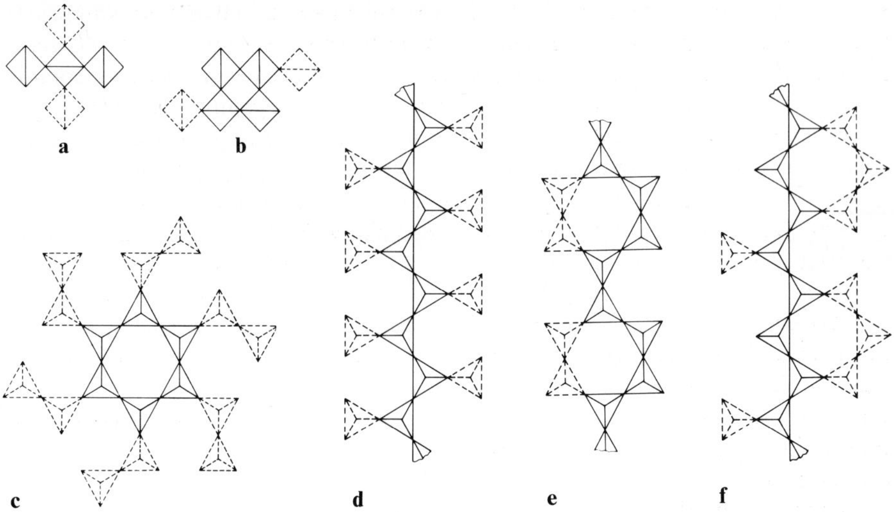

Fig. 4.4a–f. Several branched fundamental anions formed by condensation of tetrahedra. *Solid lines* indicate the linear part of the anion, *broken lines* indicate the branches.

a Open-branched triple tetrahedron; **b** and **c** open-branched single rings; **d** open-branched single chain; **e** loop-branched single chain; **f** mixed-branched single chain

The ensemble of multiple polyhedra, single chains, and single rings, irrespective of whether they are branched or unbranched, form the group of *fundamental anions. Complex anions* are formed if fundamental anions are linked in a topologically nonlinear way. Different complex anions are possible depending on the kind of fundamental anions connected together. The scheme

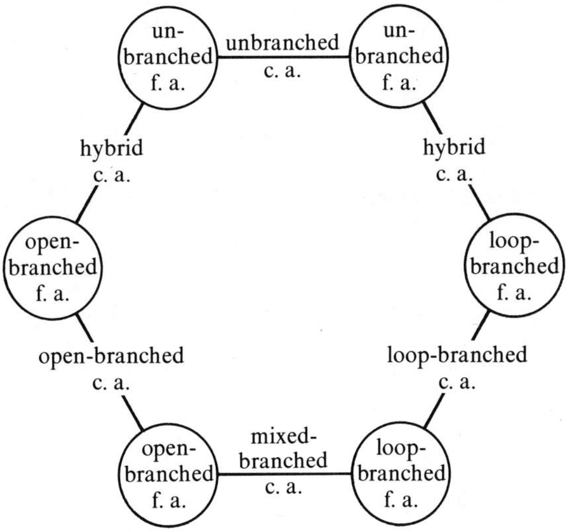

illustrates the six possible combinations between the three kinds of fundamental anions. The different complex anions are called *unbranched, open-branched, loop-branched, mixed-branched,* and *hybrid,* where in the latter category no distinction is made between open- and loop-branched components of the complex anions.

Figure 4.6 shows the various kinds of double chains which can be formed by connecting unbranched, open-branched, and loop-branched single chains.

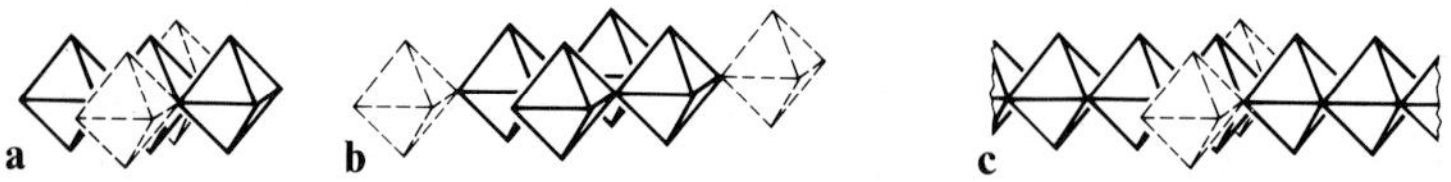

Fig. 4.5 a – c. Several branched fundamental anions formed by condensation of octahedra. *Solid lines* indicate the linear part of the anion, *broken lines* indicate the branches.

a Open-branched triple octahedron; **b** open-branched vierer single ring; **c** open-branched single chain

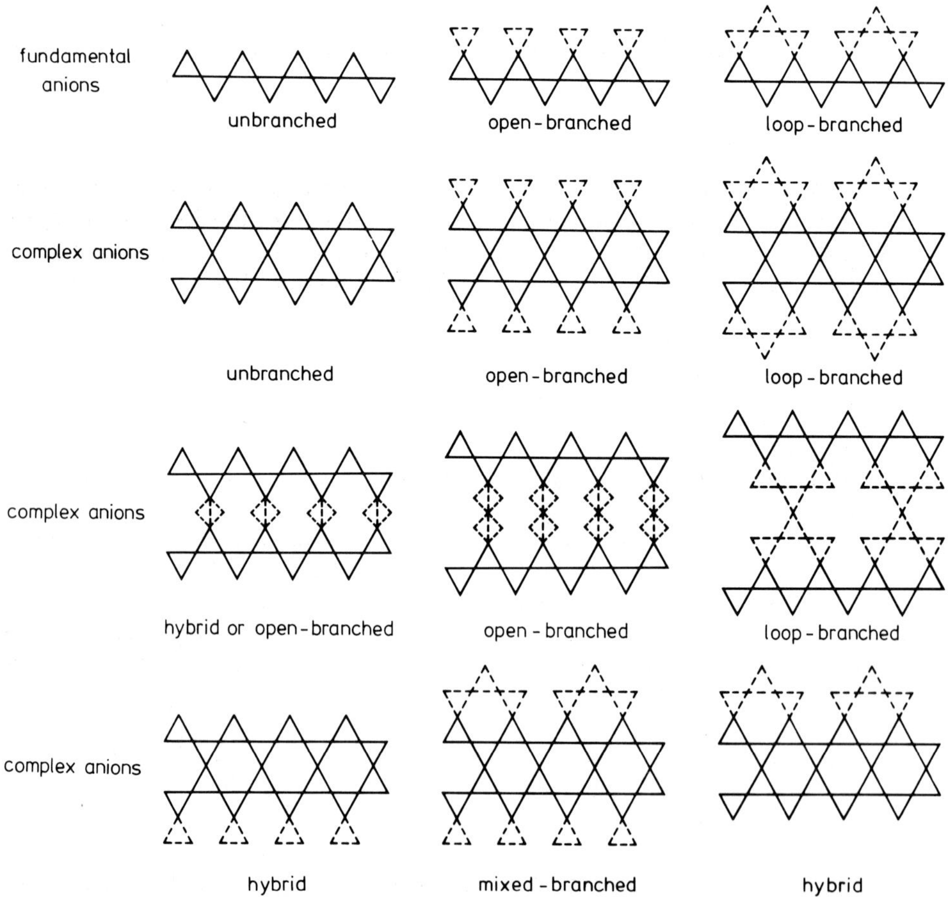

Fig. 4.6. Formation of complex anions by connecting fundamental anions containing only tetrahedra

It can be seen that a complex anion is called open-branched (loop-branched) if its fundamental chains are open-branched (loop-branched), regardless of whether the branches are involved in the linking or not.

With this procedure the fundamental chain can easily be determined in chain silicate anions since its direction is determined by the overall chain direction. In silicate anions with dimensionalities $\boldsymbol{D} > 1$, i.e., in layer and framework anions, tetrahedral chains run parallel to any direction within the layer or the framework. If an unambiguous description is required, the rules described on p. 59 must be followed.

The *branchedness* of a silicate anion, i.e., the property which describes whether the anion is unbranched, open-branched, loop-branched, mixed-branched, or hybrid is described by the symbols

B = ***uB, oB, lB, olB,*** and ***hB,***

respectively. The ***oB, lB, olB,*** and ***hB*** anions are collectively designated as *branched* anions and given the symbol ***br.***

(5) Dimensionality of Silicate Anions. Silicate anions may extend to infinity in zero, one, two, or three dimensions and their *dimensionality* is then said to be

$\boldsymbol{D} = 0, 1, 2,$ or $3,$

respectively. Single $[SiO_n]$ polyhedra, along with terminated and cyclic arrays of finite numbers of polyhedra, have dimensionality $\boldsymbol{D} = 0$, chains have $\boldsymbol{D} = 1$, layers have $\boldsymbol{D} = 2$, and frameworks have $\boldsymbol{D} = 3$. To distinguish between terminated multiple polyhedra and cyclic anions, both having $\boldsymbol{D} = 0$, the symbols

t for multiple polyhedra and

r for cyclic anions (ring anions)

are introduced.

(6) Multiplicity of Silicate Anions. Connecting limited numbers of $[SiO_n]$ polyhedra, single chains, single rings, or single layers leads to the formation of multiple anions of the same dimensionality as the generating ones. The number of single anions connected to form such a multiple anion is called its *multiplicity* ***M***. According to the value of ***M***, silicates are designated single, double, triple, fourfold (or quadruple), fivefold (or quintuple), etc. anions. Although the multiplicity of silicate anions can in principle be very large, in crystalline silicates it is seldom higher than five.

(7) Periodicity of Silicate Anions. In a crystal the atoms are arranged periodically. This implies that, for example, in a crystalline silicate containing single chains of corner-sharing $[SiO_4]$ tetrahedra, the structural motif of the chain repeats after several tetrahedra.

The *periodicity* ***P*** of a single chain, the *chain periodicity*, is defined as the number of polyhedra within one repeating unit of the linear part of the chain. For unbranched single chains, ***P*** is equal to all the polyhedra within one period of the

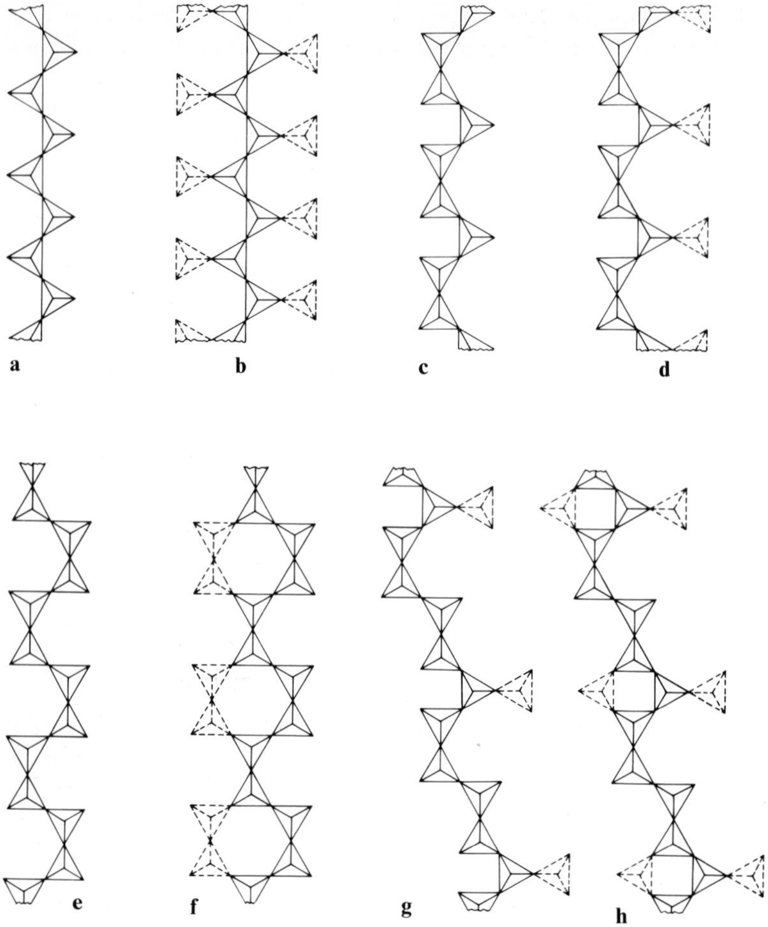

Fig. 4.7 a–h. Several tetrahedral chains which have different periodicities. **a** Unbranched zweier single chain; **b** open-branched zweier single chain; **c** unbranched dreier single chain; **d** open-branched dreier single chain (hypothetical); **e** unbranched vierer single chain; **f** loop-branched vierer single chain; **g** open-branched fünfer single chain; **h** mixed-branched fünfer single chain (hypothetical)

chain, for branched single chains, the polyhedra forming the branches are not included in $\boldsymbol{P}$.

To denote a silicate anion according to its periodicity, the terms einer, zweier, dreier, vierer, fünfer, sechser, siebener, achter, neuner, zehner, elfer, zwölfer, etc. for periodicities $\boldsymbol{P}=1, 2, 3, \ldots$ have been widely accepted. These terms are derived from the German numerals, eins, zwei, drei, vier, fünf, sechs, sieben, acht, neun, zehn, elf, zwölf, etc. by suffixing "er" to the numeral. For higher periodicities, such as $\boldsymbol{P}=24$ the shortened form 24er is preferable to the longer term vierundzwanziger. These terms indicate that the following noun is in some way related to the numeral in question. They should not be mixed up with the terms single, double, triple, fourfold, etc. which describe the multiplicity of an item.

Figure 4.7 illustrates a few different chain types and should help in assigning a chain its proper name. Theoretically the periodicity of a silicate chain can have any value.

Complex silicate anions of infinite extension, such as multiple chains, single and multiple layers, and frameworks can be subdivided according to their fundamental chains. As pointed out earlier (see p. 54 ff.), for silicate layers and frameworks the selection of fundamental chains is unambiguous only if certain rules are followed. In practice the following rules have proven to be suitable.

(1) The fundamental chains are chosen as chains of lowest periodicity which run parallel to the shortest identity period within the anion, regardless of their branchedness, and from which the anion can be generated by successive linkage[3].

(2) If more than one chain is derived in agreement with rule (1) the fundamental chains are chosen such that their number is lowest.

(3) If more than one chain is derived in agreement with rules (1) and (2) the fundamental chains are chosen in the order of preference: unbranched > loop-branched > open-branched > mixed-branched > hybrid.

In Fig. 4.8 a few tetrahedral single layers are shown with their fundamental chains indicated by heavily outlined tetrahedra. According to rule (1) the layer presented in Fig. 4.8c and d should be classified as a loop-branched dreier layer (Fig. 4.8c) rather than as an unbranched vierer layer (Fig. 4.8d).

In analogy to the practice used for multiple chains (see Fig. 4.6), the branchedness of a layer or framework is the same as that of its fundamental chain. If the branches of the fundamental chains are involved in linking these chains to the layer, then the tetrahedra of the branches are an essential part of the layer and do not stick out from the layer (Fig. 4.8c, e). If, however, the branch tetrahedra do not take part in the chain linking, then they stick out from the layer as, for example, in zussmanite (Fig. 10.29b).

In silicate frameworks it is often not obvious which of the numerous chains has been chosen as the fundamental one. Therefore, for silicate frameworks it is advisable to indicate the crystallographic direction of the fundamental chain whenever a deeper structural insight is required. For example, high temperature tridymite would be specified as an "unbranched zweier [100] framework", whereas the feldspar framework would be identified as a "loop-branched dreier [001] framework" (Fig. 4.9).

In the same way that unbranched tetrahedral single chains are subdivided according to the number of tetrahedra in the identity period of the chain (i.e., the chain periodicity $\boldsymbol{P}$), unbranched single rings are subdivided according to the num-

3 This definition of the fundamental chain differs somewhat from that given previously (Liebau 1962a, 1972) in that rather than choosing the chain of absolute lowest periodicity, the chain which is selected is that with lowest periodicity running parallel to the shortest identity period within the silicate anion. Although it is theoretically possible that the chain of lowest absolute periodicity does not run parallel to the shortest identity period within a silicate layer or framework, no such case has been observed. Therefore, previous assignments of layer and framework silicates into categories according to the periodicity of the fundamental chains need not to be changed. The new definition not only facilitates the fixing of the fundamental chain, but seems also to be crystallochemically more reasonable.

a b c d e

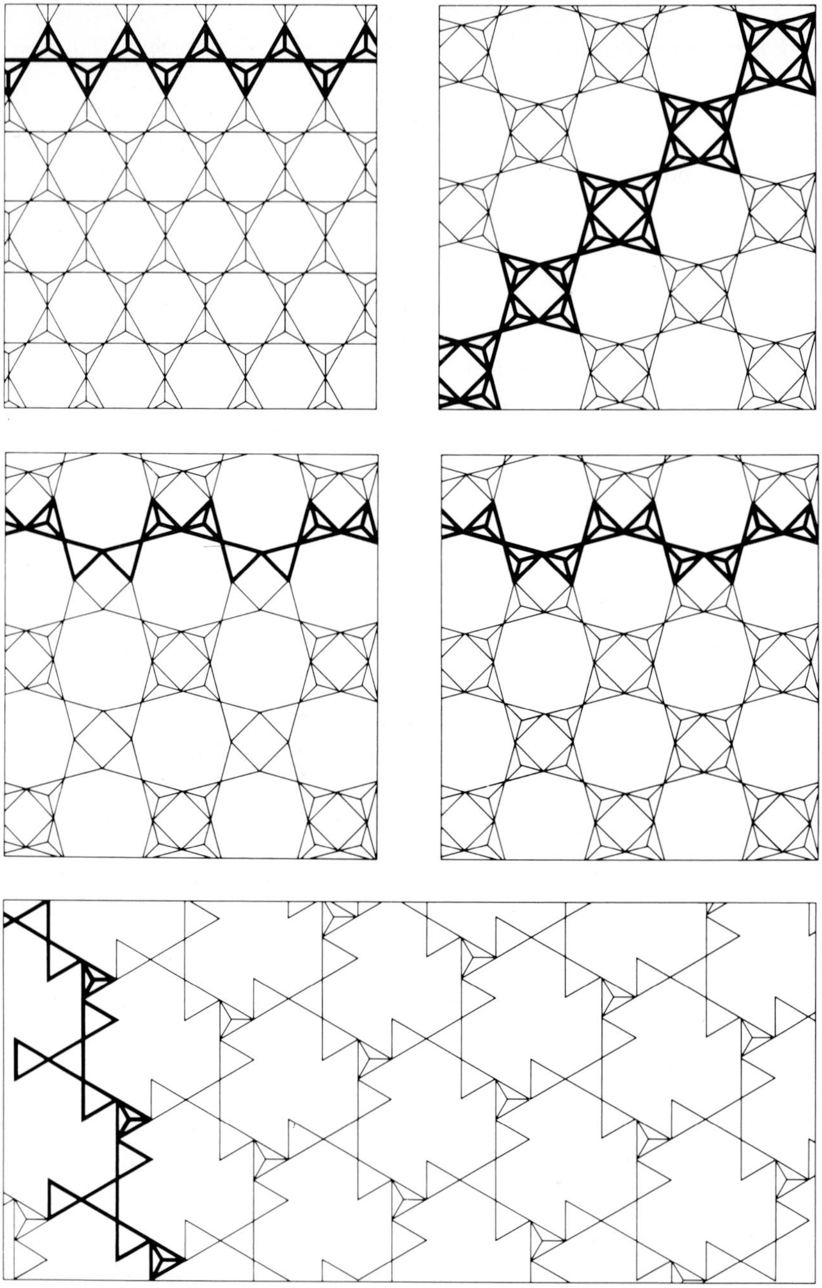

Fig. 4.8 a – e. Several tetrahedral single layers of various periodicities. In each layer one fundamental chain is indicated by *heavy lines.*

a Unbranched zweier single layer; **b** unbranched vierer single layer; **c** loop-branched dreier single layer (hypothetical); **d** the same layer as **c**, but denoted as unbranched vierer single layer [in contradiction to rule (i) of Sec. 4.2.(7)]; **e** open-branched vierer single layer

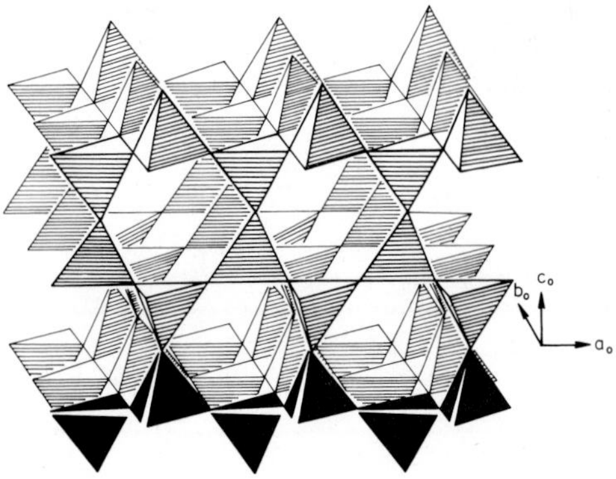

a

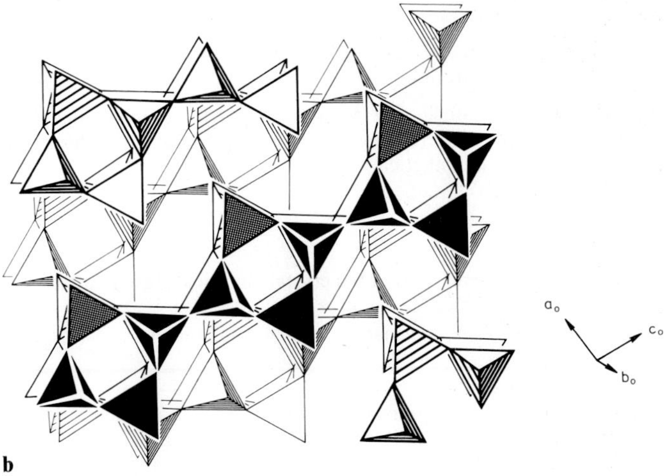

b

Fig. 4.9 a, b. Tetrahedral frameworks. In each framework one fundamental chain is indicated by *black* tetrahedra.

a Unbranched zweier [100] framework of high temperature tridymite, SiO_2; **b** loop-branched dreier [001] framework of potassium feldspar, $K[AlSi_3O_8]$

ber of tetrahedra forming the ring. In a somewhat formalistic way this number can be considered to be the "inner periodicity" of the ring; it is, therefore, called the *ring periodicity*. The symbol $\boldsymbol{P^r}$ is, therefore, used for the subdivision of the cyclic anions. By analogy with the definition of the chain periodicity of branched chains, the tetrahedra of the branches take no part in the definition of the ring periodicity $\boldsymbol{P^r}$ of a branched ring. The designation of cyclic silicate anions is shown in Fig. 4.10.

Table 4.2 is a glossary of English, German, French, and Russian terms used in the crystal chemical classification of silicate anions.

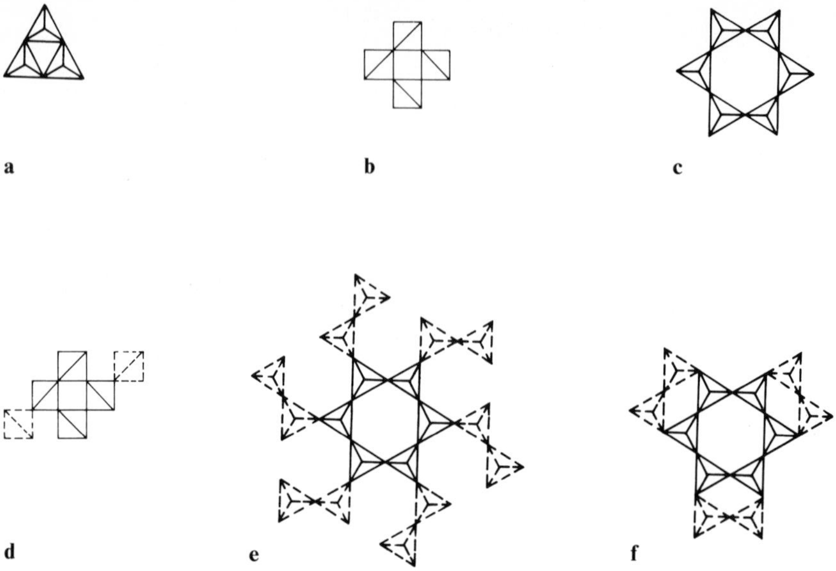

Fig. 4.10. Several kinds of tetrahedral single rings.

a unbranched dreier single ring; **b** unbranched vierer single ring; **c** unbranched sechser single ring; **d** open-branched vierer single ring; **e** open-branched sechser single ring; **f** loop-branched sechser single ring (hypothetical)

Table 4.2 Glossary of terms used in the crystal chemical classification of silicate anions

Symbol	English	German	French	Russian
CN	Coordination number of silicon	Koordinationszahl des Siliziums	Coordinence du silicium	координационное число кремния
L	Linkedness	Verknüpfungstyp	Type de liaison	тип сочленения
s	Connectedness	Verknüpfungszahl	Nombre de liaisons	число сочленений
B	Branchedness	Verzweigtheit	mode de ramification	разветленность
uB	Unbranched	Unverzweigt	Non ramifié	неразветленный
br	Branched	Verzweigt	Ramifié	разветленный
oB	Open-branched	Offen-verzweigt	À ramification ouverte	открыто разветленный
lB	Loop-branched	Geschlossen-verzweigt	À ramification annulaire	закрыто разветленный
olB	Mixed-branched	Gemischt-verzweigt	À ramification mixte	смешанно разветленный
hB	Hybrid	Hybrid	Hybride	гибридный
D	Dimensionality	Dimensionalität	Dimensionalité	размерность
t	Terminated	Mehrfachpolyeder	Polyèdre multiple	многократный полиэдр
r	Cyclic	Ringförmig, zyklisch	Cyclique	цикличный
M	Multiplicity	Multiplizität	Multiplicité	кратность
P	Chain periodicity	Kettenperiodizität	Periodicité de la chaîne	периодичность цепочки
P^r	Ring periodicity	Ringperiodizität	Periodicité cyclique	периодичность кольца
	Fundamental anion	Fundamental-Anion	Anion fondamental	фундаментальный анион
	Complex anions	Komplexe Anionen	Anions complexes	комплексные анионы
	Fundamental chain	Fundamental-Kette	Chaîne fondamentale	фундаментальная цепочка

4.3 The Crystal Chemical System of Silicate Anions

In this section the parameters defined in Sec. 4.2 are applied to all hypothetical silicate anions, irrespective of whether they exist or whether they are likely to exist. This will provide a comprehensive system of silicate anions which will aid in deriving relationships between the structures and chemical properties of silicates. At a later stage it will become clear that for most practical purposes this comprehensive system can be considerably shortened.

The various silicate anions can be filed into a hierarchical order of categories according to the following seven parameters:

coordination number	***CN***:	classes
linkedness	***L***:	subclasses
branchedness	***B***:	branches
multiplicity	***M***:	orders
dimensionality	***D***:	groups
rings or multiple polyhedra	***r*** or ***t***:	subgroups
periodicity	***P***, $\boldsymbol{P^r}$:	families

With this procedure a very broad division of silicate anions into classes and subclasses is obtained, as described in Table 4.3. In addition to the anions containing $[SiO_n]$ polyhedra with one coordination number of Si and one value of the linkedness of the polyhedra, Table 4.3 allows for the existence of polymer anions having $[SiO_n]$ groups with more than one ***CN*** value and for others in which the $[SiO_n]$ polyhedra are connected via corners, edges, and/or faces. Two such anions are illustrated in Fig. 4.11.

Within each subclass, with the exception of those with ***L*** = 0, i.e., with isolated $[SiO_n]$ polyhedra, the silicate anions can be further divided into branches according to their branchedness (Table 4.4). Once again a category for anions of mixed branchedness, i.e., anions which are open- as well as loop-branched, is provided.

Table 4.3 Very broad division of silicate anions into classes and subclasses

L \ *CN*	...	4	5	6	...	Mixed *CN*
0						╳
1						
2						
3						
Mixed ***L***						

CN = coordination number of Si towards O; ***L*** = linkedness of $[SiO_n]$ polyhedra. Each column represents one class, and each field within a column is a subclass

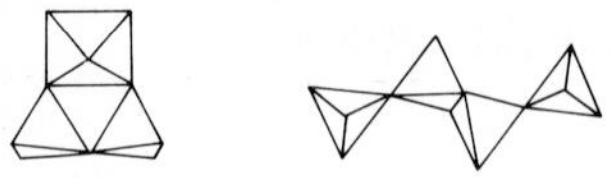

a b

Fig. 4.11. Hypothetical silicate anions with **a** mixed coordination numbers and **b** mixed linkedness

Table 4.4 Division into branches of the anions of each subclass of Table 4.3 with $\boldsymbol{L} > 0$

	oB	
uB	***lB***	***hB***
	olB	

uB = unbranched anions; ***oB*** = open-branched anions; ***lB*** = loop-branched anions; ***olB*** = mixed-branched anions; ***hB*** = hybrid anions

Table 4.5 Subdivision of the anions of each branch into orders and groups according to their multiplicity, ***M***, and dimensionality, ***D***, respectively

D \ ***M***	1	2	3	4	5	...
0	***t*** / ***r***	***t*** / ***r***	***t*** / ***r***	***t*** / ***r***	***t*** / ***r***	***t*** / ***r***
1	***P***	***P***	***P***	***P***	***P***	***P***
2	***P***	***P***	***P***	***P***	***P***	***P***
3	***P***					

Each group with $\boldsymbol{D} = 0$ is further subdivided into one subgroup for cyclic anions, ***r***, and one for noncyclic, terminated anions, ***t***. Each group with $\boldsymbol{D} > 0$ is subdivided into families according to the periodicity ***P*** of the fundamental chains of its anions

These anions are illustrated in Figs. 4.4f and 4.6. Silicate anions belonging to the same branch are subdivided according to their multiplicity and dimensionality into orders and groups, respectively. This leads to the scheme given in Table 4.5. Here it should be mentioned that silicate anions with $\boldsymbol{D} = 0$ can be formulated for all values of $\boldsymbol{L}$, whereas anions with $\boldsymbol{D} > 0$ must have $\boldsymbol{L} > 0$, i.e., they cannot be isolated polyhedra.

Each group with dimensionality $\boldsymbol{D} = 0$ is subdivided into two subgroups, one for cyclic anions, ***r***, and one for noncyclic anions, ***t***. Within each group with $\boldsymbol{D} > 0$, the anions are subclassified according to the periodicity ***P*** of their fundamental chains.

Table 4.6 Crystal chemical system of silicate anions

	CN													
	3	4					5	6					...	Mixed ***CN***
L \ ***B***		***uB***	***oB***	***lB***	***olB***	***hB***		***uB***	***oB***	***lB***	***olB***	***hB***		
0		○	×	×	×	×		○	×	×	×	×	...	×
1		*	*	*	*	*		*	*	*	*	*	...	
2		*	*	*	*	*		*	*	*	*	*	...	
3		*	*	*	*	*		*	*	*	*	*	...	
Mixed ***L***		*	*	*	*	*		*	*	*	*	*	...	

CN = coordination number of silicon towards oxygen; ***B*** = branchedness; ***uB*** = unbranched; ***oB*** = open-branched; ***lB*** = loop-branched; ***olB*** = mixed-branched; ***hB*** = hybrid; ***L*** = linkedness; × = such anions are topologically impossible; ○ = no further topological subdivision is necessary; * = further subdivision according to multiplicity ***M***, periodicity ***P***, $\boldsymbol{P^r}$, dimensionality ***D***, and (for ***D*** = 0) into terminated ***t*** and cyclic ***r*** anions

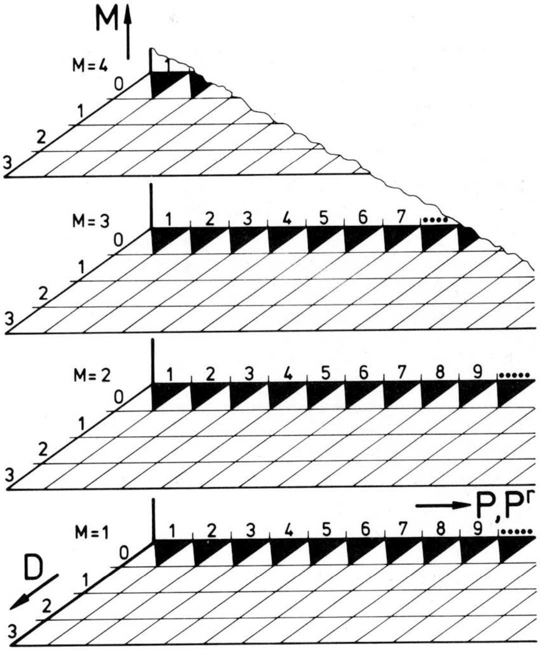

Fig. 4.12. Three-dimensional representation of the subdivision of a branch of silicate anions. Each cell represents a family of anions. Families with ***D*** = 0 are further subdivided into cyclic and terminated (noncyclic) anions by means of *black* and *white shading*

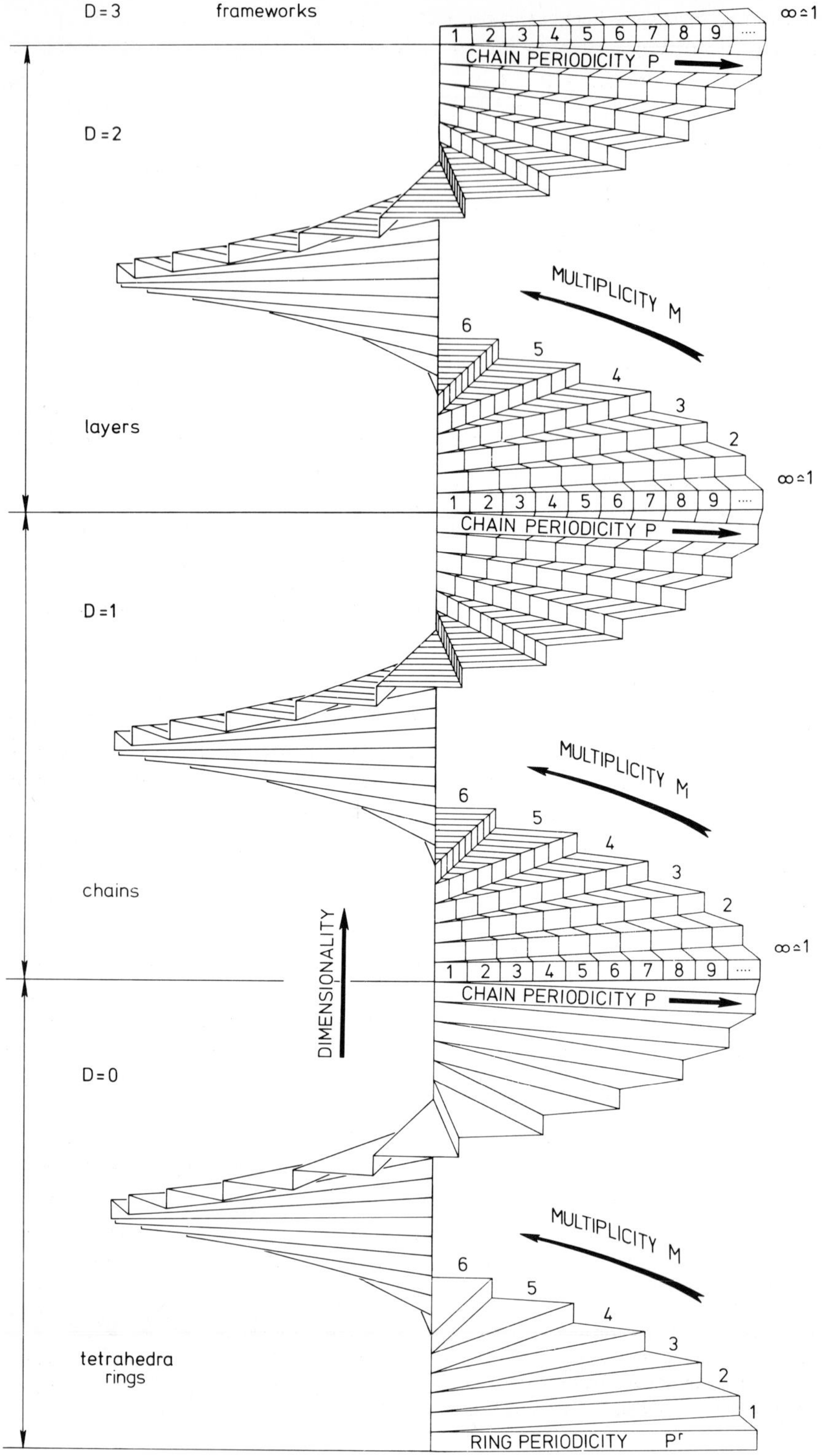
D=3
frameworks
∞≙1
1 2 3 4 5 6 7 8 9 ....
CHAIN PERIODICITY P
D=2
MULTIPLICITY M
6
5
4
3
2
layers
∞≙1
1 2 3 4 5 6 7 8 9 ....
CHAIN PERIODICITY P
D=1
MULTIPLICITY M
6
5
4
3
2
chains
DIMENSIONALITY
∞≙1
1 2 3 4 5 6 7 8 9 ....
CHAIN PERIODICITY P
D=0
MULTIPLICITY M
6
5
4
3
2
1
tetrahedra
rings
RING PERIODICITY P^r

Due to the relatively large number (seven) of parameters used to distinguish the huge variety of silicate anions it is not easy to present the content of Tables 4.3 to 4.5 in a single table. In Table 4.6 an attempt has been made to combine the first three parameters of the crystal chemical classification into a single two-dimensional table.

Each field of this table constitutes one branch of silicate anions. Those branches in Table 4.6 which have been denoted with a * require further subdivision according to multiplicity ***M***, dimensionality ***D***, and chain periodicity ***P***. In addition, each group having $\boldsymbol{D}=0$ can be further discriminated into the two subgroups for cyclic and noncyclic anions.

The existence of silicate anions in a branch marked with an × is topologically impossible.

The subdivision of the branches of silicate anions can be illustrated by the three-dimensional representation in Fig. 4.12. Each of the cells in this illustration represents a family of anions. Subdivision of the anion families with finite extension ($\boldsymbol{D}=0$) into cyclic and noncyclic anions is indicated by the black and white partition of the corresponding cells.

4.4 Periodic Character of the Crystal Chemical Classification of Silicate Anions

In Fig. 4.13 a different representation has been chosen to illustrate a remarkable property of this crystal chemical classification of silicate anions. As in Fig. 4.12 one branch of silicate anions has been selected, such as the unbranched anions (***uB***) containing corner-shared ($\boldsymbol{L}=1$) $[SiO_4]$ tetrahedra ($\boldsymbol{CN}=4$).

If the condensation of $[SiO_4]$ single tetrahedra by sharing corners is considered as a series of successive steps, then the classification of the resulting anions can be illustrated by a spiral staircase. The first step represents a single tetrahedron; the second, a double tetrahedron, $[Si_2O_7]$; the third, a triple tetrahedron, $[Si_3O_{10}]$, etc. Finite numbers ***M*** of tetrahedra corner-linked to multiple tetrahedra ($\boldsymbol{M}$ = multiplicity) are represented by the first turn of the spiral. The dimensionality of these multiple tetrahedra is $\boldsymbol{D}=0$ and their periodicity $\boldsymbol{P}=0$ is trivial. If the number of linearly linked tetrahedra becomes very large ($\boldsymbol{M}\rightarrow\infty$), the resulting anion can be considered to be an unbranched single chain. In Fig. 4.13 it is represented by the first step of the second turn of the spiral.

In crystalline silicates single chains have a periodic structure and are, therefore, subdivided according to their chain periodicity ***P***. Two or more single chains may be linked to form a double chain, a triple chain, or a chain of higher multiplicity, irrespective of their periodicity. Such multiple chains have a dimensionality $\boldsymbol{D}=1$ and form the second turn of the spiral staircase. If, once again, the multiplicity

Fig. 4.13. Spiral representation of the subdivision of a branch of silicate anions. Each field on one step represents a family of anions. In the first turn of the spiral staircase ($\boldsymbol{D}=0$), the subdivision into terminated and cyclic anions and further subdivision according to the ring periodicity $\boldsymbol{P}^r$ of the latter has been omitted for clarity.

approaches ∞ a single layer is formed that has infinite extension in two dimensions ($\boldsymbol{D}=2$) and is visualized by the first step of the third turn of the spiral. Further condensation of such single layers to multiple layers leads, at least hypothetically, step by step to three-dimensional infinitely extended frameworks ($\boldsymbol{D}=3$) of corner-shared [SiO_4] tetrahedra, i.e.; to the uppermost step of the staircase. Like the single chains, the multiple chains, layers, and frameworks are subclassified according to the periodicity of the fundamental single chains from which they can be constructed by successive linkage.

The way in which the various silicate anions of the branch with $\boldsymbol{CN}=4$, $\boldsymbol{L}=1$, and ***uB*** are formally built up in Fig. 4.13 by adding one [SiO_4] tetrahedron step by step shows some resemblance to the way in which the Periodic Table of the Elements is built by adding one proton plus one or several neutrons to each preceding atom.

Although derived by way of a completely formalistic process it will be shown in Chap. 7 that this periodic character of the system of silicate anions is revealed by the abundance of the various natural and synthetic silicates.

The periodic character of this system is, however, somewhat disturbed by the necessity to incorporate the ring anions. This is done by distinguishing, in each family with $\boldsymbol{D}=0$, between terminated noncyclic anions, ***t***, and cyclic anions, ***r***, the latter being further subdivided according to their ring periodicity $\boldsymbol{P}^r$. In the spiral staircase this means that every step on the first turn of the spiral is divided into two sections, the one for the cyclic anions being further subdivided into fields with $\boldsymbol{P}^r=3, 4, 5$, etc. In Fig. 4.13 the division into terminated and cyclic anions has been omitted for the sake of clarity.

To derive a completely comprehensive system of silicate anions, for each of the branches of silicate anions – unbranched, open-branched, loop-branched, mixed-branched, and hybrid – a similar subdivision as shown in Figs. 4.12 and 4.13 would have to be made.

5 Nomenclature and Structural Formulae of Silicate Anions and Silicates

5.1 Nomenclature

The nomenclature of silicate minerals is in a state of great confusion. Over hundreds of years the rules, if there were any, have never resulted in a systematic nomenclature like that applying to chemical substances. This is in spite of considerable effort by the responsible commissions of the International Union of Pure and Applied Chemistry (IUPAC) and of the International Mineralogical Association (IMA). Since the probability of producing a systematic nomenclature for silicates and other minerals in the near future is almost zero, the problem will not be discussed here, although the history of mineral names is an interesting and sometimes amusing subject.

The situation for synthetic silicates is also confused, despite the fact that the rules for the nomenclature of inorganic compounds as laid down by the International Union for Pure and Applied Chemistry should be used.

5.1.1 Chemical Nomenclature

As yet the number of compounds containing silicon atoms surrounded by six oxygen atoms is very small. Therefore, only a few remarks regarding the nomenclature of such hexaoxosilicates will be made here.

If, in a compound of composition $q\,M_rO \cdot s\,SiO_2$, the electronegativity of M is considerably higher than that of silicon ($\chi_{Si} = 1.74$), then SiO_2 is the basic component and Si is considered the cation which compensates for the negative charge of the $[M_tO_u]^{-v}$ anions. Such compounds are called the silicon salts of the corresponding acids. The various polymorphs of silicon diphosphate, $Si[P_2O_7]$, with $\chi_P = 2.06$ are typical examples of this kind.

If, however, the electronegativity of M is about the same as that of Si, the compound is considered to be a double oxide.

The ilmenite-type high pressure phase of $ZnSiO_3$ with Zn ($\chi = 1.66$) and Si in octahedral coordination belongs to this group.

Phases containing $[Si(OH)_6]^{-2}$ groups, such as thaumasite, $Ca_3[Si(OH)_6] \cdot [SO_4][CO_3] \cdot 12\,H_2O$, are called hexahydroxo silicates.

In the vast majority of silicon–oxygen compounds $\chi_M < \chi_{Si}$ and, therefore, SiO_2 is the acid component. Such compounds are salts of silicic acids, i.e., silicates in the proper sense of the word. In contrast to the silicon salts, silicates contain silicon in tetrahedral coordination and their nomenclature needs a more detailed discussion.

Table 5.1 Chemical nomenclature of silicates

Multiplicity / Dimensionality		1	2	3	4	...
0	Oligo-silicates	Mono-silicates	Disilicates	Trisilicates	Tetra-silicates	...
0	Cyclo-silicates	Monocyclo-silicates	Dicyclo-silicates	Tricyclo-silicates	Tetracyclo-silicates	...
1	Poly-silicates	Monopoly-silicates	Dipoly-silicates	Tripoly-silicates	Tetrapoly-silicates	...
2	Phyllo-silicates	Monophyllo-silicates	Diphyllo-silicates	Triphyllo-silicates	Tetraphyllo-silicates	...
3	Tecto-silicates	Tecto-silicates				

Silicates with single tetrahedra $[SiO_4]^{-4}$ should be called monosilicates, those with double tetrahedra $[Si_2O_7]^{-6}$ disilicates[1], those with triple tetrahedra $[Si_3O_{10}]^{-8}$ trisilicates, etc., the whole group being called oligosilicates. Silicates with tetrahedral ring anions are called cyclosilicates[2] and, depending on whether they contain single or double rings either monocyclosilicates or dicyclosilicates, respectively. Phases containing silicate chains are called polysilicates[2], and may be subdivided into monopolysilicates with single chains, dipolysilicates with double chains, etc.

The phyllosilicates with tetrahedral layers may be subdivided in a similar fashion into monophyllosilicates and diphyllosilicates. The silicates with three-dimensional tetrahedral frameworks are called tectosilicates.

This nomenclature is summarized in Table 5.1. In order to distinguish between silicates of different branchedness, the epithets, "branched" or "unbranched" may be added to the names given in Table 5.1.

1 The terminology disilicate, trisilicate, tetrasilicate is still sometimes, in particular in the technical literature, used for silicates of composition $MO \cdot 2\,SiO_2$, $MO \cdot 3\,SiO_2$, and $MO \cdot 4\,SiO_2$ ($M = Me_2^+$, Me^{+2}, $Me_{0.67}^{+3}$), respectively. This practice should definitely be abandoned since it is in contradiction to the IUPAC rules.

2 The term metasilicate for silicates of composition $MSiO_3$ ($M = Me_2^+$, Me^{+2}, $Me_{0.67}^{+3}$) should be abandoned because it includes two different types of silicates, those with ring anions (cyclosilicates) and those with chain anions (polysilicates).

5.1.2 Mineralogical Nomenclature

Minerals are usually designated by trivial names. These are often related to some conspicuous property, such as morphology, color, changes due to heating, or to their chemical composition. Here are just a few examples:

The name of actinolite, an amphibole of composition $Ca_2(Mg,Fe)_5[Si_4O_{11}]_2(OH)_2$, is derived from the Greek words *aktis* for ray and *lithos* for stone because it is often found as long prismatic crystals in radiating groups (German: *Strahlstein*). The name of the borosilicate axinite relates to the wedge-shaped crystals which often have sharp edges (Gr. *axine* = axe). The asbestos chrysotile forms delicately fibrous crystals with a silky, sometimes yellow or brownish luster and its name relates to its color (Gr. *chrysos* = gold) as well as to its morphology (Gr. *tilos* = fiber). The minerals olivine and kyanite, and the chlorites $M_{5-6}[T_2O_5]_2(OH)_8$ are named for their colors: olive-green, blue (Gr. *kyaneos*), and green (Gr. *chloros*). The white tectosilicates albite and leucite, derive their names from the Latin and Greek words *albus* and *leukos*, respectively, for white. The rare mineral melanophlogite turns black (Gr. *melas* = black) when heated with a blowpipe (Gr. *phlogos* = genitive of *phlox* = flame). The name of the clay mineral vermiculite is chosen in allusion to its exfoliation during heating into long worm-like threads (Lat. *vermiculus* = the worm). Several feldspar names relate to their characteristic cleavage properties. For example, in orthoclase the two main cleavage planes are perpendicular to each other (Gr. *orthos* = upright), and in plagioclase they are oblique (= Gr. *plagios*). On the other hand, oligoclase, calcium-poor plagioclase, is only slightly (= Gr. *oligos*) cleavable (Gr. *klasis* = cleavage). Mineral names which give an indication of their chemical composition are bafertisite for $Ba(Fe,Mn)_2Ti[Si_2O_7](O,OH)_2$, fenaksite for $Na_2K_2Fe_2[Si_8O_{20}]$, and sodalite for $Na_4[Al_3Si_3O_{12}]Cl$, a sodium-rich silicate (Gr. *lithos* = stone).

Other silicate minerals are named after the locality where they have been discovered. For example, the silica polymorph cristobalite is named after the San Cristobal mountain in Mexico, andalusite after the Spanish province Andalusia, the garnet spessartine after the German Spessart range, and vesuvianite after the Italian volcano Mount Vesuvius. Although the amphibole tremolite is named in allusion to Val Tremola, a valley near the St. Gotthard pass in the Italian Alps, this mineral has not been observed there. On the contrary, the discoverer wanted to exploit the locality himself and used its name for the sought after mineral in order to mislead other collectors.

Often minerals are named in honor of a person. Chemists will, for instance, realize that wollastonite is named in honor of the chemist W. H. Wollaston and the zeolite gmelinite in honor of L. Gmelin, the founder of the Gmelin Handbook of Inorganic Chemistry. Similarly, those people familiar with zeolites will recognize that barrerite has been named after R. M. Barrer who contributed so much to our knowledge of the chemistry of zeolites, and ceramic scientists will know that the two calcium silicate hydrates dellaite, $Ca_6[SiO_4][Si_2O_7](OH)_2$, and rustumite, $Ca_{10}[SiO_4][Si_2O_7]_2(OH)_2Cl_2$, have been named in honor of Della M. Roy and Rustum Roy, two outstanding cement scientists.

It is not only very difficult for most readers to immediately recognize a silicate mineral from its name, but it is sometimes also difficult to pronounce a mineral

Table 5.2 Nomenclature of silicates used by mineralogists

Dimensionality \ Multiplicity	1		2	3	4	...
0	Nesosub-silicates	Neso-silicates	Sorosilicates			
0	Cyclosilicates					
1	Inosilicates					
2	Phyllosilicates					
3	Tectosilicates					

name properly. For example, it is as difficult for an English or German tongue to fluently speak the name taneyamalite, derived from the Japanese locality Tane-yama, as it is for a Chinese to correctly pronounce the name of the amphibole ferroferritschermakite. More detailed information on the origin of mineral names can be found in books by Lüschen (1968) and Mitchell (1979).

Fortunately, mineralogists use names for the large groups of silicates which are quite intuitive, at least to those who have a slight knowledge of ancient Greek. These group names are presented in Table 5.2.

A great advantage of the nomenclature given in Tables 5.1 and 5.2 is that they can be applied to many other kinds of inorganic compounds, such as germanates, phosphates, etc., and that the names give an immediate indication of the structure of the silicate anion. If a more detailed structural knowledge is desired, structural chemical formulae must be used.

5.2 Structural Formulae

The structural formula of a substance should contain as much information about its structure as necessary and should be as self-explanatory as possible. This section describes a formalism which allows a maximum amount of information to be included in a silicate formula. Naturally, such a formula becomes rather complex. For practical purposes, therefore, only those details which are of interest for the problem under consideration are incorporated in the formula.

Following common practice, the silicate anions are placed between square brackets, for example, $Mg_2[SiO_4]$ for the monosilicate forsterite and $Al_2[Si_4O_{10}](OH)_2$ for pyrophyllite. If a silicate contains complex anions other than the silicate anion then they are also put between square brackets, for example, $Ca_3[Si(OH)_6][SO_4][CO_3] \cdot 12\,H_2O$ in the case of thaumasite. If desired, the coordination numbers of some or all atoms can be indicated with superscripts, as for the high pressure phase

$K_2Si^{[6]}[Si_3^{[4]}O_9]$ in which one quarter of the silicon atoms are six-coordinated and the other silicon atoms are tetrahedrally coordinated and form a cyclic anion. When, as in the following, only those silicates are considered which contain tetrahedrally coordinated silicon exclusively, the coordination number may be omitted from the formulae.

The formula can also be complemented by a suffix (lT), (mT), (hT), (lP), (mP), or (hP) to designate low, medium, or high temperature and pressure phases, respectively.

The kind and degree of condensation of the anions is indicated by indexes assembled between curly brackets preceding the anions. The general structural formula of a silicate $M_rSi_xO_y$ can then be written as

$$M_r\{\boldsymbol{B}, \boldsymbol{M}^{\boldsymbol{D}}_{\infty}\}[Si_xO_y].$$

Here the branchedness ***B*** is specified as ***uB***, ***oB***, ***lB***, ***olB***, or ***hB*** for unbranched, open-branched, loop-branched, mixed-branched, or hybrid anions, respectively; $\boldsymbol{M}=1, 2, 3, \ldots$ is the multiplicity, and $\boldsymbol{D}=0, 1, 2, 3$, is the dimensionality of the anions.

While chain, layer, and framework anions can be distinguished unequivocally by the symbols ${}^{\boldsymbol{D}}_{\infty} = {}^{1}_{\infty}, {}^{2}_{\infty}$, and ${}^{3}_{\infty}$, the symbol ${}^{0}_{\infty}$ for anions for finite extension is ambiguous. Therefore, in order to differentiate between cyclic and noncyclic anions this symbol is replaced by the index or ***t***.

For silicate anions of finite extension ($\boldsymbol{D}=0$) the numbers of silicon and oxygen atoms, x and y, are, of course, equal to the numbers of these atoms in the anion. For anions with $\boldsymbol{D}=1$ x is chosen equal to the number of silicon atoms in one period of the chain. However, for silicate anions with $\boldsymbol{D}>1$, the numbers of atoms in the formula should be chosen such that x is the smallest multiple of the number of silicon atoms in the identity period of its fundamental chain for which y is an integer.

Following this procedure, the silicate anions schematically presented in Fig. 5.1 can be described by the following structural formulae:

rosenhahnite	$H_2Ca_3\{\boldsymbol{uB}, 3\boldsymbol{t}\}[Si_3O_{10}]$;
zunyite	$Al_{12}^{[6]}Al^{[4]}\{\boldsymbol{oB}, 3\boldsymbol{t}\}[Si_5O_{16}](OH,F)_{18}O_4Cl$;
eakerite	$Ca_2Al_2Sn\{\boldsymbol{oB}, 1\boldsymbol{r}\}[Si_6O_{18}](OH)_2 \cdot 2H_2O$;
deerite	$Fe_6^{+2}Fe_3^{+3}\{\boldsymbol{lB}, 1^1_{\infty}\}[Si_6O_{17}]O_3(OH)_5$;
synthetic	$Ba_4\{\boldsymbol{uB}, 3^1_{\infty}\}[Si_6O_{16}]$;
apophyllite	$KCa_4\{\boldsymbol{uB}, 1^2_{\infty}\}[Si_4O_{10}]_2(F,OH) \cdot 8H_2O$;
hypothetic	$\{\boldsymbol{lB}, 1^2_{\infty}\}[Si_4O_{10}]^{-4}$;
feldspar, e.g.,	$K\{\boldsymbol{lB}, {}^3_{\infty}\}[(AlSi_3)O_8]$.

For silicates with anions of infinite extension it can be helpful to indicate the periodicity ***P*** of their fundamental chains, i.e., the number of $[SiO_4]$ tetrahedra in the unbranched fundamental chain of unbranched anions, or in the linear part of the branched fundamental chain of branched anions. This information can be provided by a superscript on the left side of the Si symbol. Moreover, if a cyclic anion is marked with a symbol ***r*** in the formula, the number of tetrahedra in the single ring (i.e., its ring periodicity $\boldsymbol{P}^r$) can be indicated in the same way without risk of confusion. Eakerite would then be given the formula

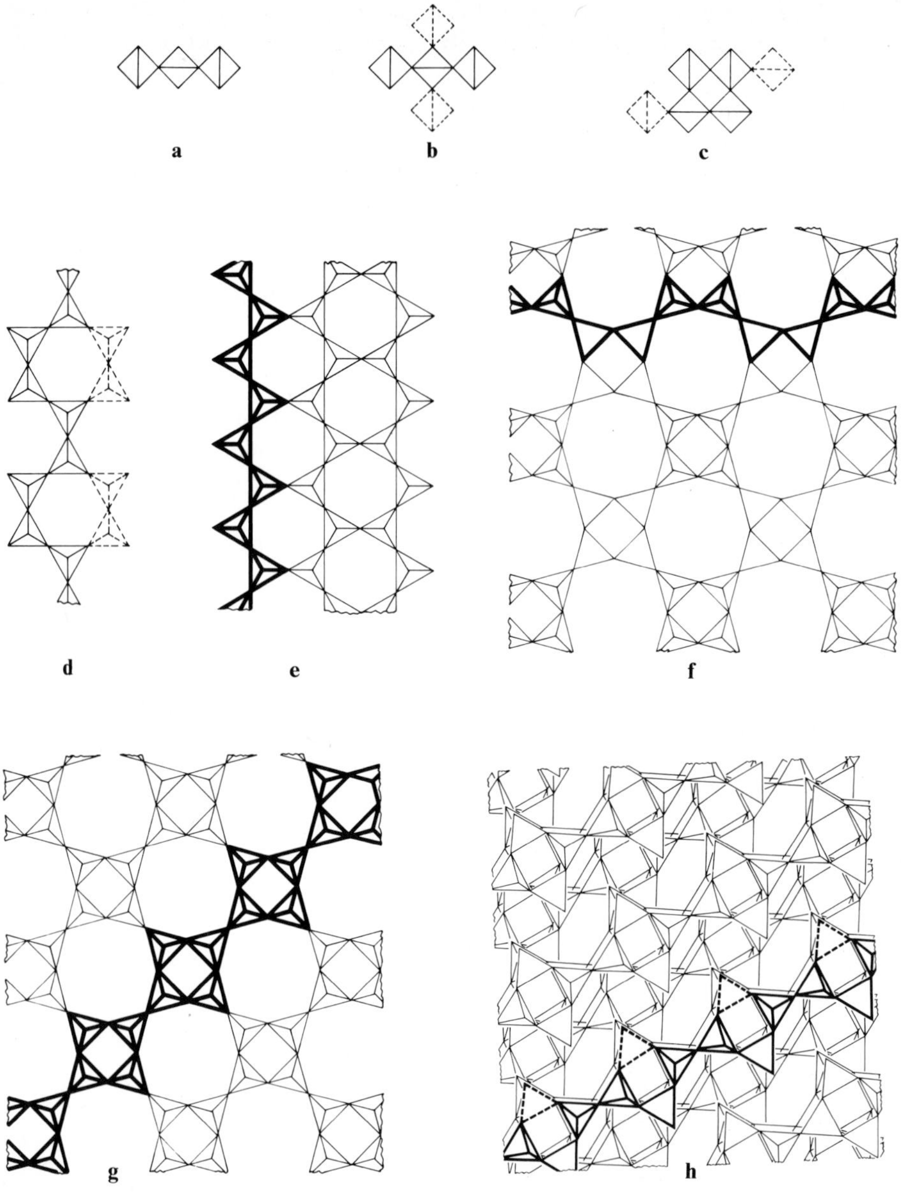

Fig. 5.1 a – h. Schematical representation of several tetrahedral silicate anions to assist in the determination of their structural formulae.

a Unbranched triple tetrahedron (rosenhahnite); **b** open-branched triple tetrahedron (zunyite); **c** open-branched vierer single ring (eakerite); **d** loop-branched vierer single chain (deerite); **e** unbranched zweier triple chain (jimthompsonite); **f** unbranched vierer single layer (apophyllite); **g** loop-branched dreier single layer (hypothetical); **h** loop-branched dreier framework (feldspar).

$Ca_2Al_2Sn\{\boldsymbol{oB}, 1\,\boldsymbol{r}\}[^4Si_6O_{18}](OH)_2 \cdot 2\,H_2O$ indicating the branched vierer single ring, whereas the formula $Ba_4\{\boldsymbol{uB}, 3^1_\infty\}[^2Si_6O_{16}]$ clearly denotes the presence of a zweier triple chain in this barium silicate.

If desired, for silicates having branched anions, the size and number of branches of an anion with $\boldsymbol{D} = 0$ or of branches in one period of the fundamental chain of an anion with $\boldsymbol{D} > 1$ can be indicated by including them between round brackets. The complete formulae of the anions are then

$\{\boldsymbol{oB}, 3\,\boldsymbol{t}\}[Si_3O_{10}(SiO_3)_2]$	for zunyite;
$\{\boldsymbol{oB}, 1\,\boldsymbol{r}\}[^4Si_4O_{12}(SiO_3)_2]$	for eakerite;
$\{\boldsymbol{lB}, 1^1_\infty\}[^4Si_4O_{12}(Si_2O_5)]$	for deerite;
$\{\boldsymbol{lB}, 1^2_\infty\}[^3Si_3O_9(SiO)]$	for the hypothetical anion in Fig. 5.1 g.

6 Crystal Chemical Classification of Silicates: General Part

6.1 Mixed-Anion Silicates

The crystal chemical classification of silicate anions described in Chap. 4 can easily be adapted to silicates by adding another parameter to the seven which have already been used.

The overwhelming majority of crystalline silicates contain only one type of silicate anion, for example, single tetrahedra $[SiO_4]^{-4}$, triple tetrahedra $[Si_3O_{10}]^{-8}$, or unbranched zweier single chains $\{\boldsymbol{uB}, 1^1_\infty\}[^2Si_2O_6]^{-4}$, and so on. However, there is a small, but steadily increasing number of silicates which have been observed to contain two different silicate anions. So far, no crystalline silicates with three or more silicate anions have been observed, although in melts, glasses, and aqueous solutions more than two types of silicate anions coexist under equilibrium conditions.

If the number of different silicate anions in a silicate is denoted $\boldsymbol{N_{an}}$, the classification of silicates should be able to distinguish between such silicates by introducing separate superclasses for $\boldsymbol{N_{an}} = 1$, $\boldsymbol{N_{an}} = 2$, etc.

Silicates with $\boldsymbol{N_{an}} = 1$ are called *uniform-anion silicates,* and those with $\boldsymbol{N_{an}} > 1$ are called *mixed-anion silicates.*

6.2 The Crystal Chemical System of Silicates

As will be shown in later Chapters the eight parameters

- $\boldsymbol{N_{an}}$: number of different silicate anions;
- ***CN***: coordination number of silicon;
- ***L***: linkedness;
- ***B***: branchedness;
- ***M***: multiplicity and
- ***D***: dimensionality of silicate anions;
- ***t***, ***r***: terminated or cyclic anions with $\boldsymbol{D} = 0$;
- ***P***, $\boldsymbol{P^r}$: periodicity of silicate anions

seem to be sufficient to arrange all known silicates and, hopefully, all silicates to be discovered, in a systematic and reasonable way.

Table 6.1 is an attempt to represent this crystal chemical system of silicates in one comprehensive table. Table 6.1, which is an extension of the corresponding system of silicate anions shown in Table 4.6, includes all of the classification parameters described above.

Table 6.1 Crystal chemical system of silicates

		CN													
		3	4					5	6					...	Mixed ***CN***
N_{an}	***B*** / ***L***		***uB***	***lB***	***oB***	***olB***	***hB***		***uB***	***lB***	***oB***	***olB***	***hB***		
1	0		○	×	×	×	×		○	×	×	×	×	...	×
	1		*	*	*	*	*		*	*	*	*	*	...	
	2		*	*	*	*	*		*	*	*	*	*	...	
	3		*	*	*	*	*		*	*	*	*	*	...	
	Mixed ***L***		*	*	*	*	*		*	*	*	*	*	...	
2	0		×	×	×	×	×		×	×	×	×	×	...	○
	1		*	*	*	*	*		*	*	*	*	*	...	
	2		*	*	*	*	*		*	*	*	*	*	...	
	3		*	*	*	*	*		*	*	*	*	*	...	
	Mixed ***L***		*	*	*	*	*		*	*	*	*	*	...	
		Mixed ***B***													
3	0		×	×	×	×	×		×	×	×	×	×	...	○
	1		*	*	*	*	*		*	*	*	*	*	...	
	2		*	*	*	*	*		*	*	*	*	*	...	
	3		*	*	*	*	*		*	*	*	*	*	...	
	Mixed ***L***		*	*	*	*	*		*	*	*	*	*	...	
		Mixed ***B***													
⋮	⋮		⋮	⋮	⋮	⋮	⋮		⋮	⋮	⋮	⋮	⋮	⋮	

N_{an} = number of different silicate anions; ***CN*** = coordination number of silicon by oxygen; ***B*** = branchedness; ***uB*** = unbranched; ***oB*** = open-branched; ***lB*** = loop-branched; ***olB*** = mixed-branched; ***hB*** = hybrid; ***L*** = linkedness; ○ = no further topological subdivision is necessary; × = such anions are topologically impossible; * = further subdivision according to multiplicity ***M***, periodicity ***P***, ***P***^{r}, dimensionality ***D***, and (for ***D*** = 0) into terminated ***t*** and cyclic ***r*** anions

Being aware that the distinction between silicates with $N_{an} = 1$ and with $N_{an} > 1$ has to be made, Tables 4.3 to 4.6 and Figs. 4.12 and 4.13, which were originally presented to illustrate the crystal chemical classification of the silicate anions, can also be used for the silicates as well.

6.3 Further Subdivision of Silicates

For several silicate anion types such a large number of silicates have been reported that a further subdivision is desirable. This can be based either on the properties of the silicate anion, or on the cation–oxygen polyhedral part of the structure.

6.3.1 Atomic Ratio Si:O of Silicate Anions

All tetrahedral single rings and single chains, with the exception of the loop-branched ones, have a silicon:oxygen atomic ratio of 1:3. When these fundamental anions are connected to form complex anions, the linkage can involve all or only part of the tetrahedra, giving rise to different Si:O ratios within the same family of ions.

This is readily visible in Fig. 6.1 where the three topologically different ways of linking two dreier single chains to a dreier double chain are shown. Depending on whether the two single chains are linked through one out of three, two out of three, or through all the tetrahedra, the composition of the double chain is $[Si_6O_{17}]^{-10}$, $[Si_6O_{16}]^{-8}$, and $[Si_6O_{15}]^{-6}$, respectively.

It is, therefore, clear that within each family of silicates with complex ions (multiple rings as well as multiple chains, single and multiple layers, and frameworks, provided their periodicity is $\boldsymbol{P} > 1$), the Si:O ratio can vary depending on the proportion of tetrahedra involved in the linkage between the fundamental anions. A more detailed description of such anions is deferred to Chap. 7.

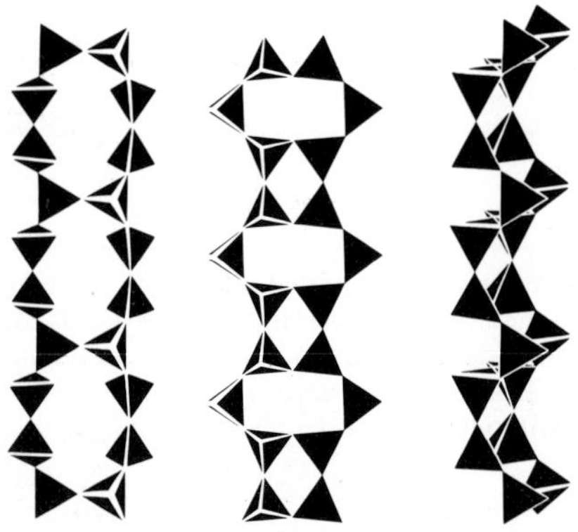

Fig. 6.1 a–c. Dreier double chains with different Si : O atomic ratios.

a $\{uB, 2^1_\infty\}[^3Si_6O_{17}]^{-10}$ in xonotlite (Kudoh and Takéuchi 1979); **b** $\{uB, 2^1_\infty\}[^3Si_6O_{16}]^{-8}$ in okenite (Merlino 1983); **c** $\{uB, 2^1_\infty\}[^3Si_6O_{15}]^{-6}$ in epididymite (Robinson and Fang 1970).

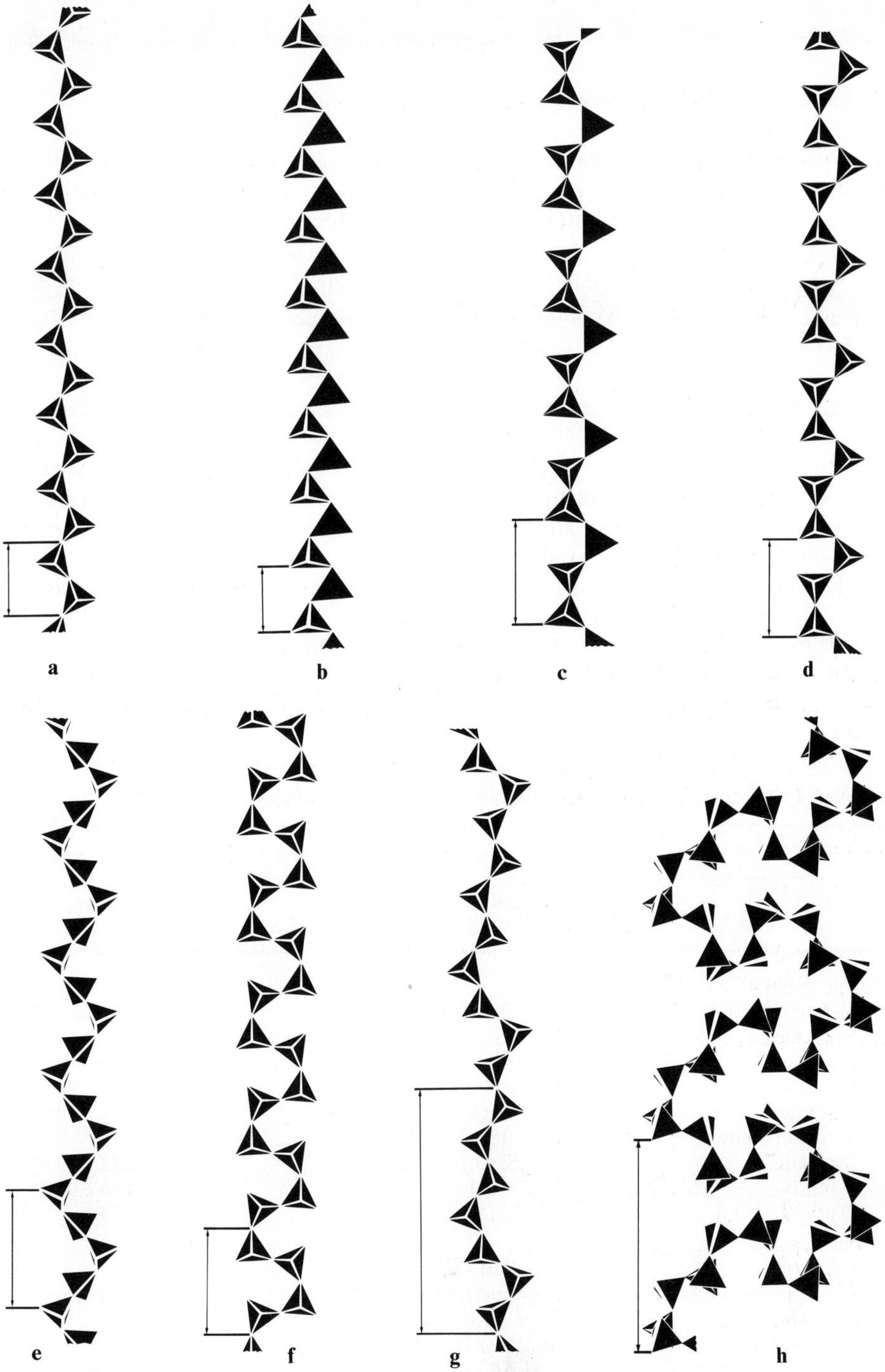

Fig. 6.2 a – h. Unbranched tetrahedral single chains of silicates illustrating the variation in the degree of chain stretching.

a Zweier single chain in enstatite, $Mg_2[Si_2O_6]$; **b** zweier single chain in synthetic $Ba_2[Si_2O_6]$(hT); **c** dreier single chain in sorensenite, $Na_4Be_2Sn[Si_3O_9]_2 \cdot 2H_2O$; **d** dreier single chain in serandite, $HNa(Mn, Ca)_2[Si_3O_9]$; **e** vierer single chain in krauskopfite, $H_4Ba_2[Si_4O_{12}] \cdot 4H_2O$; **f** vierer single chain in synthetic $Na_2Cu_3[Si_4O_{12}]$; **g** siebener single chain in pyroxferroite, $(Fe, Ca)_7[Si_7O_{21}]$; **h** 24er single chain in synthetic $Na_{24}Y_8[Si_{24}O_{72}]$.

6.3.2 Degree of Chain Stretching

In silicate anions built from chains (chain, layer, and framework silicates), the chains may either be stretched or shrunken relative to one another. As Fig. 6.2 illustrates, the degree of chain stretching varies considerably.

If a chain in the most stretched state is taken as the one with an identity period, I_{chain} (measured in Å), equal to $\boldsymbol{P}$ times the length l_T of the tetrahedron edge formed by the bridging oxygen atoms ($\boldsymbol{P}$ = periodicity of the chain) then a *stretching factor* f_s can be defined, such that

$$f_s = \frac{I_{chain}}{l_T \cdot \boldsymbol{P}}.$$

For the case of silicate chains the value of l_T may be taken as 2.70 Å, half the chain period of the zweier single chain of shattuckite, the silicate with the largest known value of $I_{chain} : \boldsymbol{P}$. With this value the stretching factors f_s of all chain silicates are ≤ 1.000 as long as no silicates with $l_T > 2.70$ Å are discovered.

From the definition of f_s it follows that $100 \cdot f_s$ is the relative stretching and $100\,(1-f_s)$ is the relative shrinkage of a chain, in percent.

We can use this stretching factor to subdivide silicates with anions of infinite extension. Since the shrinkage of the chains, which is most pronounced for chains with an even number of tetrahedra in the repeat unit, is strongly related to the

Table 6.2 Stretching factors of some unbranched tetrahedral silicate chains

Silicate		I_{chain}	f_s
Name	Formula	[Å]	
Zweier chains			
Single chains			
Shattuckite	$Cu_5[Si_2O_6]_2(OH)_2$	5.40	1.000
Synthetic	$NaIn[Si_2O_6]$	5.37	0.994
Johannsenite	$CaMn[Si_2O_6]$	5.29	0.980
Diopside	$CaMg[Si_2O_6]$	5.25	0.972
Enstatite	$Mg_2[Si_2O_6]$	5.21	0.965
Ferrocarpholite	$FeAl_2[Si_2O_6](OH)_4$	5.11	0.946
Synthetic	$Na_4[Si_2O_6]$	4.82	0.893
Synthetic	$Li_4[Si_2O_6]$	4.66	0.863
Synthetic	$Ba_2[Si_2O_6]$ (hT)	4.54	0.841
Double chains			
Grunerite	$(Fe, Mg)_7[Si_4O_{11}]_2(OH)_2$	5.35	0.991
Glaucophane	$Na_2Mg_3Al_2[Si_4O_{11}]_2(OH)_2$	5.30	0.981
Tremolite	$Mg_5Ca_2[Si_4O_{11}]_2(OH)_2$	5.28	0.978
Arfvedsonite	$(Na, Ca)_3(Fe, Al, Mg)_5$ $[(Al, Si)_4O_{11}]_2(OH)_2$	5.21	0.965
Oligo-fold chains			
Synthetic	$HNaMg_4[Si_6O_{16}](OH)_2$	5.26	0.974
Synthetic	$Ba_4[Si_6O_{16}]$	4.69	0.869
Synthetic	$Ba_5[Si_8O_{21}]$	4.70	0.870
Synthetic	$Ba_6[Si_{10}O_{26}]$	4.71	0.872

Table 6.2 (continued)

Silicate		I_{chain}	f_s
Name	Formula	[Å]	
Dreier chains			
Single chains			
Sorensenite	$Na_4Be_2Sn[Si_3O_9]_2 \cdot 2\ H_2O$	7.44	0.919
Foshagite	$Ca_4[Si_3O_9](OH)_2$	7.36	0.909
Wollastonite	$Ca_3[Si_3O_9]$	7.32	0.904
Bustamite	$(Ca, Mn)_3[Si_3O_9]$	7.16	0.884
Pectolite	$HNaCa_2[Si_3O_9]$	7.04	0.869
Serandite	$HNa(Mn, Ca)_2[Si_3O_9]$	6.89	0.851
Double chains			
Xonotlite	$Ca_6[Si_6O_{17}](OH)_2$	7.35	0.901
Elpidite	$Na_2Zr[Si_6O_{15}] \cdot 3\ H_2O$	7.14	0.881
Vierer chains			
Single chains			
Krauskopfite	$H_4Ba_2[Si_4O_{12}] \cdot 4\ H_2O$	8.46	0.783
Batisite	$Na_2BaTi_2[Si_4O_{12}]O_2$	8.08	0.748
Ohmilite	$Sr_3(Ti, Fe^{+3})[Si_4O_{12}](O, OH) \cdot 2-3\ H_2O$	7.78	0.720
Leucophanite	$Na_2Be_2Ca_2[Si_4O_{12}]F_2$	7.42	0.687
Haradaite	$Sr_2V_2[Si_4O_{12}]O_2$	7.06	0.654
Double chain			
Narsarsukite	$Na_4Ti_2[Si_8O_{20}]O_2$	8.01	0.742
Fünfer chains			
Single chains			
Synthetic	$Mn_5[Si_5O_{15}]$	12.24	0.906
Nambulite	$H(Li, Na)Mn_4[Si_5O_{15}]$	12.02	0.890
Double chain			
Inesite	$Ca_2Mn_7[Si_{10}O_{28}](OH)_2 \cdot 5\ H_2O$	11.98	0.887
Sechser chains			
Single chains			
Gaidonnayite	$Na_4Zr_2[Si_6O_{18}] \cdot 4\ H_2O$	13.51	0.834
Stokesite	$Ca_2Sn_2[Si_6O_{18}] \cdot 4\ H_2O$	11.63	0.718
Chkalovite	$Na_6Be_3[Si_6O_{18}]$	11.11	0.686
Double chains			
Synthetic	$Na_4Mg_4[Si_{12}O_{30}]$	10.21	0.630
Zektzerite	$Li_2Na_2Zr_2[Si_{12}O_{30}]$	10.16	0.627
Synthetic	$Na_4Li_2Fe_2^{+3}[Si_{12}O_{30}]$	10.11	0.624
Siebener single chains			
Pyroxmangite	$(Mn, Fe, Ca)_7[Si_7O_{21}]$	17.45	0.923
Pyroxferroite	$(Fe, Ca)_7[Si_7O_{21}]$	17.38	0.920
Neuner single chain			
Ferrosilite III	$Fe_9[Si_9O_{27}]$	22.61	0.930
Zwölfer single chain			
Alamosite	$Pb_{12}[Si_{12}O_{36}]$	19.63	0.606
24er single chain			
Synthetic	$Na_{24}Y_8[Si_{24}O_{72}]$	15.14	0.234

properties of the cations within the silicate under consideration (see Sec. 10.1.4), the use of f_s for silicate subdivision is reasonable from a structural chemical point of view. Table 6.2 compares the f_s values for some silicates with unbranched chain anions.

6.3.3 Silicate Anion Symmetry

The symmetry of a silicate anion is another property that can be used to subdivide the larger groups in the crystal chemical system of silicates. However, classification by this means is not very appropriate from a structural chemistry viewpoint, since symmetry is more a property of the crystalline state than of the chemical structure of the phase and is, therefore, strongly dependent on small changes in temperature and pressure.

6.3.4 Cation – Oxygen Polyhedra

An important factor to be considered for further subdivision of the larger silicate groups is the cation–oxygen polyhedron part of the silicate. In doing this, some compensation is made for the shortcomings of the classification given in Sec. 6.2 resulting from the neglect of the cationic part of the structure in favor of the anionic part. This is particularly important for silicates with higher cation : silicon atomic ratios M:Si, where $M = Me^{+}, Me^{+2}_{0.5}, Me^{+3}_{0.33}$, etc. (see Sec. 6.5).

6.4 Silicate Classification Procedure

In order to classify a particular silicate according to the crystal chemical systematics discussed previously, the following procedure should be applied:

Step 1: Determine which tetrahedra are occupied principally by silicon. These tetrahedra constitute the anions of the phase.

Step 2: Determine the coordination numbers ***CN*** of silicon and the atoms replacing silicon statistically in the phase. If $CN = 4$ the phase is a silicate with Si being part of the anion; if $CN = 6$ it is a silicon compound with Si as a more basic part of the structure. If the phase contains tetrahedrally and octahedrally coordinated silicon, it is one of the very rare mixed-coordination number phases. In principle, for compounds containing six-coordinated Si the same classification procedure can be applied as for silicates with $CN = 4$. However, this is perhaps unnecessary since the number of compounds with $CN = 6$ is at present rather small. Therefore, the following steps describe the procedure to classify only those silicates with $CN = 4$. For phases with $CN = 6$ or with $CN = 4$ and 6, appropriate adjustments are made.

Step 3: Determine the linkedness ***L***, i.e., whether the $[SiO_4]$ tetrahedra are singular or share corners, edges, or faces. If $L = 0$ the silicate is a monosilicate. Since only one phase with $\boldsymbol{L}=2$ and none with $\boldsymbol{L}=3$ are known, further discussion is necessary only for silicates with $\boldsymbol{L}=1$.

Step 4: Determine the number $\boldsymbol{N}_{an}$ of different types of silicate anions. If $\boldsymbol{N}_{an}=1$ the phase is a uniform-anion phase, if $\boldsymbol{N}_{an}>1$ it is a mixed-anion phase and each of the anion types should be analyzed as follows.

Step 5: Determine the dimensionality $\boldsymbol{D}$ of each anion. If the dimensionality $\boldsymbol{D}=0$, the phase is either an oligo- or a cyclosilicate, if $\boldsymbol{D}=1$, 2, or 3 it is a polysilicate, phyllosilicate, or tectosilicate, respectively. Each of these five groups of silicates is then treated in a slightly different way.

Oligosilicates

Step 6: Determine the branchedness $\boldsymbol{B}$ of the anion.

Step 7: Determine the multiplicity $\boldsymbol{M}$, i.e., the number of tetrahedra in the linear part of the anion. By analogy with the nomenclature of aliphatic hydrocarbons, the longest possible "chain" is chosen as the linear part of the anion.

Cyclosilicates

Step 6: Determine the branchedness $\boldsymbol{B}$ of the anion.

Step 7: Determine the ring periodicity $\boldsymbol{P}^r$, i.e., the number of tetrahedra in the unbranched part of the fundamental ring.

Step 8: Determine the multiplicity $\boldsymbol{M}$ of the anion.

Step 9: If further subdivision seems desirable, determine the Si:O ratio.

Polysilicates

Step 6: Determine the periodicity $\boldsymbol{P}$ of the chain.

Step 7: Determine the branchedness $\boldsymbol{B}$ of the chain.

Step 8: Determine the multiplicity $\boldsymbol{M}$ of the anion.

Step 9: Determine the Si:O ratio.

Phyllosilicates

Step 6: Determine the chain (or chains) with the lowest periodicity $\boldsymbol{P}$ which runs (run) parallel to the shortest identity period within the silicate layer.

Step 7: Determine the branchedness $\boldsymbol{B}$ of the chain (or chains). If there is only one such chain it is the fundamental chain of the silicate layer. If there are several chains of lowest periodicity running parallel to the shortest identity period within the layers, the fundamental chain is chosen such that an unbranched chain is preferred over a loop-branched chain and a loop-branched chain over an open-branched one.

Step 8: Determine the multiplicity $\boldsymbol{M}$ of the anion.

Step 9: Determine the Si:O ratio.

Tectosilicates

As for the phyllosilicates, except that step 8 is omitted since all tectosilicates have a multiplicity of $\boldsymbol{M}=1$.

Figure 6.3 is a flow chart of this procedure.

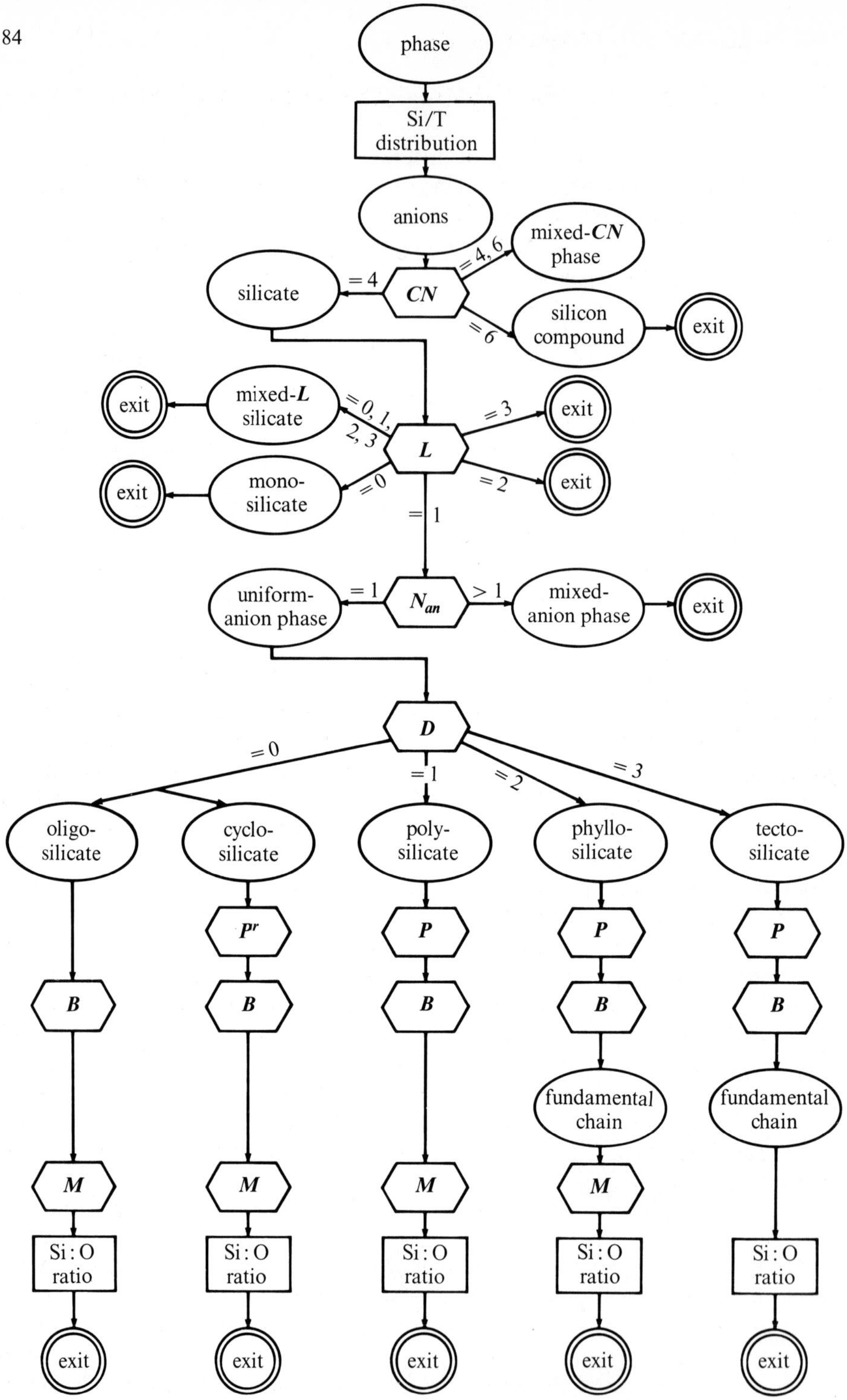

Fig. 6.3. Flow-chart diagram of the procedure to classify a silicate according to the crystal chemical classification.

6.5 Shortcomings of the Crystal Chemical Classification

The classification described above has a number of weak points that result from the principles upon which the classification is based. As pointed out earlier, this is a crystal chemical classification based on both the crystal structure and the chemical composition of the silicates. Therefore, in order to file a certain silicate into its proper category, its crystal structure must be known.

Since the system is based almost exclusively on the aggregates of silicon–oxygen polyhedra, the classification procedure is straightforward for those silicates which contain no atoms that can substitute for the silicon atoms. For other phases we follow the common practice that those atoms are regarded as part of the silicon–oxygen complex that, at least under certain thermodynamic conditions, can substitute for silicon.

This means that in some cases a very accurate structure analysis, including a determination of the site occupancy, may be necessary in order to unequivocally classify a particular silicate. However, even if complete ordering has been found under some conditions, it is often quite difficult to decide whether or not there will be some statistical replacement at higher temperatures or pressures.

As in any other classification, limits must be drawn between parameters that change more or less continuously. Such boundaries are always rather arbitrary and may be conjectural. Consequently, the decision, for example, of which minimum value of Si:Al ratio a $[TO_4]$ tetrahedron should be regarded as belonging to the silicate anion, will sometimes be disputed. Some prefer to classify the phase $(Mg_{38.46}Sc_{3.11})(Li_{1.16}Si_{0.18}Si_{40})O_{124}$, in which one of the T sites is occupied by 0.58 lithium, 0.09 silicon atoms and the rest vacant (Takéuchi et al. 1977, 1984 a, b), as a chain silicate with 22 tetrahedra in the repeat unit. However, others regard it as an oligosilicate with tenfold tetrahedra ($\boldsymbol{M}=10$) because of the low silica content (9%) of the special T site (see Sec. 7.2.2.1.1).

It has already been pointed out, and is explicitly emphasized here, that by using the silicon–oxygen anions as the basis of the classification, these anions are often regarded as the more rigid part of the structure and the cationic part is unintentionally considered to be less important. It should, however, be kept in mind that this conclusion is definitely unwarranted and should be avoided, especially for compounds with molar ratios $M:Si \geq 2$, $M = Me^{+}$, $Me^{+2}_{0.5}$, $Me^{+3}_{0.33}$, $Me^{+4}_{0.25}$. Therefore, whenever possible the classification based on silicon–oxygen anions should be supplemented by a subdivision which takes the aggregates of cation–oxygen polyhedra into account.

Ideally a classification should be able to place a phase unambiguously into just one category. Unfortunately, some two- and three-dimensional silicate anions can, at first glance, be filed into two or more categories. For example, the single layer found in the structure of prehnite, $Ca_2Al[AlSi_3O_{10}](OH)_2$ (Papike and Zoltai 1967) could be regarded either as an unbranched vierer single layer or as an open-branched zweier single layer (Fig. 6.4). However, according to the procedure described in the preceding paragraphs, the correct fundamental chain of lowest periodicity in prehnite is the branched zweier chain, filing prehnite unambiguously into the group of open-branched zweier single layer silicates: $Ca_2Al\{\boldsymbol{oB}, 1^2_\infty\}[^2(AlSi_3)O_{10}](OH)_2$.

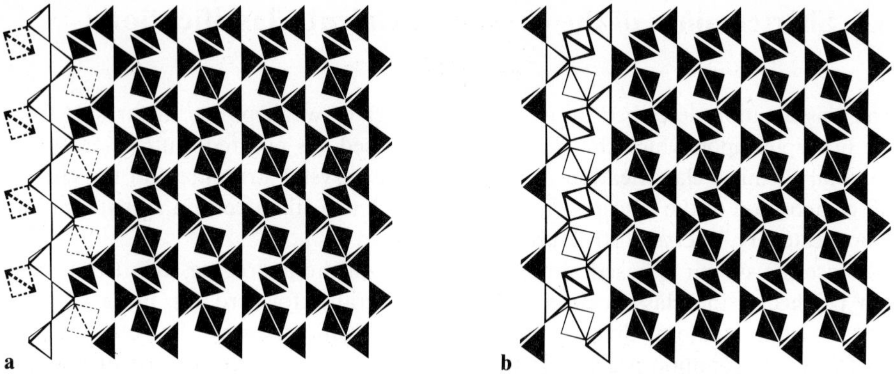

Fig. 6.4. The tetrahedral layer of prehnite, $Ca_2Al[AlSi_3O_{10}](OH)_2$, can be considered either as **a** an open-branched zweier single layer or **b** an unbranched vierer single layer. The assignment **a** with the fundamental chain of lowest periodicity is the correct one.

Latiumite, $(K_{1.7}\square_{0.3})Ca_6[(Al_{5.7}Si_{4.3})O_{22}][SO_4]_{1.4}[CO_3]_{0.6}$ (Cannillo et al. 1973a), and tuscanite, $(K,Sr,H_2O)_2(Ca,Na,Mg,Fe)_6[(Al_{3.66}Si_{6.34})O_{22}][SO_4]_{1.4}[CO_3]_{0.5}[O_4H_4]_{0.1}$ (Mellini et al. 1977), are two striking examples of the ambiguity of the silicate classification. Figure 6.5a and b give the tetrahedral part of both structures together with the silicon content in percent for each tetrahedron.

If no distinction is made between Si and Al both of the minerals should be classified as double layer silicates since each layer is linked to a symmetrically equivalent one by sharing the tetrahedral corners marked with a small circle in Fig. 6.5a and b. Within each layer there are unbranched fünfer chains with identity periods of $c_0 = 10.8$ Å (Fig. 6.5c) so that both minerals could be regarded as silicates with unbranched fünfer double layers. In each structure there is, however, a shorter period of $b_0 = 5.1$ Å parallel to the [010] direction which is characterized by unbranched zweier chains (solid tetrahedra) and open-branched zweier chains (shaded tetrahedra with white branches) (Fig. 6.5d). According to the rules for correct selection of a fundamental chain, the chain should be the one of lowest periodicity running parallel to the shortest identity period within the silicate anion. Under these circumstances both silicates would better be described as hybrid zweier double layer silicates.

There is, however, the rule which requires that only those tetrahedrally coordinated cations which are, at least under certain temperature/pressure conditions, statistically distributed over the same crystallographic sites should be regarded as part of the silicate anion. In latiumite and tuscanite two of the five T positions are occupied only by Al, although nothing is known about the Al/Si distribution at higher temperatures. If the statistical-distribution rule is applied, then both silicates contain single tetrahedra $[SiO_4]$ and double tetrahedra $[Si_2O_7]$ and should be classified as mixed-anion silicates (Fig. 6.5e).

The preferred classification of these two silicates depends on the purpose under consideration. For morphology, cleavage, and other physical properties their classification as hybrid zweier double layer silicates will be more appropriate since [010] is the direction of the physically strongest bonds. On the other hand, resis-

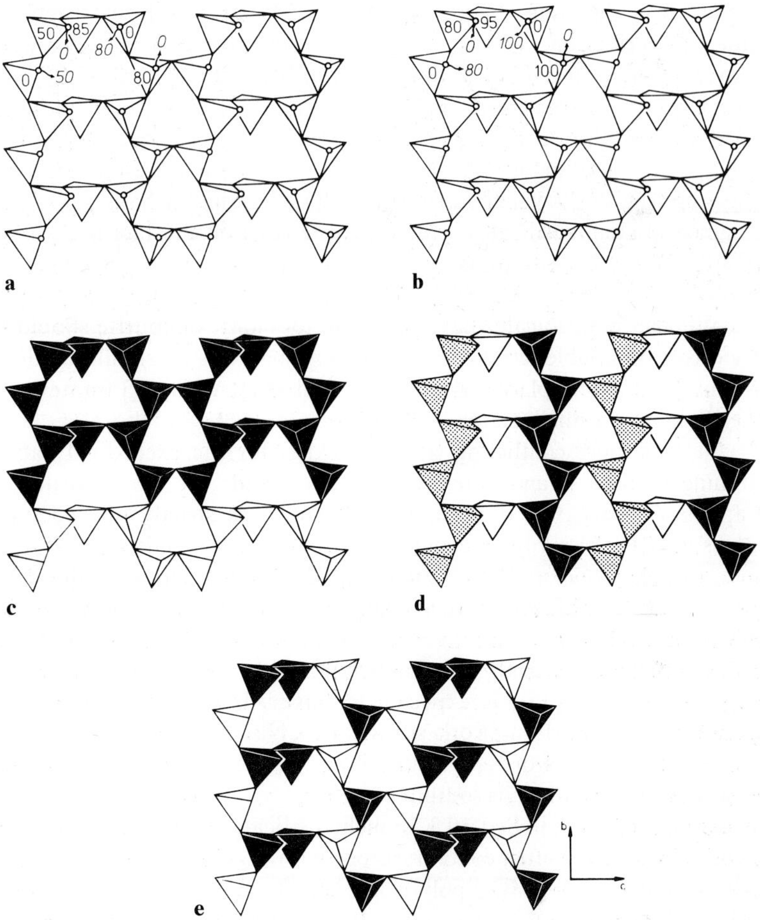

Fig. 6.5 a – e. Different possibilities for the classification of latiumite and tuscanite.

a Tetrahedral part of the latiumite structure with percentage of silicon occupation of the tetrahedral sites. Corners marked with a *small circle* are shared with tetrahedra of a second layer. The *arrows* point to the silicon content of the attached tetrahedron of the second layer; **b** the same for tuscanite. **c** Unbranched fünfer double layer; **d** hybrid zweier double layer. Unbranched zweier chains are indicated by *solid tetrahedra,* and the open-branched zweier chains by *shaded* and *white tetrahedra;* **e** mixed-anions: $[AlO_4]$ *white* and $[SiO_4]$ *solid tetrahedra*

tivity against chemical attack by acids or other solutions, and chemical reactivity in general, may distinguish between Si–O–Si, Si–O–Al, and Al–O–Al bonds and, therefore, their classification as mixed-anion silicates may be more appropriate for purely chemical purposes.

There are other cases where the definition of the T atoms as part of the silicate anion leads to an unsatisfactory separation of related structures. For example, in danburite, $CaB_2Si_2O_8$, and in paracelsian, the room temperature polymorph of $BaAl_2Si_2O_8$, the $[BO_4]$, $[AlO_4]$, and $[SiO_4]$ tetrahedra share corners to give frameworks having the same topology (Phillips et al. 1974). In both structures there is an ordered distribution of B/Si and Al/Si, respectively, over the tetrahedral sites. However, while in danburite double tetrahedra $[B_2O_7]$ and $[Si_2O_7]$ are linked together to form a framework, in paracelsian $[Al_2O_7]$ and $[Si_2O_7]$ double tetrahedra do not occur, the Si and Al atoms instead ordering into alternating $[SiO_4]$ and $[AlO_4]$ tetrahedra in accordance with Loewenstein's aluminum avoidance rule (Loewenstein 1954) which states that $[AlO_4]$ tetrahedra have no oxygen atoms in common in aluminosilicates. This principle rests on the fact that the bond energy of two Al–O–Si groups is lower than the sum of the bond energies of one Al–O–Al and one Si–O–Si group.

Although both silicates have the same topology, danburite should be placed in the class with double tetrahedra $[Si_2O_7]$, and paracelsian should be described as an aluminosilicate and placed in the class with $\{^3_\infty\}[(Al, Si)O_2]$ frameworks, provided there is at least a small degree of Al/Si disorder.

An even more conflicting situation exists for the exceedingly abundant rock-forming feldspars anorthite, $CaAl_2Si_2O_8$, and the low-sodium plagioclases $Ca_{1-x}Na_xAl_{2-x}Si_{2+x}O_8$. If Loewenstein's rule was strictly obeyed, the Al and Si atoms would necessarily be completely ordered, independent of temperature, pressure, or kinetic history of the material. The rules relating to the definition of T atoms would then classify pure anorthite as a monosilicate $CaAl_2[SiO_4]_2$ with single $[SiO_4]$ tetrahedra.

On the other hand, a plagioclase with very low albite content, say $Ca_{0.98}Na_{0.02}Al_{1.98}Si_{2.02}O_8$, is expected to have Al/Si disorder and should then be considered to be a framework silicate $(Ca, Na)\{^3_\infty\}[(Al, Si)_4O_8]$. Consequently, if Loewenstein's rule is obeyed[1], pure anorthite would be separated from a plagioclase no matter how low its sodium content.

As will be shown in Sec. 10.3, branched silicates with one- or two-dimensionally infinite anions very often contain cations with high electronegativities. In such silicates the rather rigid $[MO_n]$ polyhedra, usually with $3 \leq n \leq 6$, very often complement the silicate anion in such a way that a structure is built in which these $[MO_n]$ polyhedra, together with the $[SiO_4]$ tetrahedra, form an array of polyhedra that has a higher degree of condensation (and often a higher dimensionality) than the array of $[SiO_4]$ tetrahedra alone. Since the covalency of the M–O bond for highly electronegative M ions is similar to that of the Si–O bond, the $[MO_n]$ polyhedra may be regarded as part of a larger anion $[Si_rM_sO_t]$. Figure 6.6 demonstrates how the open-branched zweier single chain of astrophyllite (Woodrow 1967) is complemented by $[TiO_6]$ octahedra to form a single layer, how the loop-branched zehner single chains of nordite (Bakakin et al. 1970) are complemented by $[(Zn, Mg, Fe, Mn)O_4]$ tetrahedra to a single layer, and how the open-branched vierer single layer and $[SiO_4]$ single tetrahedra of the mixed-anion silicate meliphanite (Dal Negro et al. 1967) are complemented by $[BeO_3F]$ tetrahedra to form a hybrid single layer.

The difficulties that arose in the classification of the branched silicates, the closely related structures of danburite and paracelsian, and anorthite and plagio-

1 In fact, partial Al/Si disorder in pure anorthite at high temperatures has been concluded from X-ray studies of anorthite quenched from 1800 K (Bruno et al. 1976).

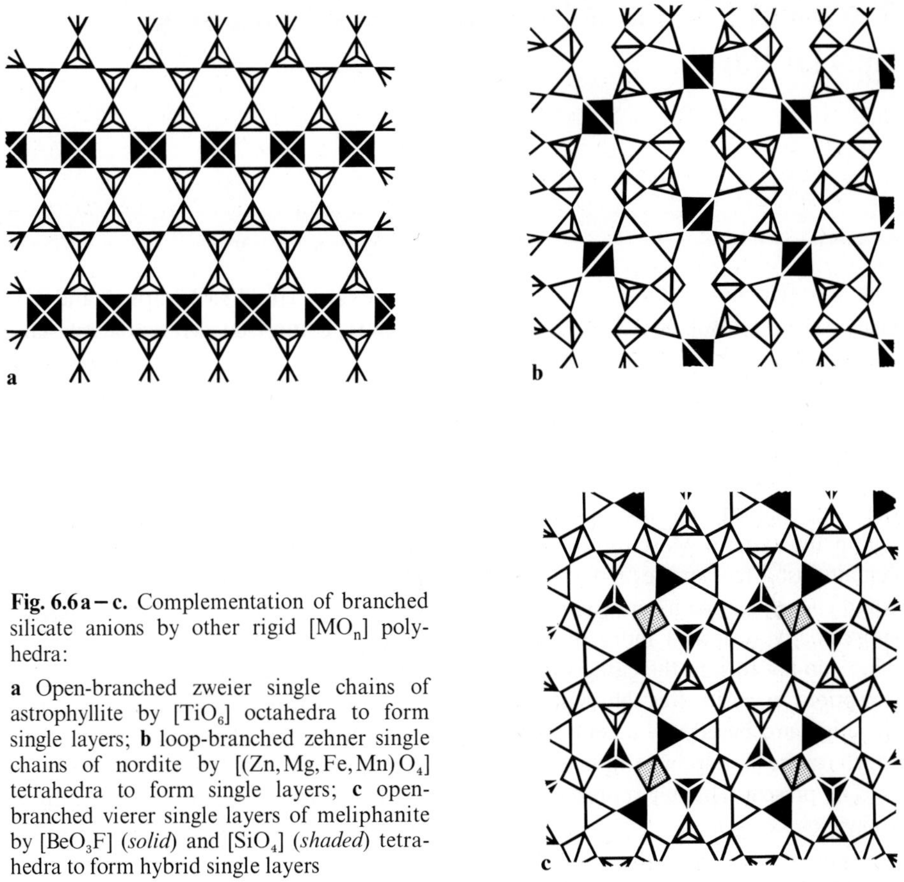

Fig. 6.6 a – c. Complementation of branched silicate anions by other rigid $[MO_n]$ polyhedra:

a Open-branched zweier single chains of astrophyllite by $[TiO_6]$ octahedra to form single layers; **b** loop-branched zehner single chains of nordite by $[(Zn,Mg,Fe,Mn)O_4]$ tetrahedra to form single layers; **c** open-branched vierer single layers of meliphanite by $[BeO_3F]$ (*solid*) and $[SiO_4]$ (*shaded*) tetrahedra to form hybrid single layers

clase resulted from the fact that it had been decided to consider as part of the silicate anion only those cations that at least under certain conditions statistically replace some of the silicon atoms. If wider limits had been set for atoms belonging to the silicate ion as, for example, were chosen by Zoltai (1960), who did not in his classification distinguish at all between different tetrahedrally coordinated cations like Li^+, Be^{+2}, Al^{+3}, Si^{+4}, P^{+5}, and even S^{+6}, then the intricacies would only have been shifted somewhere else. Such difficulties are not unique to this classification. They remind us that every system works well only within certain limits and that there may be particular areas where the system has to be adjusted by taking into account other facts and rules, or even areas in which other classification systems are more suitable. Whenever such difficulties occur they should serve as a reminder not to use rules thoughtlessly, but to remember the crystal chemistry inherent in the facts. In addition, one would perhaps do well to heed S. Butler's statement, as quoted by Keynes and Hill (1951): "The hardness of men's hearts makes an idol of classification, but they are nothing apart from our sense of our own convenience."

7 Crystal Chemical Classification of Silicates: Special Part

The classification of silicate anions and of silicates described in Chaps. 4 and 6 is clear and straightforward, but it is at the same time, somewhat sophisticated. This is also true for the chemical formulae derived in Chap. 5 since they are based on the same classification system. The desired clearness and the undesirable complexity of the system are both a result of the attempt to include all conceivable silicate anions.

A survey of the silicate anions currently known to exist, however, reveals that the large majority of silicate anions belong to a rather small number of groups within the system. Most families and even whole orders, subclasses, and classes contain just a few species or none at all. Consequently, for practical purposes, the comprehensive system represented by Tables 4.3 to 4.6 and 6.1 can be condensed substantially so that a more manageable system results. The few silicate species which do not fit into the categories of the condensed system can then be treated as exceptions or curiosities. Notwithstanding, since the parameters used for the classification are based on general crystal chemical principles it is these exceptions which can be of help in gaining a deeper understanding of the silicates in general.

The present Chapter represents a survey of the various types of silicate anions known to exist.

7.1 Silicon Compounds with $[SiO_6]$ Octahedra (Hexaoxosilicates)

At present the number of compounds known to contain $[SiO_6]$ groups is quite small. Detailed crystal structure analyses have been made for the phases listed in Table 3.4. There is about an equal number of phases which have been reported to be isostructural with one or another of the phases included in that Table, and still other phases with a statistical distribution of octahedrally coordinated silicon and aluminum, such as the high pressure phases $K[(AlSi_3)^{[6]}O_8]$ with the hollandite-type structure (Ringwood et al. 1967), and $Na[(AlSi)^{[6]}O_4]$ with the $CaFe_2O_4$-type structure (Yamada et al. 1983). In addition, there are some 20 or 30 silicon-organic compounds for which $[SiO_6]$ coordination in aqueous solution has been established by spectroscopic methods.

In Table 7.1 the crystalline silicon compounds have been arranged according to their linkedness ***L*** and dimensionality ***D***. From this compilation it is clear that isolated $[SiO_6]$ groups exist in compounds which can be synthesized at ambient pressure.

Table 7.1 Classification of silicon compounds with $[SiO_6]$ octahedra according to the linkedness ***L*** of their $[SiO_n]$ polyhedra and the dimensionality ***D*** of their anions

D \ ***L***	0	1	1 and 2	2	3	0 and 1
0	$Si_3^{[6]}Si_2^{[4]}[PO_4]_6O$ [a] $Si^{[6]}[P_2O_7]$ AI, AIII, AIV $[NH_4]_2Si^{[6]}[P_4O_{13}]$ $K_2Si^{[6]}[Si_3^{[4]}O_9]$ [a] Thaumasite $[C_5H_5NH]_2[(C_6H_4O_2)_3Si^{[6]}]$ $[Cu\,en_2][(C_6H_4O_2)_3Si^{[6]}]$ [b]					$Si_3^{[6]}Si_2^{[4]}[PO_4]_6O$ [a]
1						
2				$MgSi^{[6]}O_3$ Ilmenite-type $ZnSi^{[6]}O_3$ Ilmenite-type		
3		$MgSi^{[6]}O_3$ Perovskite-type $CaSi^{[6]}O_3$ Perovskite-type $Sc_2Si_2^{[6]}O_7$ Pyrochlore-type $In_2Si_2^{[6]}O_7$ Pyrochlore-type $K_2Si^{[6]}Si_3^{[4]}O_9$ [a]	$Si^{[6]}O_2$ Stishovite $K[(AlSi_3)^{[6]}O_8]$ Hollandite-type $Na[(AlSi)^{[6]}O_4]$ $CaFe_2O_4$-type			

[a] This phase appears at two different places in the Table, depending on whether the $[SiO_6]$ and $[SiO_4]$ polyhedra are treated separately or not.
[b] *en* = ethylenediamine $H_2N \cdot CH_2 \cdot CH_2 \cdot NH_2$

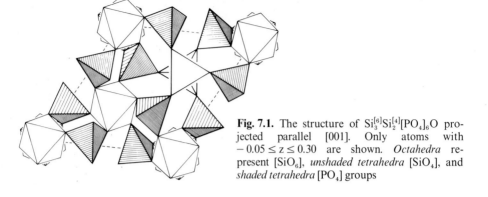

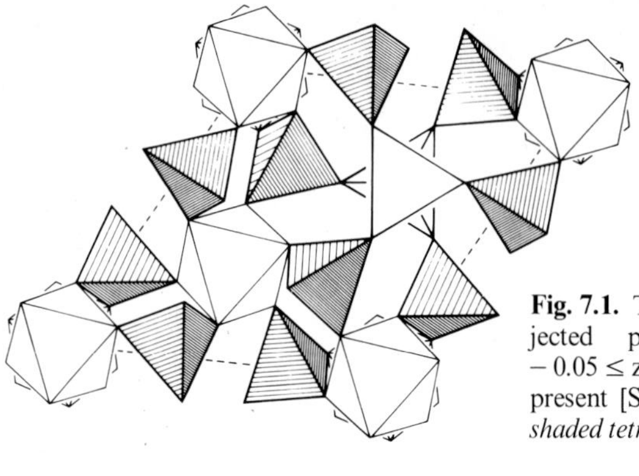

Fig. 7.1. The structure of $Si_3^{[6]}Si_2^{[4]}[PO_4]_6O$ projected parallel [001]. Only atoms with $-0.05 \leq z \leq 0.30$ are shown. *Octahedra* represent $[SiO_6]$, *unshaded tetrahedra* $[SiO_4]$, and *shaded tetrahedra* $[PO_4]$ groups

Two of the phases in Table 7.1 can be filed into two different divisions depending on whether the $[SiO_6]$ octahedra and $[SiO_4]$ tetrahedra are treated separately or not. Since the electronegativity of P is greater than that of Si, the compound described as $Si_5O[PO_4]_6$ (Mayer 1974) can be considered to be a silicon oxo-monophosphate with silicon cations partly in octahedral and partly in tetrahedral coordination: $Si_3^{[6]}Si_2^{[4]}[PO_4]_6O$. The structure is illustrated in Fig. 7.1. If the $[SiO_6]$ groups alone are considered, then the phase falls into the group with isolated octahedra, i.e., $\boldsymbol{L} = 0$ and $\boldsymbol{D} = 0$. If, however, both types of $[SiO_n]$ polyhedra are considered, then the phase is a mixed-anion silicate containing single octahedra $[SiO_6]$ and double tetrahedra $[Si_2O_7]$ and may be placed in the field with $\boldsymbol{L} = 0$ and 1, $\boldsymbol{D} = 0$. In either case, the structural formula would be written as $[Si^{[6]}O_{6/2}]_3\{2\boldsymbol{t}\}[Si_2^{[4]}OO_{6/2}][PO_{4/2}]_6$, where $O_{1/2}$ represents an oxygen atom bridging between two polyhedra.

In the $K_2Si_4O_9$ polymorph, which is stable under atmospheric conditions, all silicon atoms are four-coordinated and the silicate anion is an unbranched zweier single layer (Fig. 7.22 b) with a structural formula $K_2\{\boldsymbol{uB}, 1_\infty^2\}[^2Si_4^{[4]}O_9]$ (Schweinsberg and Liebau 1974). At moderate pressures (above ca. 2 GPa) this phase transforms to one in which one quarter of the silicon atoms are six-coordinated, with the rest in four-coordination forming dreier single rings, $K_2Si^{[6]}\{\boldsymbol{uB}, 1\,\boldsymbol{r}\}[Si_3^{[4]}O_9]$ (Swanson and Prewitt 1983). If the $[SiO_6]$ groups are considered separately, then they are single octahedra ($\boldsymbol{L} = 0$, $\boldsymbol{D} = 0$) and the structural formula could be written as $K_2[Si^{[6]}O_{6/2}]\{\boldsymbol{uB}, 1\,\boldsymbol{r}\}[Si_3^{[4]}O_3O_{6/2}]$. If, however, both kinds of $[SiO_n]$ polyhedra are included together, this phase would be described as a framework silicate, $K_2\{_\infty^3\}[Si^{[6]}Si_3^{[4]}O_9]$, ($\boldsymbol{D} = 3$), with mixed coordination number $\boldsymbol{CN} = 4$, 6 and with corner-sharing between octahedra and tetrahedra ($\boldsymbol{L} = 1$).

The high pressure phases $CaSi^{[6]}O_3$ and $MgSi^{[6]}O_3$ with the perovskite structure (Liu 1979; Yagi et al. 1978) and $Sc_2Si_2^{[6]}O_7$ and $In_2Si_2^{[6]}O_7$ with the pyrochlore structure (Reid et al. 1977) are three-dimensional frameworks of corner-shared $[SiO_6]$ octahedra. In stishovite, the rutile-type polymorph of silica (Sinclair and Ringwood 1978, Hill et al. 1983), in the high pressure phase of potassium feldspar $KAlSi_3O_8$ with the hollandite structure (Ringwood et al. 1967), and in $NaAlSiO_4$ with the $CaFe_2O_4$ structure type (Yamada et al. 1983), the octahedra share corners as well as edges to form three-dimensional frameworks. Their structural formulae are, therefore, $\{^3_\infty\}[Si^{[6]}O_2]$, $K\{^3_\infty\}[(AlSi_3)^{[6]}O_8]$, and $Na\{^3_\infty\}[(AlSi)^{[6]}O_4]$, respectively.

In the ilmenite-type high pressure polymorphs of $MgSiO_3$ (Horiuchi et al. 1982) and $ZnSiO_3$ (Ito and Matsui 1979), the $[SiO_6]$ octahedra share edges to form two-dimensional layers with the formula $\{1^2_\infty\}[Si^{[6]}O_3]$.

Further subdivision of the small number of phases with $[SiO_6]$ octahedra according to branchedness, multiplicity, and periodicity of their anions is at present unnecessary.

7.2 Silicates Containing $[SiO_4]$ Tetrahedra (Tetraoxosilicates)

7.2.1 Silicates with Edge-Sharing Tetrahedra

The overwhelming majority of oxygen compounds of silicon contain Si in tetrahedral coordination, in other words they are silicates in the normal sense. In all but one of the crystalline silicates, the $[SiO_4]$ tetrahedra are either isolated or share corners with other tetrahedra. The only phase[1] known to have edge-sharing $[SiO_4]$ tetrahedra is the fibrous polymorph of silica described by Weiss and Weiss (1954) in which the tetrahedra are linked in chains with a periodicity $\boldsymbol{P} = 2$ (Fig. 7.2). The complete classification is: $\boldsymbol{N_{an}} = 1$, $\boldsymbol{CN} = 4$, $\boldsymbol{L} = 2$, $\boldsymbol{B} = \boldsymbol{uB}$, $\boldsymbol{M} = 1$, $\boldsymbol{D} = 1$, $\boldsymbol{P} = 2$, with a structural formula $\{\boldsymbol{uB}, 1^1_\infty\}[^2Si_2^{[4]}O_4]$.

7.2.2 Silicates with Corner-Sharing Tetrahedra

7.2.2.1 Oligosilicates (Sorosilicates)

7.2.2.1.1 Unbranched Oligosilicates. Silicate anions which are the product of linear condensation of a finite number m of $[SiO_4]$ tetrahedra can be described by the general formula

$$\{\boldsymbol{uB}, \boldsymbol{Mt}\}[Si_mO_{3m+1}]^{-(2m+2)}, \quad m = \boldsymbol{M}, \text{ finite} \qquad (1)$$

where $\boldsymbol{M}$ is the multiplicity of the anion[2]. Double tetrahedra $[Si_2O_7]^{-6}$ have been known for a very long time and a large number of natural and synthetic disilicates have been described (see Table 3.1). However, the first crystal structure containing triple tetrahedra $[Si_3O_{10}]^{-8}$ was published as recently as 1968 (Donnay and All-

1 It has been claimed that edge-sharing silicate tetrahedra exist in the crystal structure of leucophoenicite, $Mn_7[SiO_4]_2[SiO_4](OH)_2$ (Moore 1970a). However, from the text of the paper it is clear that the two edge-sharing tetrahedra are actually statistically occupied by only one silicon atom, and that at no point in time and space do the $[SiO_4]$ tetrahedra share edges in this structure.

2 In this and subsequent general formulae of silicate anions, the formal charges, if given, are valid only for those anions in which Si stands for pure silicon or for silicon partially replaced by other tetravalent T atoms, such as Ge and Ti. If non-tetravalent T atoms replace some of the silicon atoms, adjustments to the charges in the formulae have to be made.

mann 1968). Since then about one dozen trisilicates have been discovered and these are listed in Table 7.2. More recently, unbranched fourfold tetrahedra have been observed in the synthetic compounds $Ag_{10}[Si_4O_{13}]$ (Jansen and Keller 1979), $Na_4Sc_2[Si_4O_{13}]$ (Maksimov et al. 1980a), and $Pb_8[Si_4O_{13}][SO_4]O_2$ (Fröhlich 1984), and fivefold tetrahedra in synthetic $Na_4Sn_2[Si_5O_{16}] \cdot H_2O$ (Safronov et al. 1983). In medaite, $HMn_6V[Si_5O_{16}]O_3$, a $[VO_4]$ tetrahedron is linked end-on to the fivefold silicate tetrahedron leading to an unbranched sixfold tetrahedron of composition $[VSi_5O_{19}]^{-12}$ (Gramaccioli et al. 1981).

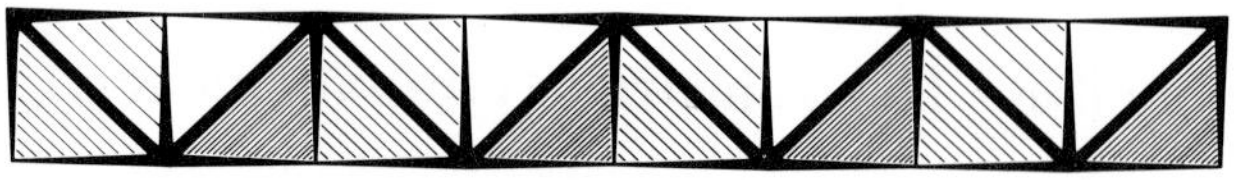

Fig. 7.2. Zweier single chain of edge-sharing $[SiO_4]$ tetrahedra in fibrous silica of Weiss and Weiss (1954)

Oligosilicates containing unbranched multiple tetrahedra with $M = 8$, 9, and 10 have been synthesized from melts in the system $MgO-Sc_2O_3-SiO_2$ (Ozawa et al. 1979; Takéuchi et al. 1977, 1984a, b). Their formulae are

$(Mg_{15.61}, Sc_{1.37})(\square_{0.68}, Mg_{0.30}, Si_{0.02})\{\boldsymbol{uB}, 8\boldsymbol{t}\}[Si_8O_{25}]_2$,
$(Mg_{17.40}, Sc_{1.49})(\square_{0.74}, Mg_{0.15}, Si_{0.11})\{\boldsymbol{uB}, 9\boldsymbol{t}\}[Si_9O_{28}]_2$, and
$(Mg_{19.60}, Sc_{1.28})(\square_{0.74}, Mg_{0.04}, Si_{0.22})\{\boldsymbol{uB}, 10\boldsymbol{t}\}[Si_{10}O_{31}]_2$.

Adjacent multiple tetrahedra within these structures are bridged by additional tetrahedra which are statistically and only partly occupied by silicon and magnesium. If these additional tetrahedra are considered to be part of the silicate anion, in spite of the fact that they contain only very little silicon, the three phases would be somewhat artificially classified as loop-branched single chain silicates with periodicities of 18, 20, and 22. Figure 7.3 shows several unbranched multiple tetrahedra which have been found in crystalline oligosilicates.

7.2.2.1.2 ***Branched Oligosilicates.*** The open-branched triple tetrahedra found in the synthetic tetrasilicate $NaBa_3Nd_3[Si_2O_7]\{\boldsymbol{oB}, 3\boldsymbol{t}\}[Si_4O_{13}]$ (Malinovskii et al. 1983) and in the natural pentasilicate, zunyite, $Al_{12}^{[6]}Al^{[4]}\{\boldsymbol{oB}, 3\boldsymbol{t}\}[Si_5O_{16}](OH, F)_{18}O_4Cl$ (Baur and Ohta 1982), are the only branched oligosilicate anions known so far. As can be deduced from Fig. 7.4 more informative formulae of these silicate anions can be given as $[Si_3O_{10}(SiO_3)]$ and $[Si_3O_{10}(SiO_3)_2]$, respectively, in which the branches are written in round brackets.

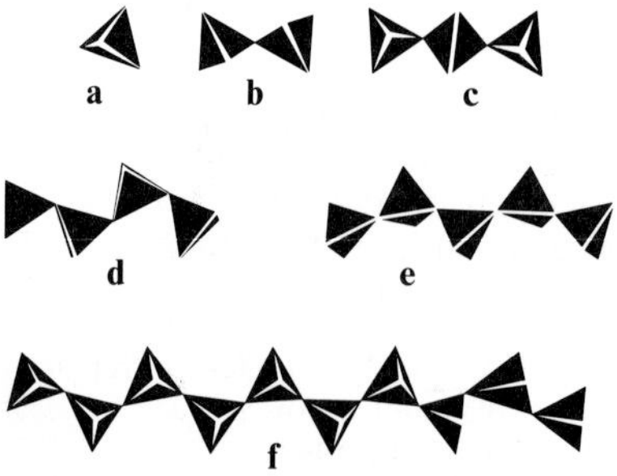

Fig. 7.3 a–f. Unbranched multiple tetrahedra found in crystalline silicates.

a Single tetrahedron in the monosilicates, e.g., olivine, $(Mg, Fe)_2[SiO_4]$; **b** double tetrahedron in barysilite, $MnPb_8[Si_2O_7]_3$; **c** triple tetrahedron in aminoffite, $Be_2Ca_3[Si_3O_{10}](OH)_2$; **d** fourfold tetrahedron in synthetic $Ag_{10}[Si_4O_{13}]$; **e** fivefold tetrahedron in medaite, $HMn_6V[Si_5O_{16}]O_3$; **f** tenfold tetrahedron in synthetic $(Mg_{19.60}Sc_{1.28})(\square_{0.74}Mg_{0.04}Si_{0.22})[Si_{10}O_{31}]_2$

Table 7.2 Oligosilicates with unbranched multiple tetrahedra and multiplicities $\boldsymbol{M} \geqq 3$ (mixed-anion oligosilicates not included)

M	Silicate		Reference
	Name	Formula	
3	Synthetic	$Na_4Mg_2[Si_3O_{10}]$	Foris et al. 1979
	Synthetic	$Na_2Ca_3[Si_3O_{10}]$	Kuznetsova et al. 1980
	Synthetic	$Na_4Cd_2[Si_3O_{10}]$	Simonov et al. 1978a
	Synthetic	$Na_2Cd_3[Si_3O_{10}]$	Simonov et al. 1977
	Synthetic	$H_2K_3Y[Si_3O_{10}]$	Pushcharovskii et al. 1981
	Rosenhahnite	$H_2Ca_3[Si_3O_{10}]$	Wan et al. 1977
	Thalenite	$Y_3[Si_3O_{10}](OH)$	Kornev et al. 1972
	Aminoffite	$Be_2Ca_3[Si_3O_{10}](OH)_2$	Coda et al. 1967a
	Tiragalloite	$HMn_4As[Si_3O_{10}]O_3$	Gramaccioli et al. 1979, 1980
	Kinoite	$Ca_2Cu_2[Si_3O_{10}] \cdot 2H_2O$	Laughon 1971
4	Synthetic	$Ag_{10}[Si_4O_{13}]$	Jansen and Keller 1979
	Synthetic	$Na_4Sc_2[Si_4O_{13}]$	Maksimov et al. 1980a
	Synthetic	$Pb_8[Si_4O_{13}][SO_4]O_2$	Fröhlich 1984
5	Synthetic	$Na_4Sn_2[Si_5O_{16}] \cdot H_2O$	Safronov et al. 1983
	Medaite	$HMn_6V[Si_5O_{16}]O_3$	Gramaccioli et al. 1981
8	Synthetic	$(Mg_{15.61}Sc_{1.37})(Mg_{0.30}Si_{0.02})[Si_8O_{25}]_2$	Ozawa et al. 1979 Takéuchi et al. 1977, 1984b
9	Synthetic	$(Mg_{17.40}Sc_{1.49})(Mg_{0.15}Si_{0.11})[Si_9O_{28}]_2$	Ozawa et al. 1979 Takéuchi et al. 1977, 1984b
10	Synthetic	$(Mg_{19.60}Sc_{1.28})(Mg_{0.04}Si_{0.22})[Si_{10}O_{31}]_2$	Ozawa et al. 1979 Takéuchi et al. 1977, 1984b

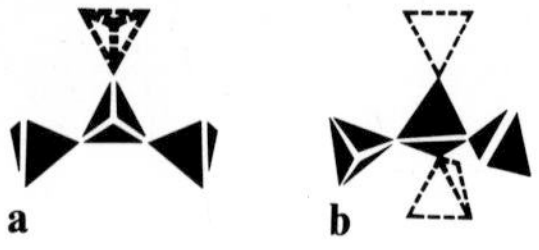

Fig. 7.4 a, b. Branched multiple tetrahedra found in crystalline silicates.

a Open-branched triple tetrahedron in synthetic $NaBa_3Nd_3[Si_2O_7]\{\boldsymbol{oB}, 3\boldsymbol{t}\}[Si_4O_{13}]$; **b** open-branched triple tetrahedron in zunyite, $Al_{13}[Si_5O_{16}](OH,F)_{18}O_4Cl$

7.2.2.2 Ring Silicates (Cyclosilicates)

7.2.2.2.1 ***Unbranched Ring Silicates.*** Unbranched single ring silicate anions formed by annular condensation of a finite number p of $[SiO_4]$ tetrahedra can be described by the general formula

$$\{\boldsymbol{uB}, 1\boldsymbol{r}\}[^{P^r}Si_pO_{3p}]^{-2p}, \quad p = \boldsymbol{P}^r \geq 3, \text{ finite.} \tag{2}$$

The number p of silicon atoms in an unbranched single ring is equal to its ring periodicity, $\boldsymbol{P}^r$. This is in contrast to branched single rings for which $\boldsymbol{P}^r$ is smaller than the number of tetrahedra in the ring.

To date, single rings with $\boldsymbol{P}^r = 3$, 4, 6, 8, 9, and 12 tetrahedra have been found in crystalline silicates (Fig. 7.5, Table 7.3). They are usually called three-membered single rings, four-membered single rings, etc. or, by analogy to the chain anions, dreier single rings, vierer single rings, etc.

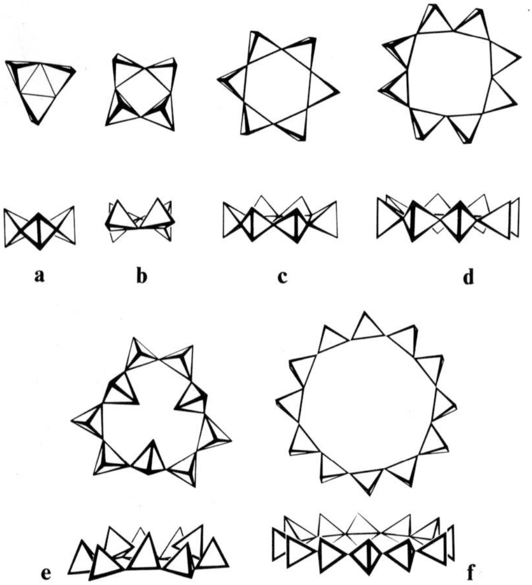

Fig. 7.5 a–f. Unbranched single rings found in crystalline silicates (for references see Table 7.3).

a Dreier single ring in benitoite, $BaTi[Si_3O_9]$; **b** vierer single ring in taramellite, $Ba_4(Fe,Ti)_4B_2[Si_4O_{12}]_2O_5Cl_x$; **c** sechser single ring in beryl, $Be_3Al_2[Si_6O_{18}]$; **d** achter single ring in muirite, $Ba_{10}(Ca,Mn,Ti)_4[Si_8O_{24}](Cl,O,OH)_{12} \cdot 4H_2O$; **e** neuner single ring in eudialyte, $Na_{12}(Ca,RE)_6(Fe,Mn,Mg)_3Zr_3(Zr,Nb)_x\{\boldsymbol{uB},\boldsymbol{r}\}[Si_3O_9]_2\{\boldsymbol{uB},\boldsymbol{r}\}[Si_9(O,OH)_{27}]_2Cl_y$; **f** zwölfer single ring in traskite, $(Ca,Sr)Ba_{24}(Mg,Mn,Fe,Al,Ti)_{16}[Si_2O_7]_6[Si_{12}O_{36}](O,OH)_{30}Cl_6 \cdot 14H_2O$

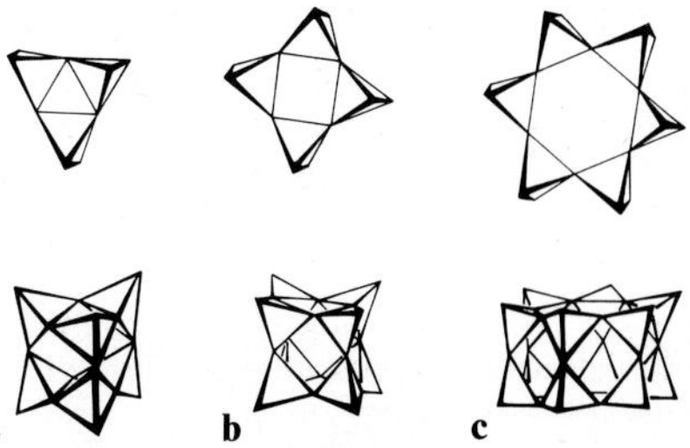

Fig. 7.6. Unbranched double rings found in crystalline silicates (for references see Table 7.4).
a Dreier double ring in synthetic $[Ni(H_2N \cdot CH_2 \cdot CH_2 \cdot NH_2)_3]_3\{\boldsymbol{uB}, 2\boldsymbol{r}\}[Si_6O_{15}] \cdot 26\,H_2O$;
b vierer double ring in steacyite, $K_{1-x}(Na, Ca)_{2-y}Th_{1-z}\{\boldsymbol{uB}, 2\boldsymbol{r}\}[Si_8O_{20}]$;
c sechser double ring in milarite, $KCa_2(Be_2Al)\{\boldsymbol{uB}, 2\boldsymbol{r}\}[Si_{12}O_{30}] \cdot 0.75\,H_2O$

Condensation of two single rings via corners of the tetrahedra results in the formation of double rings. Depending on the number of $[SiO_4]$ tetrahedra in the single rings that form the double ring, the latter may be called either a three-membered double ring, a four-membered double ring, etc. or, more concisely, a dreier double ring, a vierer double ring, and sechser double ring etc. (Fig. 7.6). Fünfer double rings have been found to exist in solution (Hoebbel et al. 1975) and their trimethylsilylester $[(CH_3)_3Si]_{10}[^5Si_{10}O_{25}]$ has been studied crystallographically (Smolin in Hoebbel et al. 1975).

Only a few crystalline silicates containing double rings have been described (Table 7.4). In every case the two single rings that constitute the double ring are linked via all their tetrahedra to give the formula

$$\{\boldsymbol{uB}, 2\boldsymbol{r}\}[^{P^r}Si_{2p}O_{5p}]^{-2p}, \quad p = P^r \geq 3, \text{ finite.} \qquad (3a)$$

It is likely that silicates with double rings will be discovered in which the two single rings use only some of their tetrahedra for linking. Therefore, a more general formula for such double rings is

$$\{\boldsymbol{uB}, 2\boldsymbol{r}\}[^{P^r}Si_{2p}O_{6p-l}]^{-(4p-2l)}, \quad p = P^r \geq 3, \text{ finite} \quad \text{and} \quad 1 \leq l \leq p \qquad (3b)$$

where l is the number of linkages between the two rings.

7.2.2.2.2 ***Branched Ring Silicates.*** If in eakerite, $Ca_2Al_2^{[4]}SnSi_6O_{18}(OH)_2 \cdot 2\,H_2O$ (Kossiakoff and Leavens 1976), the silicon and aluminum atoms retain their ordered distribution under all conditions up to its decomposition temperature, then its anions can be regarded as open-branched vierer single rings $\{\boldsymbol{oB}, 1\,\boldsymbol{r}\}[^4Si_4O_{12}(SiO_3)_2]$ (Fig. 7.7a). However, if at high temperatures there is a statistical Al/Si distribution – an unlikely situation since dehydration begins at 675 K – then eakerite should be classified as a phyllosilicate with loop-branched vierer single layers. Similarly,

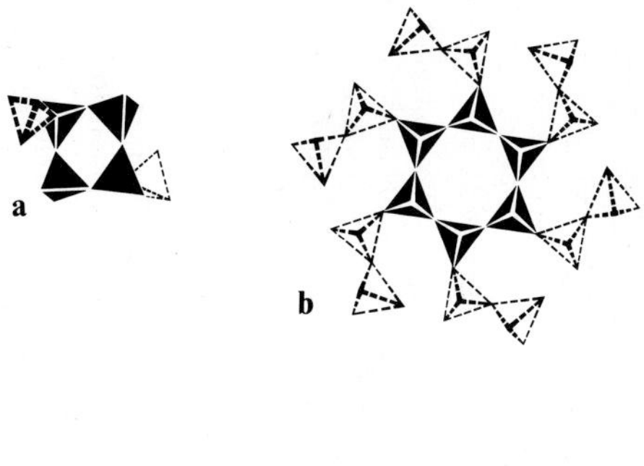

Fig. 7.7a, b. Branched single rings found in crystalline silicates.
a Open-branched vierer single ring in eakerite, $Ca_2Al_2Sn\{\boldsymbol{oB}, 1\,\boldsymbol{r}\}[^4Si_6O_{18}](OH)_2 \cdot 2\,H_2O$;
b open-branched sechser single ring in tienshanite, $Na_9KCa_2Ba_6(Mn, Fe)_6(Ti, Nb, Ta)_6B_{12}\{\boldsymbol{oB}, 1\,\boldsymbol{r}\}[^6Si_{18}O_{54}]_2O_{15}(OH)_2$

Table 7.3 Monocyclosilicates with unbranched single rings $\{\boldsymbol{uB}, 1\,\boldsymbol{r}\}\;[{}^{P^r}Si_pO_{3p}]^{-2p}$ [$p = \boldsymbol{P^r}$: number of tetrahedra within one ring (ring periodicity)]

$\boldsymbol{P^r}$	Silicate		Reference
	Name	Formula	
3	Pseudowollastonite	$Ca_3[Si_3O_9]$ (α) (hT)	Yamanaka and Mori 1981
	Synthetic	$Ca_3[Si_3O_9]$ (δ) (mP)	Trojer 1969
	Synthetic	$Sr_3[Si_3O_9]$ (α)	Machida et al. 1982
	Synthetic	$Na_2Be_2[Si_3O_9]$	Ginderow et al. 1982
	Margarosanite	$Ca_2Pb[Si_3O_9]$	Freed and Peacor 1969
	Walstromite	$Ca_2Ba[Si_3O_9]$	Dent Glasser and Glasser 1968
	Synthetic	$NaBaNd[Si_3O_9]$	Malinovskii et al. 1984
	Benitoite	$BaTi[Si_3O_9]$	Fischer 1969
	Wadeite	$K_2Zr[Si_3O_9]$	Blinov et al. 1977
	Catapleiite	$Na_2Zr[Si_3O_9] \cdot 2H_2O$	Ilyushin et al. 1981c
4	Taramellite	$Ba_4(Fe^{+3}, Ti)_4B_2[Si_4O_{12}]_2O_5Cl_x$	Mazzi and Rossi 1980
	Nagashimalite	$Ba_4(V^{+3}, Ti)_4B_2[Si_4O_{12}]_2O_3(O,OH)_2Cl$	Matsubara 1980b
	Baotite	$Ba_4Ti_4(Ti,Nb,Fe)_4[Si_4O_{12}]O_{16}Cl$	Muradyan and Simonov 1977
	M′−2PbO · SiO_2	$Pb_8[Si_4O_{12}]O_4$	Dent Glasser et al. 1981
	Synthetic	$K_4Sc_2[Si_4O_{12}](OH)_2$	Pyatenko et al. 1979
	Synthetic	$Na_2BaNd_2[Si_4O_{12}][CO_3]$	Malinovskii 1983
	Verplanckite	$Ba_{12}(Mn, Ti, Fe)_6[Si_4O_{12}]_3(O,OH)_2(OH,H_2O)_7Cl_9$	Kampf et al. 1973
6	Combeite	$Na_4Ca_4[Si_6O_{18}]$	Fischer and Tillmanns 1983
	Zirsinalite	$Na_6CaZr[Si_6O_{18}]$	Pudovkina et al. 1980
	Synthetic	$Na_8Sn[Si_6O_{18}]$	Zayakina et al. 1980
			Safronov et al. 1980
	Imandrite	$Na_{12}Ca_3Fe_2[Si_6O_{18}]_2$	Chernitsova et al. 1980
	Kazakovite	$Na_6MnTi[Si_6O_{18}]$	Voronkov et al. 1979a
	Beryl	$Be_3Al_2[Si_6O_{18}]$	Morosin 1972
	Petarasite	$Na_5Zr_2[Si_6O_{18}](Cl,OH) \cdot 2H_2O$	Ghose et al. 1980
	Baratovite	$Li_3KCa_7Ti_2[Si_6O_{18}]_2F_2$	Menchetti and Sabelli 1979
	Tourmalines	$XY_3Z_6B_3[Si_6O_{18}]O_9(O,OH,F)_4$	e.g. Nuber and Schmetzer 1981
		$X = Na, Ca$; $Z = Al, Mg$; $Y = Li, Mg, Fe^{+2}, Mn, Fe^{+3}, Al$	Fortier and Donnay 1975
	Dioptase	$Cu_6[Si_6O_{18}] \cdot 6H_2O$	Ribbe et al. 1977, Belov et al. 1978
	Scawtite	$Ca_7[Si_6O_{18}][CO_3] \cdot 2H_2O$	Pluth and Smith 1973

8	Muirite	$Ba_{10}(Ca,Mn,Ti)_4[Si_8O_{24}](Cl,O,OH)_{12} \cdot 4H_2O$	Khan and Baur 1971
9	Eudialyte[a]	$Na_{12}(Ca,RE)_6(Fe,Mn,Mg)_3Zr_3(Zr,Nb)_x[Si_3O_9]_2[Si_9(O,OH)_{27}]_2Cl_y$	Golyshev et al. 1971 Giuseppetti et al. 1971
12	Traskite	$(Ca,Sr)Ba_{24}(Mg,Mn,Fe,Al,Ti)_{16}[Si_2O_7]_6[Si_{12}O_{36}](O,OH)_{30}Cl_6 \cdot 14H_2O$	Malinovskii et al. 1976
	Synthetic	$M_{16}Ca_4[Si_{12}O_{36}]$, M = Na, K	Baumgartner and Völlenkle 1977
	Synthetic	$K_{16}Sr_4[Si_{12}O_{36}]$	Baumgartner and Völlenkle 1977
	Synthetic	$Na_{15}RE_3[Si_{12}O_{36}]$, RE = Sm, ..., Lu; Sc, Y	Merinov et al. 1978, 1980 Maksimov et al. 1979

[a] Eudialyte is a mixed-anion silicate which contains dreier as well as neuner single rings

Table 7.4 Dicyclosilicates with unbranched double rings $\{uB, 2r\}$ $[^{P^r}Si_{2p}O_{5p}]^{-2p}$ [$p = P^r$: number of tetrahedra within one fundamental ring (ring periodicity)]

P^r	Silicate		Reference
	Name	Formula	
3	Synthetic	$[Ni(H_2N \cdot CH_2 \cdot CH_2 \cdot NH_2)_3]_3[^3Si_6O_{15}] \cdot 26H_2O$	Smolin 1970
4	Synthetic	$[Cu(H_2N \cdot CH_2 \cdot CH_2 \cdot NH_2)_2]_4[^4Si_8O_{20}] \cdot 38H_2O$	Smolin et al. 1972
	Synthetic	$H_2[Co(H_2N \cdot CH_2 \cdot CH_2 \cdot NH_2)_3]_2[^4Si_8O_{20}] \cdot 16.4H_2O$	Smolin et al. 1976
	Synthetic	$[N(CH_3)_4]_8[^4Si_8O_{20}] \cdot 64.8H_2O$	Smolin et al. 1979
	Synthetic	$H_7[N(n\text{-}C_4H_9)_4][^4Si_8O_{20}] \cdot 5.33H_2O$	Bissert and Liebau 1984
	Steacyite	$K_{1-x}(Na,Ca)_{2-y}Th_{1-z}[^4Si_8O_{20}]$	Szymański et al. 1982
	Iraqite	$(K,\square)(Ca,RE,Na)_2(RE,Th)[^4Si_8O_{20}]$	Livingstone et al. 1976
5	Synthetic	$[(CH_3)_3Si]_{10}[^5Si_{10}O_{25}]$	Smolin in Hoebbel et al. 1975
6	Milarite	$KCa_2(Be,Al)_3[^6Si_{12}O_{30}] \cdot 0.75H_2O$	Černý et al. 1980
	Milarite-type phases, e.g.,	$A_xM_2M'_3[^6Si_{12}O_{30}]$; A = Na, K, Rb; M = Mg, Cu, Fe; M′ = Mg, Zn, Fe, Cu, Li	Nguyen et al. 1980
	Synthetic	$K_2Mn_5[^6Si_{12}O_{30}]$	Sandomirskii et al. 1977
	Synthetic	$K_2Mg_5[^6Si_{12}O_{30}]$	Khan et al. 1972
	Osumilite	$KMg_2Al_3[^6(Al_2Si_{10})O_{30}]$	Hesse and Seifert 1982
	Brannockite	$KSn_2Li_3[^6Si_{12}O_{30}]$	White et al. 1973

since a statistical B/Si distribution is rather unlikely, the silicate anions of tien-shanite, $Na_9KCa_2Ba_6(Mn, Fe)_6(Ti, Nb, Ta)_6B_{12}[Si_{18}O_{54}]_2O_{15}(OH)_2$ (Malinovskii et al. 1977), can be regarded as open-branched sechser single rings $\{oB, 1\,r\}[Si_6O_{18}(Si_2O_6)_6]$ (Fig. 7.7b).

Although loop-branched single rings have not as yet been proven to occur in crystalline silicates, their existence in trimethylsilylesters has been suggested by gas chromatography and ^{29}Si NMR spectroscopy (Hoebbel et al. 1976). The structural representations of two such esters are

$$
\begin{array}{ccccc}
 & M & M & M & \\
 & | & | & | & \\
M- & Q^2 & -Q^3 & -Q^2 & -M \\
 & | & | & | & \\
M- & Q^2 & -Q^3 & -Q^2 & -M \\
 & | & | & | & \\
 & M & M & M &
\end{array}
$$

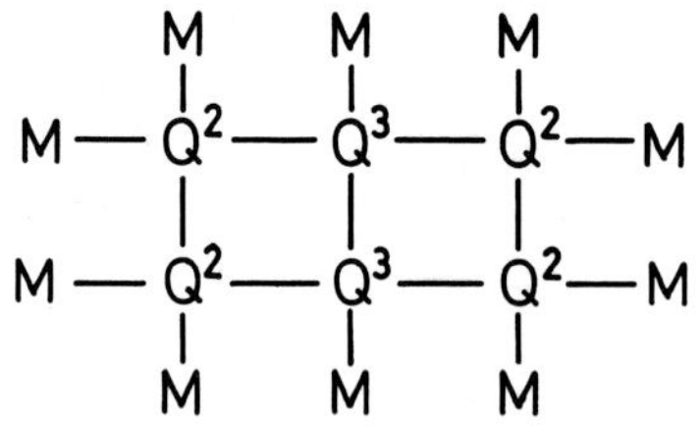

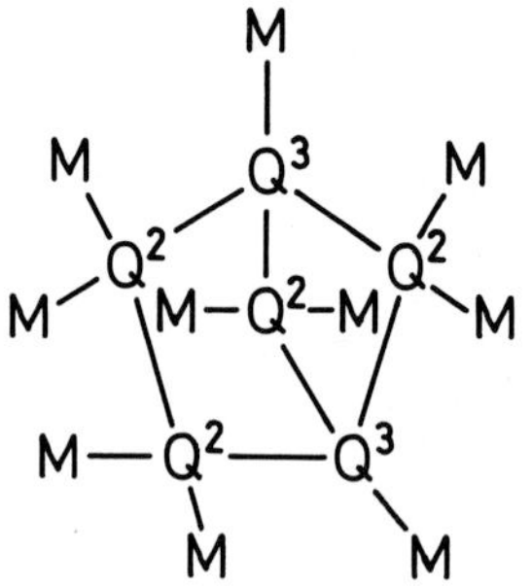

where Q^2 and Q^3 are secondary and tertiary $[SiO_4]$ tetrahedra and M is a trimethylsilyl group, $Si(CH_3)_3$. These species are not double rings because they cannot be constructed by linking two separate single rings via common oxygen atoms. One of them is, instead, a loop-branched vierer single ring $\{lB, 1\,r\}[^4Si_4O_{12}(Si_2O_5)]$ built by condensation of a vierer single ring with a double tetrahedron:

$$
\begin{array}{c}
\quad O \quad\quad O \\
\quad | \quad\quad | \\
O-Si-O-Si-O \\
\quad | \quad\quad | \\
O \quad\quad O \\
\quad | \quad\quad | \\
O-Si-O-Si-O \\
\quad | \quad\quad | \\
O \quad\quad O
\end{array}
\quad + \quad
\begin{array}{c}
\quad O \\
\quad | \\
O-Si-O \\
\quad | \\
O \\
\quad | \\
O-Si-O \\
\quad | \\
O
\end{array}
\quad \xrightarrow{-2\,O} \quad
\begin{array}{c}
\quad O \quad\quad O \quad\quad O \\
\quad | \quad\quad | \quad\quad | \\
O-Si-O-Si-O-Si-O \\
\quad | \quad\quad | \quad\quad | \\
O \quad\quad O \quad\quad O \\
\quad | \quad\quad | \quad\quad | \\
O-Si-O-Si-O-Si-O \\
\quad | \quad\quad | \quad\quad | \\
O \quad\quad O \quad\quad O
\end{array}
$$

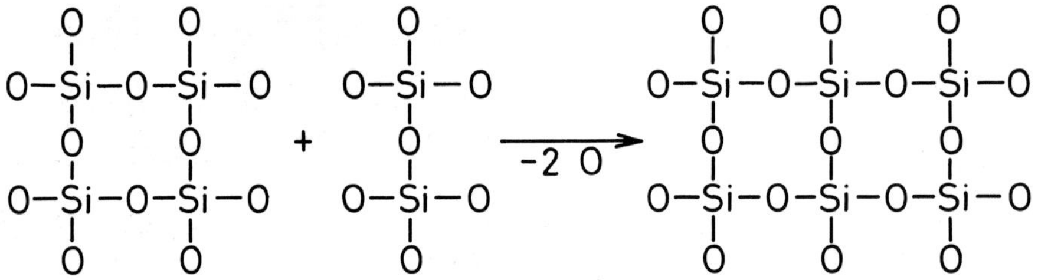

The other ester is a loop-branched fünfer single ring $\{lB, 1\,r\}[^5Si_5O_{15}(SiO_2)]$ of the same composition and can be formed by analogous condensation of an unbranched fünfer single ring with a single tetrahedron.

Fig. 7.8. The tetrahedral anion of hyalotekite, $Ca_2Ba_2Pb_2B_2[(Si_{1.5}Be_{0.5})Si_8O_{28}]F$ can be described as **a** open-branched vierer double ring $\{oB, 2\,r\}[^4T_{10}O_{28}]$, or, **b** better, as loop-branched achter single ring $\{lB, 1\,r\}[^8T_{10}O_{28}]$

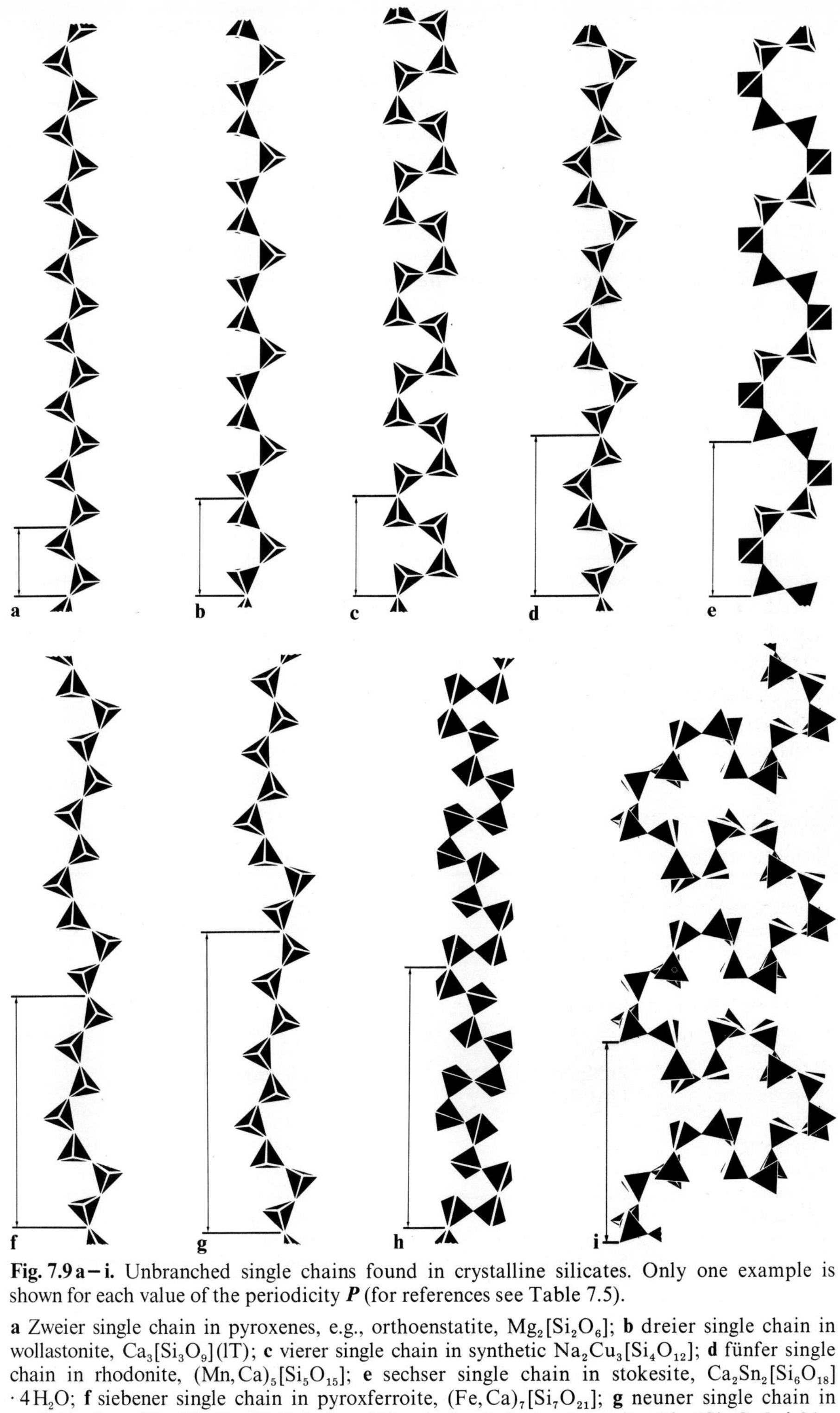

Fig. 7.9 a–i. Unbranched single chains found in crystalline silicates. Only one example is shown for each value of the periodicity ***P*** (for references see Table 7.5).

a Zweier single chain in pyroxenes, e.g., orthoenstatite, $Mg_2[Si_2O_6]$; **b** dreier single chain in wollastonite, $Ca_3[Si_3O_9]$(lT); **c** vierer single chain in synthetic $Na_2Cu_3[Si_4O_{12}]$; **d** fünfer single chain in rhodonite, $(Mn, Ca)_5[Si_5O_{15}]$; **e** sechser single chain in stokesite, $Ca_2Sn_2[Si_6O_{18}] \cdot 4H_2O$; **f** siebener single chain in pyroxferroite, $(Fe, Ca)_7[Si_7O_{21}]$; **g** neuner single chain in synthetic ferrosilite III, $Fe_9[Si_9O_{27}]$; **h** zwölfer single chain in alamosite, $Pb_{12}[Si_{12}O_{36}]$; **i** 24er single chain in synthetic $Na_{24}Y_8[Si_{24}O_{72}]$

In contrast, the cyclic anion of hyalotekite, a complex lead borosilicate of approximate composition $Ca_2Ba_2Pb_2B_2[(Si_{1.5}Be_{0.5})Si_8O_{28}]F$, has a silicate anion which can be described both as an open-branched vierer double ring or, better, as a loop-branched achter single ring (Fig. 7.8 a, b) (Moore et al. 1982).

7.2.2.3 Single Chain Silicates (Monopolysilicates)

7.2.2.3.1 Unbranched Single Chain Silicates. Unbranched tetrahedral single chains can be described by the general formula

$$\{\boldsymbol{uB}, 1^1_\infty\}[^P\mathrm{Si_pO_{3p}}]^{-2p} \quad \mathrm{p} = \boldsymbol{P}, \text{ finite} \tag{4}$$

Table 7.5 Survey of silicates containing unbranched single chains

Name	Formula	Reference
Zweier single chain silicates (excluding pyroxenes)		
Synthetic	$Li_4[Si_2O_6]$	Völlenkle 1981
Synthetic	$Na_4[Si_2O_6]$	McDonald and Cruickshank 1967
Synthetic	$Ag_4[Si_2O_6]$	Thilo and Wodtcke 1958
Synthetic	$Ba_2[Si_2O_6]$ (hT)	Grosse and Tillmanns 1974
Synthetic	$Na_2Zn[Si_2O_6]$	Belokoneva et al. 1970
Synthetic	$Na_2Ba[Si_2O_6]$	Gunawardane et al. 1973
Lorenzenite (ramsayite)	$Na_2Ti_2[Si_2O_6]O_3$	Ch'in-hang et al. 1969
Shattuckite	$Cu_5[Si_2O_6]_2(OH)_2$	Evans and Mrose 1977
Carpholite	$MnAl_2[Si_2O_6](OH)_4$	Lindemann et al. 1979; Viswanathan 1981
Pyroxenes (end members only)		
Orthoenstatite	$Mg_2[Si_2O_6]$	Hawthorne and Ito 1977
Orthoferrosilite	$Fe_2[Si_2O_6]$	Sueno et al. 1976
Synthetic	$Zn_2[Si_2O_6]$	Morimoto et al. 1975
Synthetic	$Mn_2[Si_2O_6]$	Tokonami et al. 1979
Diopside	$CaMg[Si_2O_6]$	Clark et al. 1969
Hedenbergite	$CaFe[Si_2O_6]$	Cameron et al. 1973
Johannsenite	$CaMn[Si_2O_6]$	Freed and Peacor 1967
Synthetic	$CaCo[Si_2O_6]$	Ribbe and Prunier 1977
Synthetic	$CaNi[Si_2O_6]$	Ribbe and Prunier 1977
Synthetic	$ZnMg[Si_2O_6]$	Morimoto et al. 1975
Spodumene	$LiAl[Si_2O_6]$	Clark et al. 1969
Synthetic	$LiFe[Si_2O_6]$	Clark et al. 1969
Synthetic	$LiSc[Si_2O_6]$	Hawthorne and Grundy 1977
Jadeite	$NaAl[Si_2O_6]$	Cameron et al. 1973
Acmite	$NaFe[Si_2O_6]$	Clark et al. 1969
Ureyite	$NaCr[Si_2O_6]$	Clark et al. 1969
Synthetic	$NaIn[Si_2O_6]$	Hawthorne and Grundy 1974
Synthetic	$NaSc[Si_2O_6]$	Hawthorne and Grundy 1973
Dreier single chain silicates		
Wollastonite 1T	$Ca_3[Si_3O_9]$	Ohashi and Finger 1978
Wollastonite 2M (parawollastonite)	$Ca_3[Si_3O_9]$	Hesse 1984b

Table 7.5 (cont.)

Name	Formula	Reference
Wollastonite 7T	$Ca_3[Si_3O_9]$	Henmi et al. 1978, 1983
Bustamite 1T	$(Ca, Mn)_3[Si_3O_9]$	Ohashi and Finger 1978
Bustamite 2M	$(Ca, Mn)_3[Si_3O_9]$	Hesse et al. 1982
Ferrobustamite 1T	$(Ca, Fe)_3[Si_3O_9]$	Rapoport and Burnham 1973
Pectolite	$HNaCa_2[Si_3O_9]$	Ohashi and Finger 1978
Serandite	$HNaMn_2[Si_3O_9]$	Ohashi and Finger 1978
Cascandite	$HCaSc[Si_3O_9]$	Mellini and Merlino 1982
Foshagite	$Ca_4[Si_3O_9](OH)_2$	Gard and Taylor 1960
Synthetic	$Fe_3Be[Si_3O_9](OH)_2$	Bakakin and Solov'eva 1971
Hilairite	$Na_2Zr[Si_3O_9] \cdot 3H_2O$	Ilyushin et al. 1981a
Unnamed	$K_2Zr[Si_3O_9] \cdot H_2O$	Ilyushin et al. 1981b
Sorensenite	$Na_4Be_2Sn[Si_3O_9]_2 \cdot 2H_2O$	Metcalf-Johansen and Hazell 1976
Vierer single chain silicates		
Synthetic	$Na_2Cu_3[Si_4O_{12}]$	Kawamura and Kawahara 1976
Haradaite	$Sr_2V_2[Si_4O_{12}]O_2$	Takéuchi and Joswig 1967
Suzukiite	$Ba_2V_2[Si_4O_{12}]O_2$	Matsubara et al. 1982
Batisite	$Na_2BaTi_2[Si_4O_{12}]O_2$	Schmahl 1981
Synthetic	$Ca_3Mn_2[Si_4O_{12}]O_2$	Moore and Araki 1979
Leucophanite	$Na_2Be_2Ca_2[Si_4O_{12}]F_2$	Cannillo et al. 1967
Krauskopfite	$H_4Ba_2[Si_4O_{12}] \cdot 4H_2O$	Coda et al. 1967b
Ohmilite	$Sr_3(Ti, Fe^{+3})[Si_4O_{12}]$ $(O, OH) \cdot 2{-}3H_2O$	Mizota et al. 1983
Fünfer single chain silicates		
Synthetic	$Mn_5[Si_5O_{15}]$	Narita et al. 1977
Rhodonite	$(Mn, Ca)_5[Si_5O_{15}]$	Ohashi and Finger 1975
Mg rhodonite	$(Mn, Mg)_5[Si_5O_{15}]$	Murakami and Takéuchi 1979
Babingtonite	$HCa_2Fe_2[Si_5O_{15}]$	Araki and Zoltai 1972
Synthetic	$HLiMn_4[Si_5O_{15}]$	Murakami et al. 1977
Nambulite	$H(Li, Na)Mn_4[Si_5O_{15}]$	Narita et al. 1975
Marsturite	$HNaCaMn_3[Si_5O_{15}]$	Peacor et al. 1978
Santaclaraite	$HCaMn_4[Si_5O_{15}](OH) \cdot H_2O$	Ohashi and Finger 1981
Sechser single chain silicates		
Chkalovite	$Na_6Be_3[Si_6O_{18}]$	Simonov et al. 1975/76
Gaidonnayite	$Na_4Zr_2[Si_6O_{18}] \cdot 4H_2O$	Chao 1973
Georgechaoite	$Na_2K_2Zr_2[Si_6O_{18}] \cdot 4H_2O$	Ghose and Thakur 1985
Stokesite	$Ca_2Sn_2[Si_6O_{18}] \cdot 4H_2O$	Vorma 1963
Siebener single chain silicates		
Synthetic	$Mn_7[Si_7O_{21}]$	Narita et al. 1977
Pyroxmangite	$(Fe, Mn)_7[Si_7O_{21}]$	Ohashi and Finger 1975; Pinckney et al. 1981
Pyroxferroite	$(Fe, Ca)_7[Si_7O_{21}]$	Burnham 1971
Neuner single chain silicate		
Ferrosilite III	$Fe_9[Si_9O_{27}]$	Weber 1983
Zwölfer single chain silicate		
Alamosite	$Pb_{12}[Si_{12}O_{36}]$	Boucher and Peacor 1968
24er single chain silicate		
Synthetic	$Na_{24}Y_8[Si_{24}O_{72}]$	Maksimov et al. 1980b

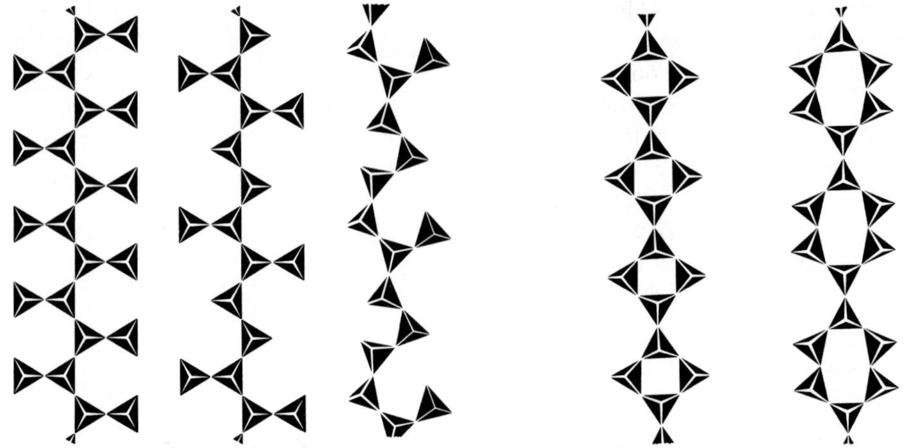

a b c d e

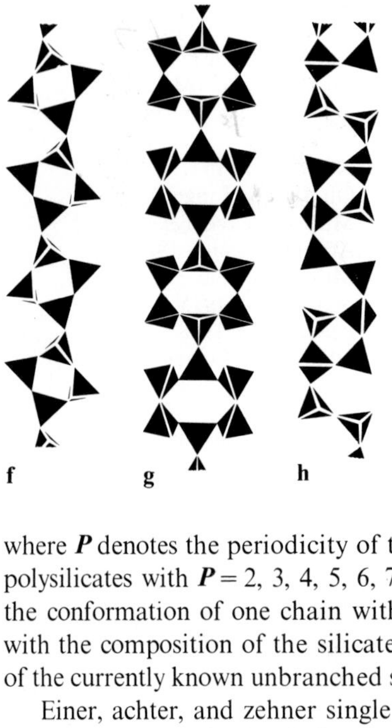

f g h

Fig. 7.10 a – h. Branched single chains found in crystalline silicates (for references see Table 10.8).

a Open-branched zweier single chain in astrophyllite, $NaK_2Mg_2(Fe,Mn)_5Ti_2\{\boldsymbol{oB}, 1^1_\infty\}[^2Si_4O_{12}]_2(O,OH,F)_7$; **b** open-branched vierer single chain in aenigmatite, $Na_2Fe_5Ti\{\boldsymbol{oB}, 1^1_\infty\}[^4Si_6O_{18}]O_2$; **c** open-branched fünfer single chain in saneroite, $HNa_{1.15}Mn_5\{\boldsymbol{oB}, 1^1_\infty\}[^5(Si_{5.5}V_{0.5})O_{18}]OH$; **d** loop-branched dreier single chain in synthetic $Li_2Mg_2\{\boldsymbol{lB}, 1^1_\infty\}[^3Si_4O_{11}]$; **e** loop-branched vierer single chain in deerite, $Fe_6^{+2}Fe_3^{+3}\{\boldsymbol{lB}, 1^1_\infty\}[^4Si_6O_{17}](OH)_5O_3$; **f** loop-branched sechser single chain in vlasovite, $Na_4Zr_2\{\boldsymbol{lB}, 1^1_\infty\}[^6Si_8O_{22}]$; **g** loop-branched achter single chain in pellyite, $Ca_2Ba_4(Fe,Mg)_4\{\boldsymbol{lB}, 1^1_\infty\}[^8Si_{12}O_{34}]$; **h** loop-branched zehner single chain in nordite, $Na_4(Na,Mn)_2(Sr,Ca)_2(Zn,Mg,Fe,Mn)RE_2\{\boldsymbol{lB}, 1^1_\infty\}[^{10}Si_{12}O_{34}]$

where $\boldsymbol{P}$ denotes the periodicity of the chain. So far, the crystal structures of monopolysilicates with $\boldsymbol{P} = 2, 3, 4, 5, 6, 7, 9, 12$, and 24 have been described. In Fig. 7.9 the conformation of one chain with each of these periodicities is shown, together with the composition of the silicate in which it has been found. Table 7.5 is a list of the currently known unbranched single chain silicates.

Einer, achter, and zehner single chains have so far not been discovered in silicates, but einer single chains have been observed in copper polygermanate, $Cu\{\boldsymbol{uB}, 1^1_\infty\}[^1GeO_3]$ (Völlenkle et al. 1967), achter single chains in a number of polyphosphates (Majling and Hanic 1980; Liebau 1981), and zehner single chains in $K_2Ba_4[P_{10}O_{30}]$ (Martin et al. 1975).

7.2.2.3.2 ***Branched Single Chain Silicates.*** Within the last two decades a number of silicate structures with open- and loop-branched single chains have been described.

Their silicate chains are illustrated in Fig. 7.10 together with their mineral names and their structural formulae. A list of these silicates is given in Table 10.8. The Si:O atomic ratio of open-branched single chains is the same as that of unbranched single chains, namely, 1:3. Therefore, it is desirable that a distinction be made between their structural formulae even when a very detailed formula is not wanted. This can easily be achieved by using the periodicity ***P*** of the chain as left superscript to the symbol Si. For example, the comprehensive formula $\{\boldsymbol{oB}, 1^1_\infty\}[{}^4Si_4O_{12}(SiO_3)_2]$ of the open-branched vierer single chain of aenigmatite can be shortened to $\{\boldsymbol{oB}, {}^1_\infty\}[{}^4Si_6O_{18}]$.

For loop-branched single chains the atomic ratio Si:O is higher than for the unbranched and open-branched single chains.

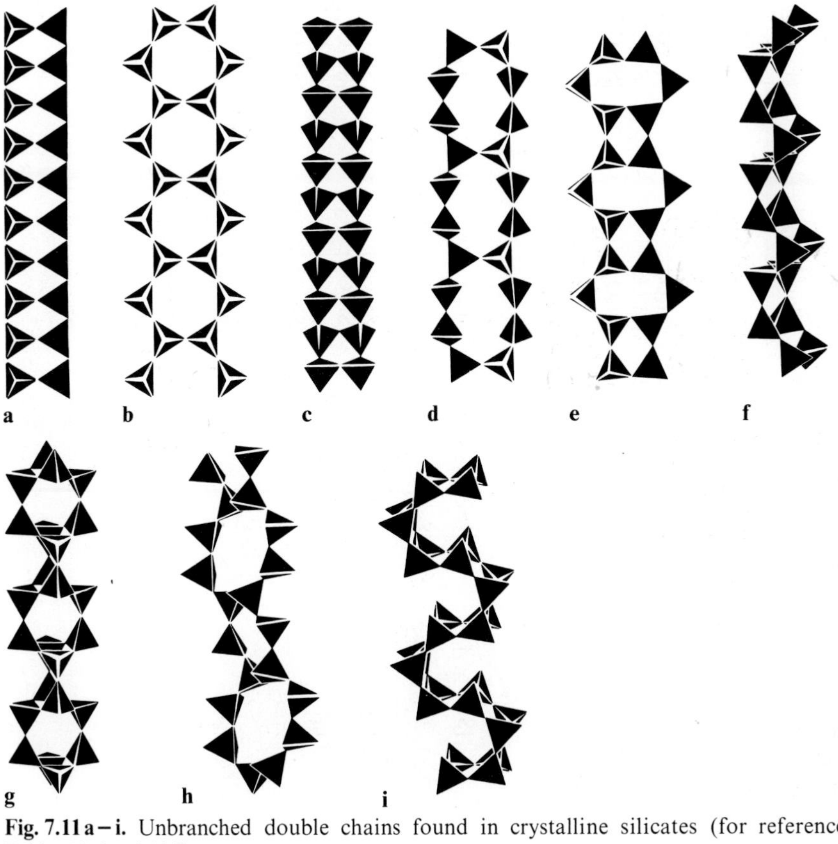

Fig. 7.11 a–i. Unbranched double chains found in crystalline silicates (for references see Tables 10.6 and 10.7).

a Unbranched einer double chain in hypothetical high-sillimanite, $Al^{[6]}[AlSiO_5](hT)$; **b** unbranched zweier double chain in amphiboles, e.g., tremolite, $Mg_5Ca_2[Si_4O_{11}]_2(OH)_2$; **c** unbranched zweier double chain in synthetic $Li_4[SiGe_3O_{10}]$; **d** unbranched dreier double chain in xonotlite, $Ca_6[Si_6O_{17}](OH)_2$; **e** unbranched dreier double chain in okenite, $Ca_{10}\{\boldsymbol{uB}, 2^1_\infty\}[Si_6O_{16}]\{\boldsymbol{uB}, 1^2_\infty\}[Si_6O_{15}]_2 \cdot 18H_2O$; **f** unbranched dreier double chain in epididymite, $Na_2Be_2[Si_6O_{15}] \cdot H_2O$; **g** unbranched vierer double chain in narsarsukite, $Na_4Ti_2[Si_8O_{20}]O_2$; **h** unbranched fünfer double chain in inesite, $Ca_2Mn_7[Si_{10}O_{28}](OH)_2 \cdot 5H_2O$; **i** unbranched sechser double chain in tuhualite, $(Na,K)_2Fe_2^{+2}Fe_2^{+3}[Si_{12}O_{30}] \cdot H_2O$

Table 7.6 Subdivision of unbranched multiple chains according to periodicity ***P***, multiplicity ***M***, and to the Si : O ratio, which is correlated with the number l of linkages between the subchains (m and p in the general formulae are numbers equal to ***M*** and ***P***, respectively)

P \ M	1 (Single)	2 (Double)	3 (Triple)	4 (Fourfold)	5 (Fivefold)	M (M-fold)
1 (Einer)	$\{1^1_\infty\}[SiO_3]$	$\{2^1_\infty\}[Si_2O_5]$	$\{3^1_\infty\}[Si_3O_7]$	$\{4^1_\infty\}[Si_4O_9]$	$\{5^1_\infty\}[Si_5O_{11}]$	$\{M^1_\infty\}[Si_mO_{2m+1}]$
2 (Zweier)	$[Si_2O_6]$	$[Si_4O_{11}]$ $[Si_4O_{10}]$	$[Si_6O_{16}]$ $[Si_6O_{15}]$ $[Si_6O_{14}]$ $[Si_6O_{13}]$ $[Si_6O_{12}]$	$[Si_8O_{21}]$ $[Si_8O_{20}]$ $[Si_8O_{19}]$ $[Si_8O_{18}]$ $[Si_8O_{17}]$ $[Si_8O_{16}]$	$[Si_{10}O_{26}]$ $[Si_{10}O_{25}]$ $[Si_{10}O_{24}]$ $[Si_{10}O_{23}]$ $[Si_{10}O_{22}]$ $[Si_{10}O_{21}]$ $[Si_{10}O_{20}]$	$[Si_{2m}O_{5m+1}]$ $[Si_{2m}O_{5m}]$ $[Si_{2m}O_{5m-1}]$ ⋮ $[Si_{2m}O_{4m+2}]$ $[Si_{2m}O_{4m+1}]$ $[Si_{2m}O_{4m}]$
	$\{1^1_\infty\}[Si_2O_6]$	$\{2^1_\infty\}[Si_4O_{12-l}]$ $l = 1, 2$	$\{3^1_\infty\}[Si_6O_{18-\sum_{i=1}^{2} l_i}]$ $2 \leq \Sigma l_i \leq 6$	$\{4^1_\infty\}[Si_8O_{24-\sum_{i=1}^{3} l_i}]$ $3 \leq \Sigma l_i \leq 8$	$\{5^1_\infty\}[Si_{10}O_{30-\sum_{i=1}^{4} l_i}]$ $4 \leq \Sigma l_i \leq 10$	$\{M^1_\infty\}[Si_{2m}O_{6m-\sum_{i=1}^{m-1} l_i}]$ $m-1 \leq \Sigma l_i \leq m\,p$
3 (Dreier)	$[Si_3O_9]$	$[Si_6O_{17}]$ $[Si_6O_{16}]$ $[Si_6O_{15}]$	$[Si_9O_{25}]$ $[Si_9O_{24}]$ $[Si_9O_{23}]$ $[Si_9O_{22}]$ ⋮ $[Si_9O_{18}]$	$[Si_{12}O_{33}]$ $[Si_{12}O_{32}]$ $[Si_{12}O_{31}]$ $[Si_{12}O_{30}]$ ⋮ $[Si_{12}O_{25}]$ $[Si_{12}O_{24}]$	$[Si_{15}O_{41}]$ $[Si_{15}O_{40}]$ $[Si_{15}O_{39}]$ $[Si_{15}O_{38}]$ ⋮ $[Si_{15}O_{32}]$ $[Si_{15}O_{31}]$ $[Si_{15}O_{30}]$	$[Si_{3m}O_{8m+1}]$ $[Si_{3m}O_{8m}]$ $[Si_{3m}O_{8m-1}]$ $[Si_{3m}O_{8m-2}]$ ⋮ $[Si_{3m}O_{6m+2}]$ $[Si_{3m}O_{6m+1}]$ $[Si_{3m}O_{6m}]$
	$\{1^1_\infty\}[Si_3O_9]$	$\{2^1_\infty\}[Si_6O_{18-l}]$ $1 \leq l \leq 3$	$\{3^1_\infty\}[Si_9O_{27-\sum_{i=1}^{2} l_i}]$ $2 \leq \Sigma l_i \leq 9$	$\{4^1_\infty\}[Si_{12}O_{36-\sum_{i=1}^{3} l_i}]$ $3 \leq \Sigma l_i \leq 12$	$\{5^1_\infty\}[Si_{15}O_{45-\sum_{i=1}^{4} l_i}]$ $4 \leq \Sigma l_i \leq 15$	$\{M^1_\infty\}[Si_{3m}O_{9m-\sum_{i=1}^{m-1} l_i}]$ $m-1 \leq \Sigma l_i \leq m\,p$

4 (Vierer)	$[Si_4O_{12}]$	$[Si_8O_{23}]$ $[Si_8O_{22}]$ ⋮ $[Si_8O_{20}]$	$[Si_{12}O_{34}]$ $[Si_{12}O_{33}]$ ⋮ $[Si_{12}O_{24}]$	$[Si_{16}O_{45}]$ $[Si_{16}O_{44}]$ ⋮ $[Si_{16}O_{40}]$ ⋮ $[Si_{16}O_{33}]$ $[Si_{16}O_{32}]$	$[Si_{20}O_{56}]$ $[Si_{20}O_{55}]$ ⋮ $[Si_{20}O_{42}]$ $[Si_{20}O_{41}]$ $[Si_{20}O_{40}]$	$[Si_{4m}O_{11m+1}]$ $[Si_{4m}O_{11m}]$ ⋮ $[Si_{4m}O_{8m+2}]$ $[Si_{4m}O_{8m+1}]$ $[Si_{4m}O_{8m}]$
	$\{1^1_\infty\}[Si_4O_{12}]$	$\{2^1_\infty\}[Si_8O_{24-l}]$ $1 \leq l \leq 4$	$\{3^1_\infty\}[Si_{12}O_{36-\sum_{i=1}^{2} l_i}]$ $2 \leq \Sigma\, l_i \leq 12$	$\{4^1_\infty\}[Si_{16}O_{48-\sum_{i=1}^{3} l_i}]$ $3 \leq \Sigma\, l_i \leq 16$	$\{5^1_\infty\}[Si_{20}O_{60-\sum_{i=1}^{4} l_i}]$ $4 \leq \Sigma\, l_i \leq 20$	$\{\boldsymbol{M}^1_\infty\}[Si_{4m}O_{12m-\sum_{i=1}^{m-1} l_i}]$ $m-1 \leq \Sigma\, l_i \leq m\,p$
⋮	⋮	⋮	⋮	⋮	⋮	⋮
P	$\{1^1_\infty\}[Si_pO_{3p}]$	$\{2^1_\infty\}[Si_{2p}O_{6p-l}]$ $1 \leq l \leq p$	$\{3^1_\infty\}[Si_{3p}O_{9p-\sum_{i=1}^{2} l_i}]$ $2 \leq \Sigma\, l_i \leq 3\,p$	$\{4^1_\infty\}[Si_{4p}O_{12p-\sum_{i=1}^{3} l_i}]$ $3 \leq \Sigma\, l_i \leq 4\,p$	$\{5^1_\infty\}[Si_{5p}O_{15p-\sum_{i=1}^{4} l_i}]$ $4 \leq \Sigma\, l_i \leq 5\,p$	$\{\boldsymbol{M}^1_\infty\}[Si_{pm}O_{3pm-\sum_{i=1}^{m-1} l_i}]$ $m-1 \leq \Sigma\, l_i \leq m\,p$

7.2.2.4 Multiple Chain Silicates (Oligopolysilicates)

7.2.2.4.1 Unbranched Multiple Chain Silicates. The condensation of two unbranched single chains via corners of the tetrahedra results in the formation of an unbranched double chain. By analogy with the cyclic anions, all, or only some, of the tetrahedra of each chain may take part in the linking of the chains. If l is the number of linkages between two single chains within one repeat unit, the possible unbranched double chains are described by the general formula

$$\{\boldsymbol{uB}, 2^1_\infty\}[^{\boldsymbol{P}}\mathrm{Si}_{2p}\mathrm{O}_{6p-l}]^{-(4p-2l)}, \quad p=\boldsymbol{P}, \quad \text{finite}, \quad 1 \le l \le p. \qquad (5)$$

Of the numerous types of unbranched double chains that could exist only those given in Fig. 7.11 have as yet been shown to form thermodynamically stable phases. The highest periodicity found so far is that of a sechser double chain.

The chain of Fig. 7.11a is a hypothetical einer double chain. The double chain observed in sillimanite, $Al^{[6]}[(AlSi)^{[4]}O_5]$, differs from this hypothetical chain in that the tetrahedral Al and Si atoms are almost completely ordered with each $[AlO_4]$ group linked to three $[SiO_4]$ tetrahedra and vice versa (in agreement with Loewenstein's rule). Strictly speaking, therefore, sillimanite contains zweier double chains. With increasing Al/Si disorder in the tetrahedral positions the structure of the sillimanite chain would approach that of a true einer double chain. In fact, however, the ordering between Al and Si in sillimanite is almost complete even at high temperatures (Winter and Ghose 1979).

The number of possible topologically distinct anions increases rapidly if condensation of three or more single chains to a higher-fold chain takes place. This becomes obvious when consideration is given to the effect of the following variables:

P: the periodicity of the single chain,
M: the multiplicity of the multiple chain, i.e., the number of single chains linked to an ***M***-fold chain,
$l \le \boldsymbol{MP}$: the number of linkages between the chains in one repeat unit of anions.

As Table 7.6 demonstrates, the number of topologically different unbranched multiple chains increases rapidly as the periodicity and/or multiplicity of the chains increases. Their general formula can be given as

$$\{\boldsymbol{uB}, \boldsymbol{M}^1_\infty\}[^{\boldsymbol{P}}\mathrm{Si}_{mp}\mathrm{O}_{3mp-\sum_1^{m-1} l_i}]^{-2(mp-\sum_1^{m-1} l_i)} \quad \text{with} \quad m=\boldsymbol{M}, \qquad (6)$$

finite, p = ***P***, finite, and $(m-1) \le \sum_1^{m-1} l_i \le mp$, where l_i is the number of linkages between two adjacent subchains per chain period.

Of the possible multiple chains listed in Table 7.6 only the "einer" double chain $\{\boldsymbol{uB}, 2^1_\infty\}[^1(AlSi)O_5]$ of sillimanite, the zweier double chain $\{\boldsymbol{uB}, 2^1_\infty\}[^2Si_4O_{11}]$ of the amphiboles, and the dreier double chain $\{\boldsymbol{uB}, 2^1_\infty\}[^3Si_6O_{17}]$ of xonotlite were known before 1960. Since then several other types of multiple chains have been found (Table 7.7), and their number is steadily increasing.

So far, crystalline silicates with multiple zweier ($\boldsymbol{P}=2$) chains with $\boldsymbol{M}=3$, 4, and 5 and $\sum l_i = 3$, 4, and 5 have been observed. They are called zweier triple

Table 7.7 Types of unbranched single and multiple chains which have been found in crystalline silicates up to 1984

P \ *M*	1	2	3	4	5
1		Si_2O_5			
2	Si_2O_6	Si_4O_{11} Si_4O_{10}	Si_6O_{16}	Si_8O_{21}	$Si_{10}O_{26}$
3	Si_3O_9	Si_6O_{17} Si_6O_{16} Si_6O_{15}		$Si_{12}O_{32}$ $Si_{12}O_{30}$	
4	Si_4O_{12}	Si_8O_{20}		$Si_{16}O_{40}$	
5	Si_5O_{15}	$Si_{10}O_{28}$			
6	Si_6O_{18}	$Si_{12}O_{30}$			
7	Si_7O_{21}				
9	Si_9O_{27}				
12	$Si_{12}O_{36}$				
24	$Si_{24}O_{72}$				

M = multiplicity; ***P*** = chain periodicity

chains, zweier fourfold (quadruple) chains, and zweier fivefold (quintuple) chains (Fig. 7.12a, b, c)[3], the corresponding phases are tripolysilicates, tetrapolysilicates, and pentapolysilicates, respectively.

A dreier fourfold chain with $\boldsymbol{M} = 4$ and $\sum l_i = 6$ and formula $[Si_{12}O_{30}]^{-12}$ has been found in miserite and a vierer fourfold chain with $\boldsymbol{M} = 4$ and $\sum l_i = 8$ in synthetic K_8Cu_4 $\{\boldsymbol{uB}, 4^1_\infty\}[{}^4Si_{16}O_{40}]$ (Kawamura and Iiyama 1981) (Fig. 7.12d, e).

The term *tubular chain* or *columnar chain* is sometimes used for multiple chains which could also be constructed by condensation of an infinite number of single rings in such a way that the planes of these rings are not parallel to the chain direction. Examples of such chains occur in narsarsukite (Fig. 7.11g), miserite and $K_8Cu_4[Si_{16}O_{40}]$ (Fig. 7.12d, e). In narsarsukite, unbranched vierer rings $\{\boldsymbol{uB}, 1\,\boldsymbol{r}\}[Si_4O_{12}]$ with $[SiO_4]$ tetrahedra alternately pointing up and down, are

3 In Fig. 7.12 and in a number of those following, one of the fundamental chains defining the type of multiple chain, single layer, double layer, or framework is distinguished by light tetrahedra.

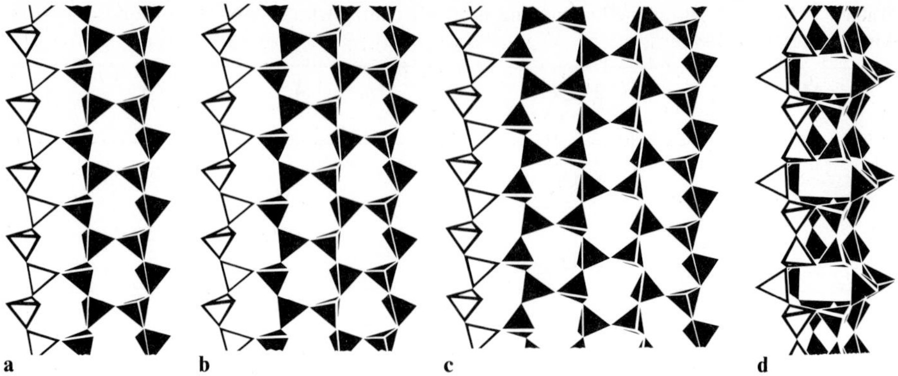

a b c d

e

Fig. 7.12 a – e. Unbranched multiple chains found in crystalline silicates. **a** Unbranched zweier triple chain in synthetic $Ba_4[^2Si_6O_{16}]$ (Hesse and Liebau 1980a); **b** unbranched zweier fourfold chain in synthetic $Ba_5[^2Si_8O_{21}]$ (Hesse and Liebau 1980a); **c** unbranched zweier fivefold chain in synthetic $Ba_6[^2Si_{10}O_{26}]$ (Hesse and Liebau 1980a); **d** unbranched dreier fourfold chain in miserite, $K_2Ca_{10}(Y,RE)_2[Si_2O_7]_2[^3Si_{12}O_{30}](OH)_2F_2$ (Scott 1976); **e** unbranched vierer fourfold chain in synthetic $K_8Cu_4[^4Si_{16}O_{40}]$ (Kawamura and Iiyama 1981)

linked while in miserite fundamental rings are loop-branched achter rings $\{\boldsymbol{lB}, 1\,\boldsymbol{r}\}[^8Si_8O_{24}(\breve{S}i_2O_5)_2]$, and in the synthetic potassium copper silicate they are unbranched achter rings $\{\boldsymbol{uB}, 1\,\boldsymbol{r}\}[Si_8O_{24}]$ (Fig. 7.13a, b, c). In these three minerals the planes of the fundamental rings are perpendicular to the chain direction.

The double chains of synthetic $Li_4[SiGe_3O_{10}]$ and epididymite (Fig. 7.11c, f and Fig. 7.13d, e) can be built by linking unbranched and open-branched vierer rings, respectively, the planes of which are not perpendicular to the chain direction. A comparison of these examples shows that the tubular character of these multiple chains becomes more pronounced the closer the angles between the chain direction and the planes of the rings approach 90°.

Although the term tubular chain describes the conformation of these anions quite clearly, it should be mentioned that this term is not consistent with the crystal chemical classification of silicates since it is not based on the classification parameters used. However, it is a term well-suited to be used as an additional descriptor in the particular designation of the crystal chemical system.

Multiple chains with multiplicities $\boldsymbol{M} > 5$ have been observed as lattice faults [chain multiplicity faults (Czank and Liebau 1980)] in high resolution electron microscopy of crystals of the silica-rich members of polysomatic series of chain silicates defined by Thompson (1978). One such series is the barium silicates $Ba_{m+1}\{\boldsymbol{uB}, \boldsymbol{M}^1_\infty\}[^2Si_{2m}O_{5m+1}]$, in which Czank and Buseck (1980) observed sixfold and sevenfold chains in crystals of the quintuple chain phase $Ba_6[Si_{10}O_{26}]$ (Fig. 7.14).

Another series is the pyriboles with an idealized general formula [if members having only one kind of silicate anions each ($N_{an}=1$) are considered] $M_{(3m+1)/2}\{\boldsymbol{uB}, \boldsymbol{M}^1_\infty\}[^2Si_{2m}O_{5m+1}](OH)_{m-1}$, where M = Mg, Fe^{+2}, Ca. The simplest pyriboles are the well-known rock-forming minerals pyroxene (such as enstatite $Mg_2[Si_2O_6]$ and diopside $CaMg[Si_2O_6]$), amphibole (such as tremolite $Mg_5Ca_2[Si_4O_{11}]_2(OH)_2$ and anthophyllite $(Mg, Fe)_7[Si_4O_{11}]_2(OH)_2$), and the recently discovered triple chain silicate jimthompsonite $(Mg, Fe)_5[Si_6O_{16}](OH)_2$. In

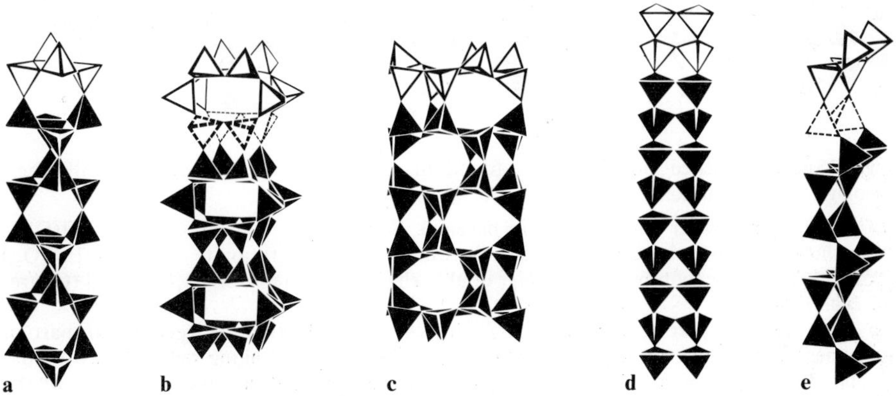

Fig. 7.13 a–e. Tubular chains found in crystalline silicates.
a Narsarsukite, $Na_4Ti_2[Si_8O_{20}]O_2$; **b** miserite, $K_2Ca_{10}(Y,RE)_2[Si_2O_7]_2[Si_{12}O_{30}](OH)_2F_2$; **c** synthetic $K_8Cu_4[Si_{16}O_{40}]$; **d** synthetic $Li_4[SiGe_3O_{10}]$; **e** epididymite, $Na_2Be_2[Si_6O_{15}]\cdot H_2O$

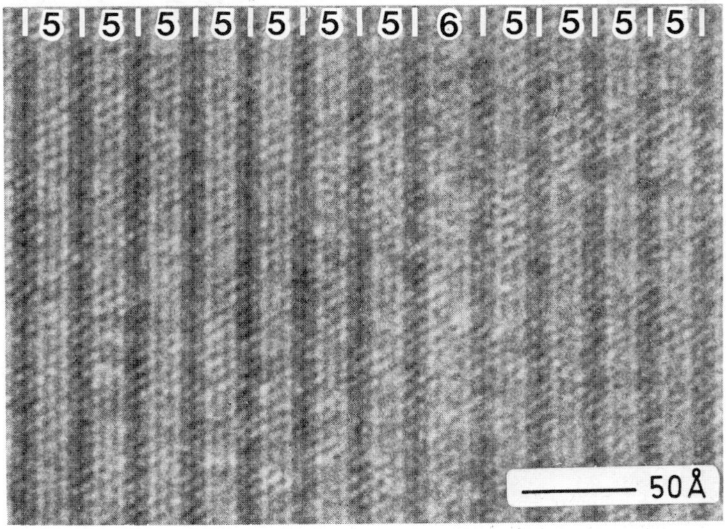

Fig. 7.14. Sixfold chains as lattice faults in the ideal sequence of fivefold chains of a $Ba_6[^2Si_{10}O_{26}]$ crystal (from Czank and Buseck 1980)

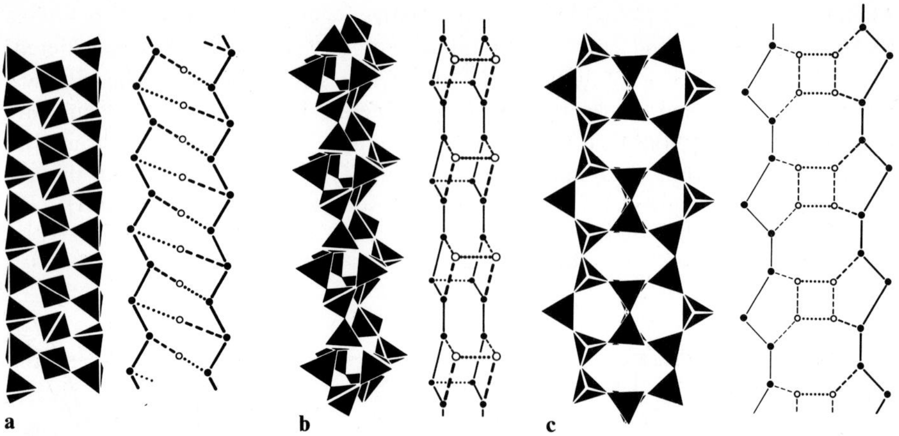

a b c

Fig. 7.15 a–c. Branched double chains found in crystalline silicates together with their topological representations.

a Open-branched zweier double chain in bavenite, $Be_2Ca_4Al_2[Si_3O_{10}]\{\boldsymbol{oB}, 2^1_\infty\}[^2Si_6O_{16}](OH)_2$ (Cannillo et al. 1966 b); **b** loop-branched dreier double chain in litidionite, $Na_2K_2Cu_2[^3Si_8O_{20}]$ (Pozas et al. 1975); **c** loop-branched dreier double chain in synthetic $Li_4Ca_4[^3Si_{10}O_{26}]$ (Castrejón et al. 1983).

–●——●– Linear part of fundamental chains; ●----○ linking between linear part of fundamental chain and branch; ●·····●, ●·····○, and ○·····○ linking between two fundamental chains. ● Si atom of the linear part of the fundamental chains, ○ Si atoms of the branches

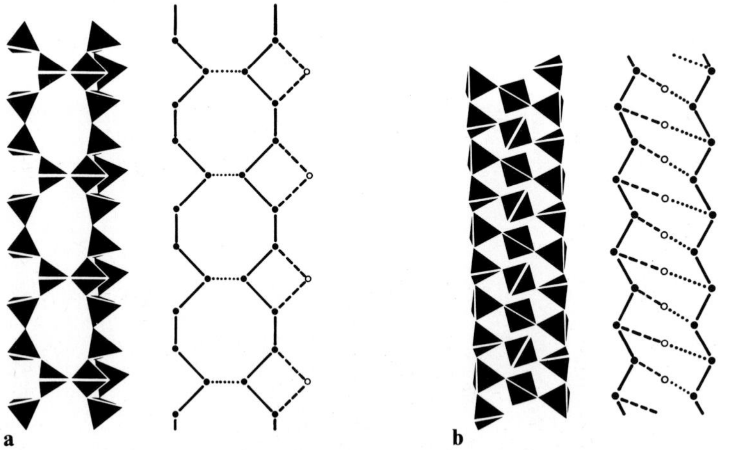

a b

Fig. 7.16 a, b. Hybrid double chains found in crystalline silicates together with their topological representations.

a Hybrid dreier double chain in tinaksite, $HNaK_2Ca_2Ti[^3Si_7O_{19}]O$ (Bissert 1980); **b** chain in bavenite, $Be_2Ca_4Al_2[Si_3O_{10}][Si_6O_{16}](OH)_2$, regarded as hybrid zweier double chain (cf. Fig. 7.15 a).

–●——●– Linear part of fundamental chains; ●----○ linking between linear part of fundamental chain and branch; ●·····●, ●·····○, and ○·····○ linking between two fundamental chains. ● Si atom of the linear part of the fundamental chains, ○ Si atom of the branches

such pyriboles Veblen and Buseck (1979), among others, obtained electron microscopic images of multiple chains with multiplicities as high as $\boldsymbol{M} = 333$.

While in these silicates multiple chains with $\boldsymbol{M} \geq 6$ were present as lattice faults within crystals of lower periodicity, stable crystalline chain silicate phases have so far been reported only with $\boldsymbol{M} \leq 5$.

7.2.2.4.2 ***Branched Multiple Chain Silicates.*** The number of branched multiple chain silicates documented to date is still quite small. They include six silicates having loop-branched dreier double chains and one silicate with open-branched zweier double chains (Fig. 7.15, Table 10.8).

7.2.2.4.3 ***Hybrid Multiple Chain Silicates.*** The silicate anion found in tinaksite can be constructed by linking an unbranched dreier single chain with a loop-branched dreier single chain (Fig. 7.16a). It is described by the formula $\{\boldsymbol{hB}, 2^1_\infty\}[^3Si_6O_{17}(SiO_2)]$.

In Fig. 7.15a the anion found in bavenite was regarded a condensation product of two topologically equivalent open-branched zweier single chains $\{\boldsymbol{oB}, 1^1_\infty\}[Si_2O_6(SiO_3)]$ and was, therefore, classified as a branched double chain $\{\boldsymbol{oB}, 2^1_\infty\}[^2Si_6O_{16}]$. As shown in Fig. 7.16b this chain can also be described as the product of the condensation of an unbranched zweier single chain $\{\boldsymbol{uB}, 1^1_\infty\}[Si_2O_6]$ with an open-branched zweier single chain $\{\boldsymbol{oB}, 1^1_\infty\}[Si_2O_6(SiO_3)_2]$ and, therefore, it can be classified as a hybrid double chain $\{\boldsymbol{hB}, 2^1_\infty\}[^2Si_6O_{16}]$. It seems reasonable, however, to prefer the former description with higher symmetry for the anion topology.

7.2.2.5 Single Layer Silicates (Monophyllosilicates)

Although crystalline silicates containing multiple chains with $\boldsymbol{M} \geq 6$ have not as yet been found, the lattice faults described in Sec. 7.2.2.4.1 indicate that a more or less continuous transition series from single chain to single layer silicates is structurally possible, even if such materials are not stable in the thermodynamic sense. Indeed, in nature, alteration of pyroxenes and amphiboles to the single layer sili-

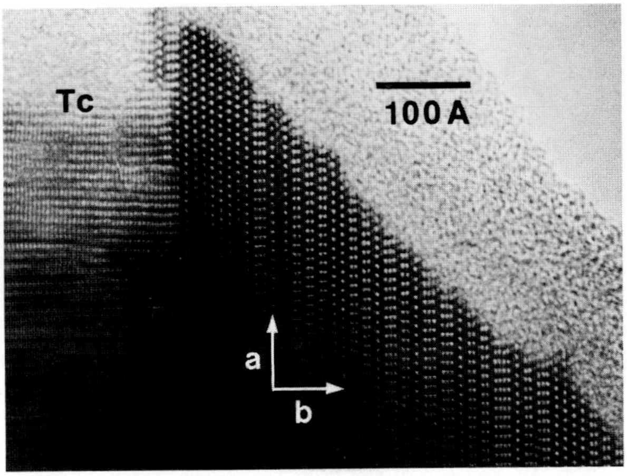

Fig. 7.17. High resolution electron micrograph of the boundary region between orthopyroxene and talc in an oriented intergrowth. The boundary region contains chains of single, double, triple, and higher multiplicity (from Veblen and Buseck 1981)

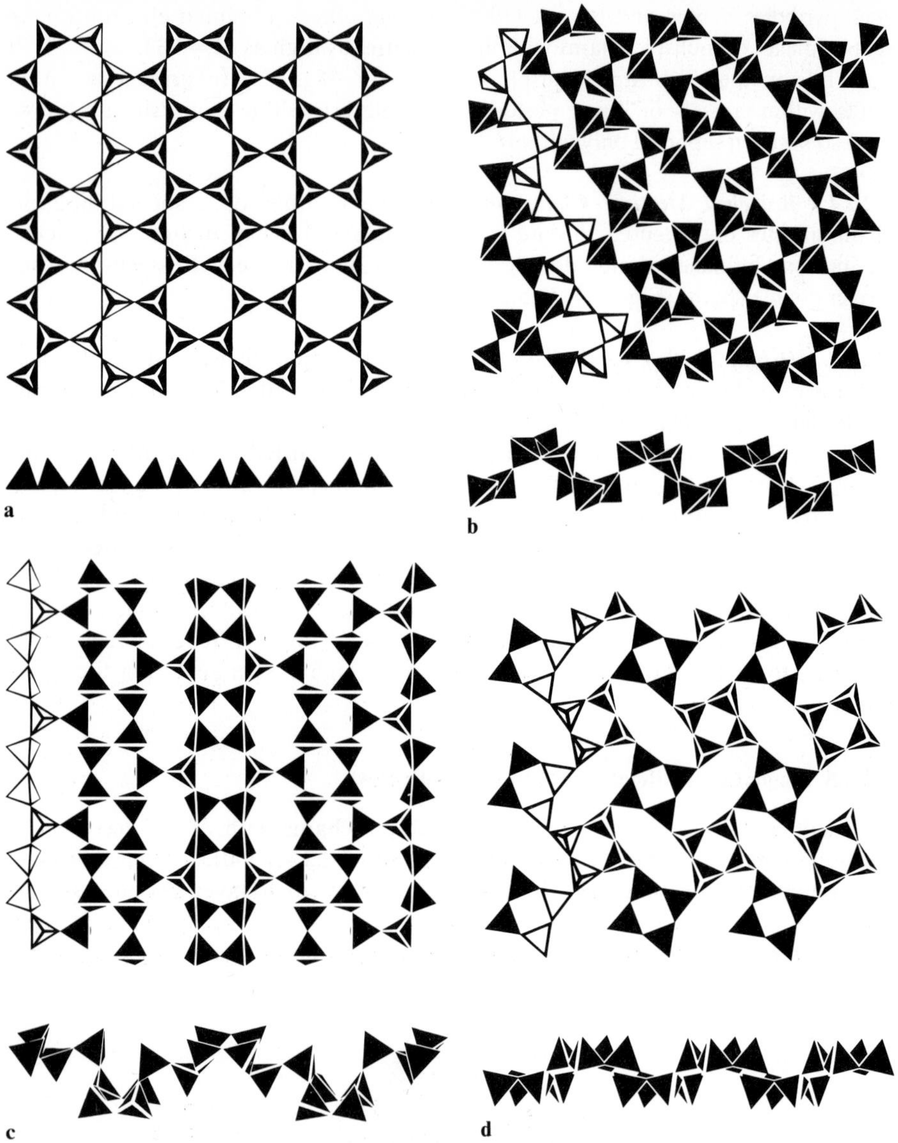

Fig. 7.18 a–d

cate talc, $Mg_3\{\boldsymbol{uB}, 1^2_\infty\}[^2\,Si_2O_5]_2(OH)_2$, can take place by stepwise condensation of single chains to multiple chains of increasing multiplicity (Veblen and Buseck 1981; Akai 1982). Therefore, the stepwise linking of silicate single chains as represented in the second turn of the spiral staircase of Fig. 4.13 is not merely hypothetical, but rather an actual reaction path. A high resolution electron microscopic image (Fig. 7.17) of a region where an orthopyroxene crystal is intergrown with talc in exact crystallographic orientation and in which single, double, triple, and quadruple chains can be seen near the boundary pyroxene-talc strongly suggests such an interpretation.

e f

Fig. 7.18 a–f. Unbranched single layers of different periodicity containing only tertiary tetrahedra found in crystalline silicates (for references see Table 10.11).

a Unbranched zweier single layer in micas, e.g., muscovite, $KAl_2[^2(AlSi_3)O_{10}](OH)_2$; **b** unbranched dreier single layer in dalyite, $K_2Zr[^3Si_6O_{15}]$; **c** unbranched dreier single layer in synthetic $H_2NaNd[^3Si_6O_{15}] \cdot nH_2O$; **d** unbranched vierer single layer in apophyllite, $KCa_4[^4Si_4O_{10}]_2(F,OH) \cdot 8H_2O$; **e** the unbranched vierer single layer in the mixed-anion silicate reyerite, $(Na,K)Ca_7\{\boldsymbol{uB}, 1^2_\infty\}[^4Si_4O_{10}]\{\boldsymbol{uB}, 2^2_\infty\}[^4(AlSi_7)O_{19}](OH)_4 \cdot 3H_2O$ (Merlino 1972a); **f** unbranched sechser single layer in manganpyrosmalite, $Mn_8[^6Si_6O_{15}](OH,Cl)_{10}$

7.2.2.5.1 ***Unbranched Single Layer Silicates.*** According to the definition of layer types based on the periodicity of the fundamental chain from which the layer can be generated by successive linking of these chains, the layers pictured in Fig. 7.18 are unbranched zweier, dreier, vierer, and sechser single layers. Unbranched fünfer single layers have recently been found in synthetic $LiBa_9[^5Si_{10}O_{25}][CO_3]Cl_7$ (Il'inets et al. 1983).

If we consider single layers as multiple chains with multiplicity $\boldsymbol{M} \rightarrow \infty$, then the general formula

$$\{\boldsymbol{uB}, \boldsymbol{M}^1_\infty\}[^{\boldsymbol{P}}\mathrm{Si}_{mp}\mathrm{O}_{3mp-\sum_1^{m-1} l_i}]^{-2(mp-\sum_1^{m-1} l_i)} \quad \text{with} \quad m = \boldsymbol{M} \text{ and } p = \boldsymbol{P}$$
$$\text{and} \quad m-1 \leq \sum_1^{m-1} l_i \leq mp, \qquad (6)$$

derived for unbranched multiple chains approaches

$$\{\boldsymbol{uB}, 1^2_\infty\}[^{\boldsymbol{P}}\mathrm{Si}_{np}\mathrm{O}_{3np-\sum_1^{n} l_i}]^{-2(np-\sum_1^{n} l_i)}. \qquad (7)$$

Here l is the number of linkages between adjacent fundamental chains per repeat units of these chains, and $p = \boldsymbol{P}$, finite, with $n \leq \sum_1^n l_i \leq np$ and $3np - \sum_1^n l_i$ is the smallest possible integer.

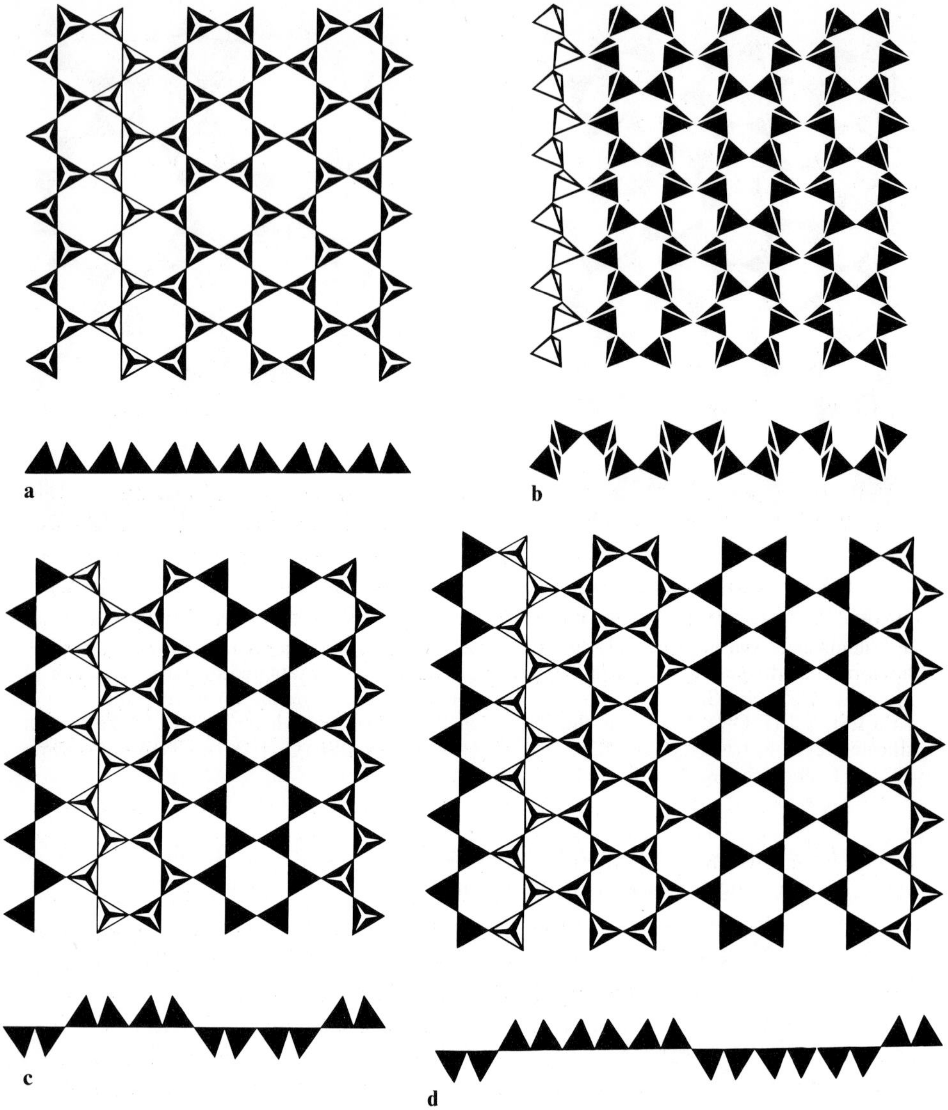

Fig. 7.19 a – d. Several unbranched zweier single layers found in crystalline silicates (cf. Fig. 7.20).

a Mica, e.g., muscovite, $KAl_2[^2(AlSi_3)O_{10}](OH)_2$; **b** sanbornite, $Ba[^2Si_2O_5]$(lT); **c** palygorskite, $Mg_5[^2Si_2O_5]_4(OH)_2 \cdot 8\,H_2O$; **d** sepiolite, $Mg_4[^2Si_2O_5]_3(OH)_2 \cdot 4\,H_2O$

From formula (7) it is clear that the number of single layers with different composition increases rapidly as the periodicity of the fundamental chain increases. In much the same way as Table 7.6 was derived for the various unbranched multiple chains, a corresponding table could be prepared for unbranched single layers. However, almost all of the silicate single layers known to exist have an Si:O ratio of 2:5 so that a table of this kind would largely be hypothetical.

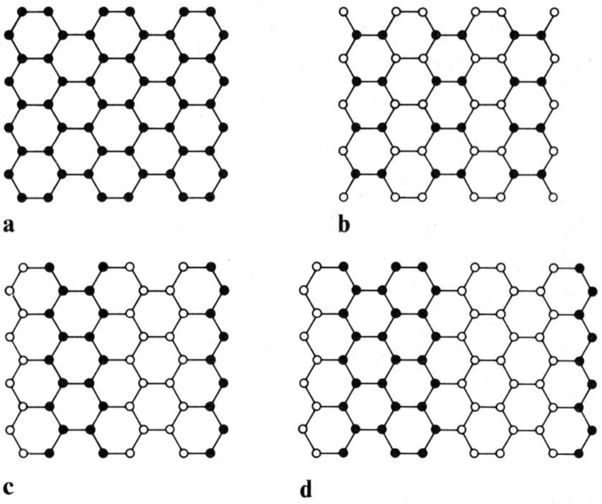

Fig. 7.20 a – d. Idealized representation of the topology of the unbranched zweier single layers in Fig. 7.19. The directedness Δ of each tetrahedron is indicated by either ● for $\mathbb{U}$ or ○ for $\mathbb{D}$.

a Mica; **b** sanbornite; **c** palygorskite; **d** sepiolite

P, n, and *l* are not the only parameters which can be used to distinguish between topologically different tetrahedral single layers, as indicated in Fig. 7.19 for various zweier single layers. In the mica layer the terminal oxygen atoms of all the $[SiO_4]$ tetrahedra are on the same side of the tetrahedral layer, whereas in sanbornite, the room temperature polymorph of $Ba[Si_2O_5]$, the terminal oxygen atoms within each fundamental zweier chain alternate between both sides of the layer. In palygorskite all terminal oxygen atoms of every two adjacent zweier chains are on one side, while those of the next two chains lie on the other side of the layer. In sepiolite all tetrahedra of every three zweier chains point to the same side of the layer.

If we assign a dot to a tetrahedron in the layers of Fig. 7.19, if its apical oxygen atom, O_{term}, points up, and assign a circle if it points down, then these layers can be represented schematically by the idealized diagrams in Fig. 7.20.

This new parameter, which we might call *directedness*, Δ, can have only two values, $\mathbb{U}$ for upwards and $\mathbb{D}$ for downwards. Following the condensation sequence in the staircase model of Fig. 4.13, the directedness comes into effect, at least theoretically, as soon as tertiary $[SiO_4]$ tetrahedra occur in cyclic and one-dimensionally infinite anions, i.e., in multiple rings and chains. In multiple ring anions $\mathbb{U}$ and $\mathbb{D}$ would distinguish between tetrahedra which point away from, and those which point towards, the interior of the cage formed by the tetrahedra. Since, in general, the "surface" of a multiple ring cage is convex it is very unlikely that terminal oxygen atoms will be located on the inner surface of the cage and, in fact, no example of $\mathbb{D}$-type multiple rings is known. However, for very large cages it is possible that, under special conditions, some tetrahedra may point to the center of the cage.

The directedness parameter Δ is, of course, also applicable in all tetrahedral layers, single as well as multiple, and in all tetrahedral frameworks. It adds considerably to the variety of possible silicate anions, the number of which is already immense due to the different combinations of the parameters coordination number ***CN***, linkedness ***L***, connectedness ***s***, branchedness ***B***, multiplicity ***M***, dimensionality ***D***, periodicity ***P*** and $\boldsymbol{P^r}$ and distinction between cyclic, ***r***, and linear, ***t***, anions.

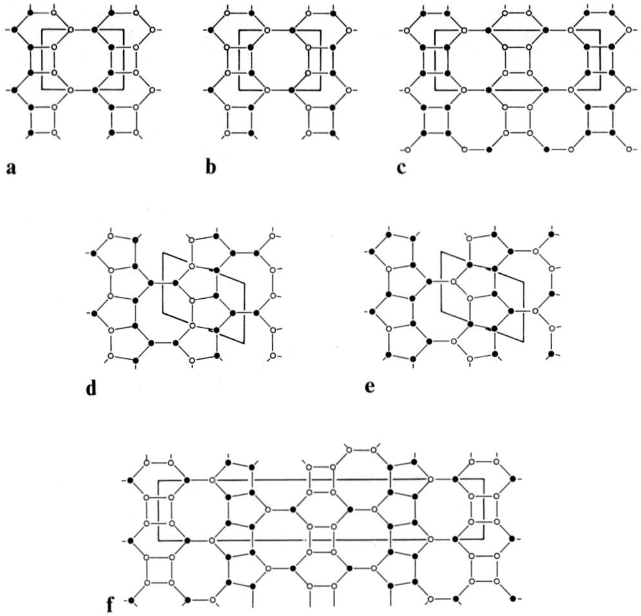

Fig. 7.21 a – f. Topology of unbranched dreier single layers of different directedness Δ. The directedness of each tetrahedron is indicated by either ● for $\mathbb{U}$ or ○ for $\mathbb{D}$.

a Dalyite, $K_2Zr[Si_6O_{15}]$ (Fleet 1965); **b** armstrongite, $CaZr[Si_6O_{15}] \cdot 2.5\,H_2O$ (Kashaev and Sapozhnikov 1978); **c** sazhinite, $HNa_2Ce^{+3}[Si_6O_{15}] \cdot n\,H_2O$ (Shumyatskaya et al. 1980); **d** nekoite, $Ca_3[Si_6O_{15}] \cdot 7H_2O$ (Alberti and Galli 1980); **e** okenite, $Ca_{10}\{\boldsymbol{uB}, 2^{1}_{\infty}\}[Si_6O_{16}]\{\boldsymbol{uB}, 1^{2}_{\infty}\}[Si_6O_{15}]_2 \cdot 18\,H_2O$ (Merlino 1983); **f** synthetic $H_2NaNd[Si_6O_{15}] \cdot n\,H_2O$ (Karpov et al. 1977b)

Figure 7.21 shows several different dreier single layers in order to further demonstrate the influence of the directedness of $[SiO_4]$ tetrahedra. All these silicates are members of the family specified by $\boldsymbol{N_{an}} = 1$, $\boldsymbol{CN} = 4$, $\boldsymbol{L} = 1$, $\boldsymbol{uB}$, $\boldsymbol{M} = 1$, $\boldsymbol{D} = 2$, and $\boldsymbol{P} = 3$, that is, the family with the formula $\{\boldsymbol{uB}, 1^{2}_{\infty}\}[{}^{3}Si_6O_{15}]$. In three of these silicates, dalyite, armstrongite and sazhinite, even the way of linking the fundamental dreier chains is identical, with the result that the single layers have the same pattern of vierer, sechser, and achter rings. They differ, however, in the distribution of tetrahedra pointing up and down, i.e., in the directedness of the tetrahedra. Similarly, the dreier single layers of okenite and nekoite contain silicate anions with an identical pattern of fünfer, sechser, and achter rings (different from the ring pattern of the other three silicates), but the directedness of the tetrahedra is different. In synthetic $H_2NaNd[Si_6O_{15}] \cdot n\,H_2O$ the topology of the layer is a rather complicated pattern of vierer, fünfer, sechser, and achter rings.

In the silicates known to date only unbranched zweier, dreier, vierer, fünfer, and sechser single layers have been discovered. Einer single layers built from $[ZnO_4]$ tetrahedra have, however, been found in $Sr[ZnO_2]$ (Schnering and Hoppe 1961).

Almost all the known single layers are composed only of tertiary $[SiO_4]$ tetrahedra. Such layers necessarily have the composition $\{\boldsymbol{uB}, 1^{2}_{\infty}\}[Si_{2n}O_{5n}]^{-2n}$ and have been given in Figs. 7.18 to 7.21. However, in recent years, a few examples of

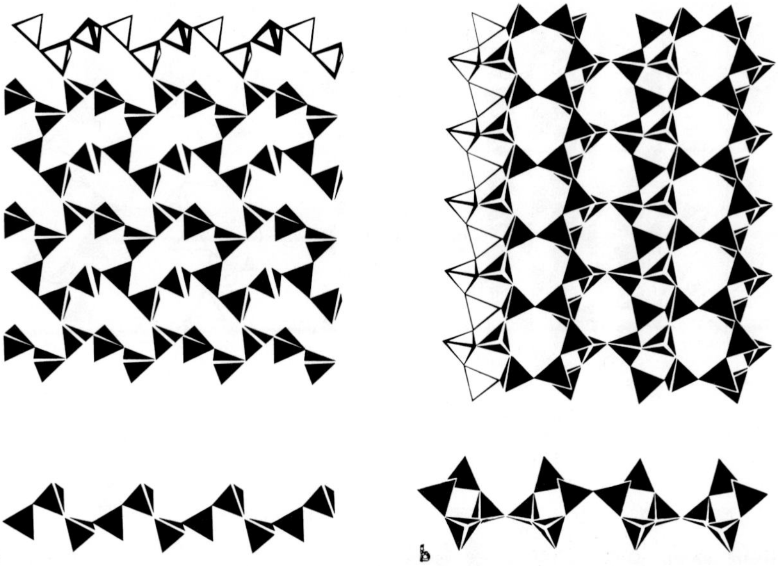

a b

Fig. 7.22 a, b. Unbranched silicate single layers containing nontertiary $[SiO_4]$ tetrahedra. **a** Unbranched dreier single layer in synthetic $Na_2Zn[Si_3O_8]$; **b** unbranched zweier single layer in synthetic $K_2[^2Si_4O_9]$

silicate single layers have been discovered which contain more than one type of $[SiO_4]$ tetrahedra. The composition of the layers is then determined by the connectedness s of the tetrahedra in the layer and the ratio between the different types of Q^s groups. For example, the single layer of $Na_2Zn\{\boldsymbol{uB}, 1^2_\infty\}[^3Si_3O_8]$ (Fig. 7.22a) is composed of secondary and tertiary $[SiO_4]$ tetrahedra in the ratio 1:2 (Hesse et al. 1977) while that of synthetic $K_2\{\boldsymbol{uB}, 1^2_\infty\}[^2Si_4O_9]$ (Fig. 7.22b) is composed of tertiary and quaternary tetrahedra Q^3 and Q^4 in the ratio 1:1 (Schweinsberg and Liebau 1974), resulting in a silicon : oxygen ratio in the layer different from the usual ratio of 2:5.

7.2.2.5.2 Branched Single Layer Silicates. Within the last two decades a number of open- and loop-branched silicate single layers have been discovered, several of which are illustrated in Fig. 7.23. In theory their atomic ratio Si : O can vary considerably: in the layers found so far it has values between 2:5.71 (meliphanite) and 2:4.67 ($NaPr[Si_6O_{14}]$).

Whether a branched single layer is considered to be open-branched, loop-branched, mixed-branched, or hybrid is often a subjective decision. For example, the single layer in semenovite (Mazzi et al. 1979) (Fig. 7.24), may be considered to belong to any one of these four categories, depending on which of the tetrahedra are regarded as branches to which linear chain. Sechser chains parallel to $\vec{a}$ and $\vec{b}$ are the chains of lowest periodicity within the layer. Since the lattice constant $b_0 = 13.835$ Å is smaller (although only slightly) than $a_0 = 13.879$ Å, the sechser chain parallel to $\vec{b}$ is chosen as the fundamental chain rather than the one parallel to $\vec{a}$. Two fundamental chains run through the unit cell, but due to the mirror

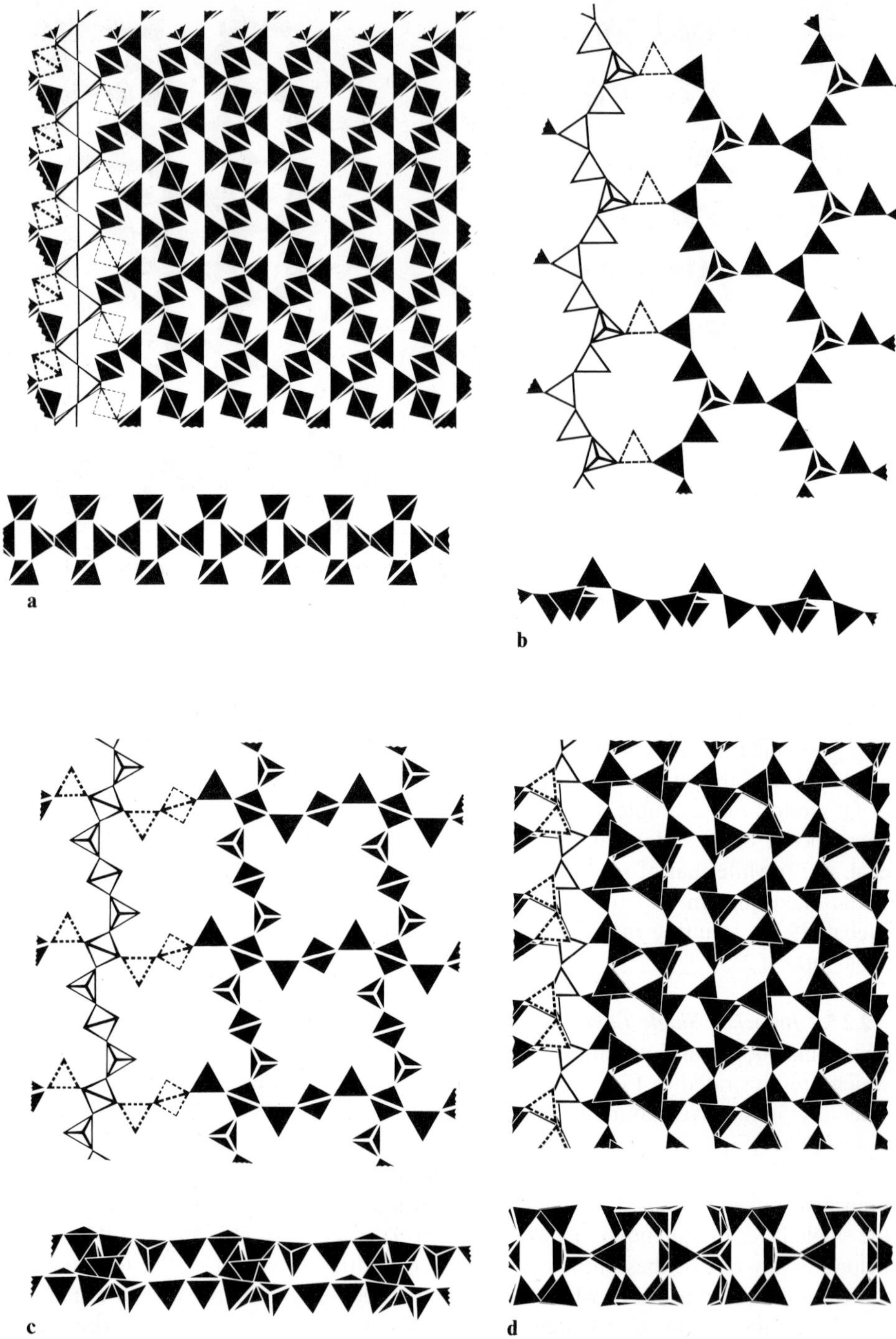

Fig. 7.23 a – d

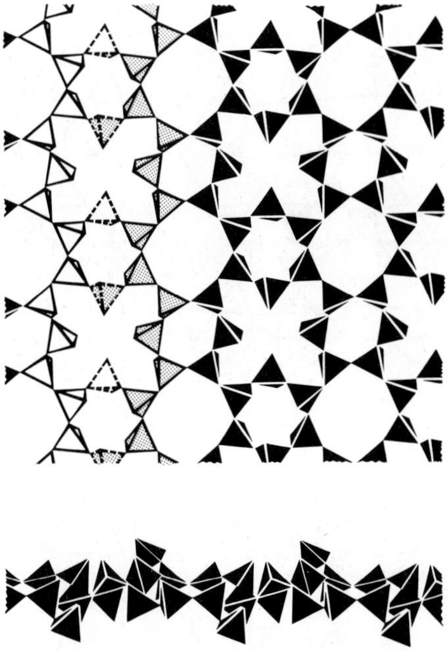

e

Fig. 7.23 a – e. Branched single layers found in crystalline silicates (for references see Table 10.8).

a Open-branched zweier single layer in prehnite, $Ca_2Al[^2(AlSi_3)O_{10}](OH)_2$; **b** open-branched vierer single layer in zeophyllite, $Ca_{13}[^4Si_5O_{14}]_2F_8(OH)_2 \cdot 6H_2O$; **c** open-branched vierer single layer in meliphanite, $Ca_4(Na,Ca)_4Be_4[SiO_4][^4Si_7O_{20}]F_4$; **d** loop-branched vierer single layer in synthetic $NaPr[^4Si_6O_{14}]$; **e** open-branched fünfer single layer in synthetic $K_8Yb_3[^5Si_6O_{16}]_2(OH)$

planes *m* perpendicular to $\vec{a}$ these chains are crystallographically equivalent, independent of whether open-branched or loop-branched chains are chosen (Fig. 24a, b, c).

Classifying semenovite as a hybrid single layer silicate requires two different fundamental chains: a branched one (white tetrahedra in Fig. 7.24f) and an unbranched one (stippled tetrahedra), thus ignoring the crystallographic symmetry. This is clearly not acceptable. If the semenovite layer is considered to be mixed-branched, i.e., to have open as well as loop branches, there are two essentially different possibilities: while in Fig. 7.24e the two fundamental mixed-branched chains are symmetrically equivalent, in Fig. 7.24d they are not, so that this latter possibility is ruled out. Keeping in mind the rule that a loop-branched fundamental chain is preferred over an open-branched one (step 7 for phyllosilicates in the silicate classification procedure, p. 83), semenovite would then be classified as the loop-branched sechser single layer silicate represented in Fig. 7.24a.

No single layer silicate has yet been found which, according to these rules, should be classified as a hybrid silicate.

General formulae for silicate branched single layers, and both branched and unbranched double layers and frameworks, become very complicated and, therefore, no attempt is made here to deduce such general formulae.

7.2.2.6 Double Layer Silicates (Diphyllosilicates)

A small number of compounds have been determined to contain silicate anions which can be regarded as a product of the condensation of two single layers. In

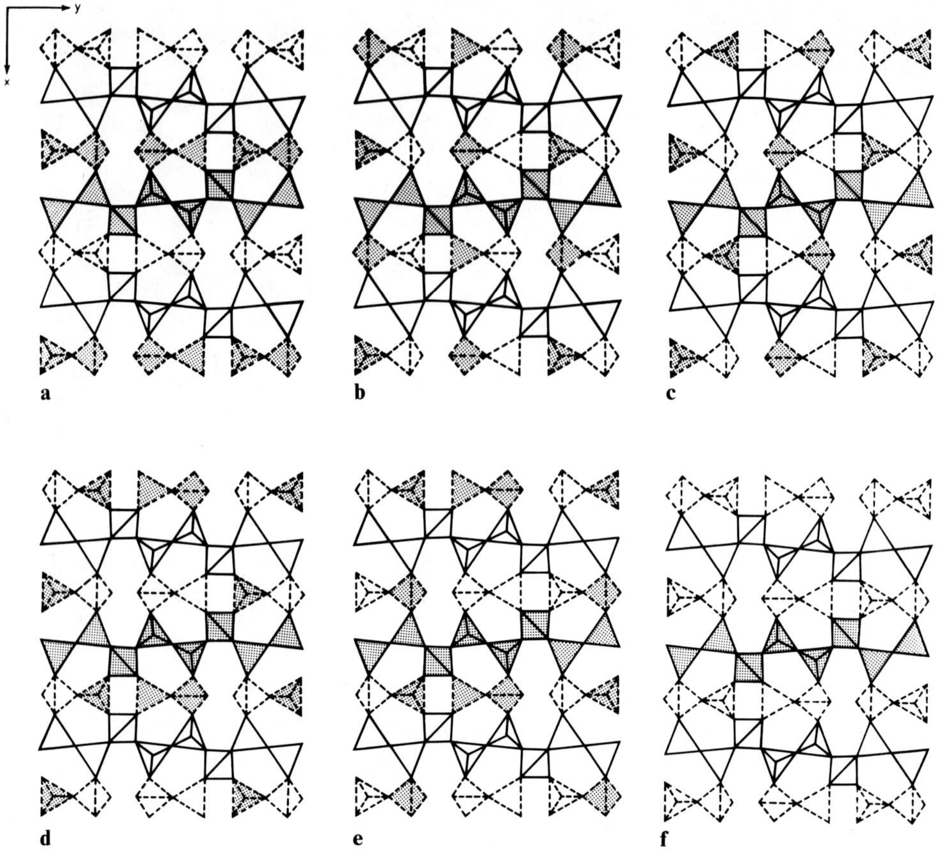

Fig. 7.24 a – f. Different ways to describe the single layer found in semenovite, $H_xNa_{0-2}(Ca,Na)_8(Fe,Mn,Zn,Ti)RE_2[^6(Si,Be)_{10}(O,F)_{24}]_2$ (Mazzi et al. 1979).

a Loop-branched sechser single layer $\{\boldsymbol{lB}, 1^2_\infty\}[^6T_{10}O_{24}]$; **b** and **c** open-branched sechser single layers $\{\boldsymbol{oB}, 1^2_\infty\}[^6T_{10}O_{24}]$; **d** and **e** mixed-branched sechser single layers $\{\boldsymbol{olB}, 1^2_\infty\}[^6T_{10}O_{24}]$; **f** hybrid sechser single layer $\{\boldsymbol{hB}, 1^2_\infty\}[^6T_{10}O_{24}]$

Fig. 7.25 a – d. Unbranched double layers found in crystalline silicates.

a Unbranched zweier double layer in hexacelsian, $Ba[^2(AlSi)O_4]_2$(hT). Two single layers of the kind shown are linked via the apical oxygen atoms marked by *dots* and lying on a mirror plane. **b** Unbranched vierer double layer in naujakasite, $Na_6Fe[^4(Al_4Si_8)O_{26}]$; **c** the unbranched vierer double layer in the mixed-anion silicate reyerite, $(Na,K)Ca_7\{\boldsymbol{uB}, 1^2_\infty\}[^4Si_4O_{10}]$ $\{\boldsymbol{uB}, 2^2_\infty\}[^4(AlSi_7)O_{19}](OH)_4 \cdot 3H_2O$. There is a pseudo mirror plane between the two sublayers. **d** Unbranched neuner double layer in stilpnomelane, $KFe_{24}[^9(AlSi_{35})O_{84}]O_6(OH)_{24} \cdot nH_2O$. As in **a** two such single layers are linked via the apical oxygen atoms pointing upwards in the layer. The side view shows only the tetrahedra within the slab that is indicated by the *double arrow* in the vertical projection. The scale is reduced by a factor of 2 compared to **a**, **b**, and **c**

a

b

c

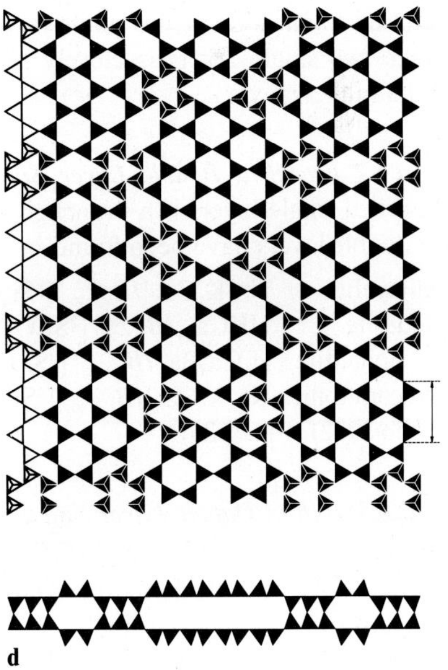

d

general, there is a mirror plane or a pseudo mirror plane between the two single layers. Depending on the type of fundamental chain from which the layers can be generated they are subdivided into unbranched and branched double layers. Hybrid double layer silicates are still unknown. Again, the periodicity of the fundamental chains is used for classification of the double layer and the framework silicates.

7.2.2.6.1 ***Unbranched Double Layer Silicates.*** When the apical oxygen atoms of a single layer all point to the same side of the layer (as in the case of the so-called mica layer of Fig. 7.18a) then two layers can be linked via oxygen atoms from all tetrahedra. If the single layers have the composition $[T_{2n}O_{5n}]$, then the resulting double layer has a T:O ratio of 1:2. Such unbranched zweier double layers (Fig. 7.25a) have been described for the hexacelsian-type high temperature phases of feldspar composition, $CaAl_2Si_2O_8$(hT), $SrAl_2Si_2O_8$(hT), and $BaAl_2Si_2O_8$(hT). In these compounds half of the silicon atoms are replaced by aluminum atoms, and the negatively charged double layers are held together by cations (see Sec. 10.6 and Table 10.17).

When the two single layers are linked via only a fraction of their tetrahedra, then the T:O ratio of the double layer is smaller than 1:2. This explains why the unbranched vierer and neuner double layers found in naujakasite $Na_6Fe\{\boldsymbol{uB}, 2^2_\infty\}[^4(Al_4Si_8)O_{26}]$ (Basso et al. 1975), reyerite $(Na,K)Ca_7\{\boldsymbol{uB}, 1^2_\infty\}[^4Si_4O_{10}]\{\boldsymbol{uB}, 2^2_\infty\}[^4(AlSi_7)O_{19}](OH)_4 \cdot 3\,H_2O$ (Merlino 1972a), and stilpnomelane $KFe_{24}\{\boldsymbol{uB}, 2^2_\infty\}[^9(AlSi_{35})O_{84}]O_6(OH)_{24} \cdot n\,H_2O$ (Eggleton 1972) (Fig. 7.25) have T:O ratios of 1:2.167, 1:2.375, and 2.333, respectively, instead of 1:2 in the hexacelsian-type phases $Ba\{\boldsymbol{uB}, 2^2_\infty\}[^2(AlSi)O_4]_2$ and $Ca\{\boldsymbol{uB}, 2^2_\infty\}[^2(AlSi)O_4]_2$.

In the double layers observed to date, the number of linkages per tetrahedron between two single layers is equal to the number of tetrahedra per tetrahedron pointing to one side of the single layer. In other words, the number of linkages between two single layers is closely related to the directedness of the $[SiO_4]$ tetrahedra in these layers.

7.2.2.6.2 ***Branched Double Layer Silicates.*** The result of the condensation of two branched single layers is a branched double layer. Loop-branched dreier and sechser double layers have been found in

delhayelite $Na_3K_7Ca_5\{\boldsymbol{lB}, 2^2_\infty\}[^3Si_6O_{17}((Al,Si)O)_2]_2F_4Cl_2$ (Cannillo et al. 1969),
rhodesite $HKCa_2\{\boldsymbol{lB}, 2^2_\infty\}[^3Si_6O_{17}(SiO)_2] \cdot (6-n)H_2O$ (Hesse 1979b),
macdonaldite $H_2Ca_4Ba\{\boldsymbol{lB}, 2^2_\infty\}[^3Si_6O_{17}(SiO)_2]_2 \cdot (8+x)H_2O$ (Cannillo et al. 1968),
and carletonite $Na_8K_2Ca_8\{\boldsymbol{lB}, 2^2_\infty\}[^6Si_{12}O_{33}(SiO)_2(SiO_{0.5})_2][CO_3]_8(OH,F)_2 \cdot 2\,H_2O$ (Chao 1972) (Fig. 7.26).

7.2.2.6.3 ***Hybrid Double Layer Silicates.*** Although hybrid anions are extremely rare, latiumite $K_{3.4}\square_{0.6}Ca_{12}\{\boldsymbol{hB}, 2^2_\infty\}[^2(Al_{10.8}Si_{5.2})O_{36}(Al_{0.15}Si_{0.85}O_2)_4][SO_4]_{2.8}[CO_3]_{1.2}$ and tuscanite $(K, H_2O, Sr)_4(Ca, Na, Fe^{+3}, Mg)_{12}\{\boldsymbol{hB}, 2^2_\infty\}[^2(Al,Si)_{16}O_{36}((Al,Si)O_2)_4][SO_4]_{2.8}[CO_3]_{1.0}[O_4H_4]_{0.2}$, which have overall Si:Al ratios of 1:1.33 and 1:0.58, respectively, may be described as containing hybrid zweier double layers (Fig. 7.27) if the fact is disregarded that some of the tetrahedral sites are almost completely occu-

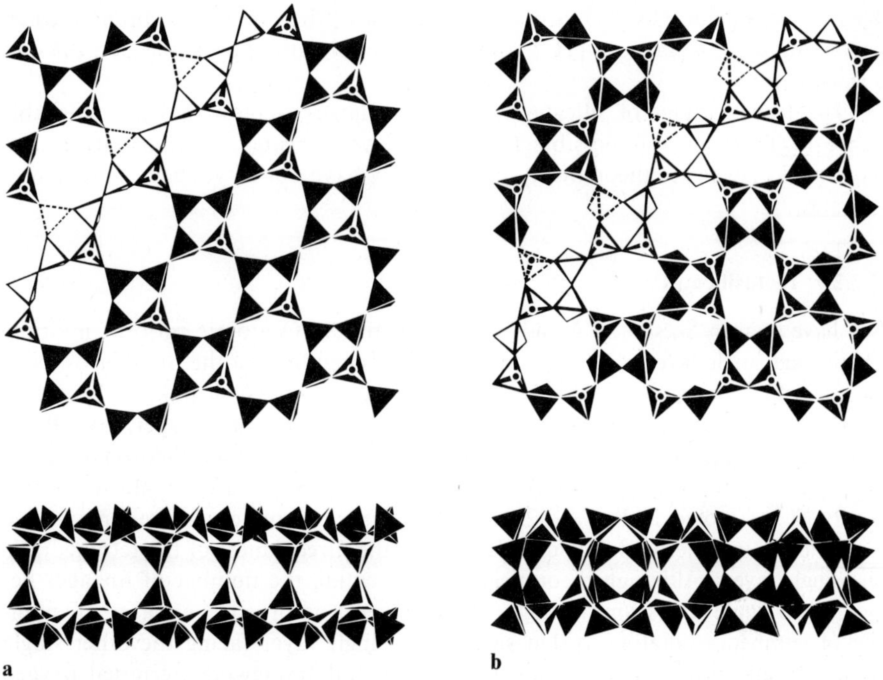

Fig. 7.26 a, b. Branched double layers found in crystalline silicates. Two single layers of the type shown are linked via the apical oxygen atoms marked by *dots* and lying on a mirror or pseudo mirror plane.

a Loop-branched dreier double layer in delhayelite, $Na_3K_7Ca_5[{}^3Si_6O_{17}((Al,Si)O)_2]_2F_4Cl_2$; **b** loop-branched sechser double layer in carletonite, $Na_8K_2Ca_8[{}^6Si_{12}O_{33}(SiO)_2(SiO_{0.5})_2][CO_3]_8(OH,F)_2 \cdot 2H_2O$.

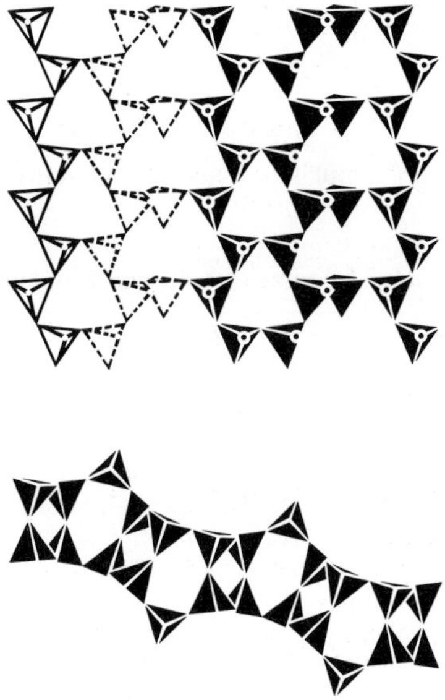

Fig. 7.27. The hybrid zweier double layer in latiumite, $(K_{1.7}\square_{0.3})Ca_6[(Al_{5.7}Si_{4.3})O_{22}][SO_4]_{1.4}[CO_3]_{0.6}$ (Cannillo et al. 1973a) and tuscanite, $(K,Sr,H_2O)_2(Ca,Na,Mg,Fe)_6[(Al_{3.66}Si_{6.34})O_{22}][SO_4]_{1.4}[CO_3]_{0.5}[O_4H_4]_{0.1}$ (Mellini et al. 1977). Two single layers of the kind shown are linked via the apical oxygen atoms marked by *dots*

pied by aluminum. Otherwise, they would be described as mixed-anion silicates containing single tetrahedra $[TO_4]$ and double tetrahedra $[T_2O_7]$ (see Sec. 6.5 and Fig. 6.5).

The few examples of silicates with double layers suggest that a remarkable range of T:O values is possible for double layer silicates. Although triple and higher-fold layers are theoretically possible they have not as yet been observed in silicates.

7.2.2.7 Tectosilicates

We have seen in Secs. 7.2.2.4 and 7.2.2.5 that the Si:O atomic ratio of multiple chains and single layers can vary substantially, depending on the number of tetrahedra in the repeat unit of the fundamental chain involved in linking the chains into a complex anion. Similarly during formation of a double layer, not all the tetrahedra in each single layer necessarily become linked to a tetrahedron of an adjacent layer. This gives rise to further variation in the Si:O ratio of silicate double layers. It was noted in Sec. 7.2.2.6.1 that the fraction of tetrahedra linking two single layers together is closely correlated with the directedness of the tetrahedra in the single layers. Although theoretically not essential, the number of linkages between two single layers constituting a double layer is, in practice, equal to the number of tetrahedra pointing to that side of the single layer facing the other single layer. During further condensation to a tetrahedral framework, terminal oxygen atoms which point away from the double layer are used to add more layers to the nucleus. In the same way as for double layer silicates, also for tectosilicates in general, all tetrahedra are involved in bonds between adjacent layers, even though this is not absolutely necessary.

The T:O ratio of a silicate framework can vary in the range $1:3 < T:O \leq 1:2$ (Table 9.2) due to variation of the fraction of tetrahedra forming linkages between the single chains and single layers, respectively, or, in other words, the ratio between secondary, tertiary, and quarternary $[TO_4]$ tetrahedra.

Almost all silicate frameworks contain only quaternary tetrahedra and, therefore, have a formula $[T_nO_{2n}]$. The few silicate frameworks reported to have T:O $<1:2$ are given in Table 7.8. They have sometimes been called *interrupted frameworks*, but since no special name has been given to double chain silicates with

Table 7.8 Framework silicates with atomic ratios T:O < 1:2 (interrupted framework silicates)

Silicate		T:O	Reference
Name	Formula		
Wenkite	$H_2(Ba,Ca)_{10}\{\boldsymbol{hB},{}^{3}_{\infty}\}[{}^{3}(Al,Si)_{20}O_{43}][SO_4]_3\cdot H_2O$	1:2.150	Wenk 1973
Leifite	$Na_6Be_2\{\boldsymbol{oB},{}^{3}_{\infty}\}[{}^{2}(Al_2Si_{16})O_{39}](OH)_2\cdot 1.5H_2O$	1:2.167	Coda et al. 1974
Roggianite	$H_4Ca_4\{\boldsymbol{oB},{}^{3}_{\infty}\}[{}^{4}(Al_4Si_8)O_{26}](OH)_4\cdot\sim 6.5H_2O$	1:2.167	Galli 1980
Ussingite	$HNa_2\{\boldsymbol{uB},{}^{3}_{\infty}\}[{}^{4}(AlSi_3)O_9]$	1:2.250	Rossi et al. 1974
Synthetic	$Rb_6\{\boldsymbol{oB},{}^{3}_{\infty}\}[{}^{4}Si_{10}O_{23}]$	1:2.300	Schichl et al. 1973
Synthetic	$Cs_6\{\boldsymbol{oB},{}^{3}_{\infty}\}[{}^{4}Si_{10}O_{23}]$	1:2.300	Schichl et al. 1973
Synthetic	$K_2Ce\{\boldsymbol{oB},{}^{3}_{\infty}\}[{}^{4}Si_6O_{15}]$	1:2.500	Karpov et al. 1977a

Fig. 7.28 a – d. Several silicate frameworks. The scale of **a**, **b**, and **d**, is enhanced by a factor of 1.75 compared to **c** and to the majority of silicate anions shown.

a Unbranched zweier framework of tridymite; **b** unbranched dreier framework of keatite, SiO_2; **c** projection of the open-branched zweier framework of leifite; **d** loop-branched dreier framework of the feldspars, e.g., orthoclase, $K[AlSi_3O_8]$

a

b

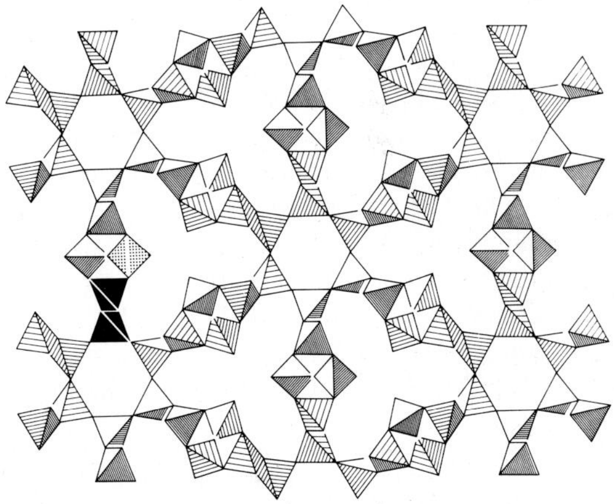

c

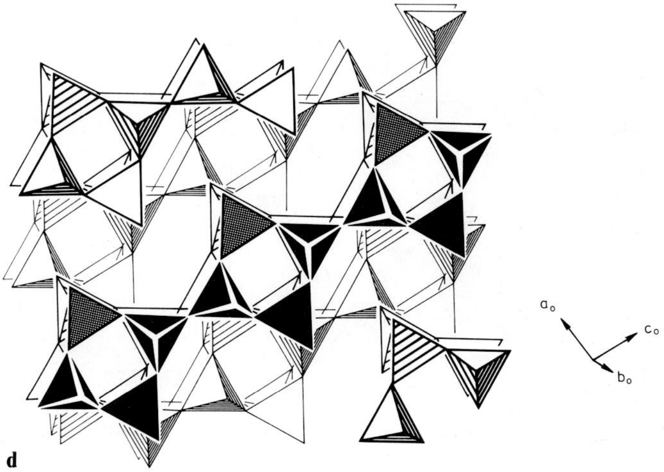

d

$T:O < 1:2.5$ or to single layer silicates with $T:O < 1:2.5$, etc., such a name is unnecessary. Nevertheless, the term " interrupted framework" immediately indicates that such a framework does not contain only quaternary $[TO_4]$ tetrahedra, i.e., that its $T:O$ ratio is less than $1:2$.

Unbranched frameworks with fundamental chain periodicities of $\boldsymbol{P} = 2$, 3, 4, and 6 exist, for example, in the silica polymorphs cristobalite and tridymite $\{\boldsymbol{uB}, {}^{3}_{\infty}\}[{}^{2}Si_2O_4]$ (Fig. 7.28 a), quartz and keatite $\{\boldsymbol{uB}, {}^{3}_{\infty}\}[{}^{3}Si_3O_6]$ (Shropshire et al. 1959) (Fig. 7.28 b), in the zeolites laumontite $Ca\{\boldsymbol{uB}, {}^{3}_{\infty}\}[{}^{3}(AlSi_2)O_6]_2 \cdot 4\,H_2O$ (Bartl and Fischer 1967), gismondine $Ca\{\boldsymbol{uB}, {}^{3}_{\infty}\}[{}^{4}(Al_2Si_2)O_8] \cdot 4\,H_2O$ (Fischer 1963), harmotome $Ba\{\boldsymbol{uB}, {}^{3}_{\infty}\}[{}^{4}(Al_2Si_6)O_{16}] \cdot 6\,H_2O$ (Sadanaga et al. 1961), and chabazite $Ca\{\boldsymbol{uB}, {}^{3}_{\infty}\}[{}^{6}(Al_2Si_4)O_{12}] \cdot 6\,H_2O$ (Mazzi and Galli 1983), and in silicalite-1 $\{\boldsymbol{uB}, {}^{3}_{\infty}\}[{}^{6}Si_6O_{12}]$, another silica polymorph (Flanigen et al. 1978; Kokotailo et al. 1978).

In comparison, leifite $Na_6Be_2\{\boldsymbol{oB}, {}^{3}_{\infty}\}[{}^{2}(Al_2Si_{16})O_{39}](OH)_2 \cdot 1.5\,H_2O$ (Coda et al. 1974) has an open-branched zweier framework (Fig. 7.28 c), and cordierite $Mg_2\{\boldsymbol{oB}, {}^{3}_{\infty}\}[{}^{4}(Al_4Si_5)O_{18}]$ has an open-branched vierer framework (Hochella et al. 1979). On the other hand, coesite $\{\boldsymbol{lB}, {}^{3}_{\infty}\}[{}^{3}Si_4O_8]$ (Gibbs et al. 1977), and the feldspars $M\{\boldsymbol{lB}, {}^{3}_{\infty}\}[{}^{3}(Al, Si)_4O_8]$ (Fig. 7.28 d) have loop-branched dreier frameworks, sodalite $Na_4\{\boldsymbol{lB}, {}^{3}_{\infty}\}[{}^{4}(Al_3Si_3)O_{12}]Cl$ (Löns and Schulz 1967) has a loop-branched vierer framework, and thomsonite $NaCa_2\{\boldsymbol{lB}, {}^{3}_{\infty}\}[{}^{6}(Al, Si)_{10}O_{20}] \cdot 6\,H_2O$ (Alberti et al. 1981) a loop-branched sechser framework.

Only a few silicate frameworks are to be regarded as hybrid frameworks. For example, wenkite $H_2(Ba, Ca)_{10}\{\boldsymbol{hB}, {}^{3}_{\infty}\}[{}^{3}(Al, Si)_{20}O_{43}][SO_4]_3 \cdot H_2O$ (Merlino 1974) is a hybrid dreier framework, the zeolite A $Na_{12}\{\boldsymbol{hB}, {}^{3}_{\infty}\}[{}^{5}(Al_{12}Si_{12})O_{48}] \cdot 27\,H_2O$ (Gramlich and Meier 1971) is a hybrid fünfer framework, and melanophlogite $\{\boldsymbol{hB}, {}^{3}_{\infty}\}[{}^{6}Si_{23}O_{46}] \cdot n\,M$, where M are small organic or inorganic molecules such as CH_4, CH_3NH_2, N_2, CO_2, Kr (Gies 1983), is a hybrid sechser framework.

7.2.2.8 Silicates with Interpenetrating Anions

Neptunite, $LiNa_2K(Fe, Mg, Mn)_2Ti_2[Si_8O_{22}]O_2$, has an interesting structure in which two identical open-branched vierer frameworks of low density are inter-

Fig. 7.29. The two interpenetrating tetrahedral frameworks of neptunite, $LiNa_2K(Fe, Mg, Mn)_2Ti_2[Si_8O_{22}]O_2$ (after Cannillo et al. 1966 a)

penetrating (Cannillo et al. 1966a) (Fig. 7.29). In this respect the neptunite structure resembles those of $Zn(CN)_2$, $Cd(CN)_2$, and three of the modifications of crystalline H_2O, tetragonal ice VI (Kamb 1965) and cubic ice VII and VIII (Kamb and Davis 1964). Since three of the polymorphs of silica are isostructural with other ice polymorphs: tridymite with hexagonal ice I, cristobalite with cubic ice Ic, and keatite with tetragonal ice III, it seems possible that a silica polymorph isostructural with ice VI (containing two interpenetrating cristobalite-like frameworks), or with ice VII and VIII or a derivative of such a silica modification, exists under certain high pressure conditions. The range of experimental conditions – pressure, temperature, starting material, etc. – for a synthesis of silica or silicates with interpenetrating tetrahedral frameworks would, however, probably be much smaller than the corresponding range for ice since silicon usually has octahedral rather than tetrahedral coordination at very high pressure, whereas octahedral coordination is unlikely for water.

Silicates with interpenetrating tetrahedral anions can be categorized in a special group of their own or, more conveniently, placed into the group appropriate for their single anions.

7.3 Mixed-Anion Silicates

Epidote $Ca_2(Fe,Al)_3[SiO_4][Si_2O_7]O(OH)$ (Carbonin and Molin 1980) and zoisite and clinozoisite $Ca_2Al_3[SiO_4][Si_2O_7]O(OH)$ (Dollase 1968) are very common rock-forming minerals which contain both single tetrahedra and double tetrahedra in the ratio 1:1, while vesuvianite $Ca_{10}(Mg, Fe)_2Al_4^{[6]}[SiO_4]_5[Si_2O_7]_2(OH)_4$ (Rucklidge et al. 1975) has single and double tetrahedra in the ratio 5:2. In contrast to these well-known minerals, other natural silicates with $[SiO_4]$ and $[Si_2O_7]$ groups, or with single and triple tetrahedra $[Si_3O_{10}]$, such as kilchoanite $Ca_6[SiO_4][Si_3O_{10}]$ (Taylor 1971), are very rare. In recent years, the complicated crystal structures of several other mixed-anion silicates have been determined to contain single or double tetrahedra together with single chains – joesmithite $Be_2Ca_2(Ca,Pb)(Fe,Mg)_5[SiO_4]_2\{\boldsymbol{uB}, 1^1_\infty\}[Si_2O_6]_2O_2(OH)_2$ (Moore 1969) – quadruple chains – miserite $K_2Ca_{10}(Y,RE)[Si_2O_7]_2\{\boldsymbol{uB}, 4^1_\infty\}[^3Si_{12}O_{30}](OH)_2F_2$ (Scott 1976) – or single layers – meliphanite $Ca_4(Na,Ca)_4Be_4[SiO_4]\{\boldsymbol{oB}, 1^2_\infty\}[^4Si_7O_{20}]F_4$ (Fig. 7.23c) (Dal Negro et al. 1967). Two kinds of single rings exist in eudialyte $Na_{12}(Ca, RE)_6(Fe, Mn, Mg)_3(Zr, Nb)_{3+x}\{\boldsymbol{uB}, 1\boldsymbol{r}\}[Si_3O_9]_2\{\boldsymbol{uB}, 1\boldsymbol{r}\}[Si_9(O, OH)_{27}]_2Cl_y$ (Golyshev et al. 1971), two types of multiple chains in chesterite $(Fe, Mg)_{17}\{\boldsymbol{uB}, 2^1_\infty\}[^2Si_4O_{11}]_2\{\boldsymbol{uB}, 3^1_\infty\}[^2Si_6O_{16}]_2(OH)_6$ (Veblen and Burnham 1978), a double chain and a single layer in okenite $Ca_{10}\{\boldsymbol{uB}, 2^1_\infty\}[^3Si_6O_{16}]\{\boldsymbol{uB}, 1^2_\infty\}[^3Si_6O_{15}]_2 \cdot 18\,H_2O$ (Merlino 1983), and even single and double layers in reyerite $(Na, K)Ca_7\{\boldsymbol{uB}, 1^2_\infty\}[^4Si_4O_{10}]\{\boldsymbol{uB}, 2^2_\infty\}[^4(AlSi_7)O_{19}](OH)_4 \cdot 3\,H_2O$ (Merlino 1972a). The synthetic silicon phosphate of composition $5\,SiO_2 \cdot 3\,P_2O_5$ is an exceptional example since it contains single $[SiO_6]$ octahedra and double tetrahedra $[Si_2O_7]$ with the structural formula $Si_3^{[6]}Si_2^{[4]}[PO_4]_6O$ (Mayer 1974) see (Sec. 7.1).

7.4 Estimated Frequency Distribution of Silicate Species

In this Chapter examples of the various families of the crystal chemical system of silicates have been presented as far as examples are known. For families with a large number of well-known species only a few typical ones have been mentioned; for other families with only two or three known examples probably all of them have been described. Consequently, a very unbalanced survey has resulted and this needs to be corrected.

In this section, therefore, tables are presented which indicate the frequency distribution of crystalline silicate species within the different categories of the system. The frequency distribution is a rough estimate only and is far from quantitative. This is due not only to the difficulties in deciding whether, for example, end members and intermediates of solid solution series should be treated as separate species, but also due to, for instance, the question of the weight to assign to a rock-forming mineral like quartz, amounting to about 18 volume percent of the earth's crust (see Table 1.3), and the weight to assign to a synthetic phase as exotic as, for example, $Cs_6Si_{10}O_{23}$. Nevertheless, a qualitative frequency distribution, however inaccurate it may be, will at least correct some of the distorted impressions obtained from the preceding sections and will give a vague idea of the relative stability of the various silicate types.

To prepare such a frequency distribution 696 crystalline silicates were categorized. For polymorphic substances separate treatment was given to those polymorphs which are topologically different. For example, quartz, cristobalite, tridymite, etc. were each considered as one silicate, whereas high and low temperature quartz were counted together as one phase only. To allow for incompleteness of the data file another 220 "ghost silicates" were added, particularly in those categories where the data file was known to be more incomplete than in

Table 7.9 Estimated frequency distribution according to coordination number of silicon, ***CN***, and linkedness, ***L***, of 874 crystalline silicates having only one type of silicate anion ($N_{an} = 1$)

L \ *CN*	...	4	5	6	...	mixed *CN*	Σ
0		160		7		×[a]	167
1		696		4		1	701
2		1		2			3
3							
mixed *L*				3			3
Σ		857		16		1	874

[a] × = theoretically not possible

Table 7.10 Estimated frequency distribution according to multiplicity ***M***, dimensionality ***D***, and branchedness of 856 crystalline silicates having only one kind of silicate anion ($N_{an} = 1$) with single or corner-shared $[SiO_4]$ tetrahedra ($L = 0, 1$; $CN = 4$)

M / *D*		1		2		3		4		5		6		7		8		9		10		...	Σ		
D		***uB***	***br***	***uB***	***br***	***uB***	***br***	***uB***	***br***	***uB***	***br***	***uB***	***br***	***uB***	***br***	***uB***	***br***	***uB***	***br***	***uB***	***br***		***uB***	***br***	Total
0	***t***	160		90		12	1	2		1						1		1		1			268	1	269
	r	73	4	27	1																		100	5	105
1		100	10	58	7	9		4		1													172	17	189
2		131	12	12	9																		143	21	164
3		89	40																				89	40	129
Σ		553	66	187	17	21	1	6		2						1		1		1			772	84	856
		619		204		22		6		2						1		1		1			856		

uB = unbranched; ***br*** = branched, including hybrid

others. This total of 916 entries includes only those silicates with structures which can reliably be filed into one category of the system. There is, therefore, a large number of silicates with unknown or insufficiently accurately known structures for which no allowance was made.

The result of this very crude survey is presented in Tables 7.9 to 7.12. In these Tables the various categories of the system are identified by the parameters $\boldsymbol{N_{an}}$, ***CN***, ***L***, ***B*** (given as ***uB*** for unbranched and as ***br*** for the entirety of branched and hybrid silicates), ***M***, ***D***, ***P***, $\boldsymbol{P^r}$, and ***r*** or ***t***. Silicates containing anions which are theoretically impossible are denoted by an ×. Categories for which no species have been reported, even though their existence is theoretically possible, are left blank.

Table 7.9 is the estimated frequency distribution obtained for crystalline uniform-anion silicates, i.e., for silicates containing only one kind of silicate anion ($\boldsymbol{N_{an}}=1$). It clearly demonstrates the numerical dominance (857:16) of the silicates with tetrahedrally coordinated silicon (tetraoxosilicates) over the phases with six-coordinated silicon (hexaoxosilicates). Of the 857 silicates with tetrahedral silicon, 160 are monosilicates with isolated $[SiO_4]$ tetrahedra, one contains edge-sharing tetrahedra (fibrous silica, Fig. 7.2), and in 696 others the $[SiO_4]$ polyhedra share corners. An additional two phases containing both tetrahedral and octahedral silicon are not included in any of the four Tables; one of the two has $\boldsymbol{N_{an}} = 2$.

Of the 16 silicon compounds known to contain silicon only in octahedral coordination, 7 have isolated $[SiO_6]$ groups, whereas in 4 the octahedra share corners, in 2 they share edges, and in another 3 corner- as well as edge-sharing takes place (see Table 7.1).

Table 7.10 presents the estimated frequency distribution for the 856 silicates with only one type of tetrahedral silicate anion and in which the $[SiO_4]$ tetrahedra are either isolated or corner-linked. It can be seen that:

(1) within each group with the same anion dimensionality there is a sharp decrease of frequency with increasing multiplicity; this frequency decrease is least pronounced for oligosilicates and, with the exception of the cyclic silicates, becomes more pronounced as the dimensionality of the silicate anions increases.
(2) The number of unbranched silicates is considerably higher than the number of branched ones.
(3) The ratio of unbranched silicates to branched silicates decreases as the dimensionality of the silicate anions increases.
(4) There is a decrease in the overall number of silicates as the dimensionality increases from $\boldsymbol{D}=0$ (374 species) to $\boldsymbol{D}=3$ (129 species).

Table 7.11 and Fig. 7.30 give the estimated frequency distribution of 587 crystalline silicates containing only one kind of silicate anion built from corner-linked tetrahedra according to chain periodicity, ring periodicity, and branchedness.

There is a general tendency for the frequency of these silicates to decrease as the periodicities ***P*** and $\boldsymbol{P^r}$ increase. This effect is most pronounced in the phyllosilicates; it is, however, somewhat masked by a rather high frequency of silicates with periodicities ***P*** and $\boldsymbol{P^r}=6$.

Table 7.12 gives the estimated frequency distribution of 42 crystalline silicates containing two different types of tetrahedral silicate anions according to the dimensionality of their silicate anions. It is evident that the number of silicate spe-

Table 7.11 Estimated frequency distribution of 587 crystalline silicates having only one kind of silicate anion with corner-linked $[SiO_4]$ tetrahedra (except multiple tetrahedra)

D	0 (Only ring silicates)						1												2						3				
M	1		2		...		1		2		3		4		5		...		1		2		...		1		Σ		
P^r, ***P***	***uB***	***br***	***uB***	***br***	***uB***	***br***	***uB***	***br***	***uB***	***br***	***uB***	***br***	***uB***	***br***	***uB***	***br***	***uB***	***br***	***uB***	***br***	***uB***	***br***	***uB***	***br***	***uB***	***br***	***uB***	***br***	Total
1	×	×	×	×	×	×			2												×	×	×	×	×	×	2		2
2	×	×	×	×	×	×	35	1	34		9		1		1				104	2	5				35	1	224	4	228
3	20		1				23		10	7			2						10	5		7			15	21	81	40	121
4	9	2	8	1			12	5	2				1						12	2	4	1			13	15	61	26	87
5		1	1				6	1	1										1	2	2					2	11	6	17
6	28	1	17				7	1	9										4	1		1			25	1	90	5	95
7							3																				3		3
8	1							1																			1	1	2
9							1														1				1		3		3
10								1																				1	1
11																													
12	15						12																				27		27
⋮																													
24							1																				1		1
⋮																													
Σ	73	4	27	1			100	10	58	7	9		4		1				131	12	12	9			89	40	504	83	587

D = dimensionality; ***M*** = multiplicity; P^r = ring periodicity (for ***D*** = 0); ***P*** = chain periodicity (for ***D*** ≠ 0); ***uB*** = unbranched; ***br*** = branched, including hybrid; × = theoretically not possible

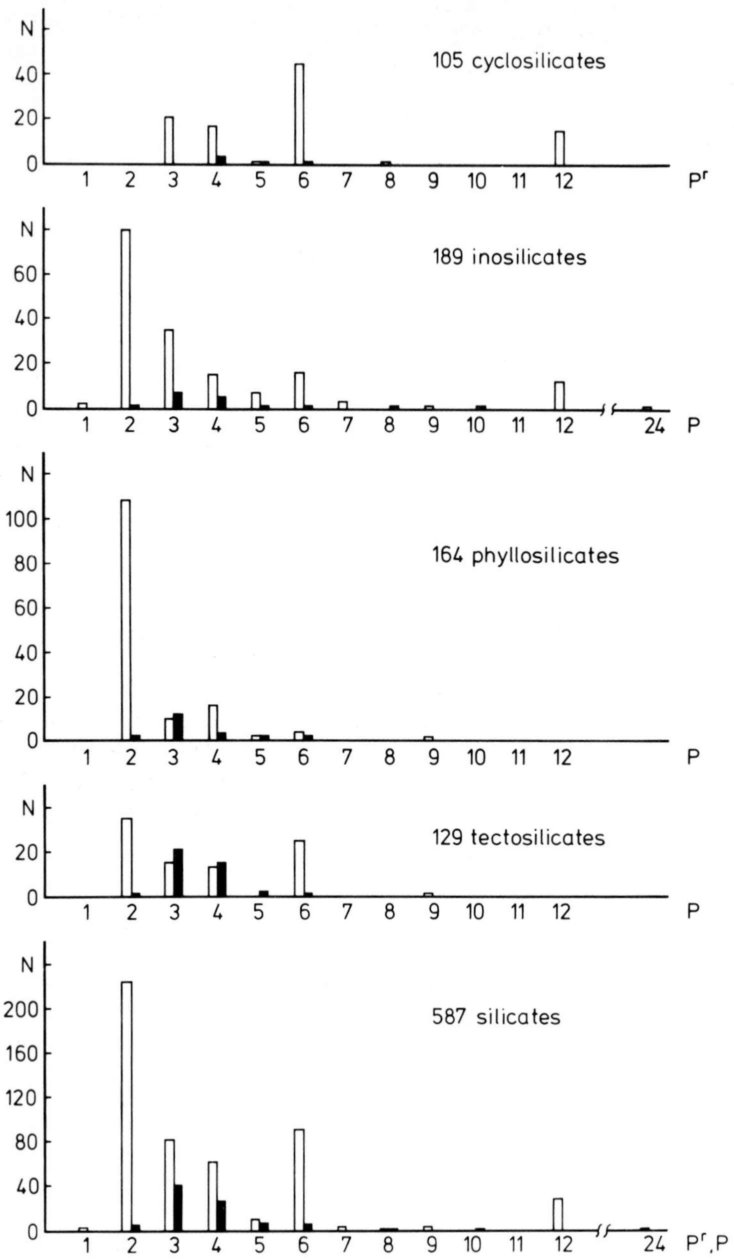

Fig. 7.30. Histograms of the estimated frequency distribution of 587 crystalline silicates with $N_{an} = 1$, $CN = 4$, and $L = 1$, according to ring periodicity P^r, chain periodicity P, and branchedness. *White bars:* unbranched silicates; *black bars:* branched silicates

Table 7.12 Estimated frequency distribution of 42 crystalline mixed-anion silicates ($N_{an} = 2$) containing only tetrahedral silicate anions ($CN = 4$)

D_2 \ D_1		0	1	2	3	Σ
0	*t*	30				30
	r	1				1
1		5	3			8
2		1	1			2
3				1		1
Σ		37	4	1		42

D_1, D_2 = dimensionality of the two types of silicate anions; t = terminal anions; r = cyclic anions

cies decreases drastically as the dimensionality of the anions increases. Half of these mixed-anion silicates contain both single and double tetrahedra.

It is stressed once again that the estimated frequency distributions of crystalline silicates presented in these Tables are estimated only and that care should, therefore, be taken when conclusions are drawn from the figures quoted. However, the general trends in Tables 7.9 to 7.12 seem to be reliable enough to enable speculation about the average stability of certain types of silicate anions and categories of silicates (see Chap. 11).

8 Other Classifications of Silicates

8.1 Early Classifications of Silicates

Machatschki (1928) seems to have been the first to realize that the kind and the degree of linkage of $[SiO_4]$ tetrahedra is a suitable property for the classification of silicates. Since at that time only a few silicate structures had been determined, this was an extremely far-sighted suggestion. Two years later Bragg (1930) and Náray-Szabó (1930), having access to the structures of the most important rock-forming silicate minerals, followed Machatschki's suggestion and created what is now usually referred to as "Bragg's classification of silicates" (Table 8.1). This classification was based on the $[SiO_4]$ tetrahedra and the dimensionality of their linkage, and took account of the partial replacement of silicon by tetrahedrally coordinated aluminum. Berman (1937) extended this partial replacement of silicon to beryllium, and Strunz (1938) extended it further to include phosphorus, arsenic, germanium, titanium, and iron.

The Bragg classification was so powerful that it was not until 1956, following the discovery of several new chain types, that an extension was suggested for silicates with infinite anions which took into account the periodicity of the tetrahedral chains (Liebau 1956). Further subdivisions of the silicates based on the multiplicity of their anions and into classes containing $[SiO_4]$ tetrahedra and $[SiO_6]$ octahedra, and into silicates having branched, unbranched, and hybrid anions have been introduced more recently (Liebau 1962a, 1972, 1978), ultimately leading to the classification described in Chaps. 4 and 6. This new classification is nothing

Table 8.1 Bragg's classification of silicates

Silicate type	Anions	T : O
Monosilicates	$[SiO_4]$	1 : ≥ 4
Group silicates		
Disilicates	$[Si_2O_7]$	1 : 3.5
Cyclosilicates	$[Si_3O_9]$, $[Si_6O_{18}]$	1 : 3
Chain silicates		
Single chain silicates	$[TO_3]$	1 : 3
Double chain silicates	$[T_4O_{11}]$	1 : 2.75
Layer silicates	$[T_4O_{10}]$	1 : 2.5
Framework silicates	$[TO_2]$	1 : 2

but an extension of the classical Bragg classification of silicates which was based solely on the type and degree of linkage of $[TO_n]$ polyhedra.

8.2 Kostov's Classification of Silicates

The extended Bragg classification described in detail in Chap. 6 is based solely on crystal chemical principles and is, therefore, suitable for finding relations between the chemical composition and crystal structure of natural and synthetic silicates and for obtaining a better understanding of stability relations between various structures. On the other hand, the classification of silicate minerals developed by Kostov (1975) is designed to fulfill the needs of mineralogists and geologists. To this end, Kostov based his classification partly on crystal structure and partly on the other equally essential properties of chemical composition and crystal morphology.

In a manner which is somewhat different from the crystal chemical classification, Kostov uses the degree of silification, i.e., the ratio (Si, Al) : M′, where $M' = Me_2^+$, Me^{+2}, $Me_{0.67}^{+3}$, and $Me_{0.5}^{+4}$, for the first broad division of the silicates. This ratio is inversely proportional to the Si : O ratio and, therefore, is approximately inversely proportional to the degree of condensation of the $[(Al, Si)O_4]$ tetrahedra. Silicates with $(Si, Al) : M' \geq 4:1$ are silica and framework silicates, while phyllosilicates have $3:1 \geq (Si, Al) : M' \geq 1:1$, chain and cyclosilicates have mainly $(Si, Al) : M' = 1:1$, and oligo- and monosilicates have $(Si, Al) : M' < 1:1$. On this basis Kostov creates only three major groups:

silica-rich silicates with $(Si, Al) : M' \geq 4:1$;
intermediate silicates with $3:1 \geq (Si, Al) : M' \geq 1:1$; and
silica-poor silicates with $(Si, Al) : M' < 1:1$.

A fourth group is formed by the borosilicates since boron is sometimes tetrahedrally coordinated like silicon, and sometimes three-coordinated.

To this point Kostov's classification is not very different from the crystal chemical one.

The first subdivision is based on the fact that in silicate minerals the following cationic elements usually appear together and very often replace each other isomorphously:

(1) Be, Al, Mg (and to a lesser content Fe);
(2) Zr, Ti (Sn), Nb;
(3) Ca (rare earth elements), Mn, Ba; and
(4) Zn, Cu, Pb (U).

This is a purely geochemical principle and, therefore, makes this classification especially suitable for earth scientists.

Since the morphology of silicate minerals is a very important property for identification purposes, the next subdivision is into types with

(a) axial habit;
(b) planar habit;
(c) isometric and pseudosisometric habit.

Table 8.2 Kostov's classification of silicates

Cations \ (Si, Al) : M′	≥ 4 : 1	3 : 1 to 1 : 1	< 1 : 1	Borosilicates
Be, Al, Mg (Fe)	Axial Planar Isometric	Axial Planar Isometric	Axial Planar Isometric	Axial
Zr, Ti (Sn), Nb	Axial Planar Isometric	Axial Planar Isometric	Axial Planar Isometric	Planar
Ca (RE), Mn, Ba	Axial Planar Isometric	Axial Planar Isometric	Axial Planar Isometric	
Zn, Cu, Pb (U)	Axial Planar Isometric	Axial Planar Isometric	Axial Planar Isometric	Isometric

The morphology of a crystal depends primarily on the relative strength and number of bonds per unit volume parallel to the various directions in the crystal. Since all bonds, i.e., M–O bonds as well as Si–O bonds, contribute to this ratio, the Kostov classification avoids those shortcomings of the crystal chemical classification discussed in Sec. 6.5 arising from the fact that the Si–O bonds are overvalued by not giving sufficient consideration to the M–O bonds.

The Kostov classification is given in concise form in Table 8.2.

8.3 Zoltai's Classification of Silicates

In 1960 Zoltai suggested a classification with a first broad division based on the dimensionality of the silicate anions. This resulted in groups containing silicates with:

(1) isolated clusters of tetrahedra;
(2) one-dimensionally nonterminated structures of tetrahedra, including cyclosilicates;
(3) two-dimensionally nonterminated structures of tetrahedra;
(4) three-dimensionally nonterminated structures of tetrahedra.

This system is complemented by a fifth group containing silicates with

(5) mixed anions.

The first three of these groups are subdivided according to the multiplicity of the silicate anions. So far this classification is identical with Bragg's.

For further subdivision of all five groups Zoltai introduced a parameter describing the degree of condensation of the tetrahedra. This parameter, which he called the *sharing coefficient*, is the average number of tetrahedra participating in the sharing of a corner in the structure under consideration, regardless of whether the tetrahedra are linked via common corners, edges, or faces. The numerical value of the sharing coefficient of a silicate anion can be obtained from its chemical formula by using the equation

$$f_{sh} = 2n + 1 - \frac{A}{4T}(n^2 + n)$$

where A is the number of oxygen atoms, T is the number of tetrahedrally coordinated cations in the silicate anion of formula $[(Al, Si, \ldots)_T O_A]$, and n is the integer equivalent of $\frac{4T}{A}$.

When silicates with edge- or face-sharing tetrahedra are omitted, this classification reduces to that presented in Table 8.3. Also included are the corresponding values of the ratios T:A, i.e., the atomic ratios T:O for the silicate anions of the group.

It is clear from this Table that all single rings and all unbranched and open-branched single chains have the same numerical value of the sharing coefficient, and that the very large number of silicates with double chains and single layers have ranges of the sharing coefficient which are so small that a further subdivision is necessary.

Table 8.3 Zoltai's classification applied to silicates containing only single and corner-sharing tetrahedra

Types	Subtypes	f_{sh}	T : O
1. Isolated groups of tetrahedra	a) Single tetrahedron	1.00	1:4
	b) Double tetrahedron	1.25	1:3.5
	c) Triple tetrahedron	1.33	1:3.33
	d) Large groups	1.33 – 1.50	1:3.33 – 1:3
	e) Mixed groups	1.00 – 1.50	1:3.67 – 1:3
2. 1-Dimensionally nonterminated structures of tetrahedra	a) Single chains	1.50	1:3
	b) Single rings	1.50	1:3
	c) Double chains	1.50 – 1.75	1:3 – 1:2.5
	d) Double rings	1.50 – 1.75	1:3 – 1:2.5
	e) Multiple chains	1.50 – 2.00	1:3 – 1:2
	f) Multiple rings	1.50 – 2.00	1:3 – 1:2
	g) Mixed chains and rings	1.50 – 2.00	1:3 – 1:2
3. 2-Dimensionally nonterminated structures of tetrahedra	a) Single layers	1.50 – 2.00	1:3 – 1:2
	b) Double layers	1.50 – 2.00	1:3 – 1:2
	c) Multiple layers	1.50 – 2.00	1:3 – 1:2
	d) Mixed layers	1.50 – 2.00	1:3 – 1:2
4. 3-Dimensionally nonterminated structures of tetrahedra	Frameworks	1.50 – 2.00	1:3 – 1:2
5. Mixed types			

f_{sh} = sharing coefficient

Table 8.4 Classification of silicates according to Zoltai

Type	Subtype	f_{sh}	Repeat unit /Loops	Unit Formula	Examples with chemical formula	
1	a (Single)	1.000	1/–	TO_4	Zircon	$Zr[SiO_4]$
	b (Double)	1.250	2/–	$TO_{3\,1/2}$	Thortveitite	$Sc_2[Si_2O_7]$
	c (Triple)	1.333	3/–	$TO_{3\,1/3}$	Rosenhahnite	$H_2Ca_3[Si_3O_{10}]$
	d (Larger)	1.333	1+5/–	$TO_{3\,1/3}$	Zunyite	$Al_{12}[AlSi_5O_{20}](OH,F)_{18}Cl$
	e (Mixed)	1.167	1+2/–	$TO_{3\,2/3}$	Epidote	$Ca_2Al_2Fe[SiO_4][Si_2O_7]O(OH)$
		1.167	1+1+1+3/–	$TO_{3\,2/3}$	Ardennite	$Mn_4Al_6[(As,V)O_4][SiO_4]_2[Si_3O_{10}](OH)_6$
		1.200	1+2+2/–	$TO_{3\,3/5}$	Rustumite	$Ca_{10}[SiO_4][Si_2O_7]_2(OH)_2Cl_2$
		1.250	1+3/–	$TO_{3\,1/2}$	Kilchoanite	$Ca_6[SiO_4][Si_3O_{10}]$
2	a (Single chains)	1.500	2/–	TO_3	Enstatite	$Mg[SiO_3]$
		1.500	3/–	TO_3	Wollastonite	$Ca[SiO_3]$
		1.500	4/–	TO_3	Batisite	$Na_2BaTi_2[SiO_3]_4O_2$
		1.500	5/–	TO_3	Rhodonite	$(Mn,Ca)[SiO_3]$
		1.500	6/–	TO_3	Stokesite	$CaSn[SiO_3]_3\cdot 2H_2O$
		1.500	7/–	TO_3	Pyroxmangite	$(Fe,Mn,Ca)[SiO_3]$
		1.500	9/–	TO_3	Ferrosilite III	$Fe[SiO_3]$
		1.500	12/–	TO_3	Alamosite	$Pb[SiO_3]$
		1.500	24/–	TO_3	Synthetic	$Na_3Y[SiO_3]_3$
		1.625	4/4	$TO_{2\,3/4}$	Vlasovite	$Na_4Zr_2[Si_8O_{22}]$
		1.750	8/4+6+8	$TO_{2\,1/2}$	Litidionite	$NaKCu[Si_4O_{10}]$
	b (Single rings)	1.500	3/3	TO_3	Benitoite	$BaTi[Si_3O_9]$
		1.500	4/4	TO_3	Baotite	$Ba_4(Ti,Nb)_8[Si_4O_{12}]ClO_{16}$
		1.500	6/6	TO_3	Dioptase	$Cu_6[Si_6O_{18}]\cdot 6H_2O$
		1.500	8/8	TO_3	Muirite	$Ba_{10}(Ca,Mn,Ti)_4[Si_8O_{24}](Cl,OH,O)_{12}\cdot 4H_2O$
		1.500	12/12	TO_3	Synthetic	$Na_{15}Y_3[Si_{12}O_{36}]$
	c (Double chains)	1.583	3/8	$TO_{2\,5/6}$	Xonotlite	$Ca_6[Si_6O_{17}](OH)_2$
		1.600	5/6+8	$TO_{2\,4/5}$	Inesite	$Ca_2Mn_7[Si_{10}O_{28}](OH)_2\cdot 5H_2O$
		1.625	2/6	$TO_{2\,1/4}$	Anthophyllite	$(Mg,Fe)_7[Si_4O_{11}]_2(OH)_2$

		1.750	1/4	$TO_{2\ 1/2}$	Sillimanite	$Al[AlSiO_5]$
		1.750	2/4	$TO_{2\ 1/2}$	Synthetic	$Li_4[SiGe_3O_{10}]$
		1.750	3/4	$TO_{2\ 1/2}$	Elpidite	$Na_2Zr[Si_6O_{15}]\cdot 3H_2O$
		1.750	4/4	$TO_{2\ 1/2}$	Caysichite	$(Ca_3RE)Y_4[Si_8O_{20}][CO_3]_6(OH)\cdot 7H_2O$
	d (Double rings)	1.750	3/3+4	$TO_{2\ 1/2}$	Synthetic	$[Ni\,en_3]_3[Si_6O_{15}]\cdot 26H_2O$
		1.750	4/4+4	$TO_{2\ 1/2}$	Steacyite	$K_{1-x}(Na,Ca)_{2-y}Th_{1-z}[Si_8O_{20}]$
	e (Multiple chains)	1.667	2/6	$TO_{2\ 2/3}$	Jimthompsonite	$Mg_5[Si_6O_{16}](OH)_2$
		1.688	2/6	$TO_{2\ 5/8}$	Synthetic	$Ba_5[Si_8O_{21}]$
		1.700	2/6	$TO_{2\ 3/5}$	Synthetic	$Ba_6[Si_{10}O_{26}]$
	f (Multiple rings)	–	–	–	–	–
	g (Mixed chains	1.500	3+9/3+9	TO_3	Eudialyte	$Na_{12}(Ca,RE)_6(Fe,Mn,Mg)(Zr,Nb)_{3+x}$ $[Si_3O_9]_2[Si_9(O,OH)_{27}]_2Cl_y$
	and rings)	1.650	2+3/6	$TO_{2\ 7/10}$	Chesterite	$(Mg,Fe)_{17}[Si_4O_{11}]_2[Si_6O_{16}]_2(OH)_6$
3	a (Single layers)	1.600	–/12	$TO_{2\ 3/5}$	Zeophyllite	$Ca_{13}[Si_5O_{14}]_2(F,OH,O)_{10}\cdot 6H_2O$
		1.750	–/4+6+8	$TO_{2\ 1/2}$	Dalyite	$K_2Zr[Si_6O_{15}]$
		1.750	–/4+6+12	$TO_{2\ 1/2}$	Pyrosmalite	$(Mn,Fe)_8[Si_6O_{15}](OH,Cl)_{10}$
		1.750	–/4+8	$TO_{2\ 1/2}$	Apophyllite	$KCa_4[Si_4O_{10}]_2(F,OH)\cdot 8H_2O$
		1.750	–/4+8	$TO_{2\ 1/2}$	Datolite	$HCa[BSiO_5]$
		1.750	–/6	$TO_{2\ 1/2}$	Kaolinite	$Al_2[Si_2O_5](OH)_4$
		1.833	–/5+5	$TO_{2\ 1/3}$	Gehlenite	$Ca_2[Al_2SiO_7]$
		1.875	–/4+6	$TO_{2\ 1/4}$	Synthetic	$K_2[Si_4O_9]$
	b (Double layers)	1.813	–/4+8	$TO_{2\ 3/8}$	Delhayelite	$Na_3K_7Ca_5[AlSi_7O_{19}]_2F_4Cl_2$
		1.833	–/3+6+6+8	$TO_{2\ 1/3}$	Zussmanite	$K(Fe,Mg,Mn,Al)_{13}[(Al,Si)_9O_{21}]_2(OH)_{14}$
		1.875	–/4+4+6+8	$TO_{2\ 1/4}$	Carletonite	$Na_4KCa_4[Si_8O_{18}][CO_3]_4(OH,F)\cdot H_2O$
		1.900	–/4+5+6+8	$TO_{2\ 1/5}$	Latiumite	$(Ca,K)_4[(Al,Si)_5O_{11}][SO_4,CO_3]$
		1.917	–/4+6+6+6	$TO_{2\ 1/6}$	Naujakasite	$Na_6Fe[Al_4Si_8O_{26}]$
		2.000	–/4+6	TO_2	Cymrite	$Ba[AlSiO_4]_2\cdot H_2O$
	c (Multiple layers)	–	–	–	–	–
	d (Mixed layers)	1.792	–/6+6+8+10+12	$TO_{2\ 5/12}$	Reyerite	$(Na,K)_2Ca_{14}[Si_8O_{20}][Al_2Si_{14}O_{38}](OH)_8\cdot 6H_2O$

Table 8.4 cont.

Type	Subtype	f_{sh}	Repeat unit /Loops	Unit Formula	Examples with chemical formula	
4	Frameworks	1.750	–/6+8+10+12	$TO_{2\ 1/2}$	Synthetic	$K_2Ce[Si_6O_{15}]$
		1.850	–/6+12	$TO_{2\ 3/10}$	Synthetic	$Rb_6[Si_{10}O_{23}]$
		1.875	–/4+6+8+10	$TO_{2\ 1/4}$	Ussingite	$HNa_2[AlSi_3O_9]$
		1.929	–/4+4+6+6+8+10	$TO_{2\ 1/7}$	Tuhualite	$(Na, K)_2Fe_2[Fe_2Si_{12}O_{30}] \cdot H_2O$
		1.975	–/4+5+6+6+7	$TO_{2\ 1/20}$	Leifite	$H_2Na_6[Be_2Al_2Si_{16}O_{41}] \cdot 1.5\ H_2O$
		2.000	–/4+6+8+8	TO_2	Quartz	$[SiO_2]$
		2.000	–/4+6+8+9	TO_2	Coesite	$[SiO_2]$
		2.000	–/6	TO_2	Cristobalite	$[SiO_2]$
		2.000	–/4+6+8+10	TO_2	Orthoclase	$K[AlSi_3O_8]$
		2.000	–/4+6+9	TO_2	Beryl	$Al_2[Be_3Si_6O_{18}]$
		2.000	–/4+6+12	TO_2	Sodalite	$Na_4[Al_3Si_3O_{12}]Cl$
		2.000	–/4+8+9	TO_2	Natrolite	$Na_2[Al_2Si_3O_{10}] \cdot 2\ H_2O$
		2.000	–/4+6+8+12	TO_2	Chabazite	$(Ca, Na_2)[Al_2Si_4O_{12}] \cdot 6\ H_2O$
		2.000	–/5+6	TO_2	Melanophlogite	$46[SiO_2] \cdot 8(N_2, CH_4, CO_2)$
		2.375	–/3+4+6	$TO_{1\ 3/4}$	Barylite	$Ba[Be_2Si_2O_7]$
		2.750	–/3+4+6	$TO_{1\ 1/2}$	Bertrandite	$H_2[Be_4Si_2O_9]$
		2.750	–/3+4+6+8	$TO_{1\ 1/2}$	Hemimorphite	$H_2[Zn_4Si_2O_9] \cdot H_2O$
		3.000	–/3+4+5	$TO_{1\ 1/3}$	Phenakite	$[Be_2SiO_4]$
5	Mixed types	1.250	2/– and	$TO_{2\ 3/4}$	Miserite	$K_2Ca_{10}(Y,RE)_2[Si_2O_7]_2[Si_{12}O_{30}](OH)_2F_2$
		+1.750	–/4+6+8		(double tetrahedra + quadruple chain)	
		1.000	1/– and	TO_3	Holdenite	$Mn_6[Zn(OH)_4][Zn_2SiAs_2O_{12}(OH)_2](OH)_2$
		+1.600	–/4		(single tetrahedra + double layer)	
		1.667	3/4+6 and	$TO_{2\ 5/9}$	Okenite	$Ca_{10}[Si_6O_{16}][Si_6O_{15}]_2 \cdot 18\ H_2O$
		+1.750	–/5+8		(double chain + single layers)	

f_{sh} = sharing coefficient

For such a subdivision of the oligosilicates Zoltai used the number of tetrahedra in their anions, and for the single chains he used the periodicity of the chain, i.e., the number of tetrahedra in its repeat unit. This is identical to the subdivision used in the crystal chemical classification. However, whereas in the latter classification the multiple chain, layer, and framework silicates are also classified according to the periodicities of their fundamental chains, Zoltai subdivided these silicates, and also the cyclosilicates, according to the number of tetrahedra in the loops contained in their anions. Therefore, for example, the zweier double chains of Fig. 7.11 b and c are described in the Zoltai classification as containing sechser rings and vierer rings, respectively, and the dreier double chain of Fig. 7.11 e is described as containing vierer plus sechser rings.

In two-dimensional anions the number of loops of different size usually increases and in tetrahedral frameworks it becomes formally infinite. Therefore, the list of different loop sizes is limited to four loops and the size of the largest loop in the list is restricted to twelve.

In order to separate as few chemically related structures as possible and to give his classification a wider range of applicability, Zoltai regarded as part of the silicate anion all of the tetrahedral cations which distribute half or more of their electrostatic bond strength within the silicate anion, in other words, all tetrahedra that are linked to at least two other tetrahedra. Here it does not matter whether the tetrahedral cation is Li^{+}, Be^{+2}, Al^{+3}, Ge^{+4}, P^{+5}, etc., or Si^{+4}.

This enabled Zoltai to apply his classification not only to silicates, but to structures with tetrahedral groups in general by making special allowance for structures in which one tetrahedral corner belongs to more than two tetrahedra. Table 8.4 presents a number of examples of the various categories of the Zoltai classification in order to demonstrate the ranges of sharing coefficients and its wide applicability.

The Zoltai classification is convenient to use and has a wider range of applicability than that of both Kostov and the crystal chemical classification. However, unlike the Kostov classification, it is not suited to the solution of geochemical problems since it does not adequately account for chemical and morphological differences between silicates. Moreover, the Zoltai classification is inferior to the crystal chemical one for the determination of correlations between chemical composition and silicate structure. This is because the periodicity of silicate chains, the parameter used to subdivide multiple chain, layer, and framework silicates in the latter classification, is very closely related to the nature of the cations present in the silicate, while the loop size used in Zoltai's classification is not.

Keeping in mind these different aspects, it is not appropriate to conclude that one classification is more powerful than the other unless it is stated for which particular purpose the classification is to be used.

8.4 Geometrical Classification of Tectosilicates

As the dimensionality $\boldsymbol{D}$ of the tetrahedral silicate anions increases it becomes progressively more difficult to visualize and to classify such anions. This is particularly true for the tetrahedral frameworks of the tectosilicates. In addition, since all bonds between neighboring atoms in such frameworks are of approximately equal strength (Si–O, Al–O, etc.), several quite different groupings of tetrahedra can equally well be chosen as building units to construct the frameworks. Indeed, the fundamental chains used in Chap. 4 to classify silicate frameworks are just one of several different kinds of possible building units. The extraordinary complexity of tetrahedral framework structures, therefore, makes it necessary to discuss the various types of framework classifications.

Since only one silicate framework contains octahedral silicon (stishovite, $Si^{[6]}O_2$), and none with edge- or face-shared $[SiO_4]$ tetrahedra are known, only the three-dimensional silicate frameworks exclusively built from corner-linked tetrahedra will be treated here.

8.4.1 Connectedness

In Sec. 4.2 the *connectedness* $\boldsymbol{s}$ of an $[SiO_n]$ polyhedron was defined as the number of other $[SiO_n]$ polyhedra with which it is connected. For $[SiO_4]$ tetrahedra the connectedness can be $s = 0, 1, 2, 3$, and 4.

If all tetrahedra have the same value of the connectedness then only $s = 3$ and $s = 4$ are possible for tectosilicates. The vast majority of framework silicates have $s = 4$, that is, their tetrahedral frameworks are 3-dimensional 4-connected nets in the sense of Wells (1975). In contrast, the silicate anion of $K_2Ce[Si_6O_{15}]$ (Karpov et al. 1977a) is a 3-dimensional 3-connected net, i.e., it contains only tertiary tetrahedra Q^3 (see Table 4.1).

If a framework of corner-linked tetrahedra has tetrahedra of differing connectedness, the values $s = 1, 2, 3$, and 4 are theoretically possible, i.e., the framework can contain Q^1, Q^2, Q^3, and Q^4 groups. However, the mean value $\langle s \rangle$ of the connectedness must be in the range $2 < \langle s \rangle < 4$.

So far, only one tectosilicate has been reported for which $\langle s \rangle < 3$: neptunite $LiNa_2K(Fe, Mg, Mn)_2Ti_2[Si_8O_{22}]O_2$ (Cannillo et al. 1966a). It contains equal numbers of secondary and tertiary tetrahedra Q^2 and Q^3. These tetrahedra are

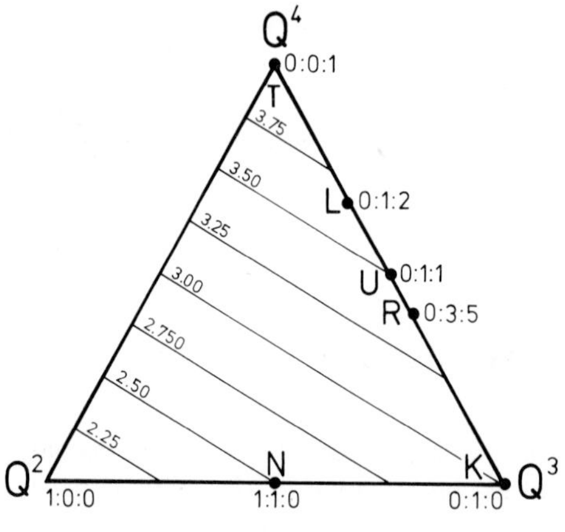

Fig. 8.1. The ratios $Q^2 : Q^3 : Q^4$ and the average connectedness $\langle s \rangle$ of the tectosilicates. Lines of constant $\langle s \rangle$ are indicated.
N neptunite, $LiNa_2K(Fe,Mg,Mn)_2Ti_2\{\boldsymbol{oB},{}^3_\infty\}[Si_8O_{22}]O_2$;
K synthetic $K_2Ce\{\boldsymbol{oB},{}^3_\infty\}[Si_6O_{15}]$;
R synthetic $Rb_6\{\boldsymbol{oB},{}^3_\infty\}[Si_{10}O_{23}]$;
U ussingite, $HNa_2\{\boldsymbol{oB},{}^3_\infty\}[AlSi_3O_9]$;
L leifite, $Na_6Be_2\{\boldsymbol{oB},{}^3_\infty\}[Al_2Si_{16}O_{39}](OH)_2 \cdot 1.5H_2O$;
T majority of tectosilicates with frameworks $\{{}^3_\infty\}[TO_2]$.

linked to form a very open framework of composition $[Si_8O_{22}]$, which, according to Wells, would be called a 2,3-connected net. Since the chemical composition of neptunite does not provide enough cations and additional anions and water to fill the voids of such a framework, two such crystallographically equivalent frameworks penetrate each other to attain higher density (Fig. 7.29).

Three silicates with 3,4-connected frameworks have so far been described. They contain Q^3 and Q^4 groups in the ratio 3:2 in synthetic $Rb_6[Si_{10}O_{23}]$ (Schichl et al. 1973), 1:1 in ussingite $HNa_2[AlSi_3O_9]$ (Rossi et al. 1974), and 1:2 in leifite $Na_6Be_2[Al_2Si_{16}O_{39}](OH)_2 \cdot 1.5\,H_2O$ (Coda et al. 1974).

In Fig. 8.1 the known tectosilicates are discriminated with regard to their ratios $Q^2 : Q^3 : Q^4$ and to their average connectedness $\langle s \rangle$.

8.4.2 Secondary Building Units

All silicate frameworks can be described in terms of finite component units from which they can be constructed. The smallest and, therefore, the primary building unit is the $[SiO_4]$ tetrahedron. However, larger building units consisting of several tetrahedra can be derived which have infinite extension in either zero, one, or two dimensions. For zeolites Meier (1968) derived so-called *secondary building units* (SBU) under the conditions that the entire framework is made up of *one* type of SBU only. These secondary building units are the unbranched vierer, sechser, and achter single rings, the unbranched vierer, sechser, and achter double rings, a loop-branched vierer single ring, and an open-branched and a loop-branched fünfer single ring. The topologies of these units are presented in Fig. 8.2 together with the symbols used to describe them in zeolite chemistry.

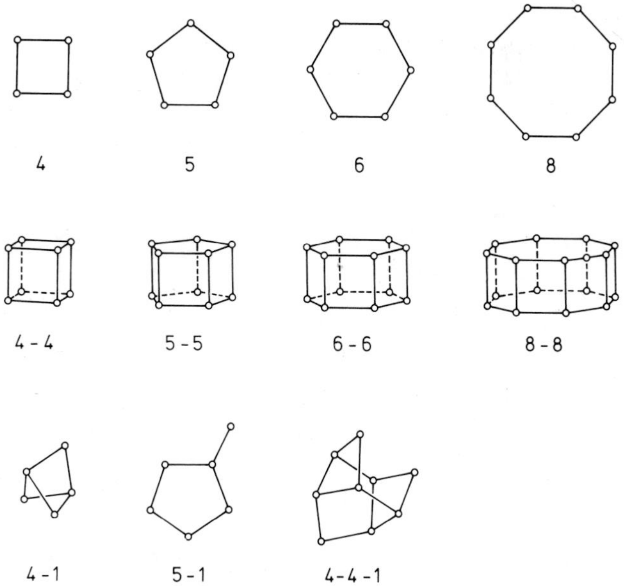

Fig. 8.2. Topology of secondary building units of framework silicates and their symbols used in zeolite science. *4:* vierer single ring; *5:* fünfer single ring; *6:* sechser single ring; *8:* achter single ring; *4-4:* vierer double ring; *5-5:* fünfer double ring; *6-6:* sechser double ring; *8-8:* achter double ring; *4-1:* loop-branched vierer single ring; *5-1:* open-branched fünfer single ring; *4-4-1:* loop-branched fünfer single ring

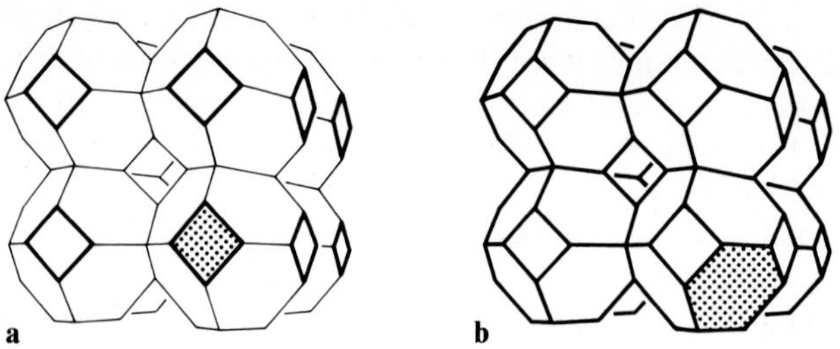

Fig. 8.3 a, b. The sodalite framework **a** built from vierer single rings as secondary building units; **b** built from sechser single rings as secondary building units. In each of the two drawings the secondary building units are drawn with *heavy lines*, one ring being *shaded*

Although originally defined for zeolites, the SBU concept can be applied to all silicate frameworks.

Most zeolite frameworks can be generated from several different SBUs. For example, the tetrahedral framework of sodalite, $Na_4[Al_3Si_3O_{12}]Cl$ (Löns and Schulz 1967) (Fig. 8.3), can be described with either a vierer single ring or a sechser single ring as the SBU. This ambiguity is analogous to the one encountered in the crystal chemical classification where different chain types could be used to generate a tetrahedral layer or framework (see Sec. 4.2). The different SBUs from which the various zeolite frameworks can be generated can be seen from Table 8.5.

8.4.3 Chain-Like Building Units

The generation of frameworks by successively linking fundamental chains has been described in detail in Chap. 4. However, sometimes it is more suitable to use other chains, often complex chains, to describe structural relations between frameworks.

The tetrahedral frameworks of coesite, the feldspars, and a series of zeolites and other tectosilicates can be generated from vierer single rings as secondary building units in such a way that the rings form ladder-like chains of various shapes. Three such ladders are presented in a very idealized way in Fig. 8.4.

The first ladder is identical to the zweier double chain in synthetic $Li_4\{\boldsymbol{uB}, 2^1_\infty\}[SiGe_3O_{10}]$, the second one is identical to the dreier double chain in epididymite $Na_2Be_2\{\boldsymbol{uB}, 2^1_\infty\}[Si_6O_{15}] \cdot H_2O$ (Figs. 7.11 c, f; 7.13 d, e), but the third

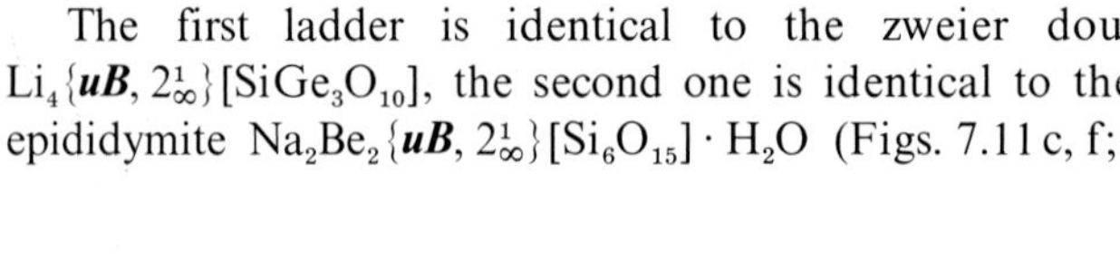

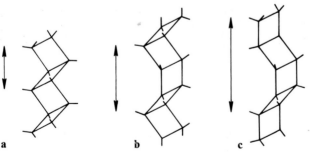

Fig. 8.4 a – c. Ladder-like chain units observed in silicate frameworks.

a In cancrinite, $Na_6Ca[AlSiO_4]_6[CO_3] \cdot 2H_2O$ (Barrer et al. 1970); **b** in offretite, $(Na_2, Ca, \ldots)[Al_2Si_7O_{18}] \cdot 7H_2O$ (Gard and Tait 1972) and zeolite L, $NaK_2[Al_3Si_9O_{24}] \cdot 7H_2O$ (Barrer and Villiger 1969); **c** in gmelinite, $(Na_2, Ca)_2[Al_4Si_8O_{24}] \cdot 12H_2O$ (Fischer 1966), feldspars, $M[(Al, Si)_4O_8]$, and coesite, SiO_2 (Levien and Prewitt 1981)

Table 8.5 Framework silicates with low framework densities and their building units (arranged in the order of decreasing framework densities) (based on Gramlich-Meier and Meier 1982)

Tectosilicate	Secondary building units [a]						Tubular building units [b]	Polyhedral building units [c]
Bikitaite						5−1		
Zeolite Li-A (BW)	4		6	8			ABW	
Melanophlogite		5						$[5^{12}6^2]$
Analcime	4		6					
Dodecasil 3C (ZSM-39]		5						$[5^{12}]$
Dodecasil 1H		5						$[5^{12}]$
Yugawaralite	4			8				
Epistilbite						5−1		
Silicalite-1 (ZSM-5)						5−1	MFI	
Ferrierite						5−1	FER	
Laumontite			6					
Brewsterite	4							
Dachiardite						5−1		
Mordenite						5−1	MOR	
Sodalite	4		6					$[4^6 6^8]$
Heulandite						4−4−1		
Stilbite						4−4−1		
Natrolite						4−1		
Thomsonite						4−1		
Edingtonite						4−1		
Cancrinite			6				CAN	$[4^6 6^5]$
Zeolite L			6				LTL	$[4^6 6^5]$
Mazzite	4					5−1	MAZ	
Merlinoite	4			8	8−8		MER	
Afghanite	4		6					
Phillipsite	4			8				
Zeolite Losod			6					
Liotite			6					
Erionite	4		6					$[4^6 6^5]$
Paulingite	4							
Offretite			6				OFF	$[4^6 6^5]$
Zeolite TMA-E (AB)	4		6					
Gismondine	4			8				
Levyne			6					
Zeolite ZK-5	4		6	8	6−6			$[4^{12}6^8 8^6]$
Chabazite	4		6		6−6			
Gmelinite	4		6	8	6−6		GME	$[4^9 6^2 8^3]$
Zeolite rho	4		6	8	8−8			$[4^{12}6^8 8^6]$
Zeolite A	4		6	8	4−4			$[4^{12}6^8 8^6]$, $[4^6 6^8]$
Faujasite	4		6		6−6			$[4^6 6^8]$

[a] Shown in Fig. 8.2;
[b] shown in Fig. 8.9;
[c] shown in Fig. 8.13

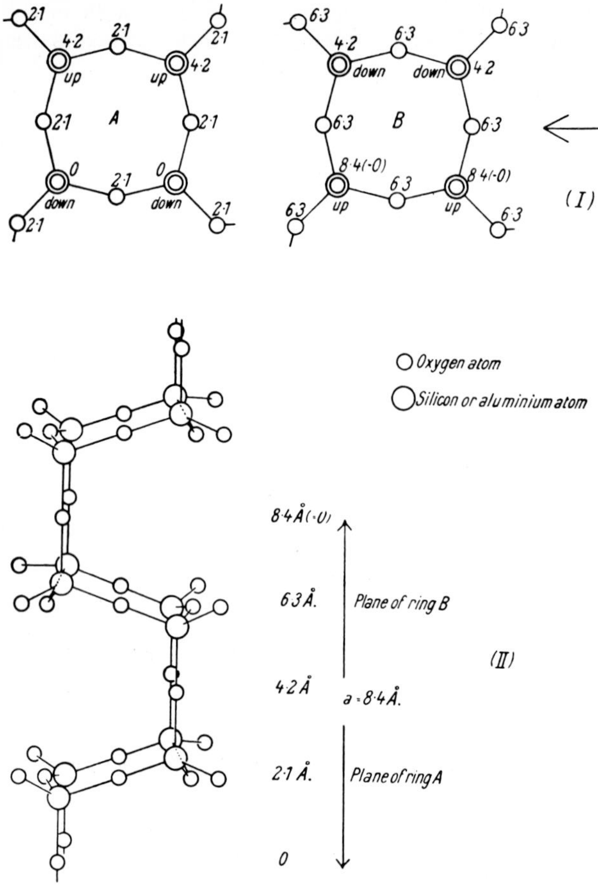

Fig. 8.5. The double crankshaft chain formed by linking vierer rings of the UUDD type (from Taylor 1933)

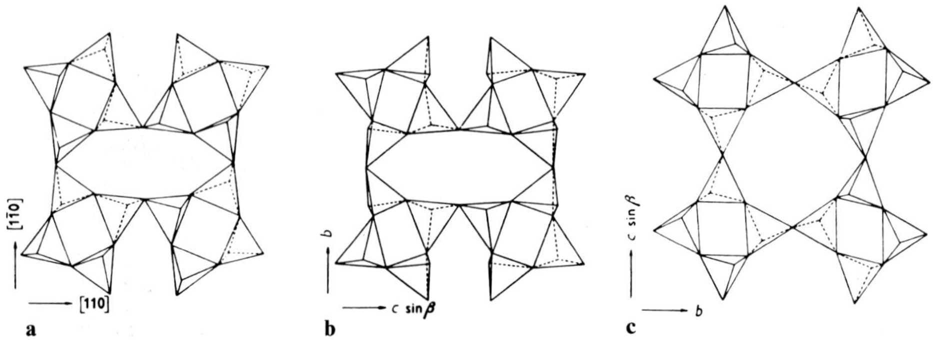

Fig. 8.6 a – c. Projections of framework structures generated from double crankshaft chains as chain-like building units (from Smith and Rinaldi 1962).

a Paracelsian, $Ba[Al_2Si_2O_8]$ (Bakakin and Belov 1961); **b** feldspars, $M[(Al,Si)_4O_8]$; **c** harmotome, $Ba[Al_2Si_6O_{16}] \cdot 6H_2O$ (Rinaldi et al. 1974)

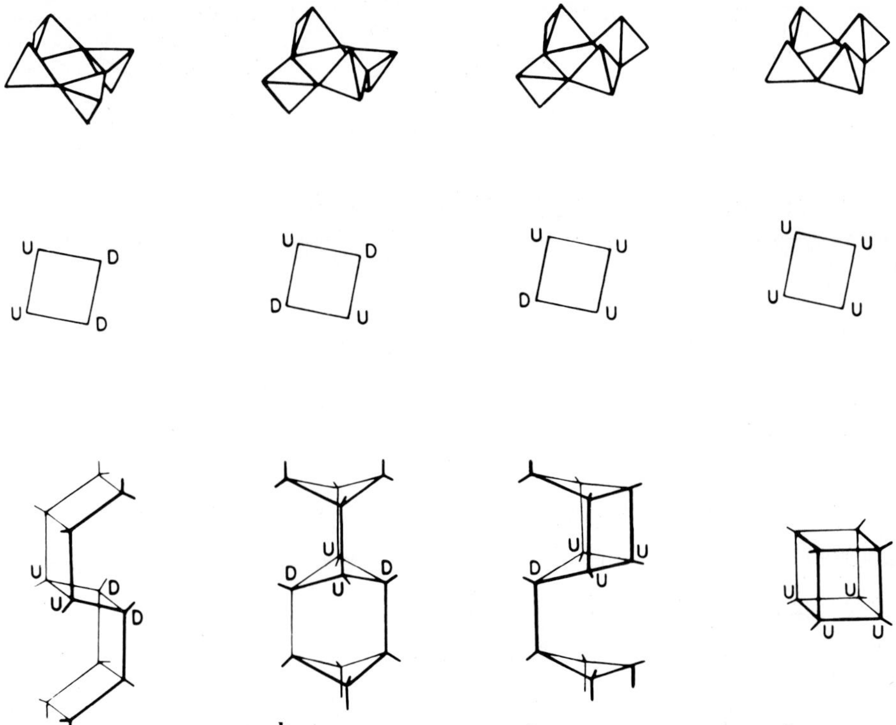

a b c d

Fig. 8.7 a – d. Silicate anions generated by linking vierer rings of different sequences of directedness of the tetrahedra (after Barrer 1978, p. 55).

a Generation of the double crankshaft chain from UUDD rings; **b** generation of the narsarsukite-type chain from UDUD rings; **c** generation of a hypothetical chain from UUUD rings; **d** generation of the vierer double ring from UUUU rings.

Fig. 8.8. Dreier quadruple chain, a chain-like building unit of mordenite, $Na_4[Al_4Si_{20}O_{48}] \cdot 12\,H_2O$ (Meier 1961), and dachiardite, $Na_5[Al_5Si_{19}O_{48}] \cdot 12\,H_2O$ (Vezzalini 1984)

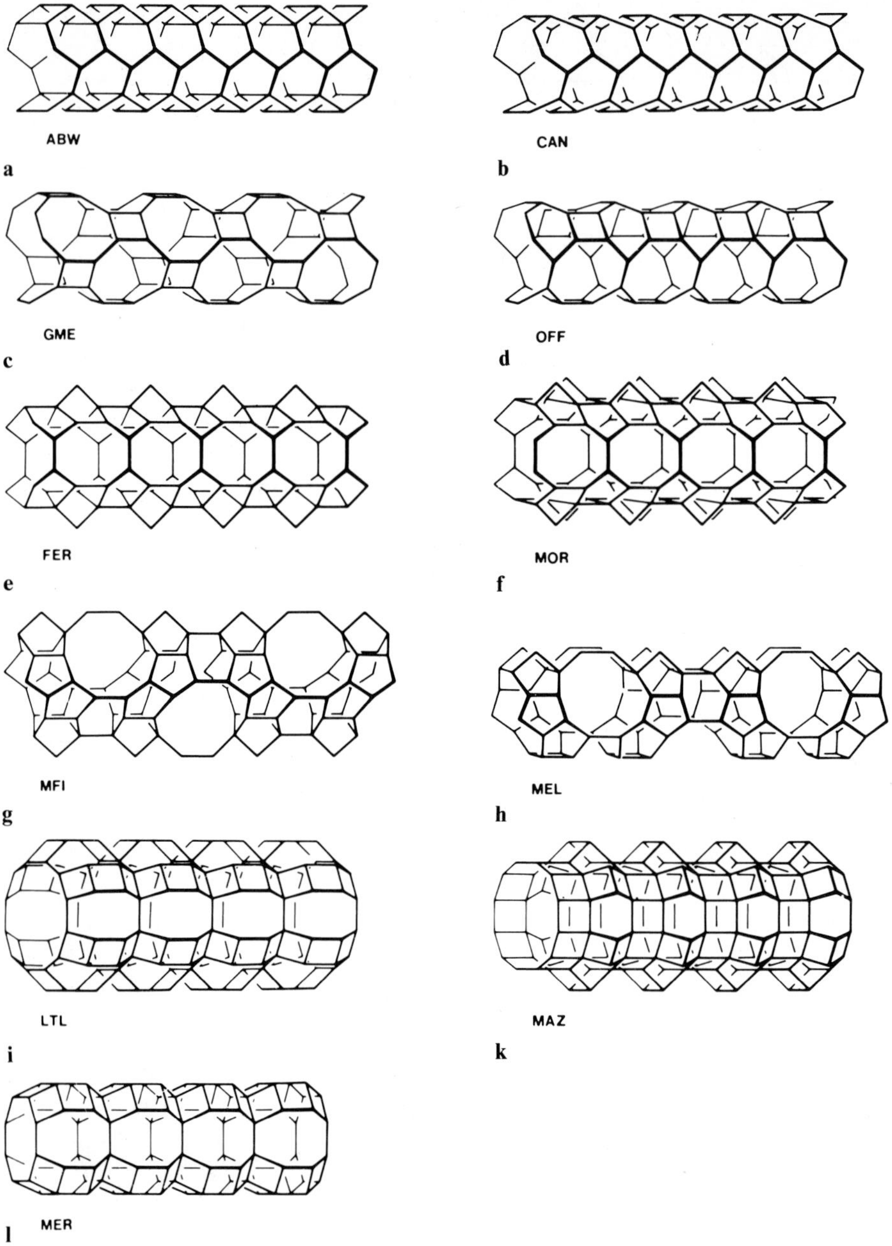

Fig. 8.9 a – l. Tubular building units of silicate frameworks which have at least achter ring apertures (from Gramlich-Meier and Meier 1982).

a Zeolite Li-A(BW) (Kerr 1974); **b** cancrinite (Barrer et al. 1970); **c** gmelinite (Fischer 1966); **d** offretite (Mortier et al. 1976); **e** ferrierite (Vaughan 1966); **f** mordenite (Meier 1961); **g** silicalite-1 (ZSM-5) (Olson et al. 1981; Price et al. 1982); **h** silicalite-2 (ZSM-11) (Kokotailo et al. 1978; Bibby et al. 1979); **i** zeolite L, $NaK_2[Al_3Si_9O_{24}] \cdot 7H_2O$ (Barrer and Villiger 1969); **k** mazzite, $(Na,K,Ca)_3Mg_2Ca[Al_{10}Si_{26}O_{72}] \cdot 28H_2O$ (Galli 1975; Rinaldi et al. 1975); **l** merlinoite, $K_5Ca_2[Al_9Si_{23}O_{64}] \cdot 24H_2O$ (Galli et al. 1979)

one has not yet been discovered in chain silicates. By linking such ladders in various ways a considerable number of frameworks can be generated, some of which have been found in real tectosilicates, but others of which are unknown.

In the ladder presented in Fig. 8.4c the two corners of adjacent tetrahedra in each vierer ring (oriented nonvertical in the Figure) which are linked to the next vierer ring point up ($\mathbb{U}$), while the other two point down ($\mathbb{D}$), forming rings with a sequence of tetrahedron directedness Δ of $\mathbb{UUDD}$. The resulting so-called double crankshaft chain can be more clearly seen in Fig. 8.5 which is a reproduction of W. H. Taylor's classical 1933 drawing. By linking such double crankshaft chains achter rings can be generated which again differ in the sequence of directedness of their tetrahedra. Of the 17 such frameworks possible (Smith and Rinaldi 1962) those of paracelsian, the feldspars, and harmotome have the sequences $\mathbb{UUDUDUUD}$, $\mathbb{UUUUDDDD}$, and $\mathbb{UUDDDDDD}$, respectively (Fig. 8.6).

As Fig. 8.7 shows, chain-like building units other than the double crankshaft chain can be formed if the sequence of directedness within the vierer ring is $\mathbb{UDUD}$ or $\mathbb{UUUD}$ instead of $\mathbb{UUDD}$. The former of these two units generates the vierer double chain observed in narsarsukite $Na_4Ti_2\{\boldsymbol{uB}, 2^1_\infty\}[Si_8O_{20}]O_2$ (Figs. 7.11g, 7.13a), but the latter unit has not been observed.

Vierer single rings with all four tetrahedra pointing up, $\mathbb{UUUU}$, form double rings when linked (Fig. 8.7d). The tubular dreier quadruple chain presented in Fig. 8.8 is another complex chain which can be described as a building unit of the tectosilicates.

Even more complex tubular building units have recently been described for zeolites and these are presented in a rather schematic way in Fig. 8.9 (Gramlich-Meier and Meier 1982).

8.4.4 Layer-Like Building Units

Instead of considering tetrahedral frameworks as being generated from zero-dimensional secondary building units or from one-dimensional chain-like building units, they may be described as being formed by linking two-dimensional layers. This can be readily demonstrated with the aid of the three structures shown in Fig. 8.6.

In the preceding section each vierer ring in this Figure was regarded as the projection of a double crankshaft chain and the framework was generated by connecting these double crankshafts. As an alternative, each vierer ring can be regarded as such and an infinite number of such single rings is then linked to form a tetrahedral layer. Such layers are the two-dimensional buildung units from which the framework can be eventually generated by successive linkage.

Since the topology of layers and the directedness of their tetrahedra can be easily and very clearly visualized, layer-like building units have proven to be very suitable not only for describing existing frameworks, but also for predicting unknown ones.

There are 17 topologically different simple ways to link $\mathbb{UUDD}$ vierer rings into a layer such that only achter rings are formed between them. Figure 8.10 is a shorthand description of these possibilities in which each ● or ○ represents a tetra-

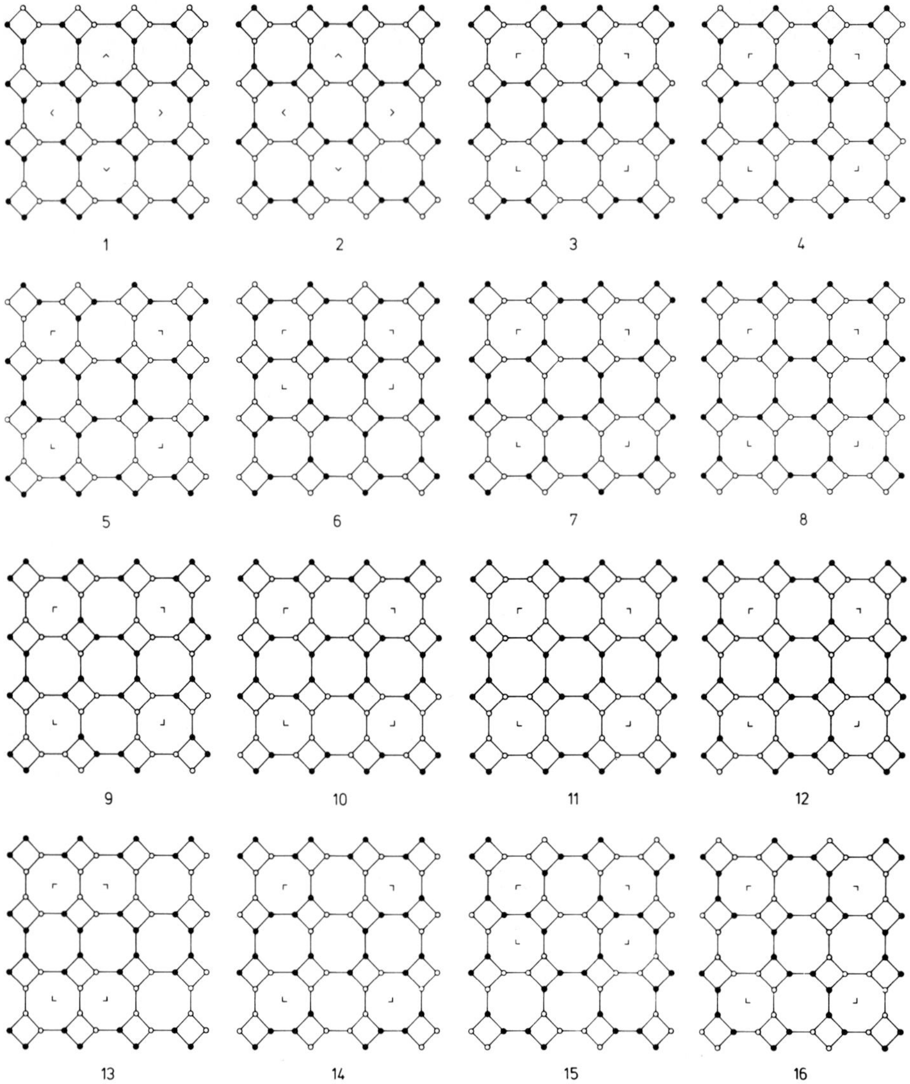

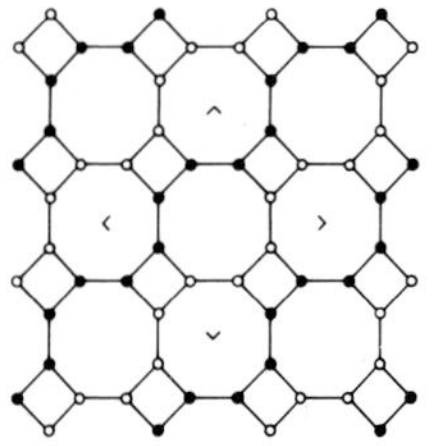

Fig. 8.10. Shorthand description of the 17 simplest ways to generate frameworks from $\mathbb{UUDD}$ vierer single rings. Tetrahedra pointing up ($\mathbb{U}$) are represented by ●; those pointing down ($\mathbb{D}$) by ○. The smallest unit cell of the ideal structure is indicated by the *rectangles* inscribed (after Smith and Rinaldi 1962).

No. 1 paracelsian, $Ba[Al_2Si_2O_8]$ (Bakakin and Belov 1961), synthetic $Na_2[ZnSi_3O_8]$ (Hesse et al. 1977); **no. 2** feldspars, $M[(Al,Si)_4O_8]$; **no. 3** harmotome, $Ba[Al_2Si_6O_{16}] \cdot 6H_2O$ (Rinaldi et al. 1974), phillipsite, $(Na,K)_5[Al_5Si_{11}O_{32}] \cdot 10H_2O$ (Rinaldi et al. 1974); **no. 11** merlinoite, $K_5Ca_2[Al_9Si_{23}O_{64}] \cdot 24H_2O$ (Galli et al. 1979), synthetic $Ba[AlSi_2O_6](OH,Cl)$ (Solov'eva et al. 1972); **no. 17** gismondine, $Ca[Al_2Si_2O_8] \cdot 4H_2O$ (Fischer 1963), garronite, $Na_2Ca_5[Al_{12}Si_{20}O_{64}] \cdot 28H_2O$ (Gottardi and Alberti 1974), zeolite P1, $Na_8[Al_8Si_8O_{32}] \cdot 16H_2O$ (Baerlocher and Meier 1972)

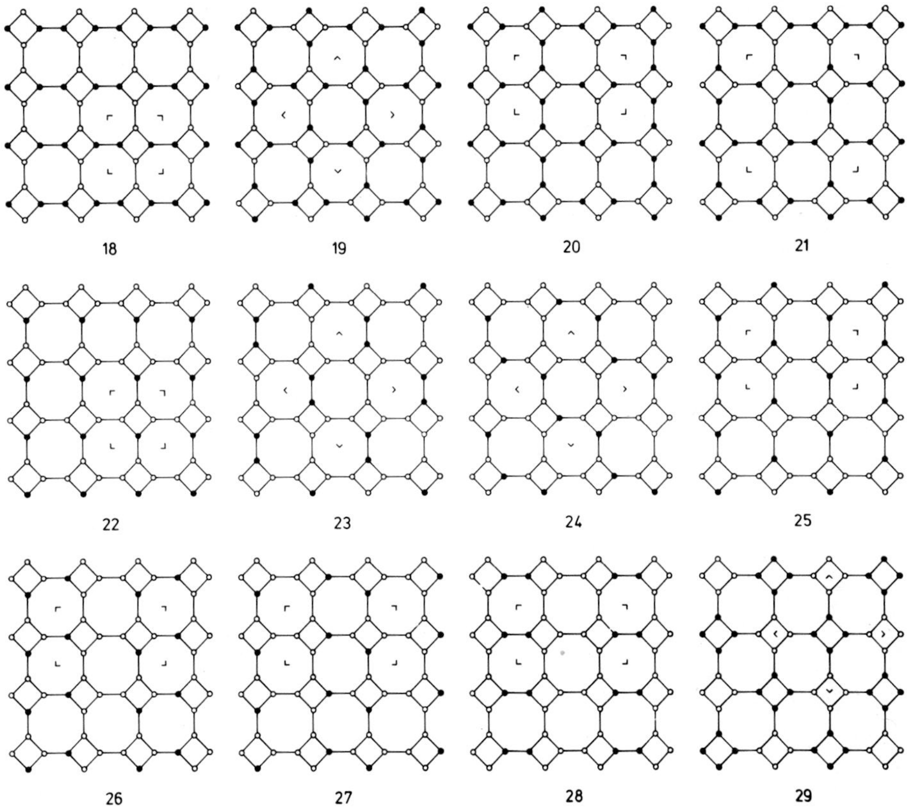

Fig. 8.11. Shorthand description of the simplest ways to generate frameworks from $\mathbb{UDUD}$ vierer rings (**18** to **21**), $\mathbb{UUUD}$ vierer rings (**22** to **28**), and $\mathbb{UUUU}$ vierer rings (**29**). Tetrahedra pointing up ($\mathbb{U}$) are represented by ●; those pointing down ($\mathbb{D}$) by ○. The smallest unit cell of the ideal structure is indicated by the *rectangles* inscribed (after Smith and Rinaldi 1962).

No. 19 banalsite, $Na_2Ba[Al_4Si_4O_{16}]$ (Haga 1973)

hedron pointing up and down, respectively. The first three of these layers correspond to the structures of harmotome, feldspar, and paracelsian shown in more detail in Fig. 8.6. Species have been observed of five of these framework types (Nos. 1, 2, 3, 11, and 17).

Figure 8.11 depicts the corresponding simplest ways of linking vierer single rings with directedness sequences $\mathbb{UDUD}$, $\mathbb{UUUD}$, and $\mathbb{UUUU}$, respectively, into layers.

Tetrahedra need not only form layers with vierer and achter rings. Several other layer-like building units are schematically presented in Fig. 8.12 without taking into account the directedness of each tetrahedron. The 29 simplest layer-like building units of Figs. 8.10 and 8.11, which have only vierer and achter rings, are represented by the one net in Fig. 8.12a. It is evident that each of the building units in Fig. 8.12 gives rise to a series of topologically different frameworks due to the fact that each tetrahedron can be either pointing up or down.

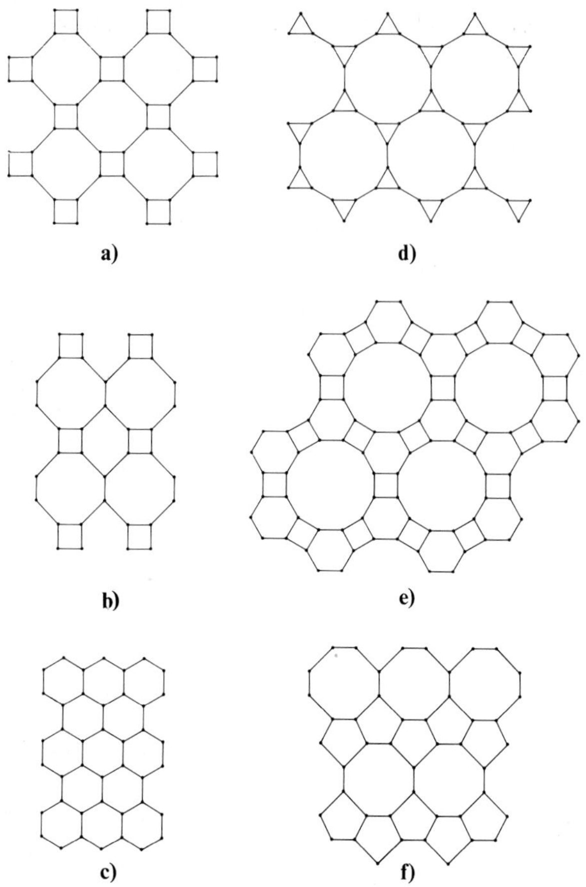

Fig. 8.12 a – f. Schematic description of several layer-like building units of silicate frameworks. Each node of these two-dimensional nets represents one tetrahedron without an indication of its directedness 𝕌 or 𝔻. The various nets are designated by their Schläfli symbols.[1]
a 4.8^2 net; **b** $(4.6.8)_2(6.8^2)_1$ net; **c** 6^3 net; **d** 3.12^2 net; **e** 4.6.12 net; **f** $(5^2.8)_2(5.8^2)_1$ net

The stereochemistry and, consequently, the classification of tectosilicates becomes even more complicated since layers can deviate in many ways from their ideal symmetry. For example, tetrahedral layers having the same topology and the same pattern of directedness can differ considerably in their actual geometrical shape. In particular, a layer can be flat, folded like a sheet of corrugated paper, or it can be curled, with small groups of tetrahedra protruding from one or the other

1 Schläfli symbols are used to describe planar networks of points which may represent atoms or, as in our case, polyhedra. Each polygon of such a network is specified by its number of corners: 3 stands for a triangle, 4 for a rectangle, 5 for a pentagon, etc. A node of the network is described by its surrounding polygons, the number of equal polygons surrounding a node being recorded by a right superscript. A network is then described by listing successively the nodes with nonequivalent polygonal surroundings and indicating the relative frequency of such nodes by right subscripts.

side of the layer (see Sec. 10.5.3). This enables adjacent layers of equivalent topology and directedness to develop different patterns of linking so that frameworks of different topology result.

In recent years J. V. Smith has made a very thorough study of silicate framework topology based on layer-like building units. This work has led to the publication of extensive lists of theoretically possible frameworks together with assignments of existing ones (Smith 1977, 1978, 1979, 1983; Smith and Bennett 1981).

8.4.5 Polyhedral Building Units; Framework Density

As shown earlier [see Sec. 2.1 (1)] silicates can be crudely described as tetrahedral arrangements of spherical oxygen atoms, each one touching three others in the same $[SiO_4]$ group. The silicon atoms are located near the centers of the oxygen tetrahedra with the other ions or molecules positioned in voids between these groups. In tectosilicates the size of the voids can vary considerably. This may be illustrated by comparing their *framework densities*, d_f, defined as the number of T atoms per 1000 Å³ of the tectosilicate, i.e.,

$$d_f = \frac{1000\,n_T}{V_{cell}} \tag{1}$$

where V_{cell} is the volume of the crystallographic unit cell and n_T is the number of T atoms in this unit cell.

The framework density is a measure of that portion of the unit cell volume that is occupied solely by the framework atoms.

Table 8.6 shows that the framework density ranges from 29.3 silicon atoms per 1000 Å³ in the high pressure silica polymorph coesite to about 12.7 (Al + Si)/1000 Å³ in the zeolite faujasite.

In tectosilicates with high framework densities the interstices between neighboring atoms are rather small. In the coesite structure, for example, the interstices are too small to permit the inclusion of other atoms or ions. On the other hand, the interstices in the quartz framework are large enough to enclose small ions, such as Li^+ and Mg^{+2} and the interstices of the tridymite and the feldspar frameworks can include larger cations, such as Na^+, K^+, Ca^{+2}, Ba^{+2}, and Pb^{+2}.

With further decrease in the framework density the voids between the framework oxygen atoms become wide enough to enclose water molecules (in bikitaite), methane and carbon dioxide molecules (in melanophlogite), larger organic molecules, such as piperidine NH or adamantylamine 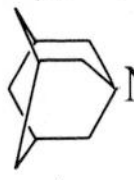NH_2 (in dodecasil 1 H), n- and iso-hydrocarbons and hydrated inorganic cations in many zeolites, and organic ions, such as $[(n\text{-}C_3H_7)_4N]^+$ (in silicalite-1) (Table 8.6).

It was mentioned previously (Secs. 8.3 and 8.4.2) that tetrahedral frameworks can be generated by linking rings of 4, 5, 6, 8 and more tetrahedra. Structural models of the frameworks can then be simplified by constructing planes of best fit through the rings, which, when allowed to intersect, will generally form concave (empty) polyhedra. A number of these polyhedra are shown in Fig. 8.13. A suitable

Table 8.6 Framework densities and types of tectosilicates

Silicate		Framework density [T/1000 Å^3]	Type	Reference
Name	Formula			
Coesite	SiO_2(hP)	29.3	P	Gibbs et al. 1977
Quartz	SiO_2	26.6	P	Wright and Lehmann 1981
Keatite	SiO_2	25.1	P	Shropshire et al. 1959
Cristobalite	SiO_2(hT)	23.0	P	Peacor 1973
Tridymite	SiO_2(hT)	22.6	P	Kihara 1978
Orthoclase	$K[AlSi_3O_8]$	22.2	P	Phillips and Ribbe 1973
Bikitaite	$Li[AlSi_2O_6] \cdot H_2O$	20.2	P	Kocman et al. 1974
Melanophlogite	$46\,SiO_2 \cdot 8\,(N_2, CH_4, CO_2)$	19.0	C	Gies 1983
Analcime	$Na[AlSi_2O_6] \cdot H_2O$	18.6	Z	Ferraris et al. 1972
Dodecasil 1H	$34\,SiO_2 \cdot (C_5NH_{11}, C_6NH_{13}, \ldots) \cdot 5\,(N_2, \ldots)$	18.4	C	Gerke and Gies 1984
Silicalite-1 (ZSM-5)	$96\,SiO_2 \cdot 4\,[N(C_3H_7)_4]F$	17.9	Z	Olson et al. 1981, Price et al. 1982
Sodalite	$Na_4[Al_3Si_3O_{12}]Cl$	17.1	C	Löns and Schulz 1967
Mordenite	$Na[AlSi_5O_{12}] \cdot 3\,H_2O$	17.0	Z	Meier 1961
Harmotome	$Ba[Al_2Si_6O_{16}] \cdot 6\,H_2O$	15.9	Z	Rinaldi et al. 1974
Chabazite	$Ca[Al_2Si_4O_{12}] \cdot 6\,H_2O$	14.6	Z	Smith et al. 1963
Zeolite rho	$(Na, Cs)_3[Al_3Si_9O_{24}] \cdot 11\,H_2O$	14.3	Z	Robson et al. 1973
Zeolite A	$Na_{12}[(Al_{12}Si_{12})O_{48}] \cdot 27\,H_2O$	12.9	Z	Gramlich and Meier 1971
Faujasite	$(Na_2, Ca, Mg)_{29}[Al_{58}Si_{134}O_{384}] \cdot 240\,H_2O$	12.7	Z	Baur 1964

P = pyknolites, C = clathrasils, Z = zeolites

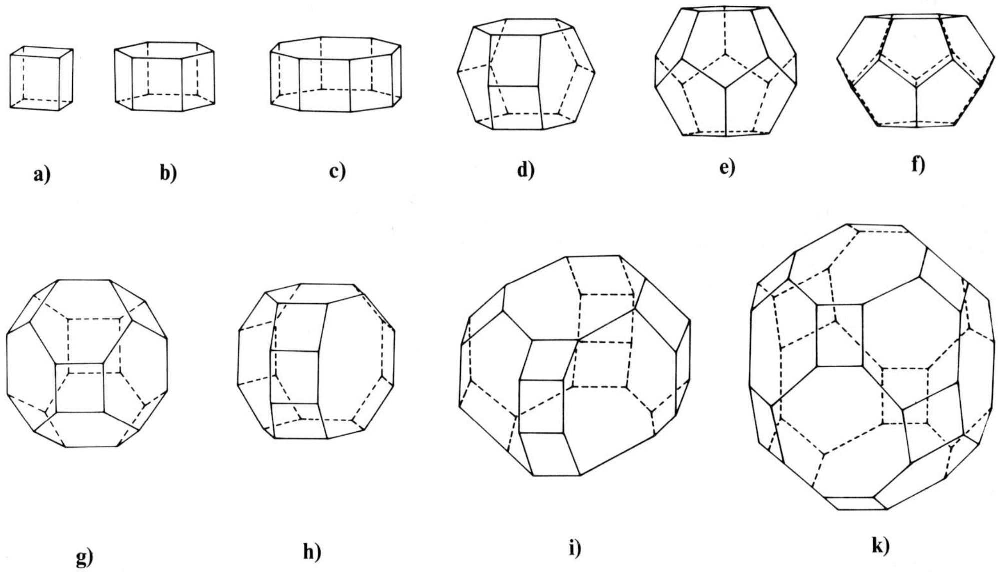

Fig. 8.13 a–k. Polyhedral building units of silicate frameworks.
a The cube $[4^6]$; **b** the hexagonal prism $[4^6 6^2]$; **c** the octagonal prism $[4^8 8^2]$. These three polyhedra are identical with the secondary building units 4-4, 6-6, and 8-8 of Fig. 8.2.
d The 11-hedron $[4^6 6^5]$, sometimes called ε-cage; **e** the 12-hedron $[5^{12}]$, pentagondodecahedron; **f** the 14-hedron $[5^{12} 6^2]$; **g** the 14-hedron $[4^6 6^8]$, called truncated octahedron or β-cage; **h** the 14-hedron $[4^9 6^2 8^3]$; **i** the 18-hedron $[4^{12} 8^6]$, called γ-cage; **k** the 26-hedron $[4^{12} 6^8 8^6]$, called truncated cuboctahedron or α-cage

way of distinguishing between different polyhedra is to use the notation $[n_i^{m_i}]$, where m is the number of n-gons defining the polyhedron and $\sum m_i$ is the number of faces of the polyhedron. For example, $[4^6 6^2]$ is the symbol for the sechser double ring circumscribed by six quadrilaterals and two hexagons, whereas $[5^{12}]$ and $[4^3 5^6 6^3]$ represent, respectively, the pentagondodecahedron having pentagons only, and a 12-hedron defined by three quadrilaterals, six pentagons, and three hexagons.

A number of silicate frameworks can be generated entirely from one kind of polyhedral void, which is then called the *polyhedral building unit.* The polyhedral building units found in tectosilicates of low framework density are given in column four of Table 8.5.

In addition to these polyhedral building units, which in clathrate science are sometimes called fundamental or basis cages, most low density tectosilicates contain other polyhedral voids. Many low density frameworks cannot be generated from a single type of polyhedral voids and, therefore, they have no polyhedral building units.

It should be mentioned that there is no clear-cut demarcation between secondary and polyhedral building units. The vierer, sechser, and achter double rings 4–4, 6–6, and 8–8 ascribed to the secondary building units by most zeolite scientists can equally well be considered to be small polyhedral building units $[4^6]$, $[4^6 6^2]$, and $[4^8 8^2]$, respectively.

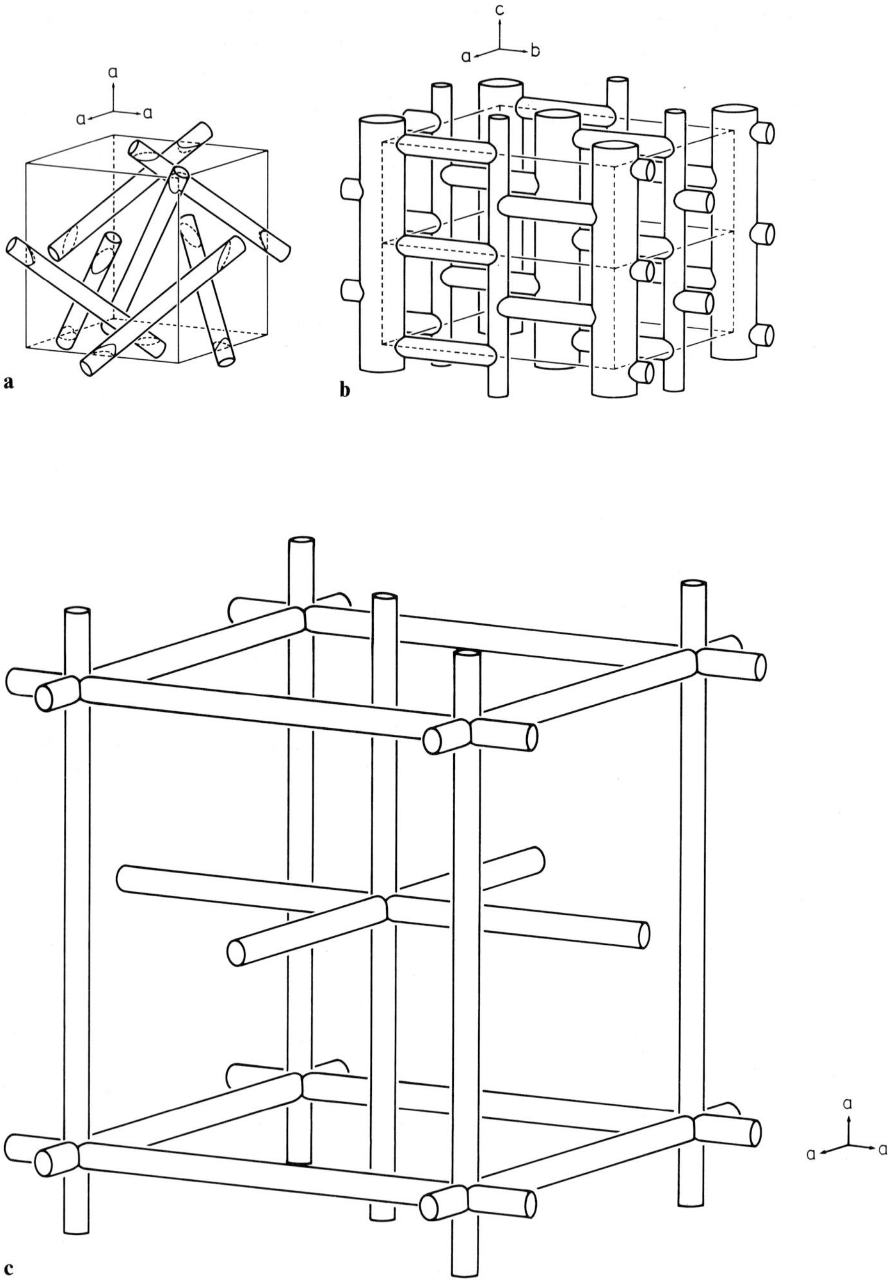

Fig. 8.14 a – c. Channel systems in zeolites.

a System of nonintersecting one-dimensional channels in analcime, $Na[AlSi_2O_6] \cdot H_2O$, running parallel to the ⟨111⟩ zone axes (Ferraris et al. 1972). **b** System of two-dimensionally intersecting channels parallel to (100) in mordenite, $Na[AlSi_5O_{12}] \cdot 3H_2O$ (Meier 1961). **c** Two interpenetrating systems of three-dimensionally intersecting channels in paulingite, $(K,Ca,Na)_2[Al_3Si_{11}O_{28}] \cdot 12H_2O$ (Gordon et al. 1966)

Polyhedral voids share corners, edges, and faces with each other. Faces shared between polyhedra are often called the "windows" of the cages. Whether or not a molecule or ion can pass from one polyhedral void to the next depends on the free diameter between the oxygen atoms of such windows. If the value of n, i.e., the number of tetrahedra forming the window, is small, for example $n = 4$ or 6, only small ions or molecules, such as Li^+ or N_2, can pass through, and even then usually only at elevated temperatures. Larger species are firmly trapped in the polyhedral voids which are, therefore, often called cages.

As the free diameter of the windows increases to $n = 8$, 10, or 12, large species can move more or less readily from one void to the next, and the cage-like or polyhedral character of the voids diminishes progressively. Instead of containing strings of concave polyhedral cavities, such frameworks contain open channels. These channels may either be separated, running in one direction only, or they may be interconnected forming a two-dimensional or even three-dimensional channel system (Fig. 8.14).

Tectosilicates can be subdivided into three main groups based on the nature of the voids between the framework oxygen atoms (Liebau 1983):

(1) Pyknolites: tectosilicates having no obvious polyhedral cavities, but rather small interstices between neighboring oxygen atoms housing cations and sometimes water;

(2) Clathrasils: tectosilicates with polyhedral cavities (cages) having windows too small to let through the encaged ions or molecules or their decomposition products;

(3) Zeolites: tectosilicates containing systems of polyhedral cavities interconnected by large windows or tunnels such that the enclosed ions or molecules can readily diffuse through the crystal.

8.4.6 Concluding Remarks

The kind of building unit – zero-, one-, or two-dimensional ones – which is most suitable for the classification of tectosilicates depends on the property under consideration. Since the chain type correlates with the properties of the cations in a silicate, the crystal chemical classification based on chain-like building units will be suitable if relationships between chemical composition and structure are of interest. If exchange, retention, or adsorption properties of ions or molecules are studied, the classification based on polyhedral cavities and channel-like voids will be preferred. Finally, if a complete enumeration of possible framework structures is desired, then layer-like building units should be used in the classification.

8.5 Silicate Classification Based on Non-Silicon Cation – Oxygen Polyhedra

It has been emphasized several times in earlier chapters that the non-tetrahedral cation–oxygen bonds should not be neglected in describing a silicate structure. Their contribution to the physical and chemical properties of a silicate is particularly prominent in silica-poor silicates and in those which contain relatively small cations with valences higher than two. Many authors have recognized the importance of the cation–oxygen bonds in determining the stability of a silicate structure and a number have, therefore, used the cation–oxygen polyhedra to describe silicate structures. Belov, in particular, has pointed out correlations between cation–oxygen polyhedra and the shape of the silicate anions in a large number of papers (Belov 1959).

However, no systematic attempt has been made to classify silicates according to the cation–oxygen polyhedra and their way of condensation.

This is not surprising since the cation–oxygen polyhedra can vary considerably in coordination number CN and in shape, especially for large values of CN. A scheme of condensation products would have to be developed for each of the different $[MO_n]$ polyhedra by analogy to the one described for the $[SiO_4]$ tetrahedra. Therefore, a classification based on $[MO_n]$ polyhedra would be much more complicated than one based on $[SiO_4]$ tetrahedra, especially since the majority of silicates contains more than one kind of cation.

An attempt at this kind of classification has been made for $[MO_6]$ octahedra, which are perhaps the most abundant polyhedra, but mainly in the case of phosphates (Moore 1970b, 1974; Keller 1972; Higgins and Ribbe 1977).

A classification optimal for interpreting and understanding the physical and chemical properties of silicates is one based on both the silicon– and cation–oxygen polyhedra, i.e., $[MO_n]$ as well as $[SiO_4]$. For this classification the priority of condensation of the various polyhedra could be chosen according to a weighting scheme in which polyhedra with smaller coordination numbers and higher valences get higher priorities than polyhedra with higher coordination numbers and lower valences. However, such a classification would probably be much too complicated to be of significant practical value.

9 General Rules for Silicate Anion Topology

A survey of the known anions containing tetrahedrally coordinated silicon shows that the topology of the vast majority of these anions obeys five general rules:

1) $[SiO_4]$ tetrahedra are linked to other $[SiO_4]$ tetrahedra via corners rather than edges or faces.
2) One oxygen atom can belong to no more than two $[SiO_4]$ tetrahedra.
3) If $\boldsymbol{s}$ is the connectedness, i.e., the number of oxygen atoms of an $[SiO_4]$ tetrahedron shared with other $[SiO_4]$ tetrahedra, then for a given silicate anion the difference, $\Delta \boldsymbol{s}$, between the $\boldsymbol{s}$ values of all $[SiO_4]$ tetrahedra tends to be small.
4) The dimensionality $\boldsymbol{D}$ of a silicate anion, i.e., the number of dimensions in which the anion is infinitely extended, tends towards the highest value possible for a given O : Si ratio.
5) Bond lengths and bond angles tend to vary as little as possible from the mean values

$$\langle d(\mathrm{Si-O})\rangle = 1.62\ \text{Å};$$
$$\langle \measuredangle(\mathrm{O-Si-O})\rangle = 109.47°;$$
$$\langle \measuredangle(\mathrm{Si-O-Si})\rangle = 140°.$$

These five rules are not natural laws with no exceptions. They only mean to imply that violations of the rules are, in general, energetically less favorable and, therefore, less likely to occur in nature. In fact, at least one exception is known to all five rules.

Ad rule 1: For purely ionic bonds between silicon and oxygen a straight Si–O–Si bond angle would be expected due to the repulsive forces between the positively charged silicon atoms in the centers of the tetrahedra. However, since the Si–O bond also has a strong covalent character, the Si–O–Si bond angle is reduced to a value near 140° (see Sec. 3.1.2.5). The Si···Si distance is thereby reduced from 2×1.62 Å $= 3.24$ Å to 3.08 Å for ideal $[SiO_4]$ tetrahedra with no statistical substitution for silicon, while the mean values of the Si–O distances and O–Si–O bond angles remain unchanged from those of rule 5.

Edge-sharing or face-sharing between two $[SiO_4]$ tetrahedra would reduce the Si···Si distance to 1.87 Å and 1.08 Å, respectively (Fig. 9.1), and would, therefore, significantly increase the repulsive forces between the silicon atoms. This suggests that edge-sharing and face-sharing of $[SiO_4]$ tetrahedra is energetically very unfavorable. In fact, the fibrous form of SiO_2 (Fig. 9.2), first synthesized by Weiss and Weiss (1954), is extremely hygroscopic and decomposes to amorphous silica in the presence of traces of water. It is the only silicon–oxygen compound in which

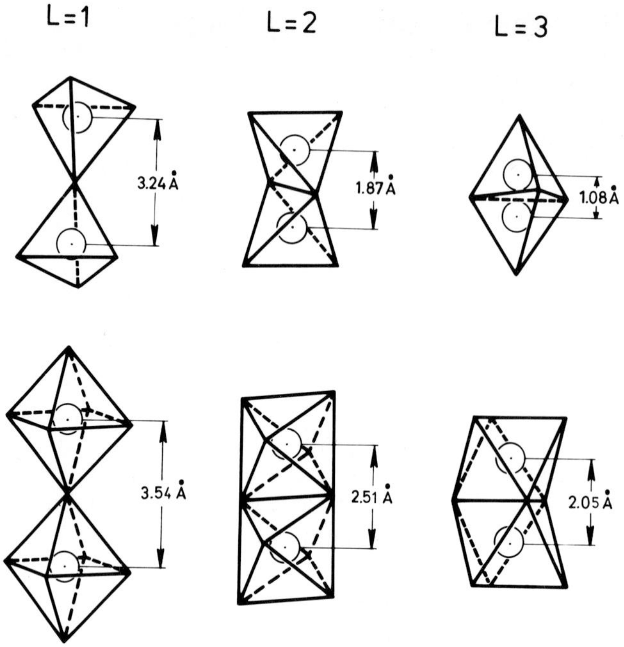

Fig. 9.1. Maximum Si···Si distances for undistorted $[SiO_4]$ tetrahedra and $[SiO_6]$ octahedra of average size for different values of linkedness ***L***

the presence of edge-shared $[SiO_4]$ tetrahedra has been established beyond doubt (see Footnote 1, Chap. 7).

For $[SiO_6]$ octahedra, an analogous rule holds. Here the Si···Si distance is reduced from 3.54 Å for ideal corner-shared octahedra with straight Si–O–Si bonds to 2.51 Å in the case of edge-sharing and 2.05 Å for face-sharing (Fig. 9.1).

By comparing these Si···Si distances and taking into account only the repulsive forces between the silicon atoms, it is clear that the stability of linked silicon–

Fig. 9.2. Chain of edge-sharing $[SiO_4]$ tetrahedra in the fibrous polymorph of silica (Weiss and Weiss 1954)

oxygen polyhedra decreases in the sequence corner-shared octahedra – corner-shared tetrahedra – edge-shared octahedra – face-shared octahedra – edge-shared tetrahedra – face-shared tetrahedra as a function of the decrease in Si···Si distance and the screening of the positive charges of the silicon ions by the negative charges of the oxygen ions. This explains why only one of the vast number of phases with $[SiO_4]$ tetrahedra has been found to contain edge-shared tetrahedra, and why none have been observed to contain face-shared tetrahedra. In contrast, of the small number of phases known to contain octahedrally coordinated silicon, about half as many phases have corner-shared octahedra relative to those which have edge-shared octahedra (Table 7.1).

Rule 1 is nothing more than Pauling's third rule (the sharing of elements rule) for ionic compounds applied to silicates (Pauling 1967, p. 559 f.).

Ad rule 2: According to Pauling's second rule (the electrostatic valence rule; Pauling 1967, p. 547), the sum of the strengths of the bonds between an anion and its adjacent cations tends to be equal to the valency of the anion as a result of an attempt by the anion to balance its electrostatic charge in its immediate neighborhood. Here the bond strength is defined as the electrostatic valency of the cation divided by its coordination number. Therefore, in silicates with $[SiO_4]$ tetrahedra, the strength of the bond between tetravalent silicon and oxygen is $4/4 = 1$.

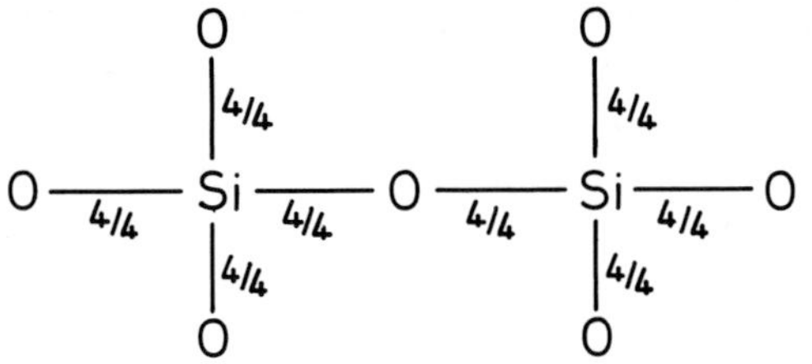

The charge of the divalent oxygen ion is then balanced by bonds to just two silicon ions. Linking to a third silicon ion would result in a high concentration of positive charge from the silicon ions bonded to this oxygen ion

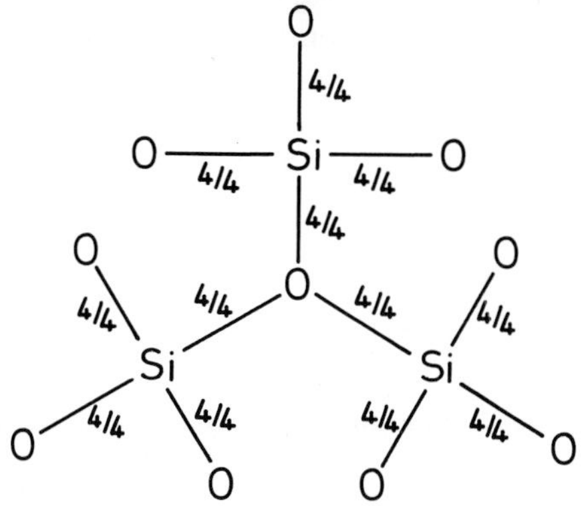

and would be energetically highly unfavorable.

Ad rule 3: This rule, first pointed out by Huggins et al. (1943) and later rediscovered by the present author (Liebau 1962b), implies that structures which contain only singular, primary, or secondary, etc., tetrahedra, should be the most stable, while those with singular and primary or with primary and secondary, etc., tetrahedra should be somewhat less favorable. Structures which contain singular and secondary $[SiO_4]$ tetrahedra are expected to be even more unstable and those with a connectedness difference $\Delta s \geq 3$ should not exist.

A survey of crystalline silicate structures shows that in terms of the number of bridging oxygen atoms, the majority of such silicates have just one kind of $[SiO_4]$ tetrahedra. These include: all the monosilicates which contain only singular tetrahedra, the disilicates which have only primary tetrahedra, the unbranched cyclosilicates and the silicates with unbranched single chains which are exclusively composed of secondary tetrahedra, the silicates with unbranched double rings and double chains of composition $[Si_{2n}O_{5n}]$, the unbranched phyllosilicates with $\{{}^{2}_{\infty}\}[Si_{2n}O_{5n}]$ layers, and all framework silicates (except those with interrupted frameworks) which have only tertiary tetrahedra.

There is only a relatively small number of silicate anions with $\Delta s = 1$. These include the unbranched oligotetrahedra having more than two tetrahedra (Fig. 7.3c–f), the loop-branched single chains (Fig. 7.10d–h), the unbranched double chains (Fig. 7.11b, d, e, and h) in which the two single chains are not linked via all their tetrahedra, the oligofold chains of Fig. 7.12a, b, and c, a few unbranched silicate single layers (Fig. 7.22), most of the double layers (except the one in Fig. 7.25a), the interrupted frameworks, and most of the branched silicate anions. Silicates containing these anions – except the amphiboles with a zweier double chain (Fig. 7.11b) – are very rare and in almost all cases only one example of the specific anion is known.

Silicates which contain tetrahedra with connectedness differing by more than one bridging oxygen atom ($\Delta s > 1$) are extremely rare; they are listed in Table 9.1.

Within a given silicate, independent of whether it is crystalline, vitreous, molten, or in solution, under certain given conditions all singular tetrahedra will have similar energy contents. This energy will, however, be distinctly different from that of primary tetrahedra under the same conditions, and so on. The lowest energy content is achieved when the silicate contains only one kind of $[SiO_4]$ tetrahedron (that is if $\Delta s = 0$), provided that is possible for the specific chemical composition and temperature and pressure conditions. Alternatively, a mixture of the two Q^s units with lowest energy content will be formed, and these will, in almost all cases, be tetrahedra that differ by just one bridging oxygen atom.

The local energy content at a particular point within a sample is dependent on the degree of electrostatic charge balance at that point. This charge balance, in turn, is strongly affected by large values of Δs in the sample.

Consider, for example, a silicate with an atomic ratio Si : O = 5 : 16, no statistical replacement of silicon by other tetrahedral cations, and highly electropositive

Table 9.1 Silicates containing $[SiO_4]$ tetrahedra which differ in their connectedness s by $\Delta s > 1$

Silicate	Connectedness s					Δs	Fig.
	0	1	2	3	4		
Uniform-anion silicates							
Eakerite		+	+	+		1, 2	7.7a
Tienshanite		+	+	+		1, 2	7.7b
Aenigmatite		+	+	+		1, 2	7.10b
Saneroite		+	+	+		1, 2	7.10c
$NaBa_3Nd_3[Si_2O_7][Si_4O_{13}]$		+		+		2	7.4a
Astrophyllite		+		+		2	7.10a
Prehnite			+		+	2	7.23a
Zunyite		+			+	3	7.4b
Mixed-anion silicates							
Ardennite	+	+	+			1, 2	
Bavenite		+	+	+		1, 2	7.15a
Miserite		+		+		2	7.12d
Meliphanite	+		+		+	2, 4	7.23c

cations (i.e., almost fully ionized silicate anions). Even if allowance is made for only two different kinds of silicate anion within the sample, a whole series of mixtures is possible, a few of which are given below:

(1)	$[SiO_4]^{-4}$ single tetrahedron	+	$\{\boldsymbol{r}\}[Si_4O_{12}]^{-8}$ single ring
(2)	$3n[SiO_4]^{-4}$ single tetrahedra	+	$\{{}^{3}_{\infty}\}[SiO_2]_{2n}$ framework
(3)	$n[Si_2O_7]^{-6}$ double tetrahedra	+	$\{{}^{1}_{\infty}\}[SiO_3]^{-6n}_{3n}$ single chain
(4)	$4n[Si_2O_7]^{-6}$ double tetrahedra	+	$\{{}^{3}_{\infty}\}[SiO_2]_{2n}$ framework
(5)	$n[Si_3O_{10}]^{-8}$ triple tetrahedra	+	$\{{}^{1}_{\infty}\}[SiO_3]^{-4n}_{2n}$ single chain
(6)	$[Si_5O_{16}]^{-12}$ unbranched fivefold tetrahedron		
(7)	$\{\boldsymbol{oB}, \boldsymbol{t}\}[Si_5O_{16}]^{-12}$ branched triple tetrahedron		

Consider as examples, cases 2, 6, and 7. While unbranched multiple tetrahedra, such as the unbranched fivefold tetrahedron (case 6) have rather stretched conformations, a branched triple tetrahedron, such as the one found in zunyite is almost spherical (Fig. 9.3). In both ions the negative charges can be considered to be localized at the terminal oxygen atoms and, therefore, rather evenly distributed over the ion surface. Due to the relatively large radius of this branched ion, there is a substantial, essentially spherical, volume in the interior of the ion containing almost no negative charges. In contrast, the "uncharged" region of the unbranched fivefold tetrahedron is a narrow cylinder. Charge balance must, therefore, be achieved over larger distances in the spherical ion than in the linear one. In case 2 the difference is even more pronounced since the silica framework is electrically neutral with no large spatial charge fluctuations within the entire crystal, while in the single tetrahedra all the positive charges of the cations have to cluster around a small highly charged anion. For cases 1, 3, 4, and 5 the situation will be somewhere between the three cases considered. From the electrostatic point of view, the unbranched fivefold tetrahedron should be the most stable of all the above anions. For reasons which will be discussed in Sec. 11.1, the branched $[Si_5O_{16}]$ group and

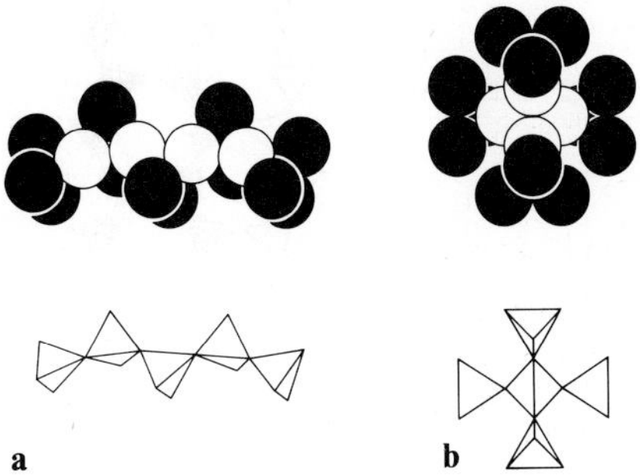

Fig. 9.3 a, b. Simplified representation of the charge distribution in $[Si_5O_{16}]^{-12}$ anions. *Black spheres:* terminal oxygen atoms as carriers of the negative charges; *white spheres:* bridging oxygen atoms carrying no formal charges.

a Unbranched quintuple tetrahedron in medaite, $HMn_6V\{\boldsymbol{uB}, 5\boldsymbol{t}\}[Si_5O_{16}]O_3$ (Gramaccioli et al. 1981); **b** open-branched triple tetrahedron in zunyite, $Al_{13}\{\boldsymbol{oB}, 3\boldsymbol{t}\}[Si_5O_{16}](OH,F)_{18}O_4Cl$ (Baur and Ohta 1982) crystalline oligosilicates with multiplicities $M > 3$ are very unstable and extremely rare. They exist only in silicates with electronegative cations in which the actual charges of the silicate anions are considerably lower than the formal charges.

Ad rule 4: By sharing corners between $[SiO_4]$ tetrahedra, the Si : O ratio of the silicate anion may be increased from 1 : 4 to 1 : 2, depending on the degree of condensation. The oligosilicate anions (see Sec. 7.2.2.1), which contain a limited number of $[SiO_4]$ tetrahedra and are, therefore, zero-dimensional, have Si:O values between 1 : 4 and 1 : 3. The value 1 : 3 is adopted by unbranched one-dimensional single chains and zero-dimensional single rings. Both types of anions contain only secondary $[SiO_4]$ tetrahedra.

As soon as the Si:O ratio becomes larger than 1 : 3, that is, as soon as the anion contains secondary and tertiary tetrahedra, i.e., Q^2 as well as Q^3 units, the existence of two-dimensionally infinite layers becomes possible. For example, Fig. 9.4 demonstrates that for Si:O = 1 : 2.67 anions are possible which are infinite in two, one, or zero dimensions. For the ratio 1 : 2.5, as already mentioned, structures may form, which contain zero-dimensional double rings or one-dimensional double chains where the single rings or chains are linked via all tetrahedra (Fig. 7.6, Fig. 7.11 a, c, f, g, i) or alternatively which can contain a number of two-dimensional single layers (Figs. 7.18, 7.19, and 7.23 a). In synthetic $K_2Ce\{\boldsymbol{oB},{}^{3}_{\infty}\}[{}^4Si_6O_{15}]$, a three-dimensional framework has been found (Strelkova et al. 1977) which contains only tertiary tetrahedra Q^3 and has a rather low framework density of 17.8 Si atoms per 1000 $Å^3$.

For Si:O > 1 : 2.5 it does not seem possible to form a zero-dimensional cluster of $[SiO_4]$ tetrahedra consistent with rule 5, although it would not be forbidden on topological grounds.

Table 9.2 gives a comparison between the dimensionalities observed and the dimensionalities possible for the various Si:O ratios when only corner-sharing between tetrahedra is allowed. As pointed out by Laves (1932), rule 4 dictates that of

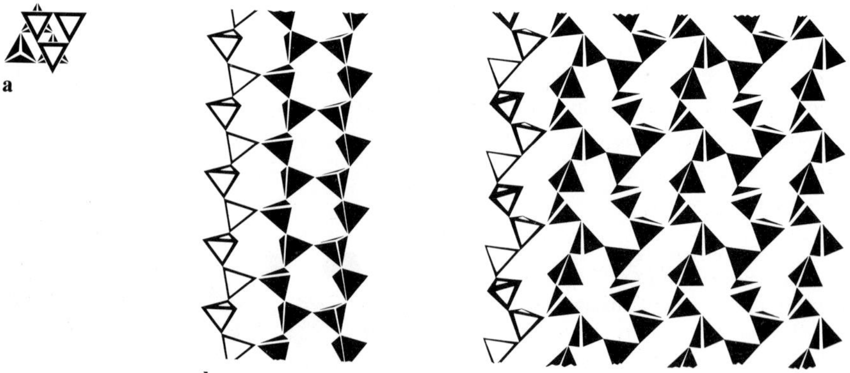

a b c

Fig. 9.4 a – c. Dimensionality of silicate anions with Si : O = 1 : 2.67.
a Zero-dimensional dreier double ring $\{\boldsymbol{uB}, 2\boldsymbol{r}\}[{}^3Si_6O_{16}]$ (hypothetical); **b** one-dimensional zweier triple chain in $Ba_4\{\boldsymbol{uB}, 3^{2}_{\infty}\}[{}^2Si_6O_{16}]$ (Hesse and Liebau 1980 a); **c** two-dimensional dreier single layer in $Na_2Zn\{\boldsymbol{uB}, 1^{2}_{\infty}\}[{}^3Si_3O_8]$ (Hesse et al. 1977)

Table 9.2 Comparison between the observed dimensionalities of silicate anions for various values of the atomic ratio Si : O and the dimensionalities topologically possible when only corner-sharing between tetrahedra is allowed

Si : O Ratio	Dimensionality	
	Possible	Observed[a]
1 : 4 ≤ r < 1 : 3	0	0
r = 1 : 3	0, 1	0, **1**
1 : 3 < r < 1 : 2.5	0, 1, 2, 3	0, 1, 2, 3
r = 1 : 2.5	0, 1, 2, 3	0, 1, **2**, 3
1 : 2.5 < r < 1 : 2	0, 1, 2, 3	2, 3
r = 1 : 2	1, 2, 3	2, **3**

[a] Bold type denotes dimensionalities most often observed for a given Si : O ratio

the various possibilities given in Table 9.2, the one with the highest dimensionality should be realized for a given Si:O ratio. This means that a definite relationship should exist between the Si:O value of the silicate anion and its dimensionality.

While this seemed to be true in the early days of silicate structure research, an increasing number of silicate anions have now been found which violate this rule. For example, the ratio Si:O = 1 : 3 applies not only to one-dimensional unbranched and open-branched single chains, but also to zero-dimensional monocyclic anions. Moreover, the ratio 1 : 2.5 has been found in two-dimensional single layers, in those zero-dimensional double rings and one-dimensional double chains in which two single chains or single rings are linked via all their tetrahedra and in the three-dimensional framework of $K_2Ce\{\boldsymbol{oB}, {}^{3}_{\infty}\}[Si_6O_{15}]$. Finally, the atomic ratio 1 : 2 is eventually not only observed in tectosilicates, but also in two-dimensional double layers of the kind shown in Fig. 7.25 a.

Although there are now quite a number of exceptions, the vast number of silicates still follow rule 4. There are probably several reasons for this. One may perhaps be seen in the fact that for a given Si:O ratio, the tetrahedral anions with the highest dimensionality can achieve charge balance over shorter distances than the others. This is clear from the models of the anions with Si:O = 1 : 2.5 shown in Fig. 9.5. Since the negative charges of the silicate anions are localized at the terminal oxygen atoms, the zero-dimensional double rings form almost circular to disk-shaped distributions of charge, while the one-dimensional multiple chains form tube-like anions, both of which have charge-free regions of considerable extension in their centers. In contrast, the corrugated two-dimensional layers in sanbornite, the room temperature polymorph of $Ba[Si_2O_5]$, have their negative charges uniformly distributed throughout the layers, while the very open framework of $K_2Ce[Si_6O_{15}]$ is completely permeated with free volume in which to homogeneously distribute the cations.

Another reason for the tendency of silicates to have anions with the highest dimensionality possible for the particular Si:O ratio may be that very often the anions with lower dimensionality form structures with rather low densities which are themselves less stable.

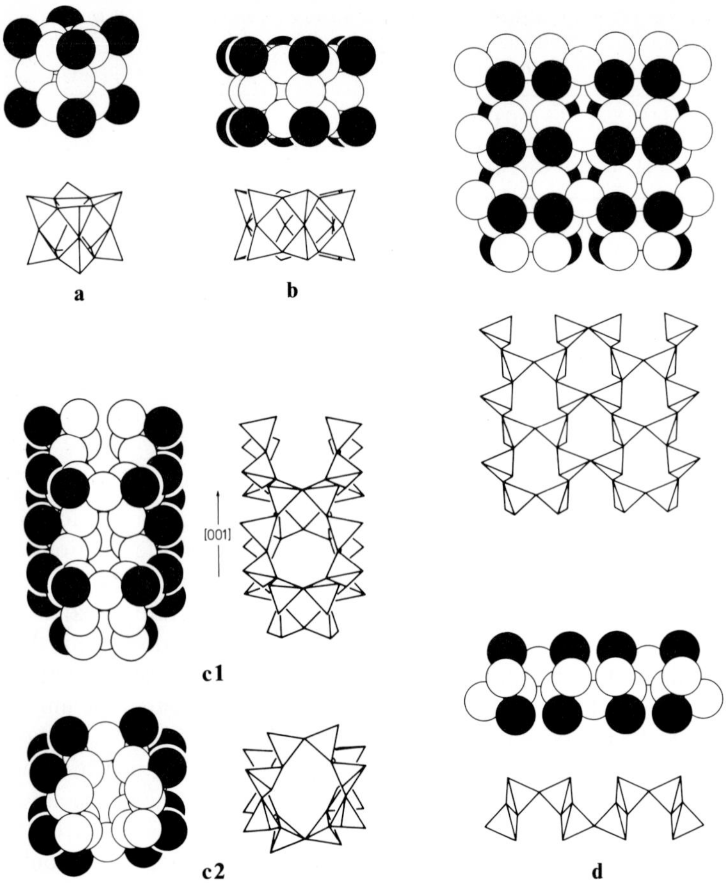

Fig. 9.5 a – d. Simplified models of the charge distribution in several anions with an Si : O ratio of 1 : 2.5.

a Zero-dimensional vierer double ring in $[N(CH_3)_4]_8[Si_8O_{20}] \cdot 64.8\,H_2O$ (Smolin et al. 1979); **b** zero-dimensional sechser double ring in $NaKMg_5[Si_{12}O_{30}]$ (Nguyen et al. 1980); **c** one-dimensional dreier quadruple chain in miserite (Scott 1976), (c1) projected approximately perpendicular to the chain direction [001], (c2) projected approximately parallel to the chain direction [001]; **d** two-dimensional zweier single layer in sanbornite, $Ba[Si_2O_5]$(lT) (Hesse and Liebau 1980 b).

Black spheres: terminal oxygen atoms as carriers of the negative charges. *Open spheres:* bridging oxygen atoms carrying no formal charges

Ad rule 5: The mean values of 1.62 Å for the Si–O bond length and 109.47° for the O–Si–O bond angle are close to the values adopted by the atoms of an isolated $[SiO_4]$ tetrahedron. The length of any particular Si–O bond and the value of a particular O–Si–O angle deviate from these "ideal" values by amounts that depend on the nature of the other atoms attached to the oxygen atom of the bond considered: in general, the length of the Si–O bond increases with increasing valence and decreasing size of these atoms (see Chap. 3). In a similar way, the mean value of the Si–O–Si bond angle, 140°, represents the bond angle at which the strain is at its minimum. It, therefore, approximates the value of the "ideal" bridging angle which the silicon and oxygen atoms attempt to attain in their structures. For a particular Si–O–Si bond this mean value may also be altered by the properties of atoms attached to the silicon atoms.

The actual values of the atomic distances and angles usually lie within the ranges:

d(Si–O): 1.57 Å to 1.68 Å;
∢(O–Si–O): 100° to 120°;
∢(Si–O–Si): 125° to 170°.

If values near or beyond the limits of these ranges are reported in a structure, it is very often a reflection of inaccuracy in the structure determination, including the choice of a wrong space group. However, in a number of silicates, such extreme values do exist and are not caused by inaccuracies or errors in the structure analysis (see Chap. 3). These structures should be carefully studied because they are most likely to enable us to obtain a deeper insight into the structural chemistry of the silicates.

10 Influence of Non-Tetrahedral Cation Properties on the Structure of Silicate Anions

The shape of a silicate anion depends on a number of factors. The most important ones are: size, valence, and electronegativity of the cations, temperature and pressure at the time when the silicate was formed, thermal and pressure history of the silicate since its formation, and temperature and pressure at the time of examination. The influence of cation properties on the shape of the silicate anions will be discussed in this Chapter.

10.1 Influence of Cation Properties on the Conformation of Unbranched Single Chain Anions

10.1.1 Conformation of Unbranched Silicate Single Chains

The selection of unbranched silicate single chains presented in Fig. 7.9 demonstrates that the shape of such chains varies greatly. Adding the branched single chains (Fig. 7.10) to the unbranched ones extends the range of variations even more. Since the vast majority of single chain silicates contain unbranched chains, these can be used to begin studying the influence of the various factors that control the shape of the silicate chains. In order to do so, a suitable way must be found to describe the chains.

One obvious property of the tetrahedral chains is their periodicity ***P*** used to classify the silicates with anions of infinite extension. Another characteristic property is the degree of stretching of the chains which can be measured by the stretching factor f_s (see Sec. 6.3.2). Instead of the stretching factor, the average value of the angles between the two tetrahedral edges formed by the bridging oxygen atoms of adjacent tetrahedra could be used. A third property is the flatness of the chains which could, for example, be defined by the average deviation of the silicon atoms of a chain from a plane of best fit through these atoms. A fourth property could be the angles of rotation of the $[SiO_4]$ tetrahedra from a position symmetrical to this plane of best fit. Finally, the distortions of the $[SiO_4]$ tetrahedra from their ideal symmetry and the deviations from their average size could also be used.

While the periodicity of known unbranched silicate single chains varies between 2 for the alkali polysilicates and 24 for synthetic $Na_{24}Y_8[Si_{24}O_{72}]$, the observed stretching factors range from 1.00 in shattuckite to 0.234 in the sodium yttrium polysilicate (see Table 6.2 and Fig. 7.9). The difference in flatness of sili-

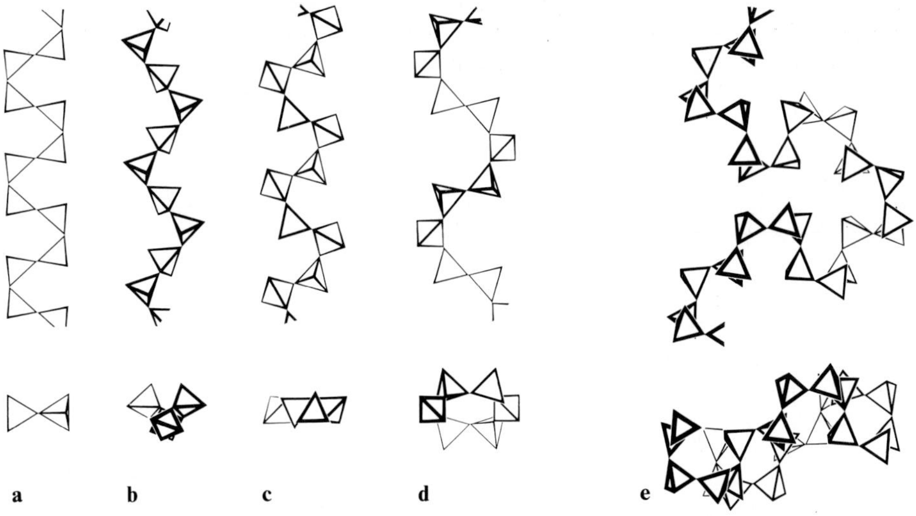

Fig. 10.1 a–e. Several single chains with different degrees of flatness and different stretching factors f_s.
Vierer single chains in **a** haradaite ($f_s = 0.654$); **b** leucophanite ($f_s = 0.687$); **c** krauskopfite ($f_s = 0.783$);
sechser single chain in **d** stokesite ($f_s = 0.718$) and
24er single chain in **e** synthetic $Na_{24}Y_8[Si_{24}O_{72}]$ ($f_s = 0.234$)

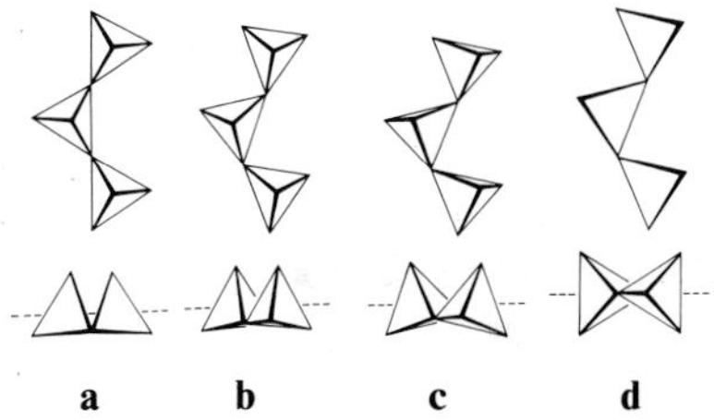

Fig. 10.2 a–d. Zweier single chains with different rotations of the tetrahedra relative to the planes of best fit through the silicon atoms (---) (after Meagher 1980).
a Synthetic pyroxene $LiFe^{+3}[Si_2O_6]$ (Clark et al. 1969); **b** orthoenstatite $Mg_2[Si_2O_6]$ (Hawthorne and Ito 1977); **c** synthetic $Na_2Ba[Si_2O_6]$ (Gunawardane et al. 1973); **d** synthetic $Na_4[Si_2O_6]$ (McDonald and Cruickshank 1967)

cate chains with constant periodicity can be seen from the examples of vierer single chains given in Fig. 10.1, while Fig. 10.2 illustrates the variations in tetrahedral rotation when the chains are flat, as they necessarily are in the zweier single chains.

Of these five properties of unbranched single chains, the periodicity and the stretching factor have been studied in more detail because they are the most conspicuous and easiest to handle.

10.1.2 Qualitative Correlations Between Chain Conformation and Cation Properties

The influence of cation size on the periodicity of silicate chains was recognized immediately after the discovery of the dreier, fünfer, and siebener single chains

Table 10.1 Relationship between cation radius, chain periodicity, and ring size for silicates of composition $M^{+2}SiO_3$

	Enstatite	Ferrosilite I	Ferrosilite III	Pyroxferroite	Rhodonite	$MnSiO_3$ (hT)	Wollastonite (lT)	Pseudowollastonite (hT)	$SrSiO_3$	Walstromite	$BaSiO_3$ (lT)	$BaSiO_3$ (hT)
Cation M^{+2}	Mg^{+2}	Fe^{+2}		$Fe^{+2}_{0.83}$, $Ca^{+2}_{0.13}$, $Mg^{+2}_{0.02}$, $Mn^{+2}_{0.02}$	Mn^{+2}		Ca^{+2}		Sr^{+2}	$Ca^{+2}_{0.67}$, $Ba^{+2}_{0.33}$	Ba^{+2}	
Mean cation radius [Å]	0.72	0.77		0.80	0.82		1.00	1.04	1.12	1.15	1.36	
Chain periodicity [tetrahedra]	2	2	9	7	5	3	3	Dreier single rings				2
Chain period [Å]	5.2	5.3	22.7	17.4	12.5	7.2	7.3					4.5

(Liebau 1956) and can be illustrated by a comparison of the anions of silicates having a general composition $M^{+2}SiO_3$ (Table 10.1). The $M^{+2}SiO_3$ silicates with small to medium size cations Mg^{+2} to Ca^{+2} are generally known as the pyroxenoids (chain periodicity $\boldsymbol{P} = 3, 5, 7, 9$) and pyroxenes ($\boldsymbol{P} = 2$). Their cations are approximately octahedrally coordinated and the $[MO_6]$ octahedra share edges to form one-dimensionally extended slabs which can be regarded as part of a hexagonal octahedral layer (Fig. 10.3). Such layers form the structure of $Mg(OH)_2$ and are also present in the trioctahedral micas (Fig. 10.19 a).

The repeat unit of the dreier chain in wollastonite (Fig. 10.3 e) consists of a tetrahedron pair in which the Si···Si direction lies parallel to the chain direction, together with a single tetrahedron protruding from the chain. The repeat unit in rhodonite (Fig. 10.3 d) consists of two pairs of tetrahedra connected by one protruding tetrahedron. The siebener chain of pyroxmangite (Fig. 10.3 c) has three pairs of tetrahedra per protruding tetrahedron, etc. The pyroxenoids, therefore, are a homologous series of single chain silicates with odd periodicities $\boldsymbol{P} = 3, 5, 7, 9, \ldots, 2n+1$. As the average size of the cations decreases from Ca^{+2} in wollastonite to Mg^{+2} in enstatite, the tetrahedral chain rotates relative to the underlying octahedral layer in order to give a good fit between these two parts of the structures. The sixfold symmetry of an ideal octahedral layer limits the asymmetric range of rotation to 30°. The pyroxenes, therefore, are the end members of the series of pyroxenoids having odd-periodic chains. In spite of their crystallographic periodicity $\boldsymbol{P}_{cryst} = 2$, they can be regarded as odd-periodic chain silicates with a formal periodicity $\boldsymbol{P}_{formal} = 2n+1, n \rightarrow \infty$.

Returning to Table 10.1, it is now clear that with increasing radius of the M^{+2} cation, the periodicity of the odd-periodic silicate chains decreases monotonously from $\boldsymbol{P} = \infty$ for the pyroxenes as one end member, to $\boldsymbol{P} = 3$ for wollastonite as the other end member of the series. Divalent cations smaller than Mg^{+2} or larger than

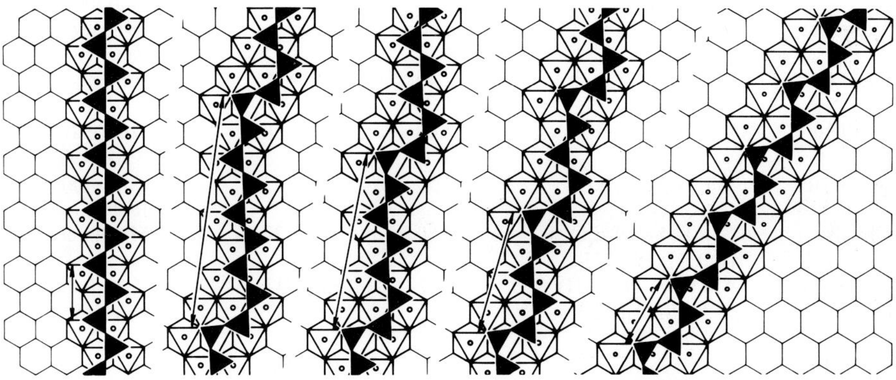

a b c d e

Fig. 10.3 a – e. Mutual adjustment between tetrahedral single chains (*black*) and slabs of edge-shared $[MO_6]$ octahedra in pyroxenoids and pyroxenes.

a Zweier chain in enstatite, $MgSiO_3$ ($f_s = 0.965$); **b** neuner chain in ferrosilite III, $FeSiO_3$(hT, mP) ($f_s = 0.930$); **c** siebener chain in pyroxmangite-type $MnSiO_3$(mP) ($f_s = 0.924$); **d** fünfer chain in rhodonite-type $MnSiO_3$(1P) ($f_s = 0.906$); **e** dreier chain in wollastonite, $CaSiO_3$(1T) ($f_s = 0.904$)

Table 10.2 Silicates with even-periodic single chains and/or an average valence of their cations $\langle v_M \rangle \neq 2$. Highly electronegative cations are printed in bold type

Silicate		$\langle v_M \rangle$	χ_M	***P***	Reference
Name	Formula				
Synthetic	$Li_4[Si_2O_6]$	1	0.97	2	Völlenkle 1981
Synthetic	$Na_4[Si_2O_6]$	1	1.01	2	McDonald and Cruickshank 1967
Synthetic	$\mathbf{Ag}_4[Si_2O_6]$	1	1.42	2	Thilo and Wodtcke 1958
Synthetic	$Na_2\mathbf{Zn}[Si_2O_6]$	1.333	1.01, 1.66	2	Belokoneva et al. 1970
Synthetic	$Na_2Ba[Si_2O_6]$	1.333	1.01, 0.97	2	Gunawardane et al. 1973
Krauskopfite	$\mathbf{H}_4Ba_2[Si_4O_{12}] \cdot 4\,H_2O$	1.333	2.1, 0.97	4	Coda et al. 1967 b
Chkalovite	$Na_6\mathbf{Be}_3[Si_6O_{18}]$	1.333	1.01, 1.47	6	Simonov et al. 1975/76
Synthetic	$Na_{24}Y_8[Si_{24}O_{72}]$	1.50	1.01, 1.11	24	Maksimov et al. 1980 b
Synthetic	$Na_2\mathbf{Cu}_3[Si_4O_{12}]$	1.60	1.01, 1.75	4	Kawamura and Kawahara 1976
Leucophanite	$Na_2\mathbf{Be}_2Ca_2[Si_4O_{12}]F_2$	1.667	1.01, 1.47, 1.04	4	Cannillo et al. 1967
Sorensenite	$Na_4\mathbf{Be}_2\mathbf{Sn}[Si_3O_9]_2 \cdot 2\,H_2O$	1.714	1.01, 1.47, 1.72	3	Metcalf-Johansen and Hazell 1976
Shattuckite	$\mathbf{Cu}_5[Si_2O_6]_2(OH)_2$	2	1.75	2	Evans, Jr. and Mrose 1977
Synthetic	$Ba_2[Si_2O_6]$(hT)	2	0.97	2	Grosse and Tillmanns 1974
Gaidonnayite	$Na_4Zr_2[Si_6O_{18}] \cdot 4\,H_2O$	2	1.01, 1.22	6	Chao and Watkinson 1974
Georgechaoite	$Na_2K_2Zr_2[Si_6O_{18}] \cdot 4\,H_2O$	2	1.01, 0.91, 1.22	6	Ghose and Thakur 1985
Alamosite	$\mathbf{Pb}_{12}[Si_{12}O_{36}]$	2	2.45	12	Boucher and Peacor 1968
Batisite	$Na_2Ba\mathbf{Ti}_2[Si_4O_{12}]O_2$	2.40	1.01, 0.97, 1.32	4	Schmahl 1981
Synthetic	$Ca_3\mathbf{Mn}_2^{+3}[Si_4O_{12}]O_2$	2.40	1.04, 1.60	4	Moore and Araki 1979
Ramsayite	$Na_2\mathbf{Ti}_2[Si_2O_6]O_3$	2.50	1.01, 1.32	2	Ch'in-hang et al. 1969
Ohmilite	$Sr_3(\mathbf{Ti};\mathbf{Fe}^{+3})[Si_4O_{12}]$ $(O,OH) \cdot 2-3\,H_2O$	2.50	0.99, 1.32	4	Mizota et al. 1983
Carpholite	$\mathbf{MnAl}_2[Si_2O_6](OH)_4$	2.67	1.60, 1.47	2	Viswanathan 1981
Ferrocarpholite	$\mathbf{FeAl}_2[Si_2O_6](OH)_4$	2.67	1.64, 1.47	2	MacGillavry et al. 1956
Stokesite	$Ca_2\mathbf{Sn}_2[Si_6O_{18}] \cdot 4\,H_2O$	3.0	1.04, 1.72	6	Vorma 1963
Haradaite	$Sr_2\mathbf{V}_2^{+4}[Si_4O_{12}]O_2$	3.5	0.99, 1.45	4	Takéuchi and Joswig 1967
Suzukiite	$Ba_2\mathbf{V}_2^{+4}[Si_4O_{12}]O_2$	3.5	0.97, 1.45	4	Matsubara et al. 1982

χ_M = electronegativity;
P = chain periodicity

Ca^{+2} cannot form chains that are capable of adjustment to octahedral layers of the kind shown in Fig. 10.3. For example, Be^{+2} does not give a stoichiometric phase $BeSiO_3$ at all, while $SrSiO_3$ and the $BaSiO_3$ phase stable below 1350 °C contain cyclic anions instead of silicate chains. Only above 1350 °C is the effective size of Ba^{+2} ions large enough to form a thermodynamically stable phase with mutual adjustment between silicate chains and corrugated layers of $[BaO_8]$ polyhedra.

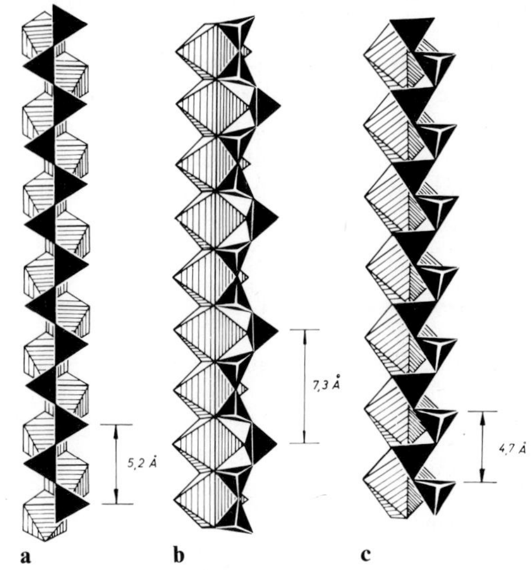

Fig. 10.4. Mutual adjustment of silicate chains (*black*) to columns of edge-shared $[MO_n]$ polyhedra in **a** enstatite $Mg_2[Si_2O_6]$; **b** wollastonite $Ca_3[Si_3O_9]$(lT); **c** synthetic $Ba_2[Si_2O_6]$(hT) (from Liebau 1972)

Figure 10.4 shows that in $MgSiO_3$, the length of two "cis" edge-shared $[MgO_6]$ octahedra is commensurable with the identity period of a stretched zweier chain, while in $CaSiO_3$ the length of two "trans" edge-shared $[CaO_6]$ octahedra is commensurable with the period of the dreier chain. In $BaSiO_3$(hT) the length of two edge-shared $[BaO_8]$ polyhedra is commensurable with the length of two tetrahedra of a zweier chain with a low stretching factor ($f_s = 0.84$ compared to $f_s = 0.965$ for enstatite).

However, cation size is not the only factor controlling the conformation of silicate chains. Indeed, it has been pointed out (Liebau 1972) that the electronegativity and the mean valence of the cations in a single chain silicate also have a large influence on the periodicity of the chains. This can be deduced from Table 10.2 which presents a survey of the silicates which contain even-periodic single chains and/or are known to have a mean cation valence $\langle v_M \rangle \neq 2$.

All but one (sorensenite) of the silicates with $\langle v_M \rangle \neq 2$ contain even-periodic chains and all but a few contain at least some cations that have a relatively high value of electronegativity, χ_M, i.e., cations that form M–O bonds with considerable covalent character. The most striking example is alamosite, $PbSiO_3$, which, despite its simple chemical formula, contains zwölfer chains, single chains with the second highest periodicity discovered so far.

10.1.3 Semiquantitative Correlations Between Chain Conformation and Cation Properties

The considerations described in the last paragraph suggest that the periodicity of unbranched single chain silicates is influenced by the size, valence, and electronegativity of their cations, but they provide no quantitative information about these relationships. On the other hand, regression analyses can lead to a better understanding of the mechanism of this influence by providing the required quantitative data. In an analysis of this kind (Liebau and Pallas 1981), correlations were obtained using the periodicity ***P*** and stretching factor f_s of 54 unbranched single chain silicates as dependent variables. The independent variables in the multiple regression analysis were the mean values of the cation radii in octahedral coordination, $\langle r_M^{[6]} \rangle$, the electronegativities, $\langle \chi_M \rangle$, and valences, $\langle v_M \rangle$, of the cations, and the number, N, of different kinds of cations in each silicate. Unfortunately, the standard deviations of the correlation coefficients b, c, d, and e of the regression equations

$$f_s = a + \sum_{k=-1}^{2} b_k \langle \chi_M \rangle^k + \sum_{l=-1}^{2} c_l \langle v_M \rangle^l + \sum_{m=-1}^{2} d_m \langle r_M \rangle^m + \sum_{n=-1}^{2} e_n N^n$$

and b′, c′, d′, and e′ of the equations

$$\boldsymbol{P} = a' + \sum_{k=-1}^{2} b'_k \langle \chi_M \rangle^k + \sum_{l=-1}^{2} c'_l \langle v_M \rangle^l + \sum_{m=-1}^{2} d'_m \langle r_M \rangle^m + \sum_{n=-1}^{2} e'_n N^n$$

were too high to make the equations useful for accurately predicting the periodicities and stretching factors of chain silicates merely from their chemical composition. This result is mainly a function of the fact that the number of silicates available for the analysis was not large enough. However, for some of the independent variables the level of significance, $l_{sign.}$, that is, the probability that the particular variable is correlated with f_s or ***P***, was more than 95%, suggesting that such correlations certainly do exist.

One of the several reliable conclusions that can be drawn from the analysis is that small to medium-sized cations capable of forming hexagonal or pseudohexagonal layers of edge-shared cation–oxygen octahedra, form single chain silicates with odd-periodic chains: the pyroxenoids and pyroxenes (Fig. 10.3). Only three other silicates with odd-periodic single chains are known in which the cations do not form slabs of edge-shared octahedra that can be considered as parts of a hexagonal layer: sorensenite, $Na_4Be_2Sn[^3Si_3O_9]_2 \cdot 2\,H_2O$ (Metcalf-Johansen and Hazell 1976), hilairite, $Na_2Zr[^3Si_3O_9] \cdot 3\,H_2O$ (Ilyushin et al. 1981a), and an unnamed mineral of composition $K_2Zr[^3Si_3O_9] \cdot H_2O$ (Ilyushin et al. 1981b).

In general, silicates containing cations that cannot form hexagonal layers of edge-shared octahedra contain even-periodic single chains. These two groups of single chain silicates are crystallochemically distinct classes that have different correlation functions. We will call them odd-periodic chain silicates (including the pyroxenes) and even-periodic chain silicates.

(1) Even-Periodic Single Chain Silicates. For the known even-periodic single chain silicates, the stretching factor lies in the range $0.23 < f_s \leq 1.00$ (Table 6.2). The correlation function obtained by regression analysis (Liebau and Pallas 1981) is

$$f_s = 1.066 - 0.125\,\langle \chi_M \rangle - 0.071\,\langle v_M \rangle$$

with levels of significance $l_{sign.} = 97\%$ for $\langle \chi_M \rangle$ and 98% for $\langle v_M \rangle$. This means that the stretching factor is negatively correlated with the average electronegativity and the average valence of the cations with a probability of 97% and 98%, respectively. In the case of the average cation radius $\langle r_M \rangle$, the corresponding $l_{sign.}$ value was too low to infer a meaningful correlation between f_s and $\langle r_M \rangle$. On the other hand, a strong inverse correlation

$$\boldsymbol{P} = -10.86 + 11.44\,f_s^{-1}$$

between chain periodicity and stretching factor was found with $l_{sign.} = 99.5\%$.

(2) Odd-Periodic Single Chain Silicates. Figures 10.5 and 10.6 give the correlations between the variables in the odd-periodic single chain silicates. For each of the subgroups of pyroxenoids with dreier ($\boldsymbol{P} = 3$) and fünfer ($\boldsymbol{P} = 5$) single chains, and for the pyroxenes ($\boldsymbol{P} = \infty$), there is a definite positive correlation between stretching factor and the average cation radius $\langle r_M \rangle$. In contrast, the correlations between f_s and the average electronegativity $\langle \chi_M \rangle$ are negative. In addition, there is a strong positive correlation between stretching factor and periodicity of the odd-periodic silicate single chains.

Table 10.3 summarizes the correlations between the stretching factor f_s and the cation properties χ_M, v_M, r_M, and the periodicity $\boldsymbol{P}$ of the unbranched silicate chains.

A positive correlation between f_s and one of the other variables is indicated by a pair of parallel arrows, ↑↑, and a negative or inverse correlation by two anti-

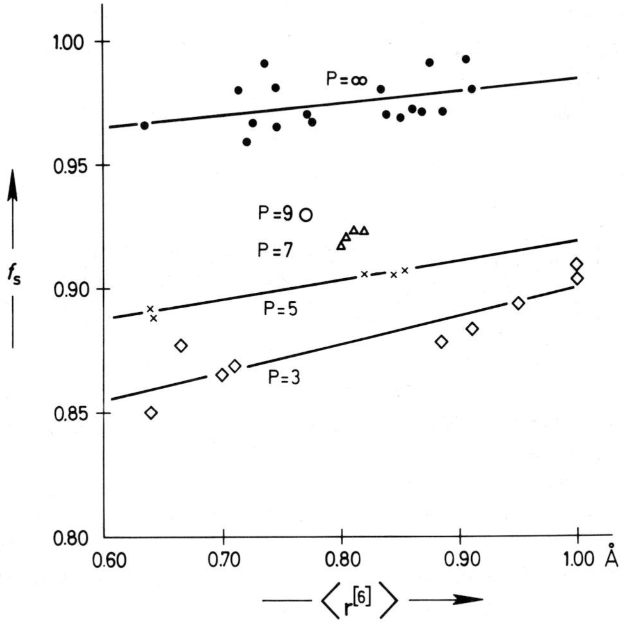

Fig. 10.5. Correlation between the stretching factor of the odd-periodic chain silicates, their average cation radius, and chain periodicity $\boldsymbol{P}$ (from Liebau and Pallas 1981)

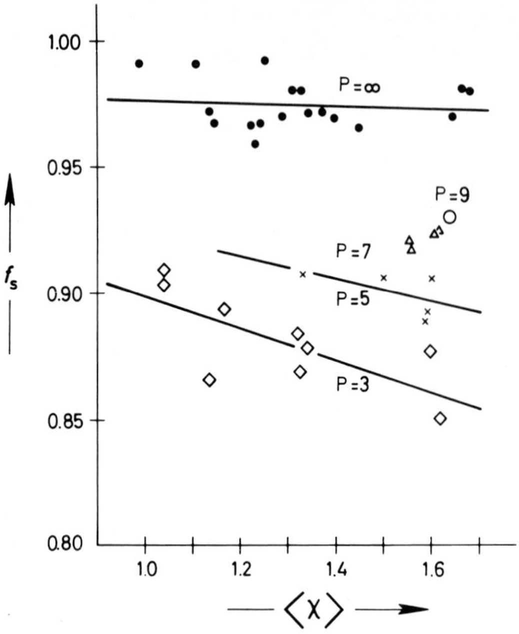

Fig. 10.6. Correlation between the stretching factor of the odd-periodic chain silicates, the average electronegativity of their cations, and their chain periodicity ***P*** (from Liebau and Pallas 1981)

Table 10.3 Comparison of the correlations between the stretching factor f_s of unbranched single chains and their periodicity ***P***, and the mean electronegativity $\langle \chi_M \rangle$, valence $\langle v_M \rangle$, and radius $\langle r_M \rangle$ of the cations in silicates and phosphates. (After Liebau 1981)

	$\langle \chi_M \rangle$	$\langle v_M \rangle$	$\langle r_M \rangle$	***P***
Odd-periodic silicates	↑↓	–	↑↑	↑↑
Even-periodic silicates	↑↓	↑↓	?	↑↓
Polyphosphates	↑↓	↑↓	↑↑	↑↓

parallel arrows, ↑↓. The length of the arrows is approximately proportional to the level of significance of the correlation. For reasons which will be discussed later, these data are complemented by the corresponding results obtained for 57 polyphosphates (Liebau 1981).

10.1.4 Crystallochemical Interpretation of the Correlations

These correlations have been obtained by unbiased mathematical treatment of experimental (f_s from crystallographic unit cell dimensions and the chain periodicity ***P***, and the chemical composition from chemical analyses) and tabulated data (χ, r, v), and can be readily interpreted with the aid of simple structural chemical considerations.

Since silicates and phosphates are chemically and crystallographically similar, interpretations given for the single chain silicates should, with the necessary adjustments, also hold for the corresponding polyphosphates. Therefore, the correlations obtained in the regression analyses of the silicates and phosphates are discussed together.

10.1.4.1 Correlation Between Stretching Factor and Average Cation Electronegativity

Atoms of highly electropositive elements, such as sodium, potassium, calcium, barium, etc., transfer their valence electrons almost completely to the neighboring anions (see Sec. 3.2.1). Therefore, in anhydrous single chain silicates $M_m[SiO_3]_n$ of such cations each $[SiO_4]$ tetrahedron carries nearly two negative charges. This high charge density causes strong repulsive forces between the tetrahedra and, as a result, the silicate chains tend to be stretched.

As the electronegativity χ_M of the cations increases, fewer electrons are transferred to the surrounding anions so that the effective charge per $[SiO_4]$ tetrahedron and the repulsive forces between them both decrease. As a consequence, in polysilicates of electronegative elements, such as Be, Mn, Cr, Cu, Sn, etc., the silicate chains are not so highly stretched.

In agreement with this hypothesis, the stretching factor of the tetrahedral chains increases as the electronegativity of the cations decreases for both groups of polysilicates and for the polyphosphates (Table 10.3).

Since hydrogen has a high electronegativity ($\chi_H = 2.1$), the effect is most pronounced if a pyroxenoid contains hydrogen bonded to an Si–O bond (Liebau 1980b). This can be seen from a comparison of the lattice constants in the chain direction of corresponding hydrogen-free and hydrogen-containing pyroxenoids (Table 10.4): there is a reduction of these lattice constants by about 0.25 Å in the acid pyroxenoids. Accurate structure analyses of several of the acid phases (e.g., Takéuchi and Kudoh 1977; Ohashi and Finger 1981) have shown that the influence of hydrogen on the silicate chain is exerted by a hydrogen bridge between terminal oxygen atoms of the two $[SiO_4]$ tetrahedra separated by the tetrahedron protruding from the chain (Fig. 10.7).

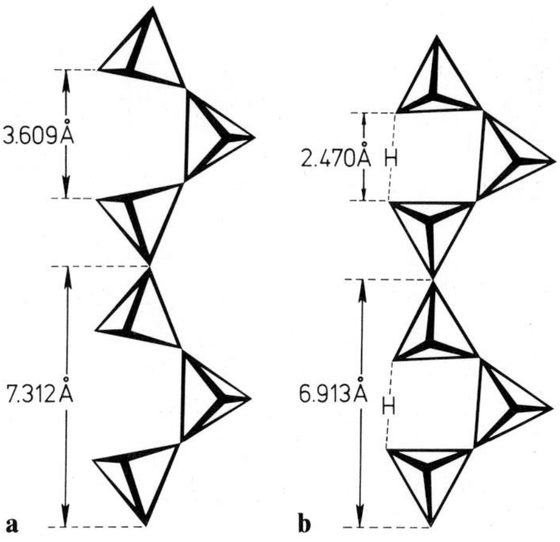

Fig. 10.7a, b. Influence of cationic hydrogen on the degree of stretching of a pyroxenoid chain.

a Dreier single chain in manganoan wollastonite, $(Ca,Mn)_3[Si_3O_9]$ and **b** dreier single chain in serandite, $HNa(Mn,Ca)_2[Si_3O_9]$ (Ohashi and Finger 1978). *Broken lines* represent hydrogen bonds

Table 10.4 Comparison of the chain periods, I_{chain}, of pyroxenoids containing cationic hydrogen with those of corresponding non-acid pyroxenoids

P	Acid compounds				ΔI_{chain} [Å]	Non-acid compounds			
	Name	Formula	Ref.	I_{chain} [Å]		I_{chain} [Å]	Formula	Name	Ref.
3	Pectolite	$HNaCa_2[Si_3O_9]$	[1]	7.040	0.28	7.320	$Ca_3[Si_3O_9]$	Parawolla-stonite	[2]
	Schizolite	$HNa(Ca, Mn)_2[Si_3O_9]$	[3]	6.978	0.25	7.231	$Ca_{2.35}(Mn, Fe, Mg)_{0.65}[Si_3O_9]$	Ca-bustamite	[3]
	Serandite	$HNa(Mn, Ca)_2[Si_3O_9]$	[4]	6.889	0.20	7.091	$Ca_{0.95}(Mn, Fe, Mg)_{2.05}[Si_3O_9]$	Mn-bustamite	[3]
	Cascandite	$HCaSc[Si_3O_9]$	[5]	7.076					
	Synthetic	$HNaCd_2[Si_3O_9]$	[6]	6.980					
	Synthetic	$HNa_2[P_3O_9]$	[7]	6.76	0.24	7.00	$Na_3[P_3O_9]$	Maddrell's salt	[8]
5	Li-hydro-rhodonite	$HLiMn_4[Si_5O_{15}]$	[9]	12.039	0.20	12.235	$Mn_5[Si_5O_{15}]$	Synthetic	[10]
	Nambulite	$H(Li, Na)Mn_4[Si_5O_{15}]$	[11]	12.016	0.22				
	Santa-claraite	$HMn_4Ca[Si_5O_{15}](OH) \cdot H_2O$	[13]	12.001	0.23	12.233	$(Mn, Fe, Mg)_{4.21}Ca_{0.79}[Si_5O_{15}]$	Rhodonite	[12]
	Marsturite	$HNaMn_3Ca[Si_5O_{15}]$	[14]	12.03	0.20				
	Saneroite	$HNa_{1.15}(Mn^{+2}, Mn^{+3})_5[^5(Si_{5.5}V_{0.5})O_{18}](OH)$	[15]	11.962	0.27				
	Babing-tonite	$HCa_2Fe^{+2}Fe^{+3}[Si_5O_{15}]$	[16]	12.18					
			[17]	12.245					
	Synthetic	$HLiCd_4[Ge_5O_{15}]$	[18]	12.673					
	Synthetic	$HNaCd_4[Ge_5O_{15}]$	[18]	12.591					

P = chain periodicity

References: [1] Prewitt 1967; [2] Trojer 1968; [3] Ohashi and Finger 1978; [4] Takéuchi et al. 1976; [5] Mellini and Merlino 1982; [6] Belokoneva et al. 1974; [7] Jost 1962; [8] Jost 1963; [9] Murakami et al. 1977; [10] Narita et al. 1977; [11] Narita et al. 1975; [12] Peacor and Niizeki 1963; [13] Ohashi and Finger 1981; [14] Peacor et al. 1978; [15] Basso and Della Giusta 1980; [16] Kosoi 1976; [17] Araki and Zoltai 1972; [18] Simonov et al. 1978 b

In the pyroxene group, the small cation – oxygen polyhedra are so rigid that the tetrahedral chains are not able to shrink when the repulsive forces between the tetrahedra become weak. Therefore, pyroxenes show no correlation, or only a very weak inverse correlation, between f_s and $\langle \chi_M \rangle$ (Fig. 10.6, Table 10.3).

Since phosphorus is pentavalent, the formal charge of a $[PO_4]$ tetrahedron in a polyphosphate $M_m[PO_3]_n$ is only -1 compared with the formal charge -2 of an $[SiO_4]$ tetrahedron of a corresponding polysilicate. Therefore, the repulsive forces between the tetrahedra of a polyphosphate chain are only about half as strong as those between the tetrahedra of a polysilicate chain.

In fact, the frequency distribution of the stretching factors for the 54 unbranched single chain silicates (plus $Na_{24}Y_8[Si_{24}O_{72}]$ published later) and 57 polyphosphates used in the regression analyses of Liebau and Pallas (1981) and Liebau (1981) clearly shows that silicate chains are, on average, more stretched than phosphate chains (Fig. 10.8).

In addition, due to the lower formal charge of the phosphate tetrahedra, the effective charges of the $[PO_4]$ groups in polyphosphate chains can only vary between -1 and 0, whereas those of the $[SiO_4]$ tetrahedra in polysilicates can vary between -2 and 0. In accordance with this hypothesis, only a very weak correlation between the stretching factor f_s and the average electronegativity $\langle \chi_M \rangle$ of the cations has been obtained in the regression analysis of the polyphosphates, whereas the corresponding correlation obtained for the polysilicates is strong (Table 10.3).

Moreover, the frequency distribution in Fig. 10.8 shows that, in general, even-periodic silicate chains are considerably less stretched than odd-periodic ones. This

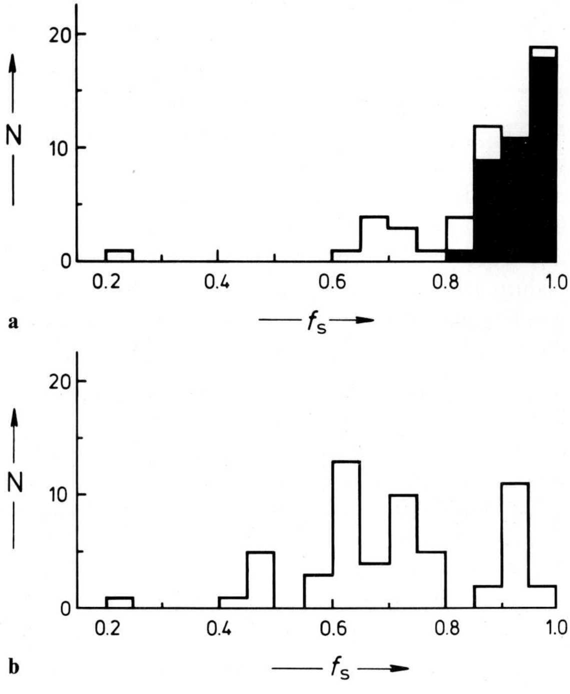

Fig. 10.8 a, b. Frequency distribution of the stretching factors of tetrahedral chains in silicates and phosphates.

a Unbranched single chain silicates. *Black:* odd-periodic chains; *white:* even-periodic chains. **b** Polyphosphates

is in good agreement with the observation that most even-periodic single chain silicates contain electronegative cations (Table 10.2) and, therefore, have a relatively low charge per $[SiO_4]$ tetrahedron.

10.1.4.2 Correlation Between Stretching Factor and Average Cation Valence

The correlation between the average cation valence and the stretching factor in the polysilicates is also easy to interpret. In a polysilicate $M_2^+[SiO_3]$, there are two monovalent cations per $[SiO_4]$ tetrahedron compensating the negative charges of the chains. In order to involve all cations in bonds to the chains, the chains themselves will have to stretch. However, as the average valence of the cations increases, the number of cations per $[SiO_4]$ tetrahedron decreases and the chain must contract, fold, or coil to involve all terminal oxygen atoms in cation–oxygen bonds. Therefore, the higher the value of $\langle v_M \rangle$, the lower f_s becomes, in agreement with the observation for the even-periodic chain silicates. For the odd-periodic chain silicates this effect is not observed since, with very few exceptions, they have an average cation valence $\langle v_M \rangle = 2$.

Table 10.5 Relationship between the atomic ratio $n_M : n_T$ and the mean cation valence, $\langle v_M \rangle$, for single chain silicates and phosphates of general formula $M_m M'_{m'} \dots [XO_3]_n$. (No phases are known where $\langle v_M \rangle$ is set in parentheses)

$\langle v_M \rangle$ for silicates	1	2	(3)	(4)	(6)	(8)
$\langle v_M \rangle$ for phosphates	–	1	1.5	2	3	4
$n_M : n_T$	2 : 1	1 : 1	0.67 : 1	0.5 : 1	0.33 : 1	0.25 : 1

In Table 10.5 the numbers of cations per $[XO_4]$ tetrahedron for a given value of $\langle v_M \rangle$ are compared for polysilicates and polyphosphates of general formula $M_m M'_{m'} \dots [XO_3]_n$. Of the silicates known to have this formula none has less than one cation per $[SiO_4]$ tetrahedron, in other words, $\langle v_M \rangle = 2$ is the highest average cation valence observed. On the other hand, polyphosphates are known to exist with $\langle v_M \rangle$ values as high as 4, i.e., with one cation per four $[PO_4]$ tetrahedra. This difference in the behavior of polysilicates and polyphosphates can also be explained by the lower repulsive forces between the $[PO_4]$ tetrahedra caused by the smaller range of negative charges per tetrahedron.

10.1.4.3 Correlation Between Stretching Factor and Average Cation Radius

To interpret the correlation between f_s and $\langle r_M \rangle$, two hypothetical single chain silicates (or phosphates) are compared which have very similar average cation radius values, but the same average values of electronegativity and valence of the cations, $\langle \chi_M \rangle$ and $\langle v_M \rangle$. Such silicates (or phosphates) will usually be isotypic or, at least, homeotypic. Within such an isotypic or homeotypic series of phases, the tetrahedral chains become more stretched with increasing cation size in order to attain satisfactory mutual adjustment between the tetrahedral chains and the arrays of cation–oxygen polyhedra of increasing size. Such a mechanism of adaptation of tetrahedral chains to $[MO_n]$ polyhedra of varying size generally works only in a rather limited range of cation radius. Beyond a certain minimum and maximum value of cation radius, no further shrinkage or stretching, respectively, of the chains is possible and a new structure type must form. As a consequence, the influence of $\langle r_M \rangle$ on the stretching factor is expected to be substantial within a group of isotypic, or at least homeotypic, phases such as the pyroxenes and pyroxenoids (Figs. 10.3 and 10.5). Polysilicates with even-periodic single chains differ greatly in their structures due to much larger variations in $\langle \chi_M \rangle$ and, in particular, $\langle v_M \rangle$. On average, therefore, the influence of $\langle r_M \rangle$ on f_s is far exceeded by that of the other two cation properties and its effect is not observed in the regression analysis (Table 10.3). In contrast, for polyphosphates the positive correlation between f_s and $\langle r_M \rangle$ is observed, in spite of the large number of different structure types, because the influence of $\langle \chi_M \rangle$ is small in this group.

10.1.4.4 Correlation Between Stretching Factor and Chain Periodicity

Table 10.3 shows that each of the correlations observed between f_s and the cation properties are of the same character for all three groups of chain compounds: negative between f_s and $\langle \chi_M \rangle$ and between f_s and $\langle v_M \rangle$, and positive between f_s and $\langle r_M \rangle$. In contrast, the correlations between f_s and the chain periodicity $\boldsymbol{P}$ is positive for the odd-periodic chain silicates, but negative for the even-periodic chain silicates and the polyphosphates.

To explain these correlations, it is necessary to consider the geometric shape of the tetrahedral chains as a function of the degree of stretching. A chain stretched to its maximum ($f_s = 1$) has all its bridging oxygen atoms on a straight line so that the lowest possible value of the periodicity is $\boldsymbol{P} = 1$. If $f_s < 1$ for a tetrahedral chain, i.e., if the bridging oxygen atoms deviate from a straight line, however little, the periodicity is necessarily higher than 1, at least $\boldsymbol{P} = 2$. If f_s is only slightly smaller than 1, the chain can still have a low periodicity, for example, $\boldsymbol{P} = 2$ or 3, but the lower the stretching factor, the more the bridging oxygen atoms have to deviate from a straight line to avoid overlapping of the bridging oxygen atoms (Fig. 7.9). A large decrease of the stretching degree is necessarily accompanied by a large increase of chain periodicity. This is particularly obvious from a comparison of the shapes of the zweier chain of enstatite ($f_s = 0.965$) and the 24er single chain of $Na_{24}Y_8[Si_{24}O_{72}]$ which is extremely folded ($f_s = 0.234$) (Fig. 7.9a and i).

While this hypothesis is in agreement with the correlations found for the polyphosphates and the even-periodic chain silicates, a different explanation must be found for the different behavior of the odd-periodic single chain silicates. Such an explanation becomes obvious by inspection of Fig. 10.3.

Within the homeotypic series of the pyroxenoids and pyroxenes, the periodicity $\boldsymbol{P} = 2n + 1$ of the chains increases in steps of two tetrahedra. Each repeat unit of the chain contains n pairs of tetrahedra with one tetrahedron set off at an angle. Within one repeat unit of a chain, the bridging oxygen atoms of the tetrahedron pairs deviate only slightly from a straight line, whereas the bridging oxygen atoms of the set-off tetrahedron deviate considerably. In this way each set-off tetrahedron produces a kink in the otherwise nearly fully stretched chain. The stretching factor, therefore, increases as the fraction of set-off tetrahedra decreases, i.e., as the periodicity of the pyroxenoid chains increases. In the pyroxenes and pyroxenoids this effect by far outweighs the general effect acting in the other two classes of single chain compounds.

10.2 Influence of Cation Properties on the Shape of Unbranched Multiple Chain Anions

The factors controlling the conformation of single chains are also effective in multiple chain silicates. In fact, many of the silicates containing even-periodic double chains contain highly electronegative cations and almost all of them contain cations which have a valence $v_M > 2$ (Table 10.6).

On the other hand, the periodicity and stretching factor of multiple chains with odd-periodic subchains is controlled by the geometric requirement of adjusting these chains to arrays of edge-shared $[MO_n]$ polyhedra, just as it is in the case for the odd-periodic single chains. In other words, the shape of these odd-periodic multiple chains is governed by the average cation radius $\langle r_M \rangle$. The examples given in Table 10.7 reveal that cations which require a certain chain type in single chain silicates tend to require the same chain type in multiple chain silicates. For example, in the so-called biopyriboles, the small Mg^{+2} ions are associated with the stretched zweier double and triple chains with $f_s > 0.95$, the slightly larger Mn^{+2} ion favors the fünfer double chain in inesite, and the medium-size Ca^{+2} ion is associated with dreier double and quadruple chains in xonotlite, okenite, and miserite. The dreier chain is also favored by Na^+ ions which have about the same size as Ca^{+2}, in devitrite, elpidite, epididymite, and canasite.

In addition to the influence of χ_M, v_M, and r_M, which is the same in single and multiple chain silicates, another factor must be considered. Unbranched single chains are flexible enough to adopt almost any shape in adjusting themselves to the $[MO_n]$ polyhedra without much deviation of the Si−O distances and O−Si−O and Si−O−Si valence angles from their mean values. They accomplish this adjustment merely by rotating the $[SiO_4]$ tetrahedra. However, linking two or more single chains to form a multiple chain reduces the freedom with which the tetrahedra can rotate to adjust the anion to the polyhedral arrays. This decrease in rotational freedom is compensated for partly by a deformation of the polyhedral arrays, and partly by an increase in the deviation from the mean values of the bond lengths and bond angles in the tetrahedral anion. This effect becomes more pronouned, the more linkages there are between adjacent subchains, and the higher the multiplicity of the chain anion. For example, for dreier double chains the freedom of tetrahedral rotation is reduced in the sequence $\{2^1_\infty\}[Si_6O_{17}]$ (xonotlite) – $\{2^1_\infty\}[Si_6O_{16}]$ (okenite) – $\{2^1_\infty\}[Si_6O_{15}]$ (epididymite), and it is higher in the dreier double chain of epididymite than in the dreier quadruple chain $\{4^1_\infty\}[Si_{12}O_{30}]$ of miserite, which has the same Si : O ratio. Loss of rotational freedom of the $[SiO_4]$ tetrahedra causes increasing difficulties in the adjustment of the silicate anion to the arrays of cation–oxygen polyhedra and, therefore, reduces the stability of the structure. On the other hand, increasing the number of linkages between adjacent subchains along

Table 10.6 Survey of silicates containing unbranched even-periodic double chains $\{\boldsymbol{uB}, 2^1_\infty\}[{}^{P}Si_{2p}O_{5p}]$

Silicate: Name	Formula	***P***	I_{chain} [Å]	f_s	Reference
[Synthetic	$Li_4\{1^1_\infty\}[Si_2O_6]$	2	4.66	0.863	Völlenkle 1981]
Synthetic	$Li_4\{2^1_\infty\}[SiGe_3O_{10}]$	2	4.94	0.915	Völlenkle et al. 1968
[Batisite	$Na_2Ba\mathbf{Ti}_2\{1^1_\infty\}[Si_4O_{12}]O_2$	4	8.087	0.749	Schmahl 1981]
Narsarsukite	$Na_4\,\mathbf{Ti}_2\{2^1_\infty\}[Si_8O_{20}]O_2$	4	8.01	0.742	Peacor and Buerger 1962
Caysichite	$(Ca_3RE)Y_4\{2^1_\infty\}[Si_8O_{20}][CO_3]_6(OH)\cdot 7\,H_2O$	4	9.73	0.901	Mellini and Merlino 1978
[Chkalovite	$Na_6\mathbf{Be}_3\{1^1_\infty\}[Si_6O_{18}]$	6	11.111	0.686	Simonov et al. 1975/76]
Synthetic	$Li_2Na_4Y_2\{2^1_\infty\}[Si_{12}O_{30}]$	6	10.375	0.640	Gunawardane et al. 1982
Synthetic	$Na_4Mg_4\{2^1_\infty\}[Si_{12}O_{30}]$	6	10.205	0.630	Cradwick and Taylor 1972
Zektzerite	$Li_2Na_2Zr_2\{2^1_\infty\}[Si_{12}O_{30}]$	6	10.164	0.627	Ghose and Wan 1978
Synthetic	$Li_4Zr_2\{2^1_\infty\}[Si_{12}O_{30}]$	6	10.146	0.626	Quintana and West 1981
Tuhualite	$(Na, K)_2\mathbf{Fe}_2^{+2}\mathbf{Fe}_2^{+3}\{2^1_\infty\}[Si_{12}O_{30}]\cdot H_2O$	6	10.11	0.624	Merlino 1969
Emeleusite	$Li_2Na_4\mathbf{Fe}_2^{+3}\{2^1_\infty\}[Si_{12}O_{30}]$	6	10.072	0.622	Johnsen et al. 1978, Sandomirskii et al. 1975
Synthetic	$Li_2Na_2\mathbf{Sn}_2\{2^1_\infty\}[Si_{12}O_{30}]$	6	10.044	0.620	Marr and Glasser 1979
Synthetic	$Li_2Na_2\mathbf{Ti}_2\{2^1_\infty\}[Si_{12}O_{30}]$	6	9.932	0.613	Marr and Glasser 1979

For each chain type a typical single chain silicate is given in square brackets. Highly electronegative cations are given in bold face.
P = Chain periodicity; f_s = Stretching factor

Table 10.7 Survey of silicates containing unbranched odd-periodic multiple chains

Silicate		***P***	***M***	I_{chain}	f_s	Reference
Name	Formula			[Å]		
[Wollastonite	$Ca_3\{1^1_\infty\}[Si_3O_9]$	3	1	7.32	0.904	Trojer 1968]
Xonotlite	$Ca_6\{2^1_\infty\}[Si_6O_{17}](OH)_2$	3	2	7.35	0.907	Kudoh and Takéuchi 1979
Okenite	$Ca_{10}\{2^1_\infty\}[Si_6O_{16}]\{1^2_\infty\}[Si_6O_{15}]_2 \cdot 18\,H_2O$	3	2	7.28	0.899	Merlino 1983
Devitrite	$Na_4Ca_6\{4^1_\infty\}[Si_{12}O_{32}]$	3	4	7.117	0.879	Ihara et al. 1984
Epididymite	$Na_2Be_2\{2^1_\infty\}[Si_6O_{15}] \cdot H_2O$	3	2	7.33	0.905	Robinson and Fang 1970
Elpidite	$Na_2Zr\{2^1_\infty\}[Si_6O_{15}] \cdot 3\,H_2O$	3	2	7.14	0.881	Cannillo et al. 1973b
Canasite	$Na_4K_2Ca_5\{4^1_\infty\}[Si_{12}O_{30}](OH, F)_4$	3	4	7.19	0.888	Chiragov et al. 1969
Miserite	$K_2Ca_{10}(Y, RE)_2[Si_2O_7]_2\{4^1_\infty\}[Si_{12}O_{30}](OH)_2F_2$	3	4	7.377	0.911	Scott 1976
[Rhodonite	$(Mn, Ca)_5\{1^1_\infty\}[Si_5O_{15}]$	5	1	12.47	0.924	Peacor and Niizeki 1963]
Inesite	$Ca_2Mn_7\{2^1_\infty\}[Si_{10}O_{28}](OH)_2 \cdot 5\,H_2O$	5	2	11.975	0.887	Wan and Ghose 1978
[Pyroxenes, e.g., enstatite	$Mg_2\{1^1_\infty\}[Si_2O_6]$	2(∞)	1	5.178	0.959	Morimoto and Koto 1969]
Amphiboles, e.g., tremolite	$Mg_5Ca_2\{2^1_\infty\}[Si_4O_{11}]_2(OH, F)_2$	2(∞)	2	5.285	0.979	Hawthorne and Grundy 1976
Synthetic	$HNaMg_4\{3^1_\infty\}[Si_6O_{16}](OH)_2$	2(∞)	3	5.257	0.974	Drits et al. 1975a
Jimthompsonite	$(Mg, Fe)_5\{3^1_\infty\}[Si_6O_{16}](OH)_2$	2(∞)	3	5.30	0.981	Veblen and Burnham 1978
Clino-jimthompsonite	$(Mg, Fe)_5\{3^1_\infty\}[Si_6O_{16}](OH)_2$	2(∞)	3	5.32	0.985	Veblen and Burnham 1978

For each chain type a typical single chain silicate is given in square brackets. For the pyroxenes and their multiple chain derivatives the crystallographic periodicity, $\boldsymbol{P}_{cryst} = 2$, is also given, along with the crystallochemically relevant formal periodicity, $\boldsymbol{P}_{formal} = \infty$ (see Sec. 10.1.2).
$\boldsymbol{P}$ = chain periodicity; $\boldsymbol{M}$ = multiplicity; f_s = stretching factor

with the multiplicity results in an increase in the Si : O ratio. Since the latter is equivalent to a decrease in the formal valence per $[SiO_4]$ tetrahedron in the silicate anion, electronegative cations are able to reduce the negative charges per tetrahedron and, in this way, decrease the repulsive forces between the tetrahedra. This can occur to such an extent that double chains become stable in which the two subchains are linked via all their tetrahedra. Inspection of Table 10.6, in fact, shows that many of these double chain silicates for which Si : O = 2 : 5 contain highly electronegative cations, such as Be, Mn, Fe, Ti, etc. $Li_4\{2^1_\infty\}[SiGe_3O_{10}]$ is one of the exceptions in that it contains no electronegative cations. In this structure the repulsive forces between the tetrahedra are reduced by the larger size of the $[GeO_4]$ tetrahedra in comparison to the smaller $[SiO_4]$ groups [$d(Ge-O) = 1.75$ Å, $d(Si-O) = 1.62$ Å].

All other multiple chain silicates which do not have cations of high electronegativity contain either "soft" cations, such as Na^+, K^+, Ca^{+2}, and Ba^{+2}, or anions and/or water molecules in addition to the silicate anions, or both. The stabilizing influence of such soft cations and additional anions or water molecules is explained in more detail in the following section.

10.3 Influence of Cation Properties on the Shape of Branched Silicate Anions

Table 10.8 is a survey of silicates known at present to contain branched anions. Inspection of the chemical compositions of these silicates reveals that almost all of them fulfill at least one of three conditions:

1) presence of *highly electronegative cations;*
2) presence of *soft cations,* such as Na^+, K^+, Ca^{+2}, Ba^{+2}, that have no pronounced preference for one or other kind of well-defined coordination polyhedron, but instead accept any site offered by the silicate anions;
3) presence of oxygen ions, hydroxyl groups, halide ions, or other nonsilicate anions and water molecules in addition to the oxygen atoms supplied by the $[SiO_4]$ tetrahedra (additional molecules and nonsilicate anions).

Most branched silicates fulfill two, and some fulfill all three of these conditions.

The first of these conditions may be explained with the help of Fig. 10.9 in which the structures of stretched and strongly shrunken unbranched single chain silicates, enstatite and haradaite, respectively, are compared with those of open- and loop-branched single chain silicates, astrophyllite and vlasovite, respectively. The two branched chains contain non-neighboring $[SiO_4]$ tetrahedra which approach each other about as near as tetrahedra of the shrunken haradaite chain do. This causes strong repulsive forces between these tetrahedra. Therefore, only cations with high electronegativities can reduce the negative charges on the terminal oxygen atoms sufficiently to stabilize silicates with branched chains in the same way as they stabilize unbranched chain silicates with low stretching factors. In fact, from the survey of branched single chain silicates given in Table 10.8, it is evident that all branched single chain silicates except agrellite, $Na_2Ca_4[Si_8O_{20}]F_2$,

Table 10.8 Survey of silicates containing branched anions. The chemical formulae are subdivided into groups containing (i) "soft" cations; (ii) electronegative cations; (iii) "hard" cations; (iv) silicate anions; and (v) additional non-silicate anions and water molecules

Name	Formula					Ref.
	"Soft" cations	Electronegative cations	"Hard" cations	Silicate anions	Additional ligands	
1. Zero-dimensional silicate anions						
Synthetic	$NaBa_3Nd_3$			{***oB***, 3***t***} $[Si_4O_{13}][Si_2O_7]$		[49]
Zunyite		Al_{13}		{***oB***, 3***t***} $[Si_5O_{16}]$	$(OH, F)_{18}O_4Cl$	[1]
Eakerite	Ca_2	Al_2Sn		{***oB***, 1***r***} $[Si_6O_{18}]$	$(OH)_2 \cdot 2\,H_2O$	[2]
Tienshanite	$Na_9KCa_2Ba_6$	$(Mn, Fe)_6(Ti, Nb, Ta)_6B_{12}$		{***oB***, 1***r***} $[Si_{18}O_{54}]_2$	$O_{15}(OH)_2$	[3]
Hyalotekite	Ca_2Ba_2	Pb_2B_2		{***oB***, 2***r***} $[(Si_{9.5}Be_{0.5})O_{28}]$	F	[4]
Synthetic esters		$[-Si(CH_3)_3]_{10}$		{***lB***, 1***r***} $[Si_6O_{17}]$		[5]
2. One-dimensional silicate anions						
Astrophyllite	NaK_2	$(Fe, Mn)_5Ti_2$	Mg_2	{***oB***, 1^{1}_{∞}} $[^2Si_4O_{12}]_2$	$(O, OH, F)_7$	[6]
Aenigmatite	Na_2	Fe_5Ti		{***oB***, 1^{1}_{∞}} $[^4Si_6O_{18}]$	O_2	[7]
Saneroite	$Na_{1.15}$	$H(Mn^{+2}, Mn^{+3})_5$		{***oB***, 1^{1}_{∞}} $[^5(Si_{5.5}V_{0.5})O_{18}]$	(OH)	[8]
Synthetic			Li_2Mg_2	{***lB***, 1^{1}_{∞}} $[^3Si_4O_{11}]$		[50]
Deerite		$Fe_6^{+2}Fe_3^{+3}$		{***lB***, 1^{1}_{∞}} $[^4Si_6O_{17}]$	$O_3(OH)_5$	[9]
Howieite	Na	$(Fe^{+2}, Fe^{+3}, Mn, Al,$	$Mg)_{12}$	{***lB***, 1^{1}_{∞}} $[^4Si_6O_{17}]_2$	$(O, OH)_{10}$	[10]
Taneyamalite	(Na, Ca)	$(Mn^{+2}, Fe^{+3}, Al,$	$Mg)_{12}$	{***lB***, 1^{1}_{∞}} $[^4Si_6O_{17}]_2$	$(O, OH)_{10}$	[11]
Vlasovite	Na_4		Zr_2	{***lB***, 1^{1}_{∞}} $[^6Si_8O_{22}]$		[12]
Pellyite	Ca_2Ba_4	(Fe,	$Mg)_4$	{***lB***, 1^{1}_{∞}} $[^8Si_{12}O_{34}]$		[13]
Nordite	$Na_4(Sr, Ca)_2RE_2(Na, Mn)_2(Zn, Fe, Mn,$		Mg)	{***lB***, 1^{1}_{∞}} $[^{10}Si_{12}O_{34}]$		[14]
Bavenite	Ca_4	Be_2Al_2		{***oB***, 2^{1}_{∞}} $[^2Si_6O_{16}][Si_3O_{10}]$	$(OH)_2$	[15]
Fenaksite	Na_2K_2	Fe_2		{***lB***, 2^{1}_{∞}} $[^3Si_8O_{20}]$		[16]
Litidionite	Na_2K_2	Cu_2		{***lB***, 2^{1}_{∞}} $[^3Si_8O_{20}]$		[17]
Synthetic	Na_4	Cu_2		{***lB***, 2^{1}_{∞}} $[^3Si_8O_{20}]$		[18]
Synthetic	Na_4	Fe_2		{***lB***, 2^{1}_{∞}} $[^3Si_8O_{20}]$		[19]
Agrellite	Na_2Ca_4			{***lB***, 2^{1}_{∞}} $[^3Si_8O_{20}]$	F_2	[20]
Synthetic	Ca_4		Li_4	{***lB***, 2^{1}_{∞}} $[^3Si_{10}O_{26}]$		[21]
Tinaksite	NaK_2Ca_2	HTi		{***hB***, 2^{1}_{∞}} $[^3Si_7O_{19}]$	O	[22]
3. Two-dimensional silicate anions						
Prehnite	Ca_2	Al		{***oB***, 1^{2}_{∞}} $[^2(AlSi_3)O_{10}]$	$(OH)_2$	[23]
Zeophyllite	Ca_{13}			{***oB***, 1^{2}_{∞}} $[^4Si_5O_{14}]_2$	$F_8(OH)_2 \cdot 6\,H_2O$	[24]
Meliphanite	$(Na, Ca)_4Ca_4$	Be_4		{***oB***, 1^{2}_{∞}} $[^4Si_4O\ ][SiO\ ]$	F	[25]

Synthetic	Na_4			$\{oB, 1^2_\infty\}[^2Si_6O_{14}]$		
Synthetic		H_4		$\{oB, 1^2_\infty\}[^2Si_6O_{14}]$		[28]
Synthetic	K_8Yb_3			$\{oB, 1^2_\infty\}[^5Si_6O_{16}]_2$	OH	[29]
Kvanefjeldite	Na_4 (Ca,	Mn) H_2		$\{oB, 1^2_\infty\}[^2Si_6O_{16}]$		[30]
Semenovite	$(Ca, Na)_8Na_{0-2}RE_2$	$H_x(Fe^{+2}$, Mn, Zn, Ti)		$\{lB, 1^2_\infty\}[^6(Si, Be)_{10}(O, F)_{24}]_2$		[31]
Eudidymite	Na_2	Be_2		$\{lB, 1^2_\infty\}[^3Si_6O_{15}]$	H_2O	[32]
Lemoynite	$(Na, K)_2Ca$		Zr_2	$\{lB, 1^2_\infty\}[^4Si_5O_{13}]_2$	$5-6\ H_2O$	[33]
Synthetic	NaPr			$\{lB, 1^2_\infty\}[^4Si_6O_{14}]$		[34]
Synthetic	NaNd			$\{lB, 1^2_\infty\}[^4Si_6O_{14}]$		[35]
Leucosphenite	Na_4Ba	Ti_2		$\{lB, 2^2_\infty\}[^4(B_2Si_{10})O_{28}]$	O_2	[36]
Rhodesite	KCa_2	H		$\{lB, 2^2_\infty\}[^3Si_8O_{19}]$	$5\ H_2O$	[37]
Macdonaldite	Ca_4Ba	H_2		$\{lB, 2^2_\infty\}[^3Si_8O_{19}]_2$	$(8+x)H_2O$	[38]
Delhayelite	$Na_3K_7Ca_5$			$\{lB, 2^2_\infty\}[^3(AlSi_7)O_{19}]_2$	F_4Cl_2	[39]
Hydro-delhayelite	KCa_2	H_2		$\{lB, 2^2_\infty\}[^3(AlSi_7)O_{19}]$	$6\ H_2O$	[40]
Carletonite	Na_4KCa_4			$\{lB, 2^2_\infty\}[^6Si_8O_{18}]$	$[CO_3]_4(OH, F) \cdot H_2O$	[41]
4. Three-dimensional silicate anions						
Feldspars	Na			$\{lB, ^3_\infty\}[^3(AlSi_3)O_8]$		
	K			$\{lB, ^3_\infty\}[^3(AlSi_3)O_8]$		
	Ca			$\{lB, ^3_\infty\}[^3(Al_2Si_2)O_8]$		
	Ba			$\{lB, ^3_\infty\}[^3(Al_2Si_2)O_8]$		
Sodalite	Na_4			$\{lB, ^3_\infty\}[^4(Al_3Si_3)O_{12}]$	Cl or OH	[42]
Thomsonite	$NaCa_2$			$\{lB, ^3_\infty\}[^6(Al, Si)_{10}O_{20}]$	$6\ H_2O$	[43]
Leifite	Na_6	Be_2		$\{oB, ^3_\infty\}[^2(Al_2Si_{16})O_{39}]$	$(OH)_2 \cdot 1.5\ H_2O$	[44]
Synthetic	K_2Ce			$\{oB, ^3_\infty\}[^4Si_6O_{15}]$		[45]
Synthetic	Rb_6			$\{oB, ^3_\infty\}[^4Si_{10}O_{23}]$		[46]
Synthetic	Cs_6			$\{oB, ^3_\infty\}[^4Si_{10}O_{23}]$		[46]
Roggianite	Ca_4	H_4		$\{oB, ^3_\infty\}[^4(Al_4Si_8)O_{26}]$	$(OH)_4 \cdot \sim 6.5\ H_2O$	[47]
Wenkite	$(Ba, Ca)_{10}$	H_2		$\{hB, ^3_\infty\}[^3(Al, Si)_{20}O_{43}]$	$[SO_4]_3 \cdot H_2O$	[48]

References: [1] Baur and Ohta 1982; [2] Kossiakoff and Leavens 1976; [3] Malinovskii et al. 1977; [4] Moore et al. 1982; [5] Hoebbel et al. 1976; [6] Woodrow 1967; [7] Cannillo et al. 1971; [8] Basso and Della Giusta 1980, Lucchetti et al. 1981; [9] Fleet 1977; [10] Wenk 1974; [11] Matsubara 1981; [12] Voronkov et al. 1974; [13] Meagher 1976; [14] Bakakin et al. 1970; [15] Cannillo et al. 1966b; [16] Golovachev et al. 1971; [17] Pozas et al. 1975; [18] Kawamura and Iiyama 1981; [19] Kovalenko et al. 1977; [20] Ghose and Wan 1979; [21] Castrejón et al. 1983; [22] Bissert 1980; [23] Papike and Zoltai 1967; [24] Merlino 1972b; [25] Dal Negro et al. 1967; [26] Lopes-Vieira and Zussman 1969; [27] Jamieson 1967; [28] Guth et al. 1977; [29] Pushcharovskii et al. 1981; [30] Johnsen et al. 1983; [31] Mazzi et al. 1979; [32] Fang et al. 1972; [33] Le Page and Perrault 1976; [34] Karpov et al. 1976; [35] Karpov et al. 1976; [36] Shumyatskaya et al. 1971; [37] Hesse 1979b; [38] Cannillo et al. 1968; [39] Cannillo et al. 1969; [40] Dorfman and Chiragov 1979; [41] Chao 1972; [42] Hassan and Grundy 1983; [43] Alberti et al. 1981; [44] Coda et al. 1974; [45] Karpov et al. 1977a; [46] Schichl et al. 1973; [47] Galli 1980; [48] Merlino 1974; [49] Malinovskii et al. 1983; [50] Czank and Bissert 1985

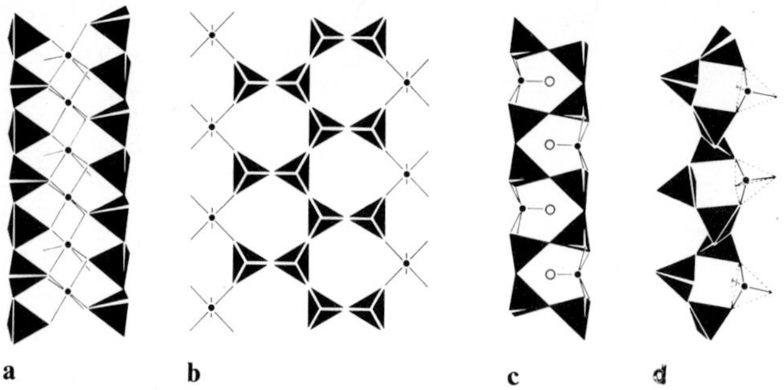

Fig. 10.9 a – d. Comparison between unbranched and branched single chains.
a Stretched unbranched zweier chain in enstatite, $Mg_2[Si_2O_6]$; **b** open-branched zweier chain in astrophyllite, $NaK_2Mg_2(Fe,Mn)_5Ti_2[Si_4O_{12}]_2(O,OH,F)_7$; **c** strongly folded unbranched vierer chain in haradaite, $Sr_2V_2[Si_4O_{12}]O_2$; **d** loop-branched sechser chain in vlasovite, $Na_4Zr_2[Si_8O_{22}]$.
● Cations not belonging to the silicate chain (Mg^{+2} in **a**, Ti^{+4} in **b**, V^{+4} in **c**, and Si^{+4} in **d**); ○ Oxygen atoms not bonded to silicon

and synthetic $Li_4Ca_4[Si_{10}O_{26}]$ and $Li_2Mg_2[Si_4O_{11}]$ contain at least some highly electronegative cations.

In strongly folded unbranched chains like those in haradaite, and in open-branched silicate chains, loops between the chain tetrahedra are closed by rigid polyhedra with the electronegative cations at their centers, e.g., $[TiO_6]$ in astrophyllite and $[VO_5]$ in haradaite. However, in loop-branched silicates closure is achieved by attaching $[SiO_4]$ tetrahedra as for example in vlasovite (Fig. 10.9 d). In other words, the silicon atoms of the loop-branched $[SiO_4]$ tetrahedra can be considered to play the same role in the loop-branched silicates as the highly electronegative cations play in open-branched and unbranched silicates.

The influence of electronegativity on the formation and shape of branched single chains is quite strong, whereas that of the size and valence of the cations is weaker and is not immediately evident.

Since loop-branched single chains have a higher Si : O ratio than unbranched ones, their negative charge per $[SiO_4]$ tetrahedron is lower, relative to that of an unbranched single chain of a silicate having the same cations. This causes a further reduction of the repulsive forces between the tetrahedra and, therefore, stabilization of the structure over and above that achieved by the presence of highly electronegative cations. With an increasing degree of condensation or dimensionality of the silicate anion, the Si : O ratio increases further, thereby reducing the negative charge per tetrahedron even more. As a result, the number of branched silicates not containing highly electronegative cations increases with increasing dimensionality (Table 10.8).

With regard to the second condition for the formation of branched silicate anions, i.e., the presence of soft cations, it should be noted that small, polyvalent cations form rather rigid $[MO_n]$ polyhedra to which the silicate anions must adjust themselves by rotation of the $[SiO_4]$ tetrahedra, distortion of the tetrahedra, and/or deviation of the Si–O–Si angle from its unstrained value of approximately 140°. In contrast, soft cations fit into almost any site and will add no more strain to a silicate anion than it would already contain without such cations. In this sense, Ca^{+2}, Na^{+}, and K^{+} ions have no destabilizing effect and can readily be incorporated into a branched silicate, while "hard" cations, like Li^{+}, Mg^{+2}, and Al^{+3}, in general, cannot.

The third condition for the existence of branched silicates, i.e., the presence of anions or water in addition to the silicate anions, also has a simple crystallochemical explanation. If only the oxygen atoms of the branched silicate anions are available to coordinate the cations, then a certain amount of strain will usually be unavoidable in achieving mutual adjustment between the silicate anions and the cation–oxygen polyhedra. However, if additional anions or water molecules are available, they can be used to complement the coordination polyhedra around the cations, thereby reducing the strain on the silicate anion and the arrays of $[MO_n]$ polyhedra.

These observations and their explanations can be summarized in the following conclusion: Silicon tends to form *any* silicate anion, branched or unbranched, that is theoretically possible within certain ranges of Si–O bond lengths, O–Si–O and Si–O–Si bond angles. Deviations from these bond lengths and angles create strain in the silicate anions and destabilize the structure. Of all the theoretically possible branched anions, the only one which exist are those which are not destabilized beyond a certain limit.

Destabilization is caused by

(1) repulsive forces between the $[SiO_4]$ tetrahedra due to a relatively high negative charge per tetrahedron. The repulsive forces are weaker, the more electronegative the cations of the silicate are (see Sec. 10.1.4.1).
(2) Small polyvalent cations destabilize the branched anions further, whereas soft cations, like Ca^{+2}, Na^{+}, and K^{+} do so to a much lesser degree.
(3) The destabilizing effect of mutual adjustment between silicate ions and the cations can be reduced by the presence of small anions and water molecules in addition to the oxygen atoms of the silicate anion.

10.4 Influence of Cation Properties on the Formation of Cyclic Silicate Anions

In silicate rings the distances between the centers of gravity of the tetrahedra are rather short. This is particularly true for dreier and vierer single rings and for all the known double rings which also contain vierer rings between the two subrings (Fig. 7.6). As in silicate chain anions, such a close approach of tetrahedra in cyclic anions produces considerable repulsive forces between the tetrahedra. Unlike unbranched single chains, in which an increase in the rather short distances between non-neighboring $[SiO_4]$ tetrahedra can take place by increasing the stretching factor of the chains, in ring anions the Si $\cdots$ Si distances of non-neighboring tetrahedra are essentially fixed. Therefore, unless prevented from doing so by incommensurability between the sizes of cation–oxygen polyhedra and the tetrahedral chains, a silicate will form chains rather than rings. However, if a chain anion is

Table 10.9 Partial survey of silicates containing unbranched single rings. The chemical formulae are subdivided into groups containing (i) "soft" cations; (ii) electronegative cations; (iii) "hard" cations; (iv) silicate anions; and (v) additional non-silicate anions plus water molecules. Branched cyclosilicates are included in Table 10.8

P^r	Name	Formula					Ref.
		"Soft" cations	Electronegative cations	"Hard" cations	Silicate anion	Additional ligands	
3	Synthetic (hT) (α)	Ca_3			$[Si_3O_9]$		[1]
	Synthetic (mP) (δ)	Ca_3			$[Si_3O_9]$		[2]
	Synthetic (mP) (α)	Sr_3			$[Si_3O_9]$		[3]
	Walstromite	Ca_2Ba			$[Si_3O_9]$		[4]
	Benitoite	Ba	Ti		$[Si_3O_9]$		[5]
	Pabstite	Ba	Sn		$[Si_3O_9]$		[6]
	Bazirite	Ba		Zr	$[Si_3O_9]$		[7]
	Margarosanite	Ca_2	Pb		$[Si_3O_9]$		[8]
	Synthetic	Na_2	Be_2		$[Si_3O_9]$		[9]
	Synthetic (hP)	K_2	$Si^{[6]}$		$[Si_3O_9]$		[10]
	Synthetic	K_2	Ti		$[Si_3O_9]$		[11]
	Wadeite	K_2		Zr	$[Si_3O_9]$		[12]
	Catapleiite	Na_2		Zr	$[Si_3O_9]$	$2\,H_2O$	[13]
	Hilairite	Na_2		Zr	$[Si_3O_9]$	$3\,H_2O$	[14]
4	Synthetic	K_4	H_4		$[Si_4O_{12}]$		[15]
	Synthetic	K_4		Sc_2	$[Si_4O_{12}]$	$(OH)_2$	[16]
	Papagoite	Ca_2	Cu_2Al_2		$[Si_4O_{12}]$	$(OH)_6$	[17]
	Kainosite	Ca_2Y_2			$[Si_4O_{12}]$	$[CO_3] \cdot H_2O$	[18]
	Joaquinite	Ba_2RE_2	$FeTi_2$		$[Si_4O_{12}]_2$	O_2	[19]
	Taramellite	Ba_4	$B_2(Fe^{+3}, Ti)_4$		$[Si_4O_{12}]_2$	O_5Cl_x	[20]
	Baotite	Ba_4	(Ti, Fe,	$Nb)_8$	$[Si_4O_{12}]$	$O_{16}Cl$	[21]
	Nagashimalite	Ba_4	$B_2(V^{+3}, Ti)_4$		$[Si_4O_{12}]_2$	$O_3(O, OH)_2Cl$	[22]
	Synthetic		Pb_8		$[Si_4O_{12}]$	O_4	[23]
6	Combeite	Na_4Ca_4			$[Si_6O_{18}]$		[24]
	Zirsinalite	Na_6Ca		Zr	$[Si_6O_{18}]$		[25]

	Scawtite	Ca_7			$[Si_6O_{18}]$	$[CO_3] \cdot 2\,H_2O$	[26]
	Synthetic	Na_6	Mn_3		$[Si_6O_{18}]$		[27]
	Synthetic	Na_6	Cd_3		$[Si_6O_{18}]$		[28]
	Kazakovite	Na_6	MnTi		$[Si_6O_{18}]$		[29]
	Imandrite	$Na_{12}Ca_3$	Fe_2		$[Si_6O_{18}]_2$		[30]
	Synthetic	Na_8	Sn		$[Si_6O_{18}]$		[31]
	Petarasite	Na_5		Zr_2	$[Si_6O_{18}]$	$(Cl, OH) \cdot 2\,H_2O$	[32]
	Dioptase		Cu_6		$[Si_6O_{18}]$	$6\,H_2O$	[33]
	Katayamalite	$(K, Na)Ca_7$	$(Ti, Fe^{+3}, Mn)_2$	Li_3	$[Si_6O_{18}]_2$	$(OH, F)_2$	[44]
	Tourmalines,						
	e.g., elbaite and	Na	$B_3Al_6(Al,$	$Li)_3$	$[Si_6O_{18}]$	$O_9(OH)_4$	[34]
	liddicoatite	Ca	$B_3Al_6(Al,$	$Li)_3$	$[Si_6O_{18}]$	$O_9(O, OH, F)_4$	[35]
8	Muirite	$Ba_{10}(Ca,$	$Mn, Ti)_4$		$[Si_8O_{24}]$	$(Cl, O, OH)_{12} \cdot 4\,H_2O$	[36]
9	Eudialyte	$Na_{12}(Ca, RE)_6$	$H_y(Fe, Mn,$	$Mg)_3Zr_3(Zr, Nb)_x$	$[Si_9O_{27}]_2$		
(+3)					$[Si_3O_9]_2$	Cl_z	[37]
12	Synthetic	$Na_{15}RE_3$			$[Si_{12}O_{36}]$		[38]
	Synthetic	$Na_{15}Gd_3$			$[Si_{12}O_{36}]$		[39]
	Synthetic	$Na_{16}Ca_4$			$[Si_{12}O_{36}]$		[40]
	Synthetic	$K_{16}Ca_4$			$[Si_{12}O_{36}]$		[41]
	Synthetic	$K_{16}Sr_4$			$[Si_{12}O_{36}]$		[42]
	Traskite	$(Ca, Sr)Ba_{24}$	$(Fe, Mn, Ti)_4$	$(Ti, Fe, Al, Mg)_{12}$	$[Si_{12}O_{36}]\,[Si_2O_7]_6$	$(O, OH)_{30}Cl_6 \cdot 14\,H_2O$	[43]

$\boldsymbol{P^r}$ = ring periodicity, i.e., number of tetrahedra in the silicate ring

References: [1] Yamanaka and Mori 1981; [2] Trojer 1969; [3] Machida et al. 1982; [4] Dent Glasser and Glasser 1968; [5] Fischer 1969; [6] Gross et al. 1965; [7] Young et al. 1978; [8] Freed and Peacor 1969; [9] Ginderow et al. 1982; [10] Swanson and Prewitt 1983; [11] Shumyatskaya et al. 1973; [12] Blinov et al. 1977; [13] Ilyushin et al. 1981c; [14] Ilyushin et al. 1981a; [15] Hilmer 1964; [16] Pyatenko et al. 1979; [17] Guillebert and Le Bihan 1965; [18] Volodina et al. 1963; [19] Cannillo et al. 1972; [20] Mazzi and Rossi 1980; [21] Muradyan and Simonov 1977; [22] Matsubara 1980b; [23] Dent Glasser et al. 1981; [24] Fischer and Tillmanns 1983; [25] Pudovkina et al. 1980; [26] Pluth and Smith 1973; [27] Pushcharovskii et al. 1976; [28] Simonov et al. 1968; [29] Voronkov et al. 1979a; [30] Chernitsova et al. 1980; [31] Zayakina et al. 1980; [31] Safronov et al. 1980; [32] Ghose et al. 1980; [32] Khomyakov et al. 1981; [33] Ribbe et al. 1977; [33] Belov et al. 1978; [34] Donnay and Barton 1972; [35] Nuber and Schmetzer 1981; [36] Khan and Baur 1971; [37] Giuseppetti et al. 1971; [37] Golyshev et al. 1971; [38] Merinov et al. 1978, 1980; [38] Maksimov and Belov 1981; [39] Shannon et al. 1977; [40] Fischer and Tillmanns 1984; [41] Baumgartner and Völlenkle 1977; [42] Baumgartner and Völlenkle 1977; [43] Malinovskii et al. 1976; [44] Murakami et al. 1983.

Table 10.10 Survey of silicates containing unbranched double rings {***uB***, 2***r***} $[^{P^r}Si_{2p}O_{5p}]^{-2p}$ with p = ***P***^{r} = number of tetrahedra in the fundamental ring

P^r	Name	Formula: Cations	Formula: Silicate anion	Formula: Additional molecules	Reference
3	Synthetic	$[Ni^{+2}(H_2N \cdot CH_2 \cdot CH_2 \cdot NH_2)_3]_3$	$[^3Si_6O_{15}]$	$26\ H_2O$	Smolin 1970
4	Synthetic	$[Cu^{+2}(H_2N \cdot CH_2 \cdot CH_2 \cdot NH_2)_2]_4$	$[^4Si_8O_{20}]$	$38\ H_2O$	Smolin et al. 1972
	Synthetic	$H_2[Co^{+3}(H_2N \cdot CH_2 \cdot CH_2 \cdot NH_2)_3]_2$	$[^4Si_8O_{20}]$	$16.4\ H_2O$	Smolin et al. 1976
	Synthetic	$[N(CH_3)_4]_8$	$[^4Si_8O_{20}]$	$64.8\ H_2O$	Smolin et al. 1979
	Synthetic	$H_7[N(n\text{-}C_4H_9)_4]$	$[^4Si_8O_{20}]$	$5.33\ H_2O$	Bissert and Liebau 1984
	Steacyite	$K_{1-x}(Na, Ca)_{2-y}Th_{1-z}$	$[^4Si_8O_{20}]$		Szymański et al. 1982; Richard and Perrault 1972; Perrault and Szymański 1982
	Iraqite	(K, □)(Ca, RE, Na)$_2$(RE, Th)	$[^4Si_8O_{20}]$		Livingstone et al. 1976
5	Synthetic ester	$[(CH_3)_3Si\text{-}]_{10}$	$[^5Si_{10}O_{25}]$		Smolin in Hoebbel et al. 1975
6	Roedderite	K_2Mg_5	$[^6Si_{12}O_{30}]$		Seifert and Schreyer 1969; Khan et al. 1972
	Eifelite	KNa_3Mg_4	$[^6Si_{12}O_{30}]$		Abraham et al. 1983
	Synthetic	K_2Mn_5	$[^6Si_{12}O_{30}]$		Sandomirskii et al. 1977
	Osumilite	KMg_2Al_3	$[^6(Al_2Si_{10})O_{30}]$		Hesse and Seifert 1982
	Milarite	$KCa_2(Be, Al)_3$	$[^6Si_{12}O_{30}]$	$0.75\ H_2O$	Černý et al. 1980
	Brannockite	KSn_2Li_3	$[^6Si_{12}O_{30}]$		White et al. 1973
	Synthetic osumilite isotypes	$A_xM_2M'_3$; A = K, Na, Rb; M = Mg, Cu, Fe; M′ = Mg, Zn, Fe, Cu, Li	$[^6Si_{12}O_{30}]$		Nguyen et al. 1980

very unfavorable, a cyclic anion can more easily be formed in the presence of highly electronegative cations than it can in the presence of more electropositive ones. In fact, most of the cyclosilicates contain electronegative cations (Tables 10.9, 10.10). In addition, almost all the single ring silicates contain soft cations, like Na^+, K^+, Ca^{+2}, and Ba^{+2} and many of them contain additional anions, such as O^{-2}, OH^-, Cl^-, F^-, $[CO_3]^{-2}$, and/or water molecules. By analogy to the situation of the branched silicates, this may be regarded as another indication that, in general, the cyclic anions are energetically less favorable than chain anions and are formed only in the absence (or near absence) of destabilizing small or/and polyvalent cations.

In agreement with the explanation given, the number of single ring silicates $M_m M'_{m'} \cdots \{1\,\boldsymbol{r}\}[Si_nO_{3n}]$ is considerably smaller than the number of single chain silicates $M_m M'_{m'} \cdots \{1^1_\infty\}[Si_nO_{3n}]$, whereas the number of cyclophosphates $M_m M'_{m'} \cdots \{1\,\boldsymbol{r}\}[P_nO_{3n}]$ is at least as large as the number of polyphosphates $M_m M'_{m'} \cdots \{1^1_\infty\}[P_nO_{3n}]$ due to the lower formal charge of the $[PO_4]$ tetrahedra compared to the $[SiO_4]$ tetrahedra (see Secs. 10.1.4.1 and 10.1.4.2).

10.5 Influence of Cation Properties on the Shape of Single Layer Silicate Anions

The number of single layer silicates is so large that it would be beyond the scope of this section to give a complete list of the known single layer silicates. This is especially true of the silicates containing zweier layers. Therefore, only a selection of typical examples for each group of silicates with unbranched single layers has been presented in Table 10.11. The silicates known to contain branched single layers have been surveyed in Table 10.8.

In Sec. 10.1.4 it was demonstrated that the shape of the silicate chains is controlled by the mutual adjustment between these chains and the arrays of cation–oxygen polyhedra. Since the tetrahedral chains are infinite in *one* dimension this adjustment has to be in *one* dimension. However, since the tetrahedral layers of the phyllosilicates are infinite in *two* dimensions, the mutual adjustment between these layers and the arrays of cation–oxygen polyhedra must also be *two*-dimensional.

The relationships developed in Sec. 10.1.4 between the size, electronegativity, and valence of cations on the one hand, and the shape of silicate chains on the other, also hold for tetrahedral layers. Thus, for example, lithium, sodium, silver, and barium ions, which form polysilicates with shrunken zweier single chains, also form anhydrous phyllosilicates with tetrahedral layers constructed by linking together these shrunken zweier single chains (Table 10.12). Moreover, the small divalent cations Mg^{+2} and Fe^{+2}, which form the pyroxenes with stretched zweier single chains, also form the micas and the large group of clay minerals containing tetrahedral layers built by linking together the same stretched zweier single chains. In addition, the medium-sized calcium and sodium ions form dreier chains in wollastonite, pectolite, serandite, and xonotlite, and also form phyllosilicates in which the layers contain dreier chains, for example in nekoite, $Ca_3[Si_6O_{15}] \cdot 7\,H_2O$ (Fig. 10.10b).

Table 10.11 Silicates containing unbranched single layers $\{uB, 1^2_\infty\}[^PSi_{2n}O_{5n}]$

P	Silicate		Reference
	Name	Formula	
2	Synthetic	$H_2[Si_2O_5]$	Liebau 1964; Le Bihan et al. 1971
	Synthetic	$Li_2[Si_2O_5]$	Liebau 1961b
	Synthetic	α-$Na_2[Si_2O_5]$	Pant and Cruickshank 1968
	Sanbornite	$Ba[Si_2O_5]$(lT)	Hesse and Liebau 1980b
	Petalite	$LiAl[Si_2O_5]_2$	Effenberger 1980; Tagai et al. 1982
	Talc	$Mg_3[Si_2O_5]_2(OH)_2$	Perdikatsis and Burzlaff 1981
	Lizardite	$Mg_3[Si_2O_5](OH)_4$	Mellini 1982
	Muscovite	$KAl_2[AlSi_3O_{10}](OH)_2$	Rothbauer 1971
	Amesite	$Mg_2Al[AlSiO_5](OH)_4$	Anderson and Bailey 1981
	Sepiolite	$(Mg,Fe,Al)_4[Si_2O_5]_3(O,OH)_2 \cdot 4H_2O$	Brauner and Preisinger 1956
	Yofortierite	$Mn_5[Si_2O_5]_4(OH)_2 \cdot 8H_2O$	Perrault et al. 1975
	Palygorskite	$Mg_5[Si_2O_5]_4(OH)_2 \cdot 8H_2O$	Bradley 1940
	Chrysotile	$Mg_3[Si_2O_5](OH)_4$	Whittaker 1956
	Antigorite	$Mg_{48}[Si_4O_{10}]_{8.5}(OH)_{62}$	Kunze 1959
	Synthetic	$K_2[Si_4O_9]$	Schweinsberg and Liebau 1974
3	Synthetic	$K_2Be_2[Si_6O_{15}]$	Naumova et al. 1976
	Synthetic	$K_3Nd[Si_6O_{15}]$	Pushcharovskii et al. 1977
	Synthetic	$K_2Ti[Si_6O_{15}]$	Gebert et al. 1983
	Dalyite	$K_2Zr[Si_6O_{15}]$	Fleet 1965
	Armstrongite	$CaZr[Si_6O_{15}] \cdot 2.5H_2O$	Kashaev and Sapozhnikov 1978
	Sazhinite	$HNa_2Ce[Si_6O_{15}] \cdot nH_2O$	Shumyatskaya et al. 1980
	Synthetic	$H_2NaNd[Si_6O_{15}] \cdot nH_2O$	Karpov et al. 1977b
	Nekoite	$Ca_3[Si_6O_{15}] \cdot 7H_2O$	Alberti and Galli 1980
	Synthetic	$Na_2Zn[Si_3O_8]$	Hesse et al. 1977
4	Synthetic	$K_2Be[Si_4O_{10}]$	Balko et al. 1980
	Synthetic	$K_2Ba_7[Si_4O_{10}]_4$	Cervantes-Lee et al. 1982
	Gillespite	$BaFe[Si_4O_{10}]$	Hazen and Finger 1983
	Synthetic	$MCu[Si_4O_{10}]$, M = Ca, Sr, Ba	Pabst 1959
	Ekanite	$Ca_2Th[Si_4O_{10}]_2$	Szymański et al. 1982
	Makatite	$H_2Na_2[Si_4O_{10}] \cdot 4H_2O$	Annehed et al. 1982
	Cavansite, pentagonite	$CaV[Si_4O_{10}]O \cdot 4H_2O$	Evans 1973
	Apophyllite	$KCa_4[Si_4O_{10}]_2(F,OH) \cdot 8H_2O$	Bartl and Pfeifer 1976
	Natroapophyllite	$NaCa_4[Si_4O_{10}]_2F \cdot 8H_2O$	Miura et al. 1981
5	Synthetic	$LiBa_9[Si_{10}O_{25}][CO_3]Cl_7$	Il'inets et al. 1983
6	Manganpyrosmalite	$Mn_8[Si_6O_{15}](OH,Cl)_{10}$	Kato and Takéuchi 1983
	Mcgillite	ca. $(Mn,Fe)_8[Si_6O_{15}](OH,Cl)_{10}$	Iijima 1982; Ozawa et al. 1983
	Bementite	$Mn_7[Si_6O_{15}](OH)_8$	Kato and Takéuchi 1980

P = Chain periodicity

Table 10.12 Identity periods parallel to corresponding chain directions of some poly- and phyllosilicates

P	Polysilicate			Phyllosilicate		
	Name	Formula	I [Å]	I [Å]	Formula	Name
2	Synthetic	$Li_4[Si_2O_6]$	4.68	4.79	$Li_2[Si_2O_5]$	Synthetic
	Synthetic	$Na_4[Si_2O_6]$	4.82	4.80	β-$Na_2[Si_2O_5]$	Synthetic
				4.90	α-$Na_2[Si_2O_5]$	Synthetic
	Synthetic	$Ag_4[Si_2O_6]$	4.55	4.79	$Ag_2[Si_2O_5]$	Synthetic
	Enstatite	$Mg_2[Si_2O_6]$	5.21	5.27	$Mg_3[Si_2O_5]_2(OH)_2$	Talc
	Orthoferrosilite	$Fe_2[Si_2O_6]$	5.24	5.30	$Mg_{48}[Si_4O_{10}]_{8.5}(OH)_{62}$	Antigorite
	Synthetic	$Ba_2[Si_2O_6]$(hT)	4.58	4.63	$Ba[Si_2O_5]$(lT)	Sanbornite
				4.66	$Ba[Si_2O_5]$(hT)	Synthetic
3	Wollastonite	$Ca_3[Si_3O_9]$(lT)	7.32	2 × 6.34	$KCa_4[Si_4O_{10}]_2(F, OH) \cdot 8\,H_2O$	Apophyllite
	Pectolite	$HNaCa_2[Si_3O_9]$	7.04	6.53	$Na_3K_7Ca_5[AlSi_7O_{19}]_2F_4Cl_2$	Delhayelite
	Serandite	$HNa(Mn, Ca)_2[Si_3O_9]$	6.89	2 × 6.59	$Na_4KCa_4[Si_8O_{18}][CO_3]_4(OH, F) \cdot H_2O$	Carletonite

P = Chain periodicity

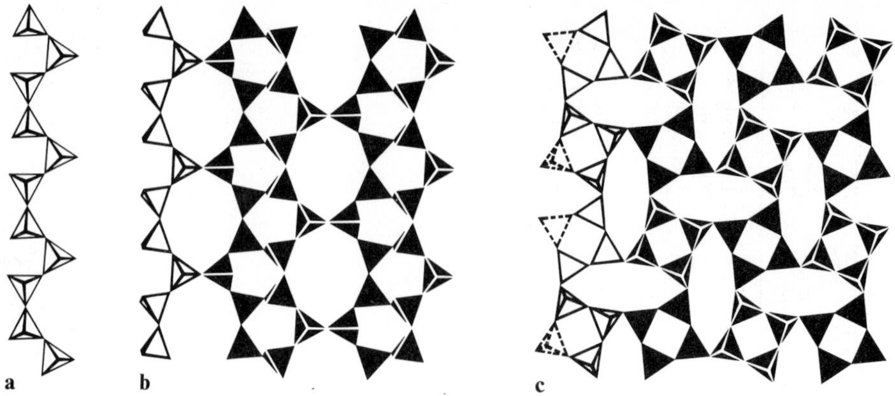

Fig. 10.10 a – c. Comparison between single chains and single layers in calcium silicates. **a** Unbranched dreier single chain in serandite, $HNa(Mn,Ca)_2[Si_3O_9]$ (Ohashi and Finger 1978); **b** unbranched dreier single layer in nekoite, $Ca_3[Si_6O_{15}]\cdot 7H_2O$ (Alberti and Galli 1980); **c** the tetrahedral layer of apophyllite, $KCa_4[Si_4O_{10}]_2(F,OH)\cdot 8H_2O$ (Chao 1971; Prince 1971) described as a loop-branched sechser single layer. [Notice that the correct fundamental chain of apophyllite is an unbranched vierer chain (Fig. 4.8 b)]

However, in some of these layers, the identity period of the chain required by a cation in a polysilicate is doubled in the layer silicate by a different directedness of the tetrahedra in alternate periods of the subchain. This is clearly visible in the single layer of apophyllite, $KCa_4\{^2_\infty\}[Si_4O_{10}]_2(F,OH)\cdot 8H_2O$ (Fig. 10.10c). According to the procedure to classify silicates described in Sec. 6.4, this is a vierer single layer. However, the layer contains the typical elements of a dreier chain. If the terminal oxygen atoms of all the $[SiO_4]$ tetrahedra were located at the same side of the layer, then apophyllite would have to be classified as a loop-branched dreier single layer with its basic chains running at 45° to the vierer chains. However, since the vierer rings point up and down alternately, the apophyllite layer could be regarded as a branched sechser single layer. Although this would not be in accordance with the procedure suggested, it would emphasize the crystal chemical similarity to the other calcium silicates better than would its formally correct description as an unbranched vierer single layer.

The few examples listed in Table 10.12 demonstrate that the one-dimensional adjustment between tetrahedral chains and polyhedral arrays still holds for layer silicates. It is only slightly altered and complemented by an additional adjustment in the second direction.

10.5.1 Anhydrous Single Layer Silicates

In order to clarify the influence of the cations on the adjustment between tetrahedral chains and polyhedral arrays in two dimensions, consider first the flat tetrahedral layer found in the micas and in many clay minerals (Fig. 10.11). In this zweier layer the free corners of all the tetrahedra point to the same side of the layer. Since the negative charge of the silicate tetrahedra is localized at these non-

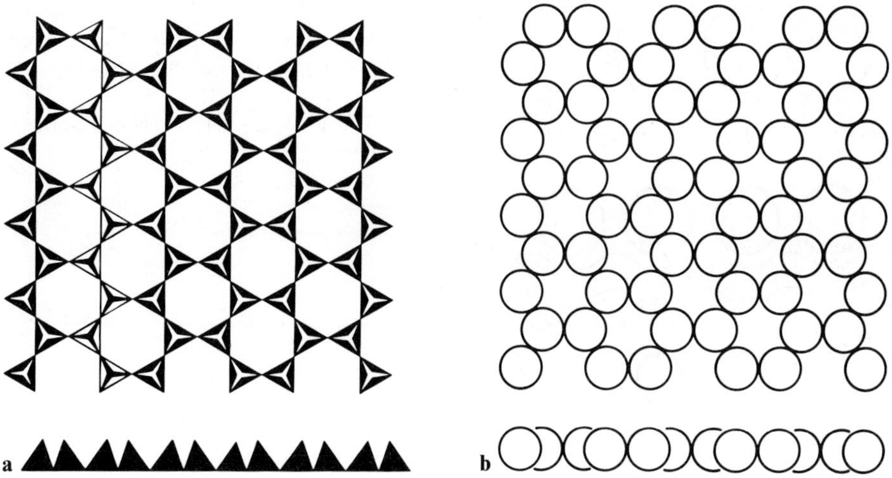

Fig. 10.11. An ideally flat $[Si_2O_5]$ tetrahedral layer of hexagonal symmetry and a layer of densely packed monovalent cations which compensate the negative charges of the polar tetrahedral layer

bridging oxygen atoms, the cations which compensate these charges are attracted to this side of the layer. In order to form an ideally flat zweier single layer of hexagonal symmetry the repulsive forces between the cations would have to be disregarded and the assumption made that the cations are approximately densely packed on top of these layers. If there is some replacement of silicon by aluminum in the tetrahedral positions, the $\{{}^{2}_{\infty}\}[Si_2O_5]$ layer would have translation periods $a_0 = b_0 \simeq 5.3$ Å.

Since each $[SiO_4]$ tetrahedron in an $[Si_2O_5]$ layer carries one negative charge, one monovalent cation is necessary to balance this charge. Moreover, in order to stabilize the ideal flat layer of Fig. 10.11, the cations should have a radius of approximately 1.7 Å. Both of these criteria are met by the cesium ion, but unfortunately, the structure of $Cs_2[Si_2O_5]$ is still unknown[1].

With smaller cations than cesium, the tetrahedral layer will cling to the cations due to electrostatic attraction between the negatively charged layers and the cations and the layer will become warped. The smaller the monovalent cations are, the higher the degree of warping. For very small cations, the layer should become significantly folded, since the nonbridging oxygen atoms of the tetrahedra in a non-flat layer will point to both sides of the layer due to the repulsive forces between the cations and between the $[SiO_4]$ tetrahedra.

---

1 Although the size of the Cs^+ ion would be large enough to stabilize this flat layer, it is unlikely that $Cs_2[Si_2O_5]$ has this structure because of strong repulsive forces between the Cs^+ ions. If $Cs_2[Si_2O_5]$ has a single layer structure at all, it is more likely that the nonbridging oxygen atoms point partly to one side and partly to the other side of the layer in order to increase the cation–cation distance. It is, however, even more probable that the hypothetical phase of composition $Cs_2[Si_2O_5]$ has a framework structure built from tertiary tetrahedra in a similar way to $K_2Ce[Si_6O_{15}]$ (see p. 205) and $Cs[Be_2F_5]$.

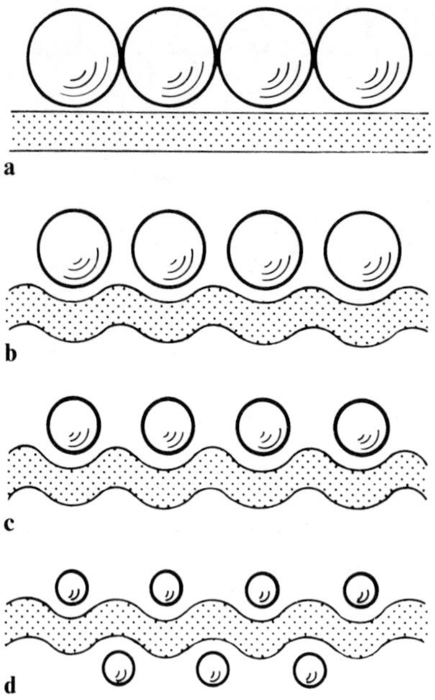

Fig. 10.12 a – d. Schematical diagram of the convolution of tetrahedral layers (*dotted bars*) due to the clinging of the layers to monovalent cations of various size.

a Cs^+, $r \simeq 1.7$ Å; **b** K^+, $r \simeq 1.4$ Å; **c** Na^+, $r \simeq 1.0$ Å; **d** Li^+, $r \simeq 0.6$ Å

This is illustrated schematically in Fig. 10.12 for monovalent cations of the size of cesium ($r \simeq 1.7$ Å), potassium ($r \simeq 1.40$ Å), sodium ($r \simeq 1.0$ Å), and lithium ($r \simeq 0.6$ Å).

The structures of the $M_2^+[Si_2O_5]$ phases determined so far suggest that this very crude picture is indeed applicable to layer silicates. In $Li_2[Si_2O_5]$ (Liebau 1961 b) and phyllosilicic acid, $H_2[Si_2O_5]$ (Liebau 1964), the tetrahedral layers, which in Fig. 10.13 are projected parallel to the zweier chains, are very strongly folded. However, in α-$Na_2[Si_2O_5]$ they are slightly less folded (Pant and Cruickshank 1968) and in β-$Na_2[Si_2O_5]$ remarkably less so (Pant 1968). Unfortunately, the structures of $K_2[Si_2O_5]$, $Rb_2[Si_2O_5]$, and $Cs_2[Si_2O_5]$ are unknown, so there is no evidence that the silicate layer becomes even flatter when the size of the monovalent cations is increased beyond the value for sodium.

In phyllosilicates $M^{+2}[Si_2O_5]$ *one* divalent cation must balance the charge of *two* $[SiO_4]$ tetrahedra. Because of this decrease in $n_M : n_{Si}$ ratio, the folding of the tetrahedral layer should be much greater for a divalent cation than for a monovalent one of the same size. In accordance with this description, the folding of the silicate layer in the barium silicates sanbornite, $Ba^{[9]}[Si_2O_5]$(1T) and its high temperature

Fig. 10.13 a – c. Convolution of tetrahedral layers in anhydrous phyllosilicates of monovalent cations. The layers are projected parallel to the fundamental zweier chains.

a $Li_2[Si_2O_5]$; **b** α-$Na_2[Si_2O_5]$; **c** β-$Na_2[Si_2O_5]$

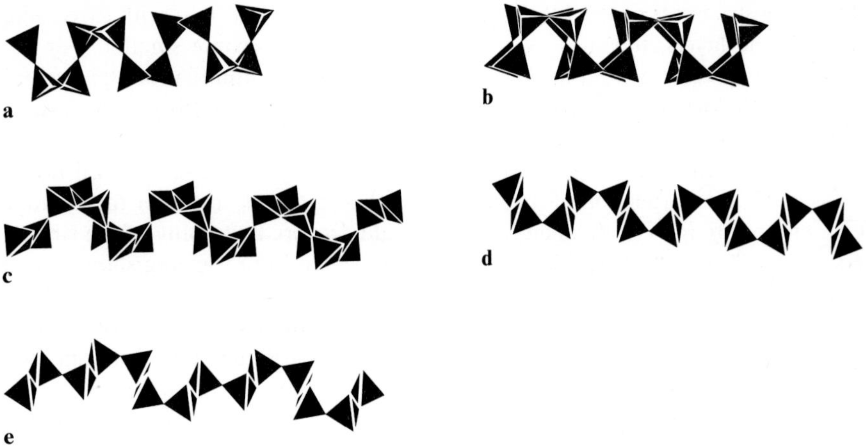

Fig. 10.14a–e. Convolution of tetrahedral layers in anhydrous phyllosilicates with $n_{cat} : n_{tetr} = 1 : 2$. The layers are projected parallel to the direction of a fundamental chain.

a Petalite, $LiAl[Si_2O_5]_2$; **b** gillespite, $BaFe[Si_4O_{10}]$; **c** dalyite, $K_2Zr[Si_6O_{15}]$; **d** sanbornite, $Ba[Si_2O_5]$(lT); **e** synthetic $Ba[Si_2O_5]$(hT)

modification $Ba^{[9]}Ba^{[10]}[Si_2O_5]_2$(hT) (Hesse and Liebau 1980b) (Fig. 10.14d, e) lies somewhere between that of α-$Na_2^{[5]}[Si_2O_5]$ and β-$Na^{[5]}Na^{[6]}[Si_2O_5]$ despite the fact that the barium ion ($r^{[9,10]} = 1.5$ Å) is considerably larger than the sodium ion ($r^{[5,6]} = 1.0$ Å) (Fig. 10.13b, c).

Although divalent cations larger than barium are not available for study, the fluoroberyllates $M^+[Be_2F_5]$ can be used as silicate models in order to deduce how the tetrahedral layer responds to larger cations. Since the fluoride ion ($r^{[4]} = 1.31$ Å) is smaller than the oxygen ion ($r^{[4]} = 1.38$ Å), the $[BeF_4]$ tetrahedron is smaller than the $[SiO_4]$ group. Therefore, the mesh size of a $[Be_2F_5]$ layer is slightly smaller than that of a corresponding $[Si_2O_5]$ layer of the same type and degree of warping. As a result, the size of the monovalent Rb^+ ion ($r^{[6]} = 1.49$ Å) in the fluoroberyllate $Rb[Be_2F_5]$ is equivalent to the size of a 12-coordinated divalent M^{+2} ion of approximately 1.80 Å radius in a silicate $M^{+2}[Si_2O_5]$. Therefore, since the tetrahedral layers in $Rb[Be_2F_5]$ (Ilyukhin and Belov 1962) and $Tl[Be_2F_5]$ (Le Fur 1972) are almost flat and deviate only slightly from trigonal symmetry, it can be inferred that a flat $[Si_2O_5]$ layer would be obtained if the silicate contained divalent cations of approximately 1.9 Å radius. No divalent monatomic cations of such size exist, so only complex divalent cations are expected to form such flat silicate layers of the mica type. Still larger cations cannot be accommodated to the tetrahedral layers as evidenced by the fact that $Cs[Be_2F_5]$ has a low density interrupted framework which contains only tertiary $[BeF_4]$ tetrahedra (connectedness $s = 3$) (Le Fur and Aléonard 1972).

Although the cations Ca^{+2}, Mg^{+2}, Fe^{+2}, Ni^{+2}, and Co^{+2}, which are smaller than Ba^{+2}, form chain silicates readily, no anhydrous layer silicates of these cations are known. This can also be attributed to the required degree of warping of the tetrahedral layers.

From the structures of the $M^{+2}[Si_2O_5]$ compounds, we find that with increasing warping of the layers, the sites available to the cations become smaller. For the phyllosilicates presented in Fig. 10.13, the coordination number of the cations decreases, with increasing warping from five and six in β-$Na_2[Si_2O_5]$, to five in α-$Na_2[Si_2O_5]$, and to four in $Li_2[Si_2O_5]$. In other words, the coordination number of the cations in phyllosilicates is a function of the warping of the tetrahedral layers and this, in turn, is a function of mean cation size and $n_M : n_{Si}$ ratio, or the average valence of the cations. The divalent Sr^{+2} ion would require a warping of the silicate layers greater than that in sanbornite, $Ba[Si_2O_5](1T)$, and the warping should be at least as large as that in α-$Na_2[Si_2O_5]$. However, strongly folded layers of this kind would lead to cation sites of coordination numbers less than six for at least some of the cations. These coordination numbers are too small for calcium and strontium ions, thus explaining why anhydrous layer silicates of composition $Ca[Si_2O_5]$ and $Sr[Si_2O_5]$ do not exist. It also explains why the small cations Li^+ and Al^{+3} in the ratio 1:1 (mean valence $\langle v_M \rangle = 2$) form a very stable anhydrous layer silicate, petalite $Li^{[4]}Al^{[4]}[Si_2O_5]_2$, with strongly folded tetrahedral layers (Fig. 10.14a).

In silicates $M_2^{+3}[Si_6O_{15}]$ *one* trivalent cation is required to balance the negative charges of *three* $[SiO_4]$ tetrahedra. This would lead to an extremely high convolution of the tetrahedral layers even for rather large cations. Indeed, a rough calculation suggests that trivalent ions of approximately 1.0 Å radius would require silicate layers as strongly folded as those of petalite (Fig. 10.14a) or of lithium phyllosilicate $Li_2[Si_2O_5]$ (Fig. 10.13a). Although such large trivalent cations exist (e.g., La^{+3}, Bi^{+3} and the large rare earth ions), they seem to be too large to fit into the small sites provided by such folded tetrahedral layers.

In summary, it is clear that the relationship between the degree of convolution of tetrahedral layers on the one hand, and the average size and valence of cations on the other, explains not only the shape of the layers in anhydrous phyllosilicates of mono- and divalent cations, but also the nonexistence of layer silicates $M_2[Si_6O_{15}]$ with trivalent cations and of $M[Si_2O_5]$ with cations M^{+2} smaller than barium.

In Fig. 10.15 the anhydrous or almost anhydrous silicates containing anions with Si : O = 2 : 5, i.e., with single layers, multiple chains, double rings, and the framework silicate $K_2Ce[Si_6O_{15}]$ (Karpov et al. 1977a) containing only tertiary $[SiO_4]$ tetrahedra, are plotted as a function of mean cation radius $\langle r_M^{[6]} \rangle$ and of the ratio $\langle r_M^{[6]} \rangle : \langle v_M \rangle$, where $\langle v_M \rangle$ is the mean valence of the cations. Although the regions containing silicate anions of different dimensionality overlap in this diagram, an explanation for the overall distribution of phases can be presented.

In Sec. 10.4 the preference of tetrahedral single chains over single rings for Si : O = 1 : 3 was explained by the fact that larger Si···Si distances between non-neighboring tetrahedra are more possible in single chains than in single rings. Therefore, because of the repulsive forces between the negative charges of the tetrahedra, single chains are, in general, energetically more favorable than single rings. The same principle holds for silicates with Si : O = 2 : 5 and is the key for an understanding of the preference of single layers over double rings and multiple chains for silicates with this atomic ratio.

A comparison of the various single layers presented in Figs. 7.18 to 7.24 clearly shows that the flat or almost flat zweier single layers in micas and many clay minerals (Fig. 7.19a, c, and d) are the most favorable in relation to the repulsive forces between the tetrahedra: they contain only sechser rings and the average distance between the centers of gravity of non-neighboring tetrahedra is a maximum, or at least nearly so. However, as the degree of warping of such zweier layers increases, e.g., in Fig. 7.19b, these distances become smaller. A further decrease of distances between non-neighboring tetrahedra occurs when a single layer contains fünfer rings (Fig. 7.21d, e, f) or vierer rings (Figs. 7.18b, c, d, f, 7.22b). The higher the proportion of fünfer, vierer, and especially dreier rings, the less energetically favorable the layer type will be.

In all the multiple chains (Figs. 7.11 to 7.13) and double rings (Fig. 7.6) with Si : O = 2 : 5, each tetrahedron participates in at least one vierer ring. In other words, these anions have a high proportion of vierer rings and are, therefore, less favorable than the majority of single layers.

In summary, the stability of silicate anions with an Si : O ratio of 2 : 5 decreases in the order: flat single layer with sechser rings exclusively – warped single layer with sechser rings exclusively – single layer with fünfer rings – single layer with increasing portion of vierer rings – multiple chains – double rings.

As a consequence, whenever possible, a silicate with Si : O = 2 : 5 tends to have flat tetrahedral zweier layers. As the average size of the cations decreases and/or their valence increases, warping becomes progressively more necessary. At the same time, the sites available for the cations become smaller. Below a certain degree of convolution at least some of the cations can only be located in sites with coordination number four. This is indicated by the broken line running NW–SE in Fig. 10.15 below which only single layer silicates with 50% of their cations in tetrahedral sites exist: $Li_2^{[4]}[Si_2O_5]$, $Li^{[4]}Al^{[4]}[Si_2O_5]_2$, $K_2^{[7]}Be_2^{[4]}[Si_6O_{15}]$, $Ca^{[8]}Cu^{[4]}[Si_4O_{10}]$, $Sr^{[8]}Cu^{[4]}[Si_4O_{10}]$, $Ba^{[8]}Fe^{[4]}[Si_4O_{10}]$, and $Ca_2^{[4]}Th^{[8]}[Si_4O_{10}]_2$ in which the Ca^{+2} ions have four nearest-neighbor oxygen atoms at a distance of 2.34 Å and four next-nearest neighbor oxygens at a distance of 2.69 Å (Szymański et al. 1982).

Silicates with Si : O = 2 : 5 for which considerable degrees of layer convolution would be required due to the presence of relatively small cations and/or, more importantly, ones with $\langle v_M \rangle > 1$, and in which there are not enough cations suitable for 4-coordination, cannot have single layer anions, but must instead form multiple chains or double rings.

At first sight, $K_8Cu_4[Si_{16}O_{40}]$ with its vierer quadruple chain seems to contradict this explanation. However, Fig. 7.13c indicates that this tubular chain with its wide cross-section may be considered to be a layer which is rolled up into a cylinder.

The higher the mean electronegativity, $\langle \chi_M \rangle$, of the cations, i.e., the lower the effective negative charge per $[SiO_4]$ tetrahedron and, therefore, the lower the repulsive forces between the tetrahedra, the more likely it is that multiple chains or double rings with Si : O = 2 : 5 will be formed when no layer can be formed (because of sterical incommensurability between tetrahedral layers and arrays of cation–oxygen polyhedra). For highly electropositive cations the strong repulsive forces between the highly charged $[SiO_4]$ tetrahedra could be counteracted by increasing the distances between non-neighboring tetrahedra by the formation of interrupted frameworks. Since such frameworks have large cavities they require large cations to gain the necessary packing density. The largest monatomic cation

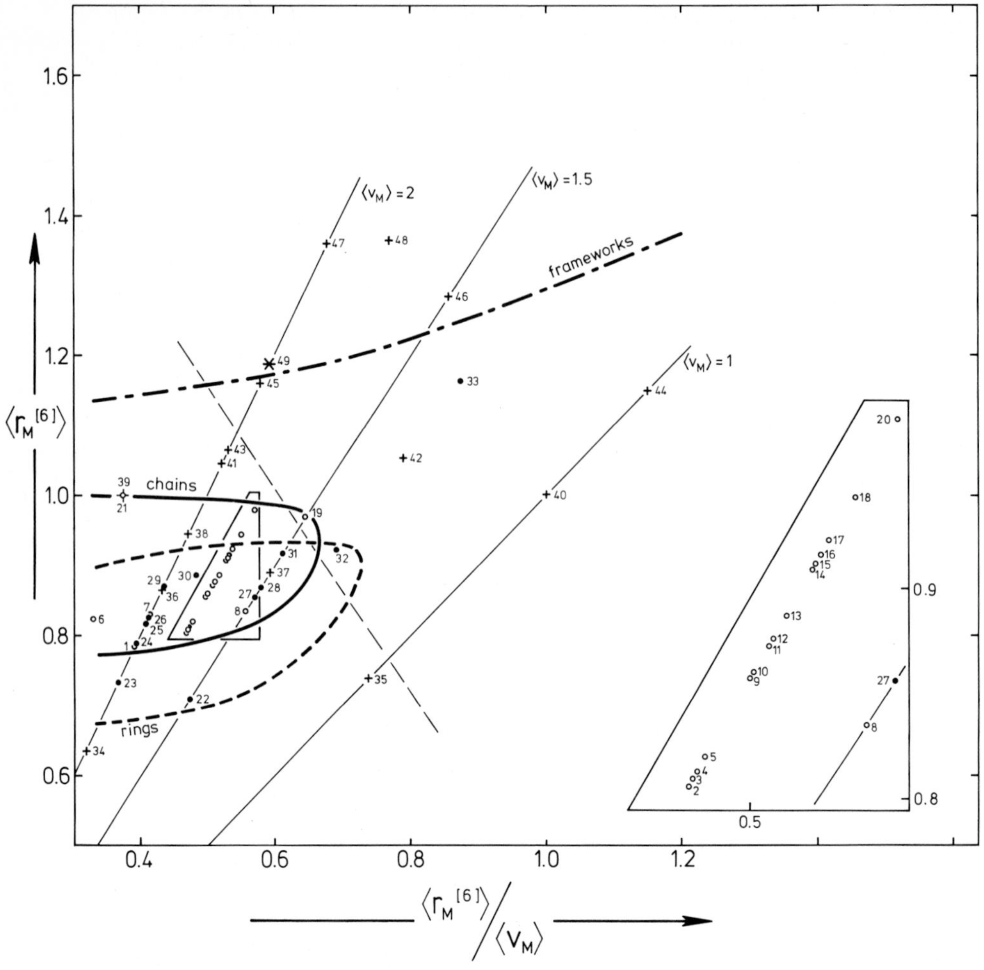

Fig. 10.15. Dimensionality of anhydrous silicates with atomic ratio Si : O = 2 : 5 in relation to mean radius, $\langle r_M^{[6]} \rangle$, and mean valence, $\langle v_M \rangle$, of their cations.

● Silicates containing double rings $\{2r\}[Si_2O_5]_p$; ○ silicates containing multiple chains $\{M^1_\infty\}[Si_2O_5]_p$; + silicates containing single layers $\{1^2_\infty\}[Si_2O_5]_n$; ∗ silicate containing a framework $\{^3_\infty\}[Si_2O_5]_n$. In the layer silicates below the *broken straight line* at least 50% of the cations are in four-coordinated positions

Key to the numbers:

Double ring silicates:

1 $KCa_2(Be_2Al)$ $[Si_{12}O_{30}] \cdot 0.75\,H_2O$, milarite (Černý et al. 1980)
2 $Na_2Mg_5[Si_{12}O_{30}]$ (Nguyen et al. 1980)
3 $Na_2Mg_3Cu_2[Si_{12}O_{30}]$ (Nguyen et al. 1980)
4 $Na_2Mg_3Zn_2[Si_{12}O_{30}]$ (Nguyen et al. 1980)
5 $Na_2Mg_3Fe_2[Si_{12}O_{30}]$ (Nguyen et al. 1980)
6 $BaCa_2Al_3[Al_3Si_9O_{30}] \cdot 2\,H_2O$, armenite (Tennyson 1960)
7 $KSn_2Li_3[Si_{12}O_{30}]$, brannockite (White et al. 1973)
8 $Na_3Mg_4Li[Si_{12}O_{30}]$ (Nguyen et al. 1980)
9 $NaKMg_5[Si_{12}O_{30}]$ (Nguyen et al. 1980)
10 $NaKMg_3Cu_2[Si_{12}O_{30}]$ (Nguyen et al. 1980)
11 $RbNaMg_5[Si_{12}O_{30}]$ (Nguyen et al. 1980)
12 $RbNaMg_3Cu_2[Si_{12}O_{30}]$ (Nguyen et al. 1980)
13 $RbNaMg_3Fe_2[Si_{12}O_{30}]$ (Nguyen et al. 1980)
14 $K_2Mg_5[Si_{12}O_{30}]$, roedderite (Khan et al. 1972)
15 $K_2Mg_3Cu_2[Si_{12}O_{30}]$ (Nguyen et al. 1980)
16 $K_2Mg_3Zn_2[Si_{12}O_{30}]$ (Nguyen et al. 1980)

available is Cs^+ ($r^{[12]} = 1.88$ Å) and although its size should be sufficient to stabilize the framework structure of $Cs_2\{^3_\infty\}[Si_2O_5]$, its existence has not definitely been established. However, $Cs_6[Si_{10}O_{23}]$ and $Rb_6[Si_{10}O_{23}]$ with Si : O = 2 : 4.6 have been shown to have interrupted frameworks (Schichl et al. 1973).

Up to now, $K_2Ce[Si_6O_{15}]$ is the only silicate with Si : O = 2 : 5 known to have an interrupted framework of tertiary $[SiO_4]$ tetrahedra. Taking account of the mean size $\langle r^{[6]}_M \rangle$ and mean valence $\langle v_M \rangle$ of the cations, a phyllosilicate of this composition would be expected to have tetrahedral layers so strongly folded that at least some of the cations would have to be located in octahedral sites. In dalyite, $K_2Zr[Si_6O_{15}]$, which has the same mean valence and only a slightly lower mean cation radius than $K_2Ce[Si_6O_{15}]$, zirconium is just small enough to fit into octahedral sites and so a layer structure, $K^{[8]}_2Zr^{[6]}\{1^2_\infty\}[Si_6O_{15}]$, is formed. In contrast, the Ce^{+3} is too large to fit into octahedral sites and consequently, a layer structure cannot be formed. Of the three alternatives – interrupted framework, multiple chains, and double rings – the K, Ce silicate has a framework structure, $K^{[8]}_2Ce^{[6]}\{^3_\infty\}[Si_6O_{15}]$, since the very low value of the mean electronegativity $\langle \chi_M \rangle = 0.96$ is incompatible with the formation of higher proportions of vierer rings as part of the ring and chain structures.

---

17 $K_2Mg_3Fe_2[Si_{12}O_{30}]$ (Nguyen et al. 1980)
18 $K_2Fe_5[Si_{12}O_{30}]$, merrihueite (Khan et al. 1972)
19 $K_3Mg_4Li[Si_{12}O_{30}]$ (Nguyen et al. 1980)
20 $K_2Mn_5[Si_{12}O_{30}]$ (Sandomirskii et al. 1977)
21 $Ca_2Th[Si_8O_{20}]$, steacyite (Szymański et al. 1982; Richard and Perrault 1972)

Multiple chain silicates:

22 $Na_2Be_2[Si_6O_{15}] \cdot H_2O$, epididymite (Robinson and Fang 1970)
23 $Li_4Zr_2[Si_{12}O_{30}]$ (Quintana and West 1981)
24 $Li_2Na_2Ti_2[Si_{12}O_{30}]$ (Marr and Glasser 1979)
25 $Li_2Na_2Sn_2[Si_{12}O_{30}]$ (Marr and Glasser 1979)
26 $Li_2Na_2Zr_2[Si_{12}O_{30}]$, zektzerite (Ghose and Wan 1978)
27 $Li_2Na_4Fe^{+3}_2[Si_{12}O_{30}]$, emeleusite (Johnsen et al. 1978)
28 $Na_4Mg_4[Si_{12}O_{30}]$ (Cradwick and Taylor 1972)
29 $(Na,K)_2Fe^{+2}_2Fe^{+3}_2[Si_{12}O_{30}] \cdot H_2O$, tuhualite (Merlino 1969)
30 $Na_4Ti_2[Si_8O_{20}]O_2$, narsarsukite (Peacor and Buerger 1962)
31 $Na_4Li_2Y_2[Si_{12}O_{30}]$ (Gunawardane et al. 1982)
32 $Na_4Cu_2[Si_8O_{20}]$ (Kawamura and Iiyama 1981)
33 $K_8Cu_4[Si_{16}O_{40}]$ (Kawamura and Iiyama 1981)

Single layer silicates:

34 $LiAl[Si_2O_5]_2$, petalite (Effenberger 1980)
35 $Li_2[Si_2O_5]$ (Liebau 1961 b)
36 $CaCu[Si_4O_{10}]$, cuprorivaite (Pabst 1959)
37 $K_2Be_2[Si_6O_{15}]$ (Naumova et al. 1976)
38 $SrCu[Si_4O_{10}]$ (Pabst 1959)
39 $Ca_2Th[Si_4O_{10}]_2$, ekanite (Szymański et al. 1982)
40 $Na_2[Si_2O_5]$ (Pant 1968; Pant and Cruickshank 1968)
41 $BaCu[Si_4O_{10}]$ (Pabst 1959)
42 $K_2Be[Si_4O_{10}]$ (Balko et al. 1980)
43 $BaFe[Si_4O_{10}]$, gillespite (Hazen and Finger 1983)
44 $Ag_2[Si_2O_5]$ (Liebau 1961 c)
45 $K_2Zr[Si_6O_{15}]$, dalyite (Fleet 1965)
46 $K_3Nd[Si_6O_{15}]$ (Pushcharovskii et al. 1977)
47 $Ba[Si_2O_5]$, sanbornite (Hesse and Liebau 1980 b)
48 $K_2Ba_7[Si_4O_{10}]_4$ (Cervantes-Lee et al. 1982)

Framework silicate:

49 $K_2Ce[Si_6O_{15}]$ (Karpov et al. 1977 a)

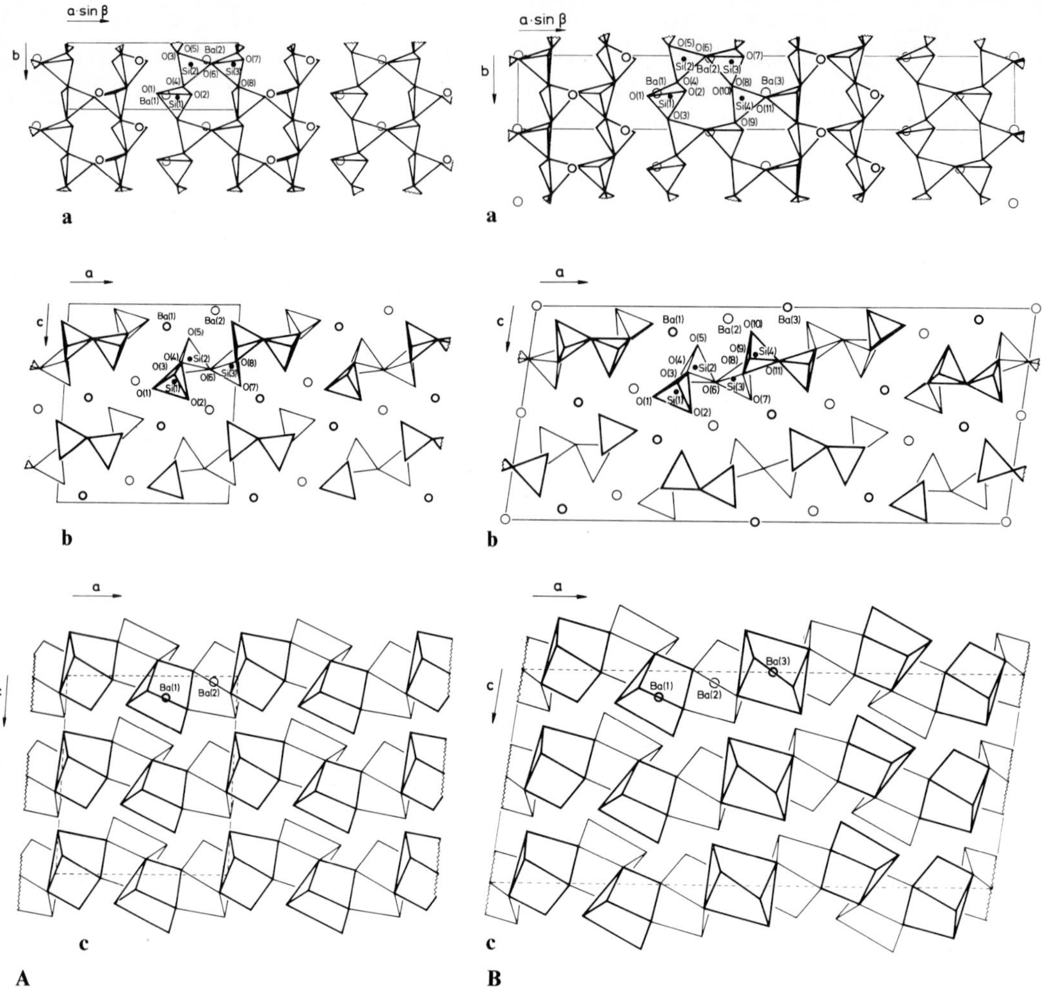

Fig. 10.16 A–C. The structures of silica-rich barium silicates (Hesse and Liebau 1980 a). **A** $Ba_4[Si_6O_{16}]$; **B** $Ba_5[Si_8O_{21}]$; **C** $Ba_6[Si_{10}O_{26}]$. **a** Silicate chains within half the unit cells projected approximately perpendicular to the chain directions; **b** silicate chains projected parallel to the chain directions; **c** corrugated layers of edge-shared barium–oxygen polyhedra projected parallel to the silicate chain directions

10.5.2 The Shape of Silicate Anions in Anhydrous Silica-Rich Barium Silicates

The various effects governing the shape of silicate ions can most readily be demonstrated with the homologous series of barium silicates of general formula $Ba_{M+1}[Si_{2M}O_{5M+1}]$. Within this series six different phases are known and their crystal structures have been studied in detail (Grosse and Tillmanns 1974; Hesse and Liebau 1980 a, b):

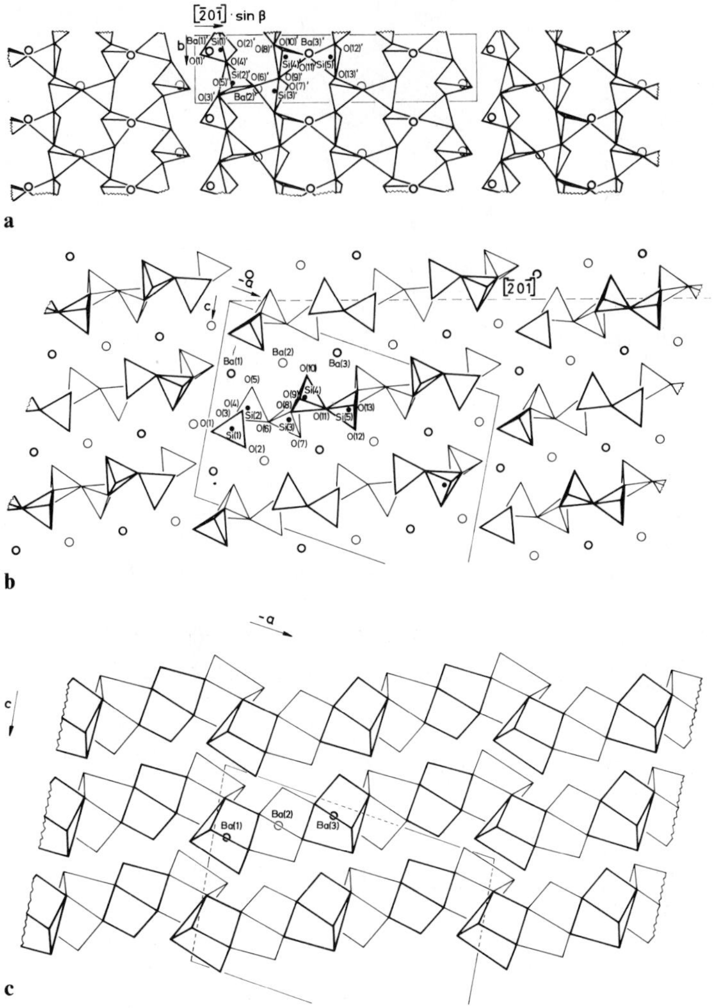

Fig. 10.16C

$Ba_2[Si_2O_6]$(hT) with $\boldsymbol{M} = 1$ is a single chain silicate;
$Ba_4[Si_6O_{16}]$ ($\boldsymbol{M} = 3$) contains triple chains;
$Ba_5[Si_8O_{21}]$ ($\boldsymbol{M} = 4$) contains fourfold chains;
$Ba_6[Si_{10}O_{26}]$ ($\boldsymbol{M} = 5$) fivefold chains.

$Ba[Si_2O_5]$ may be considered as the end member of the series with multiplicity $\boldsymbol{M} = \infty$, in other words, it is a single layer silicate of barium. Whereas two polymorphs of this phyllosilicate are known, no barium silicate $Ba_3[Si_4O_{11}]$ with double chains has been found.

Figure 10.16 shows the structures of the three multiple chain phases, Fig. 10.17 gives the corresponding projections for the two layer structures and the structure of the single chain member of the series is given in Fig. 10.4c. Relevant data obtained during the crystal structure analyses of all these phases, are summarized in Fig. 10.18 and in Table 10.13.

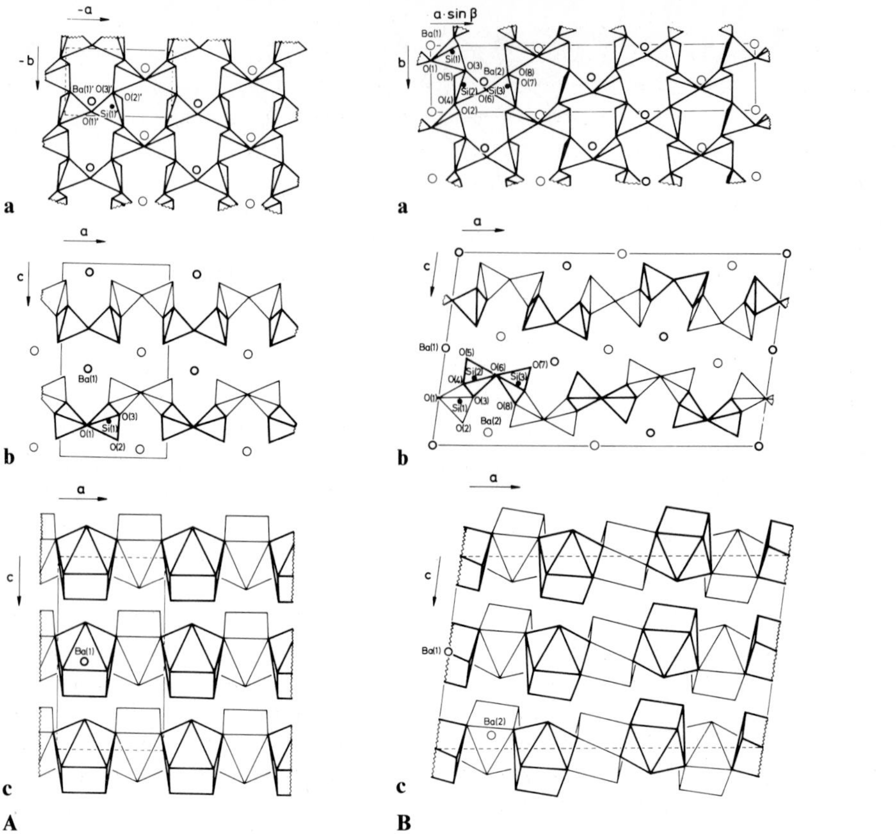

Fig. 10.17 A, B. The structures of barium phyllosilicate (Hesse and Liebau 1980b).
A Low temperature phase sanbornite, $Ba[Si_2O_5]$(lT); **B** high temperature phase $Ba[Si_2O_5]$(hT).
a One silicate layer projected perpendicular to the layer; **b** silicate layers projected parallel to the directions of their fundamental chains; **c** corrugated layers of edge-shared barium–oxygen polyhedra projected parallel to the directions of their fundamental silicate chains

The highly electropositive Ba^{+2} ions cause strong repulsive forces between the $[SiO_4]$ tetrahedra of the silicate chains so that rather stretched chains are expected. In contrast, all four barium silicates contain zweier chains with stretching factors f_s between 0.848 ($I_{chain} = 4.580$ Å) and 0.872 ($I_{chain} = 4.707$ Å). These low degrees of stretching are due to the size of the barium ions. In order to achieve reasonable mutual adjustment between the barium–oxygen polyhedra and the silicate chains, the latter have to shrink so that the height of one $[BaO_8]$ polyhedron is commensurable with that of two $[SiO_4]$ tetrahedra of the chain (Fig. 10.4c). For comparison, in enstatite, $Mg_2[Si_2O_6]$, the height of two of the smaller $[MgO_6]$ octahedra is commensurable with that of two tetrahedra of the stretched zweier chain ($I_{chain} \simeq 5.25$ Å, $f_s = 0.97$). The shrunken zweier chain in $Ba_2[Si_2O_6]$(hT) is, therefore, a compromise between the influence of low electronegativity and large radius of the barium cations and may be rather strained.

The presence of this strain is indicated by the 128° Si–O–Si bond angle within the single chain (intrachain angle). This angle is much smaller than the grand mean value of 140° found in silicates (see Sec. 3.1.2.5), a value which is regarded as typical of an unstrained Si–O–Si bond. In agreement with this explanation, $BaSiO_3$ has this chain structure only in the narrow temperature range between 1625 K and the melting point of 1695 K. Below 1625 K a structure with $[Si_3O_9]$ rings is the stable phase (Table 10.1).

Since the atomic ratio Ba : Si decreases from 1.00 to 0.60 in the series $Ba_2[Si_2O_6]$ – $Ba_4[Si_6O_{16}]$ – $Ba_5[Si_8O_{21}]$ – $Ba_6[Si_{10}O_{26}]$, the influence of the cations on the silicate chains is also expected to decrease. In fact, there is a steady increase of the chain period from 4.580 Å in the single chain phase to 4.707 Å in the fivefold chain silicate, and a slight increase of the intrachain Si–O–Si angles from 128° to 132°. The Si–O–Si angle increases still further to 136° in the two polymorphs of the layer silicate thus, indicating a considerable reduction of strain.

As shown in Fig. 10.16 there are more Ba^{+2} ions attached to the outermost subchains than to the inner ones and, therefore, their influence is lower in the inner chains. This explains the slight increase of the intrachain Si–O–Si bond angles in passing from the outermost to the innermost subchains within each multiple chain (Fig. 10.18b).

In the chain direction the repulsive forces between the negative charges on the $[SiO_4]$ tetrahedra are overcome by the size effect of the barium ions, leading to nonstretched chains. However, these forces can act quite freely in the second dimension of the multiple chain anions, and as a result, all three of these anions are almost flat (Fig. 10.16b). The stretching of the tetrahedral ribbons in their second (finite) dimension causes appreciable deviations of the Si–O–Si bond angles between the subchains (interchain Si–O–Si angles) from the 140° value of the unstrained bond. These angles reach values as high as 173° in the quadruple

Table 10.13 Chain periods, mean values of bond lengths and bond angles, and mean deviations $\frac{1}{n}\sum[140° - (\sphericalangle \text{Si–O–Si})]^2$ as a measure of angular strain in high-polymer barium silicates

	$Ba_2[Si_2O_6]$ (hT)	$Ba_4[Si_6O_{16}]$	$Ba_5[Si_8O_{21}]$	$Ba_6[Si_{10}O_{26}]$	$Ba[Si_2O_5]$ (hT)	$Ba[Si_2O_5]$ (lT)
Chain period [Å]	4.580	4.685	4.695	4.707	4.658	4.629
⟨d(Si–O)⟩ [Å]	1.623	1.627	1.613	1.621	1.618	1.614
⟨d(O···O)⟩ [Å]	2.643	2.646	2.623	2.640	2.635	2.632
Intrachain ⟨∢Si–O–Si⟩ [°]	128.2	131.4	132.7	132.3	135.8	136.0
Interchain ⟨∢Si–O–Si⟩ [°]	–	147.9	151.1	146.2	171.0	143.8
Overall ⟨∢Si–O–Si⟩ [°]	128.2	135.6	137.7	136.3	137.6	138.6
$\frac{1}{n}\sum\Delta^2$ [°²]	139	76	142	72	37	15
⟨d(Ba–O)⟩ [Å]	2.883	2.839	2.866	2.874	2.881	2.890

chain silicate (Fig. 10.18b). When there are two crystallographically nonequivalent interchain angles (in $Ba_5[Si_8O_{21}]$ and $Ba_6[Si_{10}O_{26}]$) the inner angle has the higher value, indicating a higher degree of strain in the middle of the bands. This higher strain is only partly balanced by the slight increase in the intrachain Si–O–Si angles from the outer to the inner chains.

In order to achieve a rather dense packing of the edge-sharing $[BaO_8]$ polyhedra (avoiding sharing of faces), the barium ions scatter about two levels that are 1/2 $d_{[010]}$ apart in all these structures. The $[BaO_8]$ polyhedra are linked to infinite chains parallel [010] and, in order of increasing silica content of the six phases, two, four, five, six, two, or three such chains, respectively, are linked to form bands (Figs. 10.16c, 10.17c). These bands are then connected to form stepped layers holding the silicate anions together.

From the observed barium chain silicate structures compared in Fig. 10.18, it can be seen that the chains with an odd multiplicity, $\boldsymbol{M} = 1, 3, 5$, have a twofold *screw* axis in the innermost subchain. The quadruple chain ($\boldsymbol{M} = 4$, even), and probably other such possible multiple chains with an even multiplicity, contain a twofold *rotation* axis between the two innermost subchains. Although no explanation for this observation is available, it can be assumed that the screw axes do not increase the strain at the centers of the multiple chains further, while the rotation axes do. The large Si–O–Si angle of 173° between the inner two subchains of the quadruple chain is a direct indication of this strain.

Due to the high Ba : Si atomic ratio (0.75) of the hypothetical phase $Ba_3[Si_4O_{11}]$ containing double chains, a low chain period of 4.65 Å is expected by interpolation from the chain periods of existing phases (Table 10.13). A chain compressed to this extent is so strained that it cannot cope with any additional strain associated with a rotation axis of a hypothetical double chain. This probably explains why a double chain silicate $Ba_3[Si_4O_{11}]$ does not occur.

When, in comparison with $Ba_2[Si_2O_6]$(hT) and $Ba_4[Si_6O_{16}]$, the strain revealed by the short chain period and the low intrachain Si–O–Si angle in $Ba_5[Si_8O_{21}]$ is reduced by further reduction of the Ba : Si ratio to $5:8 = 0.625$, the quadruple chain is just able to sustain the additional stress associated with the rotation symmetry. Therefore, the fourfold chain silicate exists despite the fact that the sum $\frac{1}{n}\Sigma\Delta^2$ of the differences $\Delta = |\,140° - (\sphericalangle\,\text{Si–O–Si})\,|$, which may be regarded a measure of the strain within the silicate anions, is somewhat higher for $Ba_5[Si_8O_{21}]$ than for the phase $Ba_4[Si_6O_{16}]$ containing chains of odd multiplicity.

In Sec. 10.5.1, it was shown that in order to achieve satisfactory adjustment between the arrays of cation–oxygen polyhedra and tetrahedral silicate layers, the layers become more corrugated with decreasing size and increasing valence of the cations or, in other words, with decreasing atomic ratio $n_M : n_{Si}$.

This concept not only applies to two-dimensionally infinite single layers, but also to bands that are infinite in one dimension and have an appreciable extension in the second dimension. For the series of barium multiple chain silicates, this means that with decreasing Ba : Si ratio the anions should become more and more corrugated. However, from the shape of the multiple chains, it is clear that this increase of corrugation is almost nonexistent, no doubt due to the strong repulsive forces between the negatively charged $[SiO_4]$ tetrahedra. Therefore, if the multiple

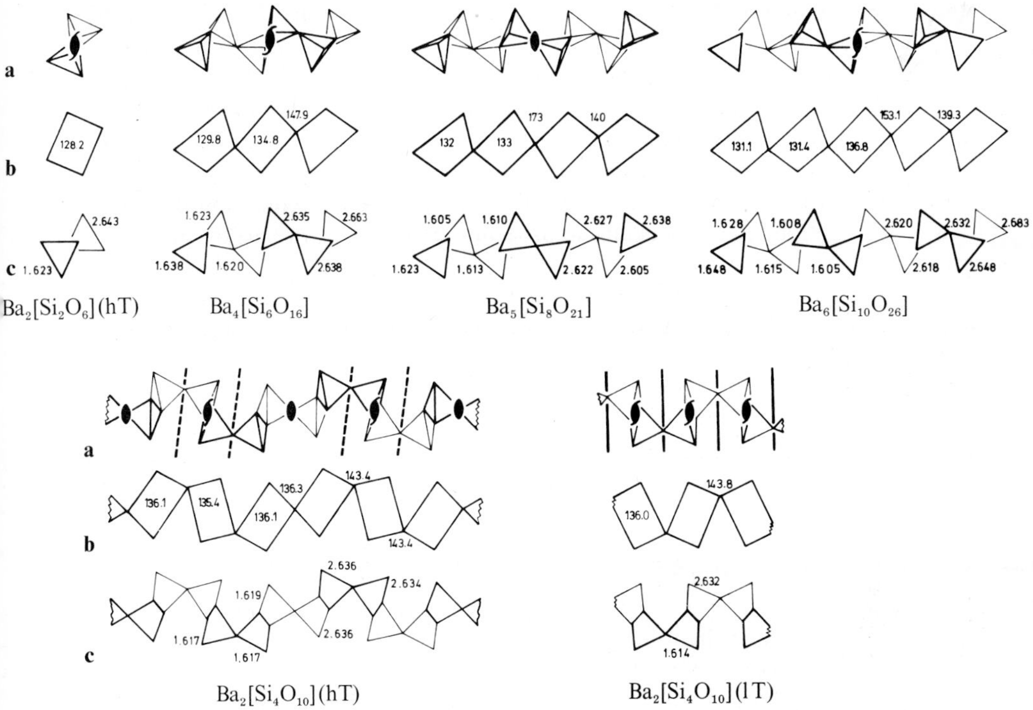

Fig. 10.18 a–c. Comparison between some of the structural details of the tetrahedral anions of silica-rich barium silicates.

a Tetrahedral anions projected parallel to the directions of their fundamental chains and their symmetry. ⬮ twofold rotation axis; twofold screw axis; —— mirror plane; ----- pseudo mirror plane. **b** Values of intrachain and interchain Si–O–Si angles in the tetrahedral anions. **c** Mean values of the Si–O distances and O···O distances for the $[SiO_4]$ tetrahedra of the silicate anions

chains are unable to corrugate, the series of multiple chain silicates of barium will terminate at some value of the multiplicity ***M***.

In fact, the quintuple chain silicate $Ba_6[Si_{10}O_{26}]$ is the silica-rich end member of the series, in which the strain caused by the shrinkage of the subchains (due to the size of Ba^{+2}) adds to that caused by the inability of the anions to corrugate (due to the repulsive forces between the negatively charged tetrahedra). At temperatures above 1575 K these contributions to the strain are tolerated so that this silicate is thermodynamically stable. However, below 1575 K it decomposes into a mixture of the quadruple chain silicate $Ba_5[Si_8O_{21}]$ and the layer silicate $Ba[Si_2O_5]$(lT) (Oehlschlegel 1971). This decomposition can be explained by a decrease in the effective cation size with decreasing temperature, thereby requiring the silicate ribbons to corrugate in opposition to the repulsive forces within the ribbons.

From the change in the average Si–O–Si angles (Fig. 10.18 b) and, in particular, their deviations from 140° ($\frac{1}{n}\Sigma\Delta^2$ in Table 10.13) with increasing multiplicity of the silicate chains, it can be expected that the hypothetical sixfold chains of the next member of this series of barium silicates would have to be even more strained than the fourfold chains of $Ba_5[Si_8O_{21}]$. In agreement with this expectation, no phase of composition $Ba_7[Si_{12}O_{31}]$ with sixfold chains has been found in the system $BaO-SiO_2$.

A further decrease of $n_M : n_{Si}$ to 0.50 reduces the formal charge per $[SiO_4]$ tetrahedron to -1 in $Ba[Si_2O_5]$ compared with -2 in $Ba_2[Si_2O_6]$. Since this is equivalent to a strong decrease in the repulsive forces between the tetrahedra, corrugation of the silicate anions becomes possible even in the case of the electropositive Ba^{+2} ions as the multiplicity of the chains approaches infinity, i.e., in barium phyllosilicate.

The reduction of the repulsive forces in the barium layer silicates relative to its chain silicates can be inferred from the increase in the intrachain angles and the decrease in the interchain angles in both $Ba[Si_2O_5]$ phases. On the other hand, the substantial reduction of angular strain, indicated by the decrease in the sum $\frac{1}{n}\Sigma\Delta^2$ of the differences $\Delta = |140° - (\sphericalangle\, Si-O-Si)|$ for the phases studied (Table 10.13), enables the subchains of the silicate layers to cope with a slight decrease in the chain period from 4.707 Å in $Ba_6[Si_{10}O_{26}]$ to 4.658 Å and 4.629 Å in the two $Ba[Si_2O_5]$ phases.

In the high temperature form of barium phyllosilicate, the effective size of the barium ions is larger than in the low temperature form, sanbornite, due to the increase in thermal vibration. Therefore, the tetrahedral layers in $Ba[Si_2O_5](hT)$ are less strongly folded than those in sanbornite, $Ba[Si_2O_5](lT)$. This decrease in layer convolution results in a reduction in symmetry from the orthorhombic space group $D_{2h}^{16}-Pmcn$ to monoclinic $C_{2h}^{6}-C2/c$ during the low → high phase transformation. Such a reduction of symmetry with increasing temperature is rather uncommon, but is readily explained by the mechanism described.

10.5.3 Hydrous Single Layer Silicates

In anhydrous single layer silicates the tetrahedral layers are increasingly corrugated as the cation size decreases and the cation valence increases. Why, then, do the small cations Mg^{+2}, Fe^{+2}, and Al^{+3} form flat layers in mica and a number of clay minerals?

The answer lies in the fact that the cations in the anhydrous silicates have at their disposal only oxygen ions which are a part of the $[SiO_4]$ tetrahedra, whereas the mica and clay minerals also contain hydroxyl groups OH. The latter silicates are, therefore, often called hydrous silicates. In these hydrous phases the cations can surround themselves not only with oxygen ions that are bonded to silicon, but also with hydroxyl groups and sometimes with water molecules which do not belong to the tetrahedral layer. The $[M(O,OH,H_2O)_n]$ polyhedra formed in this way, share edges with each other to build larger arrays. In most of these structures the M cations are small (Al^{+3}, Mg^{+2}, Fe^{+3}, Fe^{+2}, etc.) and the $[M(O,OH,H_2O)_n]$ polyhedra are octahedra and, therefore, for simplicity the polyhedra and their arrays can be referred to merely as "octahedra", "octahedral" arrays, and "octahedral" layers, disregarding the actual shape of the polyhedra.

10.5.3.1 Kaolinite-Like and Mica-Like Arrangements

Edge-shared $[M(O,OH)_6]$ octahedra of most of the small and medium-sized cations M form layers of hexagonal or pseudohexagonal symmetry. Figure 10.19 gives an idealized representation of such layers as they exist in the structure of talc, $Mg_3[Si_2O_5]_2(OH)_2$ (Perdikatsis and Burzlaff 1981), and pyrophyllite, $Al_2[Si_2O_5]_2(OH)_2$ (Lee and Guggenheim 1981). Since the octahedral layer has a positive charge and the tetrahedral layer has a negative charge, the layers attract each other. One octahedral layer attracts either one or two tetrahedral layers, forming a kaolinite-like layer or a mica-like (sometimes named talc-like) layer, respectively (Fig. 10.20).

In the field of mica and clay mineralogy, the terms 1 : 1 layer silicates and 2 : 1 layer silicates are used for the two groups of hydrous phyllosilicates. Figure 10.20 shows clearly that the 1 : 1 layer silicates, in which one layer of cation–oxygen polyhedra is linked to only one tetrahedral layer, have a higher M : T ratio than the 2 : 1 layer silicates in which one octahedral layer is sandwiched between two tetrahedral layers. For reasons to be discussed at the end of Sec. 10.5.3.2.3, the terms *cation-rich single layer silicates* and *cation-poor single layer silicates* are preferred over 1 : 1 and 2 : 1 layer silicates, respectively.

The mesh size of an unstrained octahedral layer can be described by its lattice parameter b_{oct}, that of the unstrained hexagonal tetrahedral layer by b_{tetr}. Ideal fit between a tetrahedral and an octahedral layer requires that $b_{oct} = b_{tetr}$. For a purely siliceous layer $[Si_2O_5]$, this is expected for cations with $r_M^{[6]} \simeq 0.70$ Å, i.e., for a

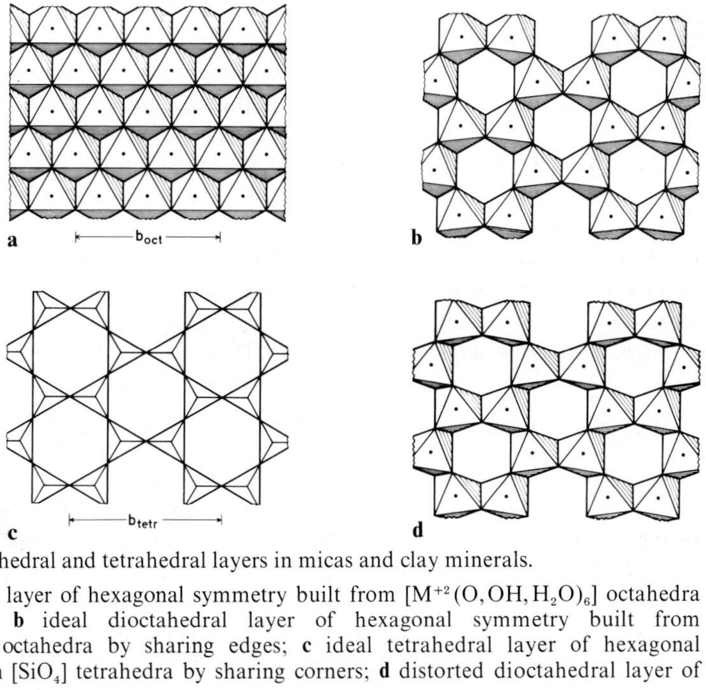

Fig. 10.19 a – d. Octahedral and tetrahedral layers in micas and clay minerals.

a Ideal trioctahedral layer of hexagonal symmetry built from $[M^{+2}(O,OH,H_2O)_6]$ octahedra by sharing edges; **b** ideal dioctahedral layer of hexagonal symmetry built from $[M^{+3}(O,OH,H_2O)_6]$ octahedra by sharing edges; **c** ideal tetrahedral layer of hexagonal symmetry built from $[SiO_4]$ tetrahedra by sharing corners; **d** distorted dioctahedral layer of pseudohexagonal symmetry

cation radius slightly smaller than that of magnesium. In this case a flat composite layer is formed.

In micas and many clay minerals there is extensive replacement of silicon in the tetrahedral layer as well as of M cations in the octahedral layers. Such replacement leads to considerable and nearly continuous variation of the mesh sizes of both layers.

Replacement of Si^{+4} by Al^{+3}, Fe^{+3}, and sometimes Mg^{+2} and Zn^{+2} causes an increase in the mesh size of the tetrahedral layers, thereby allowing the incorporation of larger cations into the octahedral layers of a given structure type. This

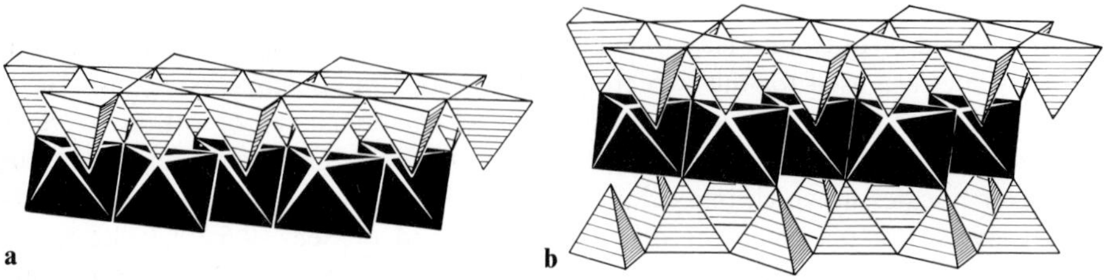

Fig. 10.20 a, b. Packages of tetrahedral and octahedral layers in micas and clay minerals. **a** Ideal kaolinite-like layer; **b** ideal talc-like or mica-like layer

explains, for example, the existence of the minerals sauconite, $Zn_6[Si_{8-x}Zn_xO_{20}](OH)_4$ (Roy and Mumpton 1956), annite, $KFe_3^{+2}[AlSi_3O_{10}](OH)_2$ (Eugster and Wones 1962), and the synthetic mica phases $KFe_3^{+2}[Si_3Co^{+3}O_{10}](OH)_2$, $KCo_3^{+2}[Si_3Co^{+3}O_{10}](OH)_2$ and $KNi_3^{+2}[Si_3Ni^{+3}O_{10}](OH)_2$ (Lindqvist 1966; Donnay et al. 1964). In dioctahedral clay minerals, in which trivalent cations fill only two out of every three octahedra, the mesh size of the octahedral layer is slightly smaller than that of the tetrahedral layer if the trivalent cation is predominantly Al^{+3} as in kaolinite, $Al_2[Si_2O_5](OH)_4$ (Zvyagin 1960), and pyrophyllite, $Al_2[Si_2O_5]_2(OH)_2$ (Lee and Guggenheim 1981). The same is true for trioctahedral species (i.e., those in which all octahedral sites are filled with divalent cations) when there is substantial substitution of the small silicon ion in the tetrahedral layer by larger ions, as in xanthophyllite, $CaMg_2Al[Al_3SiO_{10}](OH)_2$. In contrast, octahedral layers containing primarily the larger cations Mg^{+2} and Fe^{+2} have a larger mesh size than tetrahedral layers containing little or no substitution of Si^{+4} by larger cations, as in talc $Mg_3[Si_2O_5]_2(OH)_2$.

10.5.3.2 Strain Reduction Mechanisms in Cation-Poor Phyllosilicates

As the misfit between the octahedral and the tetrahedral layer increases, the mica-like layer becomes more strained. Depending on size and sign of the difference $b_{oct} - b_{tetr}$, there are different ways the composite layer responds to this strain. These different mechanisms are demonstrated for the cation-rich phyllosilicates in Table 10.14.

In this Table the silicates are arranged (with only few exceptions) according to their mean cation radius $\langle r_M^{[6]} \rangle$ independent of the actual coordination number of the cations. Since the adjustment between the "octahedral" and the tetrahedral parts of the structures is, to a first approximation, a problem of geometric fit, the acid hydrogen atoms of Si–O–H bonds have not been taken into consideration because of their very small effective size [2].

10.5.3.2.1 Tetrahedron Rotation. If the mesh size of the strain-free hexagonal tetrahedral layer is larger than that of the octahedral layer ($b_{oct} < b_{tetr}$) mutual adjustment between both layers requires a reduction of the tetrahedral mesh size. Such reduction can be achieved by rotating each $[SiO_4]$ tetrahedron about an axis normal to the tetrahedral layer by an angle α_r. Adjacent tetrahedra are rotated

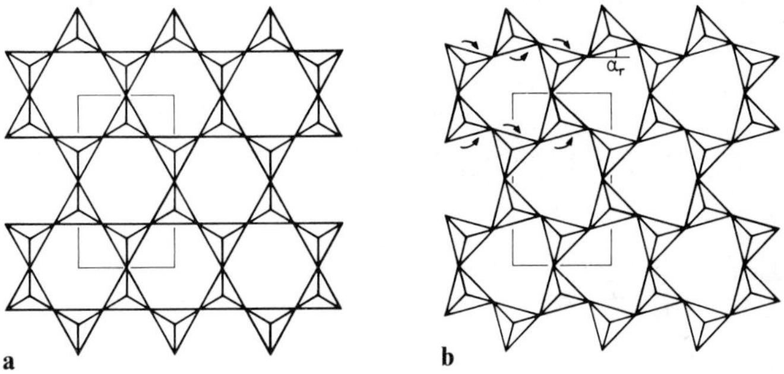

Fig. 10.21 a, b. Reduction of mesh size and symmetry of the tetrahedral layer of clay minerals by rotation of tetrahedra. **a** Ideal layer of hexagonal symmetry; **b** contracted layer of trigonal symmetry. In one ring of tetrahedra the directions of rotations by an angle α_r are indicated by *arrows*

alternately clockwise and anticlockwise, thereby reducing the symmetry from hexagonal to approximately ditrigonal (Fig. 10.21 b), and decreasing the mesh size by up to about 13% (Radoslovich and Norrish 1962; Bailey 1966). In fact, values of the rotation angles α_r between 1.1° (taeniolite $KLiMg_2[Si_2O_5]_2F_2$, Toraya et al. 1977) and about 23° (xanthophyllite $CaMg_2Al[Al_3SiO_{10}](OH)_2$, Takéuchi and Sadanaga 1966) have been observed, whereas the highest possible theoretical value is 30°.

10.5.3.2.2 Tetrahedron Tilting. In dioctahedral clay minerals the edges shared between the octahedra are shortened as a function of cation repulsion, but the non-shared edges are longer than in the more symmetrical arrangement of octahedra present in the trioctahedral layers (Fig. 10.19 b, d). In order to adjust the tetrahedral layer to a dioctahedral layer, the apical (terminal) oxygen atoms of the $[SiO_4]$ tetrahedra forming the non-shared edges of the octahedra are pushed apart, while the other oxygen atoms are slightly bent towards each other. These shifts of the apical oxygen ions are accomplished by slight tilts of the tetrahedra. The basal bridging oxygen atoms of the tetrahedral layer are then no longer coplanar and

2 The effective ionic radii of hydrogen given by Shannon and Prewitt (1969, 1970) and Shannon (1976) are, in fact, negative: $r_H^{[1]} = -0.38$ Å and $r_H^{[2]} = -0.18$ Å for hydrogen atoms having coordination numbers 1 and 2, respectively.

may deviate by up to 0.1 Å from a common plane parallel to the layer. This tilting of the tetrahedra takes place even when the tetrahedral layer has a larger mesh size than the octahedral layer. This causes a slight corrugation of the layers as, for example, in the case of margarite, $CaAl_2[AlSiO_5]_2(OH)_2$ (Fig. 10.22) (Joswig et al. 1983).

10.5.3.2.3 Tetrahedron Inversion and Octahedral Layer Splitting. In silicates with mica-like layers, where the octahedral layer is sandwiched between the tetrahedral apices of two $[Si_2O_5]$ layers, a more drastic change in structure takes place if the mesh size of the octahedral layer is larger than that of the tetrahedral layer ($b_{oct} > b_{tetr}$) as in the aluminum-free magnesium phyllosilicates having a cation

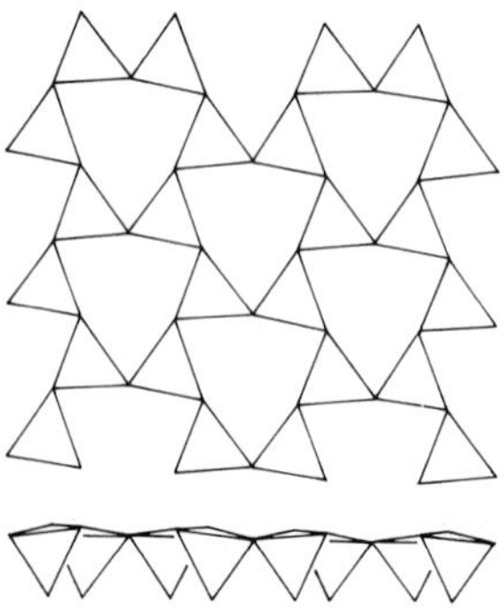

Fig. 10.22. Slight corrugation of the plane of the bridging oxygen atoms in dioctahedral margarite, $CaAl_2[AlSiO_5]_2(OH)_2$ (Takéuchi 1965)

radius $r_M^{[6]} = 0.72$ Å. Then the complete two-dimensional octahedral layer present in talc (Rayner and Brown 1973) (Fig. 10.23a) is split into one-dimensional octahedral bands eight or five octahedra wide, as in sepiolite, $(Mg, Fe, Al)_4[Si_2O_5]_3(O,OH)_2 \cdot 4\,H_2O$ (Brauner and Preisinger 1956), and palygorskite, $Mg_5[Si_2O_5]_4(OH)_2 \cdot 8\,H_2O$ (Bradley 1940), respectively.

At the same time, the tetrahedral layers are rearranged by inverting part of the $[SiO_4]$ tetrahedra, i.e., by changing their directedness. In sepiolite three adjacent zweier single chains have their free tetrahedral apices pointing to one side of the $[Si_2O_5]$ layer, while the next three chains point to the other side (Figs. 10.23b, 10.24b). In palygorskite and yofortierite, $Mn_5[Si_2O_5]_4(OH)_2 \cdot 8\,H_2O$ (Perrault et al. 1975), the *regions of equal directedness* of the $[SiO_4]$ tetrahedra are only two zweier single chains wide (Figs. 10.23c, 10.24c).

A hydrous phyllosilicate with zweier layers in which the free corners of neighboring single chains alternately point to different sides of the layer is not known. It would be expected to require a mean cation radius of about 0.76 Å, the exact value depending on the aluminum content of the tetrahedral layer and on the ratio $n_M : n_T$.

So far we have discussed cation-poor hydrous phyllosilicates whose cations M are small and which contain tetrahedral zweier layers (Table 10.14). The larger cations Ca^{+2} and Na^+ form larger $[M(O,OH,H_2O)_n]$ polyhedra which cannot adjust to the stretched zweier chains of the hexagonal or pseudohexagonal zweier

Fig. 10.23 a–c. Reduction of strain between octahedral layers (*white bars*) and tetrahedral layers (*black triangles*) in hydrous magnesium phyllosilicates with mica-like arrangements.

a Talc, $Mg_3[Si_2O_5]_2(OH)_2$ (Perdikatsis and Burzlaff 1981); **b** sepiolite, $Mg_4[Si_2O_5]_3(OH)_2 \cdot 4H_2O$ (Brauner and Preisinger 1956); **c** palygorskite, $Mg_5[Si_2O_5]_4(OH)_2 \cdot 8H_2O$ (Bradley 1940)

layer, not even if the mean cation radius is reduced by the presence of smaller cations, such as V^{+4} ($r^{[6]} = 0.58$ Å) or Zr^{+4} ($r^{[6]} = 0.72$ Å).

In pentagonite $CaV[Si_4O_{10}]O \cdot 4H_2O$ and its polymorph cavansite (Evans Jr. 1973) which have a mean cation radius of 0.79 Å, the tetrahedral layers still resemble zweier single layers, but the regions of equal directedness are one vierer single chain wide.

The structural diagrams of Table 10.14 show that up to this degree of misfit (indicated by the mesh sizes b_{oct} and b_{tetr}), the bridging oxygen atoms of a tetrahedral layer lie on a flat plane or deviate only slightly from such a plane due to slight tilting of the tetrahedra. However, if the mean cation radius increases further to values $\langle r_M^{[6]} \rangle \gtrsim 0.85$ Å, this is no more possible. Instead, the tetrahedral layers become folded in a fashion similar to the one described for the anhydrous phyllosilicates (see Sec. 10.5.1 and Figs. 10.13, 10.14). This can, for example, be seen in armstrongite, $CaZr[Si_6O_{15}] \cdot 2.5H_2O$ (Kashaev and Sapozhnikov 1978) in which the region of equal directedness is one dreier chain (Table 10.14) which is the chain type characteristic for calcium silicates.

Following the crystal chemical classification nekoite, $Ca_3[Si_6O_{15}] \cdot 7H_2O$ (Alberti and Galli 1980), a dreier single layer silicate is in agreement with a mutual adjustment between dreier chains and the larger $[Ca(O, H_2O)_n]$ polyhedra in one dimension. For $r_M^{[6]} = 1.00$ Å, however, adjustment in the second dimension is no longer possible with only single chains as regions of equal directedness. On the contrary,

Mesh sizes b_{oct}, b_{tetr}	$b_{oct} < b_{tetr}$		b_{oct}
Regions of equal directedness of $[T_2O_5]$ layer		Zweier single layer ($D_{reg} = 2$)	
Array of $[MO_n]$ polyhedra		Continuous layer	
Silicate	Hypothetic $MgAl_2[AlSiO_5]_2(OH)_2$	Margarite $CaAl_2[AlSiO_5]_2(OH)_2$	Pyrophyllite $Al_2[Si_2O_5]_2(OH)_2$
Mean cation radius $\langle r_M^{[6]} \rangle$ [Å]	0.59	0.69	0.53
$n_M : n_T$	0.75	0.75	0.75
α_r [°]		decreases	
	~ 30	21	10

Table 10.14 Structural data of cation-poor hydrous single layer silicates [a]

[a] Regions of equal directedness of $[TO_4]$ tetrahedra are indicated by black and white triangles, respectively.
The arrays of $[MO_n]$ polyhedra ("octahedral" layers) are represented by hatched bars.
$\langle r_M^{[6]} \rangle$ = mean cation radius for coordination number CN = 6 (Shannon and Prewitt 1969, 1970; Shannon 1976); n_M = number of cations M; n_T = number of $[TO_4]$ tetrahedra; α_r = angle of rotation of the $[TO_4]$ tetrahedra away from hexagonal orientation

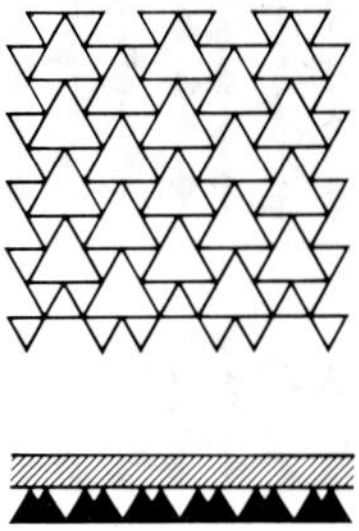

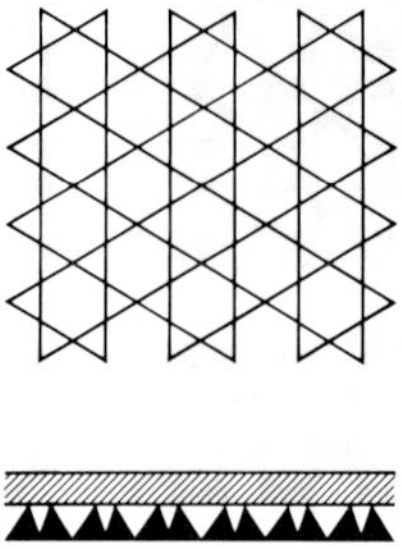

Mesh sizes $b_{oct'}$ b_{tetr}	$b_{oct} < b_{tetr}$		$b_{oct} = b_{tetr}$	
Regions of equal directedness of $[T_2O_5]$ layer			Zweier single layer ($D_{reg} = 2$)	
Array of $[MO_n]$ polyhedra				
Silicate	Hypothetic $LiAl_2[AlSiO_5](OH)_4$	Amesite $Mg_2Al[AlSiO_5](OH)_4$	Kaolinite $Al_2[Si_2O_5](OH)_4$	Liz $Mg_3[Si$
Mean cation radius $\langle r_{M^{[6]}} \rangle$ [Å]	0.60	0.66	0.53	0
$n_M : n_T$	1.50	1.50	1.00	1
α_r[°]	~ 25	15 decreases	9	
r_{curv}[Å]	infinite			
λ [Å]				

Table 10.15 Structural data of cation-rich hydrous single layer silicates [a]

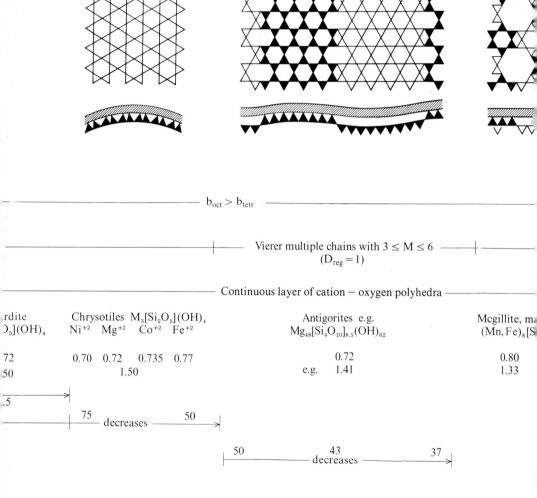

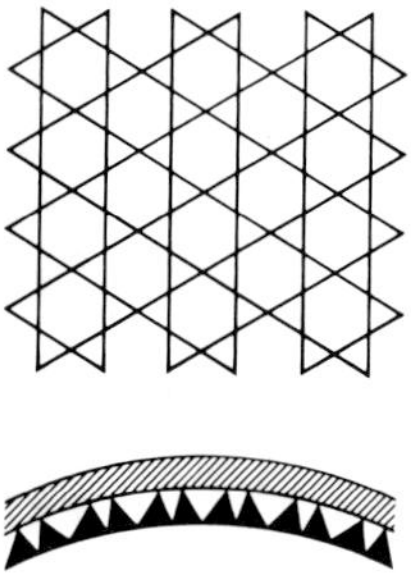

— $b_{oct} > b_{tetr}$ —

— Vierer multiple chains with $3 \leq M \leq 6$ ($D_{reg} = 1$) —

— Continuous layer of cation – oxygen polyhedra —

.rdite $)_5](OH)_4$	Chrysotiles $M_3[Si_2O_5](OH)_4$ Ni^{+2} Mg^{+2} Co^{+2} Fe^{+2}	Antigorites e.g. $Mg_{48}[Si_4O_{10}]_{8.5}(OH)_{62}$	Mcgillite, m. $(Mn,Fe)_8[S$
72	0.70 0.72 0.735 0.77	0.72	0.80
50	1.50	e.g. 1.41	1.33
.5 →			
→	75 — decreases — 50 →		
		50 — 43 decreases — 37 →	

[a] Regions of equal directedness of $[TO_4]$ tetrahedra are indicated by black and white triangles, respectively.
The arrays of $[MO_n]$ polyhedra ("octahedral" layers) are represented by hatched bars.
$\langle r_M^{[6]} \rangle$ = mean cation radius for coordination number CN = 6 (Shannon and Prewitt 1969, 1970; Shannon 1976); n_M = number of cations M; n_T = number of $[TO_4]$ tetrahedra; α_r = angle of rotation of the $[TO_4]$ tetrahedra away from hexagonal orientation; r_{curv} = radius of curvature; λ = wavelength of corrugation in antigorites

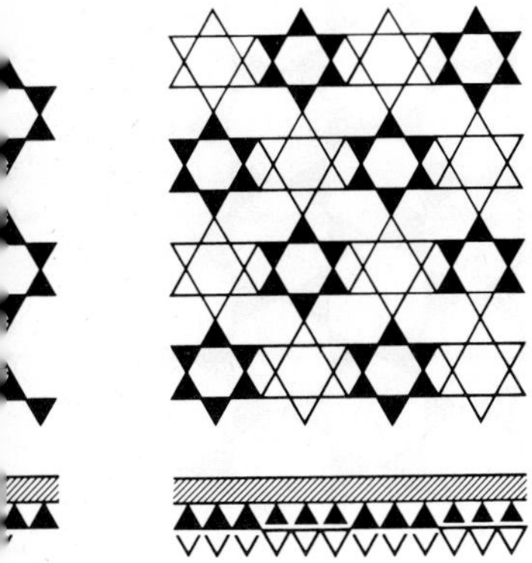

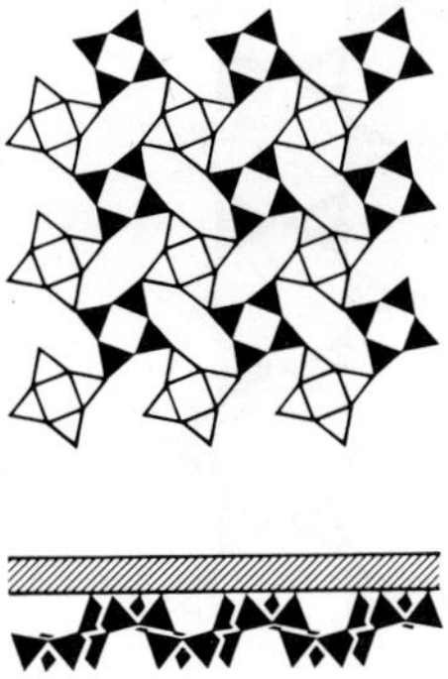

$b_{oct} \gg b_{tetr}$

Sechser single rings ($D_{reg} = 0$) | Vierer single rings ($D_{reg} = 0$)

·malite ,Cl)$_{10}$	Bementite $Mn_7[Si_6O_{15}](OH)_8$	Natroapophyllite $NaCa_4[Si_4O_{10}]_2F \cdot 8\,H_2O$	Apophyllite $KCa_4[Si_4O_{10}]_2(F,OH) \cdot 8\,H_2O$
	0.82	1.00	1.08
	1.17	1.25	1.25

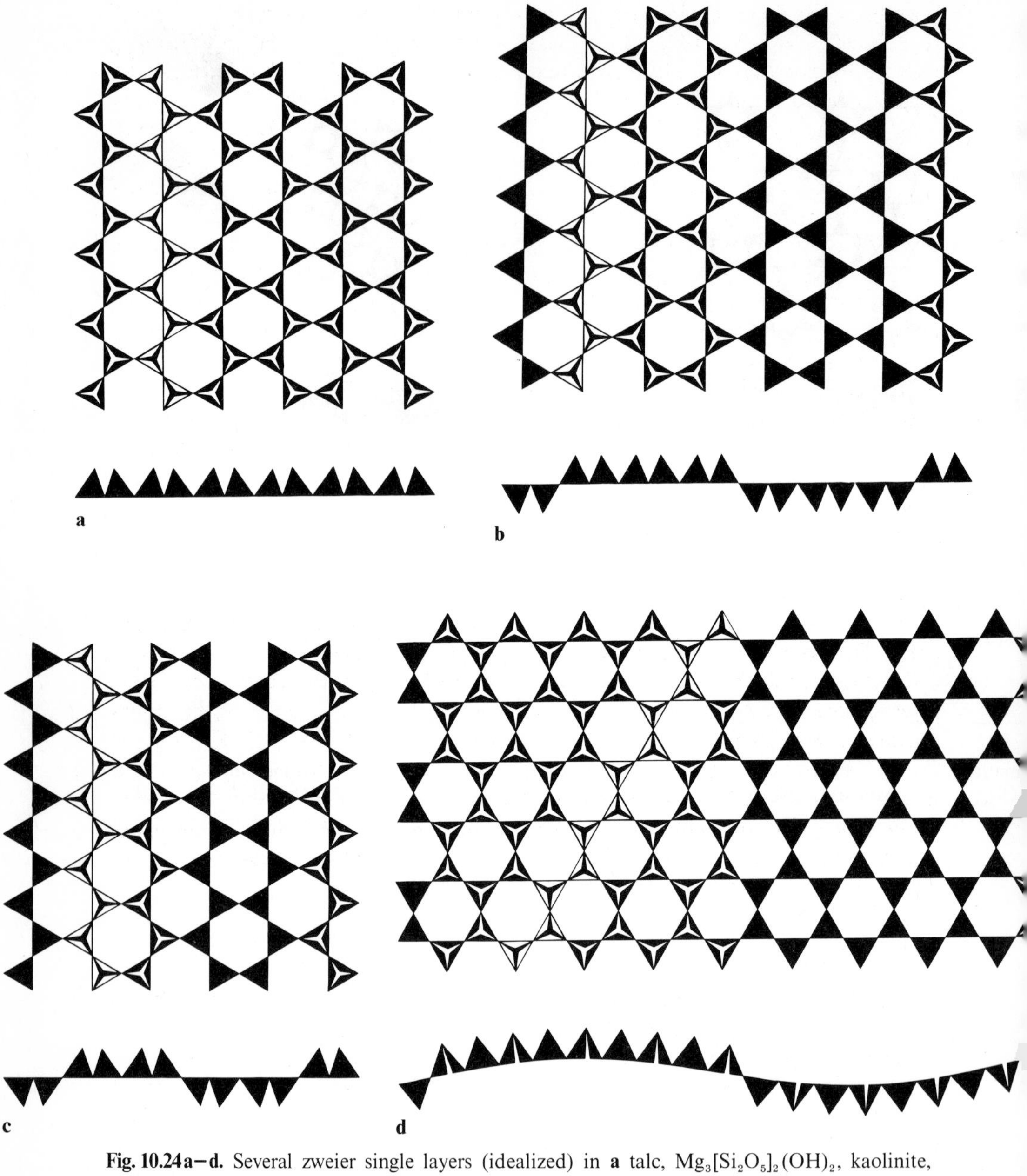

Fig. 10.24a–d. Several zweier single layers (idealized) in **a** talc, $Mg_3[Si_2O_5]_2(OH)_2$, kaolinite, $Al_2[Si_2O_5](OH)_4$; **b** sepiolite $Mg_4[Si_2O_5]_3(OH)_2 \cdot 4H_2O$; **c** palygorskite $Mg_5[Si_2O_5]_4(OH)_2 \cdot 8H_2O$; **d** antigorite $Mg_{48}[Si_2O_5]_{17}(OH)_{62}$

open-branched dreier chains alternate with double tetrahedra as regions of equal directedness. Further increase of the mean cation size in synthetic $H_2NaNd[Si_6O_{15}] \cdot nH_2O$ (Karpov et al. 1977b) leads to vierer single chains, vierer single rings, and single tetrahedra and in sazhinite, $HNa_2Ce^{+3}[Si_6O_{15}] \cdot n\,H_2O$ (Shumyatskaya et al. 1980), to vierer single rings and single tetrahedra as regions of equal directedness of the dreier layers.

Table 10.14 clearly demonstrates that, to a good approximation, a one-dimensional adjustment between tetrahedral layers and "octahedral" arrays takes place following the principles described for the chain silicates: the small cations Mg^{+2}, Fe^{+2}, Al^{+3}, and Fe^{+3} favor zweier chains; the larger cations Na^+ and Ca^{+2} dreier chains. Adjustment in the second dimension is achieved, in the order of increasing ratio $b_{oct} : b_{tetr}$, by tetrahedron rotation, by tetrahedron tilt, and by reducing the regions of equal directedness of the tetrahedra from the two-dimensional layer to one-dimensional bands of decreasing width and, eventually, to zero-dimensional rings, double tetrahedra, and single tetrahedra. This reduction of the regions of equal directedness is accompanied by splitting the two-dimensional "octahedral" layers into one-dimensional bands of decreasing width and, eventually, into zero-dimensional arrays. Table 10.14 also indicates that there is a general tendency for the atomic ratio $n_M : n_T$ to decrease as the mean cation radius increases.

In this study of the influence of cation size on the shape of tetrahedral layers in hydrous phyllosilicates, only those cations have been considered which are bonded to the so-called apical oxygen atoms of the layer, that is, to the oxygen atoms that are bonded to only one silicon atom. In some phyllosilicates there are additional cations between the octahedra–tetrahedra–octahedra layer packages. These interlayer cations are, for example, the potassium and calcium ions in the micas muscovite, $KAl_2[AlSi_3O_{10}](OH)_2$, and margarite, $CaAl_2[AlSiO_5]_2(OH)_2$. These cations are coordinated by six close and six slightly more distant bridging oxygen atoms, the latter of which are already almost electrostatically balanced by their two T–O bonds. The resultant weak bonds between K^+ or Ca^{+2} and the bridging oxygen atoms indicate that the influence of these cations on the shape of the tetrahedral layers is small and will, therefore, not be discussed here.

10.5.3.3 Strain Reduction Mechanisms in Cation-Rich Phyllosilicates

10.5.3.3.1 Tetrahedron Rotation and Tilting. In Table 10.15 it is demonstrated that in the cation-rich hydrous phyllosilicates, the mutual adjustment between the "octahedral" arrays and the tetrahedral $[Si_2O_5]$ layers takes place along the same lines as in the cation-poor ones. In particular, slight misfits between octahedral and tetrahedral parts of the kaolinite-like layer presented in Fig. 10.20a is reduced by rotating each tetrahedron about an axis normal to the layer by an angle α_r as shown in Fig. 10.21. In lizardite, $Mg_3[Si_2O_5](OH)_4$, a rotation angle of 3.5° has been observed (Mellini 1982). The value of α_r increases by replacement of Mg^{+2} by the smaller Al^{+3} ion of the octahedral layer as well as by a replacement of Si^{+4} by larger cations, such as Al^{+3} in the tetrahedral layer. Corresponding values $\alpha_r = 9°$ and 15° have been found in the dioctahedral mineral kaolinite, $Al_2[Si_2O_5](OH)_4$, and the trioctahedral amesite, $Mg_2Al[AlSiO_5](OH)_4$, respectively (Zvyagin 1960; Anderson and Bailey 1981).

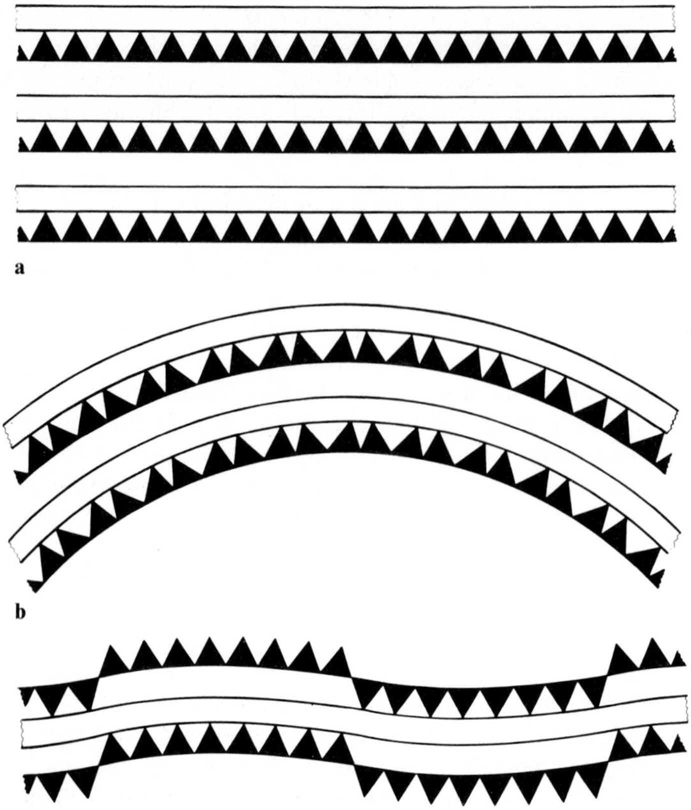

a

b

c

Fig. 10.25 a – c. Reduction of strain between octahedral layers (*white bars*) and tetrahedral layers (*black triangles*) in hydrous phyllosilicates with kaolinite-like arrangements.

a Kaolinite $Al_2[Si_2O_5](OH)_4$ (Zvyagin 1960); **b** chrysotile $Mg_3[Si_2O_5](OH)_4$ (Whittaker 1956); **c** antigorite $Mg_{48}[Si_4O_{10}]_{8.5}(OH)_{62}$ (Kunze 1959; Evans et al. 1976)

In analogy with the cation-poor hydrous phyllosilicates, the tetrahedra of the $[Si_2O_5]$ layers of the corresponding cation-rich silicates with b_{oct} only slightly smaller than b_{tetr} are tilted so that the nonbridging basal oxygen atoms of the tetrahedra deviate from a flat plane. However, since only one side of the octahedral layer is linked to a tetrahedral layer, the strain caused by the misfit between the two parts of the composite layer can be reduced by a bending of the tetrahedral–octahedral layer pair (Fig. 10.25 b).

Bending of this kind is observed in chrysotile, $Mg_3[Si_2O_5](OH)_4$, in which the layer pair is rolled up like a carpet. It has been observed not only by X-ray diffraction (Whittaker 1956), but also by high resolution electron microscopy (Yada 1971) (Fig. 10.26).

10.5.3.3.2 Tetrahedron Inversion. In antigorite, which has almost the same chemical composition as chrysotile, this tilting mechanism is accompanied by periodically inverting the directedness of the $[SiO_4]$ tetrahedra. This combination of tetra-

Table 10.16 Some crystal chemical data relevant to hydrous layer silicates of the chrysotile type

Cation	1	2	3	4	Name	Formula
Si^{+4}	–	4.98*	–	–	Cristobalite (lT)	SiO_2
Al^{+3}	0.53	5.05	650	35	Metahalloysite	$Al_2[Si_2O_5](OH)_4$
Ni^{+2}	0.70	5.40	154	9	Pecoraite	$Ni_3[Si_2O_5](OH)_4$
Mg^{+2}	0.72	5.39	156	11	Chrysotile	$Mg_3[Si_2O_5](OH)_4$
Co^{+2}	0.735	5.50	101	4	Synthetic	$Co_3[Si_2O_5](OH)_4$
Zn^{+2}	0.745	5.53	–	–	–	–
Fe^{+2}	0.77	5.63	?	?	Synthetic	$Fe_3[Si_2O_5](OH)_4$
Mn^{+2}	0.82	5.76	–	–	Manganpyrosmalite	$Mn_8[Si_6O_{15}](OH,Cl)_{10}$
Ca^{+2}	1.00	6.21	–	–	Apophyllite	$KCa_4[Si_4O_{10}]_2(F,OH)\cdot 8H_2O$
Sr^{+2}	1.16	–	–	–	–	–
Ba^{+2}	1.36	–	–	–	–	–

1 = Radius of six-coordinated cation;
2 = Lattice period (in Å) within the hexagonal or pseudohexagonal layer of the hydroxide of the cation;
3 = Mean diameter (in Å) of the tubes (Noll et al. 1960);
4 = Average number of tetrahedral–octahedral layer pairs (thickness of the tube walls) (Noll et al. 1960);
* = Dollase 1965

hedron tilting and inversion gives rise to a structure that is reminiscent of a stack of corrugated iron (Figs. 10.24d, 10.25c, 10.26d) (Kunze 1956, 1958, 1959; Yada 1984, personal communication).

Since the octahedral layers of chrysotile and antigorite are bonded to only one tetrahedral layer, the octahedral layer is not split into bands, as in the mica-like arrangements described in Fig. 10.23, but retains its infinite extension in two dimensions (Fig. 10.25b, c). The degree of misfit between octahedral and tetrahedral layers and, therefore, the amount of strain in the tetrahedral–octahedral layer pair depends again on the effective size of the octahedra, that is, on the radius of the cations in the octahedra.

Table 10.16 gives the ionic radii of medium-sized cations with coordination number six, together with the values of the translation periods in the hexagonal or pseudohexagonal layers of their hydroxides, and information about their hydrous phyllosilicates. From this Table it is clear that no hydrous zweier layer silicate with $r_M^{[6]} > r_{Mg}^{[6]}$ has yet been observed with a mica- or talc-like arrangement. This is probably due to the fact that for cations larger than Mg^{+2}, the strain between the octahedral and tetrahedral part of the structure is so high that it cannot be sufficiently reduced, even by splitting the octahedral layer into one-dimensional bands, without bending these bands along an axis parallel to the band. This, of course, is not possible because there is a tetrahedral layer on both sides of the band. With the kaolinite-like arrangement, cations larger than magnesium can be accommodated merely by bending the layer pairs into tubes as described above. The larger the misfit between the tetrahedral and octahedral layer becomes with increasing cation size, the smaller the radius of curvature of the tube becomes. This is analogous to the way a bimetal strip of a thermostat bends more with increasing temperature

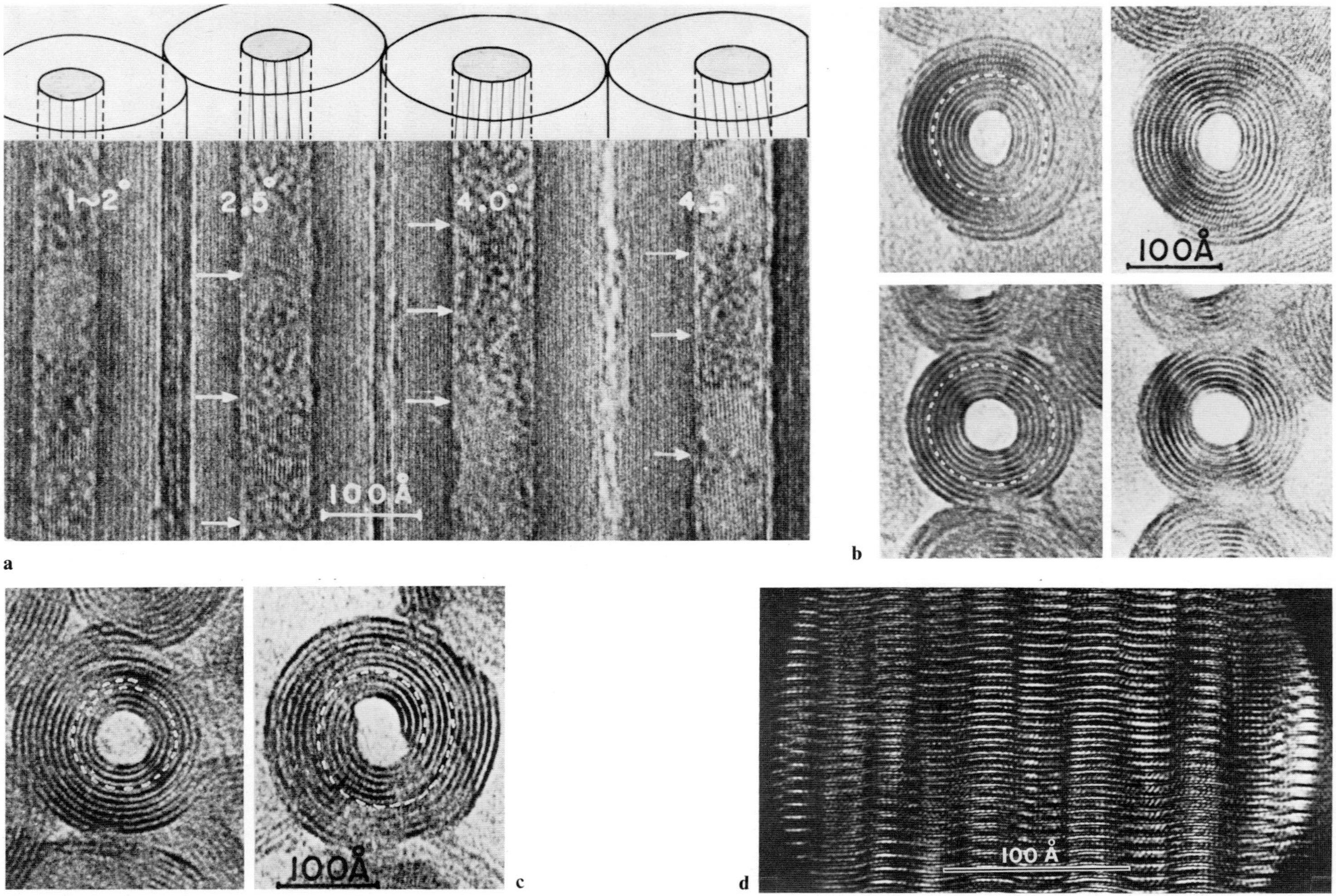

Fig. 10.26 a – d. High resolution electron micrographs of chrysotile, $Mg_3[Si_2O_5](OH)_4$ (from Yada 1971) and antigorite, $Mg_{48}[Si_4O_{10}]_{8.5}(OH)_{62}$ (Yada, personal communication).

a Four tubes of chrysotile from Coalinga, California; **b** concentric structure in the cross-section of a tube of chrysotile from Transvaal, S. Africa;

due to the difference in thermal expansion of the two metals. The relationship between layer misfit and radius of curvature for the phases isostructural with chrysotile is evident from the data given in Table 10.16. Whereas the tubes of metahalloysite, $Al_2[Si_2O_5](OH)_4$, have a mean diameter of 650 Å, this value is reduced to 156, 154, and 101 Å, respectively, for the corresponding phases of the larger cations magnesium (chrysotile), nickel (pecoraite), and cobalt, as demonstrated by the electron microscopic study of Noll, Kirchner and Sybertz (1960). This decrease in mean diameter with increasing cation size is accompanied by a decrease in the thickness of the tube walls, i.e., the number of tetrahedral–octahedral layer pairs forming the tubes. Iron chrysotile has been synthesized, but corresponding values are not yet available (Jasmund et al. 1975). In synthetic cobalt- and iron-chrysotile, the tubes are very fragile and unstable.

So far, the discussion has involved the cation-rich hydrous phyllosilicates of small cations which require zweier chains for adjustment in one direction. The slight difference in size of these cations required different adjustments to the layers in the second dimension giving rise to the structure types shown in Fig. 10.24. However, when the size of the cations bonded to the apical oxygen atoms of the $[Si_2O_5]$ layers increases up to the values for Mn^{+2} and Ca^{+2}, the adjustment of the $[M(O,OH,H_2O)_n]$ polyhedral aggregates to fit zweier chains becomes energetically unfavorable and a new tetrahedral chain type is adopted. For manganese, which forms inosilicates containing fünfer (rhodonite $Mn_5[Si_5O_{15}]$(1P)) and siebener ($Mn_7[Si_7O_{21}]$(mP)) single chains, the new chain type is a sechser chain, intermediate between those two. For calcium it is the sechser chain with two dreier units per chain period, a chain very similar to that known from a number of calcium inosilicates (see p. 186).

To achieve adjustment between tetrahedral and octahedral layers in both dimensions in these cation-rich phyllosilicates, a scheme similar to that in the cation-poor phyllosilicates (Table 10.14) is adopted. As discussed in Sec. 10.5.3.2.3, the regions of equal directedness of the tetrahedral layers of the cation-poor phyllosilicates change from two-dimensional layers to one-dimensional strips of decreasing width and, eventually, to zero-dimensional islands as the mean cation radius increases.

In cation-rich phyllosilicates an increase of cation radius beyond that of Mg^{+2} reduces the size of the regions of equal directedness of the tetrahedral layers also from two-dimensional zweier layers in, for example, kaolinite and the chrysotiles to one-dimensional strips several vierer chains wide, and eventually to zero-dimensional islands. In manganpyrosmalite, $Mn_8[Si_6O_{15}](OH,Cl)_{10}$ (Kato and Takéuchi 1983), and bementite, $Mn_7[Si_6O_{15}](OH)_8$ (Kato and Takéuchi 1980), these islands are one sechser ring wide, in apophyllite, $KCa_4[Si_4O_{10}]_2(F,OH) \cdot 8\,H_2O$ (Colville et al. 1971; Chao 1971), they have the size of a vierer ring (Fig. 10.27a, b, c). This inversion of directedness produces a change from flat and rolled layers to corrugated and eventually curled layers (Table 10.15). The rings of equal directedness may be considered as *curling units*.

Because of the very widespread isomorphous replacement between Mg^{+2} ($r^{[6]} = 0.72$ Å), Fe^{+2} ($r^{[6]} = 0.77$ Å), Mn^{+2} ($r^{[6]} = 0.82$ Å), and Ca^{+2} ($r^{[6]} = 1.00$ Å), the mean cation radius $\langle r_M \rangle$ in such layer silicates can vary almost continuously between 0.72 Å and 1.00 Å. Small changes in $\langle r_M \rangle$ probably cause changes in the

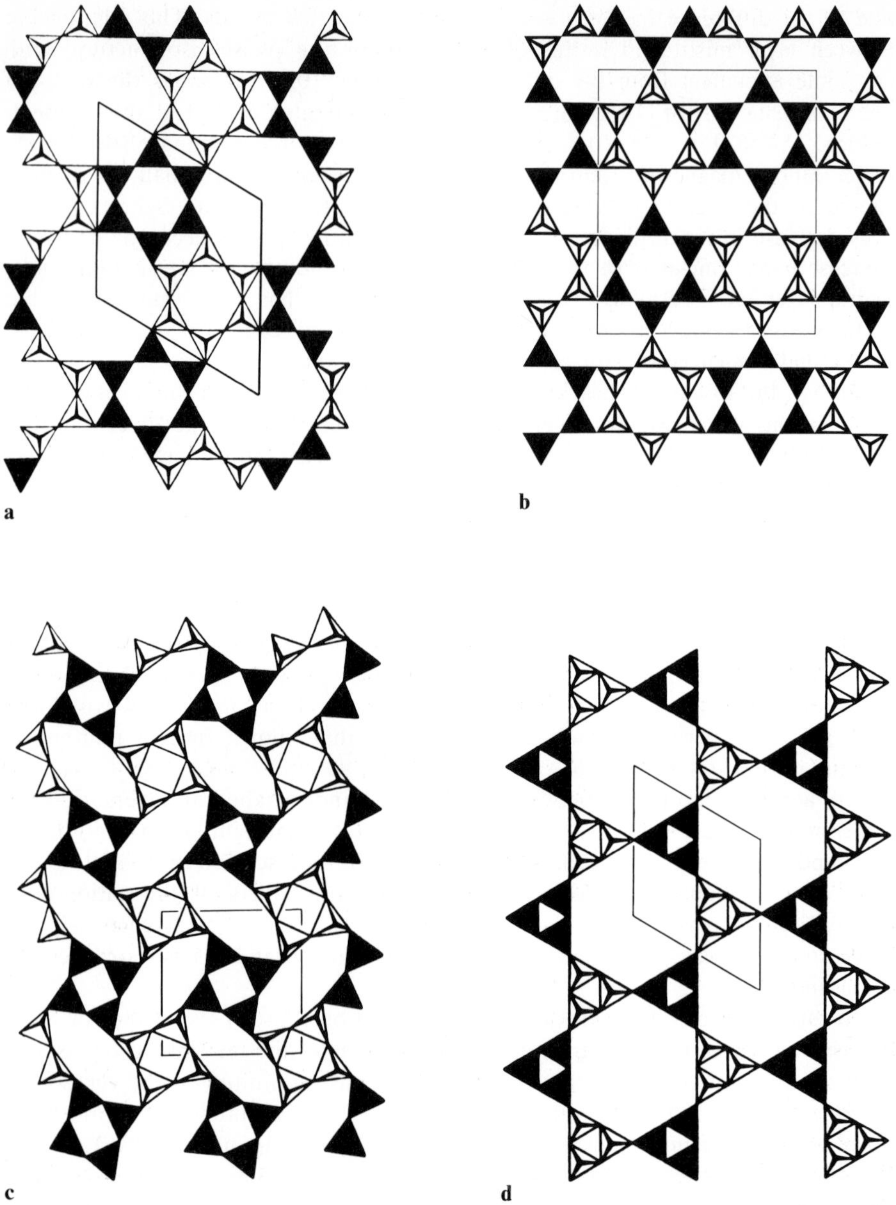

Fig. 10.27 a – d. Curled tetrahedral single layers with different curling units. Sechser rings in **a** manganpyrosmalite, $Mn_8[Si_6O_{15}](OH,Cl)_{10}$ (Kato and Takéuchi 1983) (cf. Fig. 7.18 f), and **b** bementite, $Mn_7[Si_6O_{15}](OH)_8$ (Kato and Takéuchi 1980); vierer rings in **c** apophyllite, $KCa_4[Si_4O_{10}]_2(F,OH) \cdot 8H_2O$ (Colville et al. 1971; Chao 1971); **d** hypothetical layer with dreier rings as curling unit

geometrical pattern of directedness of the tetrahedra without changing the kind of tetrahedral rings or their topological linkage. Slightly larger changes in $\langle r_M \rangle$ result in the formation of tetrahedral layers with the same kind of rings, but different topological linkage as, for example, in the case of manganpyrosmalite and bementite (Fig. 10.27a, b). Still larger changes in $\langle r_M \rangle$ eventually alter the kind of rings from sechser rings for $\langle r_M \rangle \simeq 0.80$ Å to vierer rings for $\langle r_M \rangle \simeq 1.00$ Å (Fig. 10.27c).

It is clear that isomorphous replacement of Si^{+4} ($r^{[4]} = 0.26$ Å) by Al^{+3} ($r^{[4]} =$ 0.39 Å) in the tetrahedral layer may compensate for isomorphous replacement of smaller cations by larger ones in the octahedral layer.

Caryopilite, $Mn_6[Si_4O_{10}](OH)_8$ (Kato and Takéuchi 1980),
greenalite, $(Fe^{+2}, Fe^{+3})_{<6}[Si_4O_{10}](OH)_8$ (Guggenheim et al. 1982),
mcgillite, $(Mn, Fe)_8[Si_6O_{15}](OH, Cl)_{10}$ (Iijima 1982; Ozawa et al. 1983),
ekmanite, $(Fe^{+2}, Mg, Mn, Fe^{+3})_{<9}[Si_9(Al, Si)_3O_{30}](OH)_6X_n \cdot 6\,H_2O$ (Nagy 1954),
bannisterite, $(K, Na, Ca, H_2O)(Mn, Mg, Fe^{+2}, Zn, Al)_{9.6}[Si_{12}O_{30}](OH)_{10} \cdot 4\,H_2O$ (Jefferson 1978), and
parsettensite, $(K, H_2O)_3(Mn, Fe^{+3}, Mg, Al)_{<9}[Si_{12}O_{30}](OH)_6X_n \cdot 6\,H_2O$ (Strunz and Tennyson 1970),

probably contain curled single layers which are similar to the ones in manganpyrosmalite and bementite.

In the phyllosilicates known to contain curled layers, the cation–anion polyhedra form two-dimensional layers and are not split into one-dimensional islands. In this respect they are more related to antigorite than to sepiolite and palygorskite.

With dreier rings as curling units, it might be possible to accommodate cations even larger than calcium. Such a hypothetical layer is shown in Fig. 10.27d. It has a very low density and this, together with the strong repulsive forces between the tetrahedra of a dreier ring, is probably the reason why it has not been observed.

After the originally flat two-dimensional array of apical atoms has, as a result of the increasing size of cations, been bent and then split into one-dimensional strips of various widths or into zero-dimensional islands of varying diameter, no further mechanism of reducing the strain between octahedral and tetrahedral layers seems possible. In fact, neither natural nor synthetic hydrous single layer phyllosilicates with cations larger than 1.1 Å (mean value) are known.

For a more detailed discussion of the hydrous single layer silicates with mean cation radius $\langle r_M^{[6]} \rangle \lesssim 0.85$ Å, the papers by Radoslovich and Norrish (1962), Franzini (1969), Whittaker and Wicks (1970), McCauley and Newnham (1973), Wicks and Whittaker (1975), Appelo (1978), and Baronnet (1980) should be consulted.

10.5.4 Comparison of Anhydrous and Hydrous Single Layer Silicates

In order to compare the structures of hydrous single layer silicates with those of the anhydrous varieties, a semi-quantitative relationship can be derived between the nonplanarity of the layers on the one hand, and the size and valence of the cations on the other, by using the average area parallel to the layer plane required

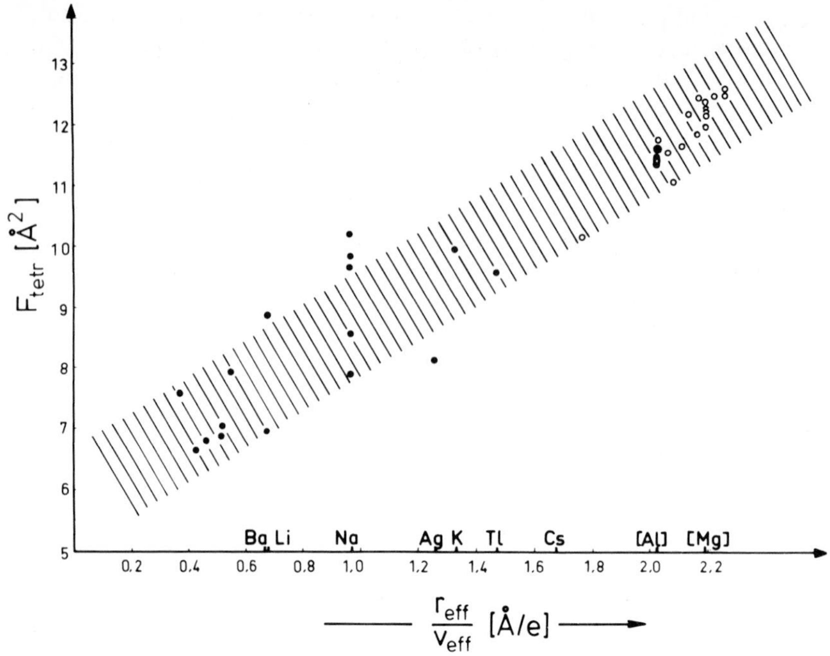

Fig. 10.28. The area F_{tetr} required per tetrahedron of a silicate single layer plotted against the ratio between the effective cation radius r_{eff} and effective valence v_{eff} of the cations of anhydrous (●) and hydrous (○) single layer silicates (after Liebau 1968)

by one [SiO_4] tetrahedron (Liebau 1968). This area, F_{tetr}, can be evaluated from the lattice constants a_1, a_2, and α_3 within the layers, the number, n_T, of [SiO_4] tetrahedra, and the number, n_S, of layers in the unit cell by using the formula

$$F_{tetr} = \frac{a_1 \cdot a_2 \cdot \sin \alpha_3 \cdot n_S}{n_T}.$$

For anhydrous phyllosilicates, the average values of the radii and the valences of all the "naked" cations bonded to the apical oxygen atoms of the layers can be used for this comparison. On the other hand, for hydrous phyllosilicates, the effective radii, r_{eff}, must be used for the cation–oxygen polyhedra which, in the case of octahedra, are given by half the length of the edge of the octahedron:

$$r_{eff} = \frac{r_{cat} + r_{O^{-2}}}{\sqrt{2}}.$$

For the valence, v_{eff}, of a polyhedron we use

$$v_{eff} = \langle v_{tetr} \rangle \cdot \frac{n_{TFU}}{n_{PFU}}$$

where $\langle v_{tetr} \rangle$ is the mean valence (charge) of the [TO_4] tetrahedra (i.e., $\langle v_{tetr} \rangle = 1$ for a tetrahedron of an [Si_2O_5] layer, and $\langle v_{tetr} \rangle = 1.25$ for the tetrahedron of an [$AlSi_3O_{10}$] layer), n_{TFU} is the number of [TO_4] tetrahedra in the formula unit, and

n_{PFU} is the number of polyhedra in the formula unit. From a comparison of the octahedral layers in, for example, trioctahedral layer silicates, such as antigorite, $Mg_{48}[Si_2O_5]_{17}(OH)_{62}$, in which all the $[M(O,OH)_6]$ octahedra contain Mg^{+2}, with the octahedral layers in dioctahedral layer silicates, such as kaolinite, $Al_2[Si_2O_5](OH)_4$, in which only two-thirds of the octahedra are filled with Al^{+3}, while the remaining are empty (Fig. 10.19), it is obvious that empty polyhedra must be considered in the same way as filled ones [3].

A plot of F_{tetr} versus r_{eff}/v_{eff} (Fig. 10.28) indicates that the influence of cation size and valence on the shape of the silicate layers, at least as far as the degree of convolution and curling is concerned, is the same for anhydrous and hydrous phyllosilicates (Liebau 1968).

10.6 Influence of Cation Properties on the Shape of Double Layer Silicate Anions

The unbranched and branched double layer silicates have been listed in Tables 10.17 and 10.18, respectively. If we neglect the fact that some of the tetrahedral sites in the structures of latiumite, $K_{1.7}Ca_6[(Al_{5.7}Si_{4.3})O_{22}][SO_4]_{1.4}[CO_3]_{0.6}$, and tuscanite, $(K,Sr,H_2O)_2(Ca,Na,Mg,Fe)_6[(Al_{3.66}Si_{6.34})O_{22}][SO_4]_{1.4}[CO_3]_{0.5}[O_4H_4]_{0.1}$, are completely filled with Al, then these two silicates should also be classified as unbranched fünfer double layer silicates (see Sec. 6.5).

In Sec. 10.5 it was demonstrated that anions which form polysilicates with a certain chain type generally form phyllosilicates with the same, or a closely related, fundamental chain. Since this is true for single layer silicates, it can also be expected to hold for double layer silicates.

In fact, the anions of the calcium silicates macdonaldite, rhodesite, and delhayelite are branched dreier double layers (Fig. 10.29d), whereas carletonite (Fig. 10.29f) contains branched sechser double layers. In contrast to the sechser chains observed in the chain silicates stokesite, chkalovite, gaidonnayite, and tuhualite and its isotypes (Figs. 7.9e, 7.11i), the sechser chains in carletonite are composed of two of the typical dreier subperiods, indicating that this sechser chain is crystal chemically more similar to the odd-periodic dreier chains of the calcium silicates than to the even-periodic sechser chains of the inosilicates containing highly electronegative cations.

Instead of classifying the sodium silicate naujakasite as a silicate with vierer double layers (Fig. 10.29e) according to the procedure described in Sec. 6.4, it could be regarded as a sechser double layer silicate with the sechser chains running horizontally in the upper part of the Figure. Again, as expected for a sodium

3 This is necessary because in hydrous phyllosilicates the empty polyhedra have about the same size as filled ones. In the case of anhydrous silicates empty polyhedra are in general smaller than filled ones and more irregular, so that their r_{eff} is difficult to evaluate. The empty polyhedra are, therefore, not taken into consideration for anhydrous phyllosilicates. This is obviously the reason why, in Fig. 10.28, the anhydrous silicates show a larger spread than the hydrous ones.

Table 10.17 Silicates containing unbranched double layers $\{\boldsymbol{uB}, 2^2_\infty\}$ $[T_mO_n]$

P	Silicate		Reference
	Name	Formula	
2	Synthetic	$Ca[AlSiO_4]_2$ (hT)	Takéuchi and Donnay 1959
	Synthetic	$Sr[AlSiO_4]_2$ (hT)	Pentinghaus 1975
	Synthetic	$Pb[AlSiO_4]_2$ (hT)	Pentinghaus 1975
	Synthetic	$Ba[AlSiO_4]_2$ (hT)	Takéuchi 1958
	Synthetic	$Rb[AlSi_3O_8]$ (hT)	Sorrell and Negas 1963
	Cymrite	$Ba[AlSiO_4]_2 \cdot H_2O$	Drits et al. 1975b
	Vertumnite	$\sim Ca_4Al_4Si_4O_6(OH)_{24} \cdot 3H_2O$	Galli and Passaglia 1978
4	Synthetic	$Cs_2Cu_2[Si_8O_{19}]$	Heinrich and Gramlich 1982
	Naujakasite	$Na_6Fe[Al_4Si_8O_{26}]$	Basso et al. 1975
	K phase	$Ca_7[Si_8O_{19}]_2(OH)_2$	Gard et al. 1981
	Reyerite	$(Na,K)Ca_7\{\boldsymbol{uB}, 1^2_\infty\}[Si_4O_{10}]$ $\{\boldsymbol{uB}, 2^2_\infty\}[AlSi_7O_{19}](OH)_4 \cdot 3H_2O$	Merlino 1972a
	Fedorite	$(K,Na)_{2.5}(Ca,Na)_7[Si_8O_{19}]_2(OH,F)_2 \cdot H_2O$	Sokolova et al. 1983
9	Stilpnomelane	$KFe_{24}[AlSi_{35}O_{84}]O_6(OH)_{24} \cdot nH_2O$	Eggleton 1972; Crawford et al. 1977

P = Chain periodicity

Table 10.18 Silicates containing branched double layers $\{\boldsymbol{lB}, 2^2_\infty\}$ $[T_mO_n]$. (No silicates have been observed which could be classified as open-branched double layer silicates)

P	Silicate		Reference
	Name	Formula	
3	Rhodesite	$HKCa_2[Si_8O_{19}] \cdot 5H_2O$	Hesse 1979b
	Hydrodelhayelite	$H_2KCa_2[AlSi_7O_{19}] \cdot 6H_2O$	Dorfman and Chiragov 1979
	Delhayelite	$Na_3K_7Ca_5[AlSi_7O_{19}]_2F_4Cl_2$	Cannillo et al. 1969
	Macdonaldite	$H_2Ca_4Ba[Si_8O_{19}]_2 \cdot (8+x)H_2O$	Cannillo et al. 1968
	Monteregianite	$(Na,K)_3Y[Si_8O_{19}] \cdot 5H_2O$	Chao 1978
4	Leucosphenite	$Na_4BaTi_2[(B_2Si_{10})O_{28}]O_2$	Malinovskii et al. 1981
6	Carletonite	$Na_4KCa_4[Si_8O_{18}][CO_3]_4(OH,F) \cdot H_2O$	Chao 1972

P = Chain periodicity

silicate, the sechser chain period is composed of two typical dreier chain subperiods, as can be seen from the side-view of Fig. 10.29e. Of all these calcium–sodium double layer silicates, reyerite is the only one in which the anions cannot be built solely from dreier or (2× dreier) chains (Fig. 10.29c).

In the preceding section and, in particular, Tables 10.14 and 10.15, we have seen that in hydrous single layer silicates, the size of the regions of equal directedness becomes progressively smaller as the mean radius of the cations increases. A similar effect is visible in the hydrous double layer silicates and zussmanite. Figures 10.29 and 10.30 show that in a double layer, the set of tetrahedra pointing to one side of a single layer act as linkages to the tetrahedra of opposite directedness in another single layer. In other words, regions of the same directedness in the single layers may either become *linking units* or *nonlinking units* of the double layers. The linking units may be two-dimensional layers, one-dimensional chains, or zero-dimensional groups of various size (between one and six tetrahedra in the structures hitherto observed) (Table 10.19). In contrast, the nonlinking units are

Table 10.19 Correlations between the mean cation radius and atomic ratio $n_M : n_T$ on the one hand, and the structure of composite tetrahedral layers on the other

Phase	$(n_T)_{nl}$ [a,c]	$(n_T)_l$ [b,c]	$\langle r^{[6]}\rangle$ [Å] [d]	$n_M : n_T$ [e]	$n_X : n_T$ [f]
Monteregianite	${}^{1}_{\infty}$	1	1.06	0.50	0.63
Macdonaldite	${}^{1}_{\infty}$	1	1.07 [g]	0.44	0.65
Rhodesite	${}^{1}_{\infty}$	1	1.13 [g]	0.50	0.38
Delhayelite	${}^{1}_{\infty}$	1	1.18	0.94	0.38
$Cs_2Cu_2[Si_8O_{19}]$	${}^{1}_{\infty}$	2	1.37	0.50	0
Stilpnomelane	24	6	0.75	0.72	0.75
Zussmanite	6	3	0.8	0.78	0.58 [h]
Reyerite	6	1	1.00	1.00 [i]	0.44 [i]
Carletonite	2	2	1.05	1.13	0.75
Naujakasite	2	${}^{1}_{\infty}$	0.98	0.58	0
Leucosphenite	–	${}^{2}_{\infty}$	0.95	0.58	0.17
$M[AlSiO_4]_2$ (hT)	–	${}^{2}_{\infty}$	1.00	0.25	0
M = Ca,			1.16		
Sr, Pb, Ba			1.18		
			1.36		
Cymrite	–	${}^{2}_{\infty}$	1.36	0.25	0.25
$Rb[AlSi_3O_8]$ (hT)	–	${}^{2}_{\infty}$	1.49	0.50	0

[a] $(n_T)_{nl}$ = Number of tetrahedra in one nonlinking unit of the composite tetrahedral layer;
[b] $(n_T)_l$ = Number of tetrahedra in one linking unit of the composite tetrahedral layer;
[c] ${}^{1}_{\infty}$, ${}^{2}_{\infty}$ = Unit has infinite extension in one and two dimensions, respectively;
[d] $\langle r^{[6]}\rangle$ = Mean cation radius for coordination number CN = 6;
[e] $n_M : n_T$ = Ratio between number of cations and number of $[TO_4]$ tetrahedra in the chemical formula;
[f] n_X = Number of H_2O molecules and O^{-2}, OH^-, and other anions which are not part of the silicate anion;
[g] H atoms not considered;
[h] Value has been adjusted by taking into account the fact that zussmanite is not a true double layer silicate;
[i] Value has been adjusted to the double layer only

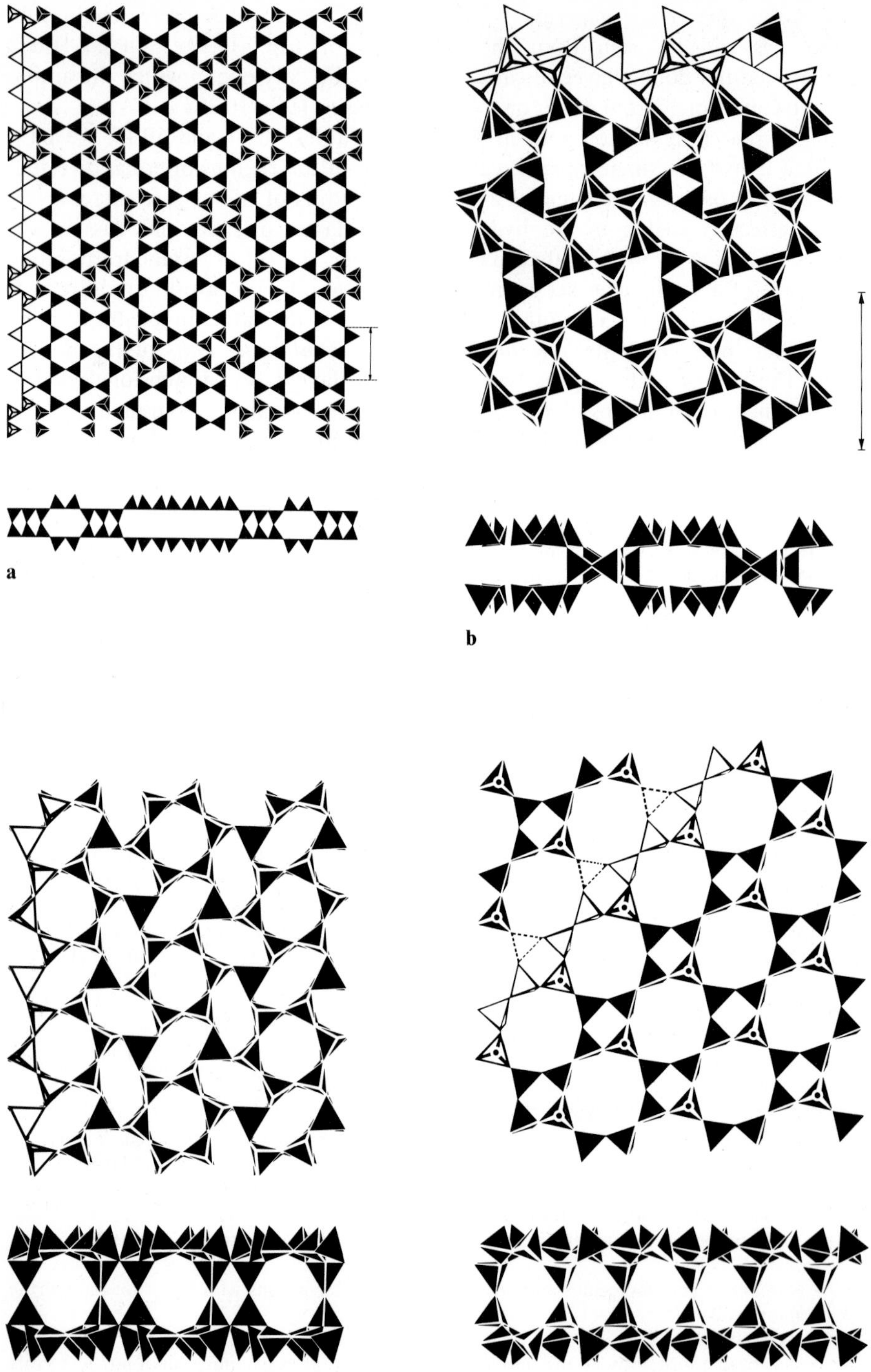

Fig. 10.29 a – d

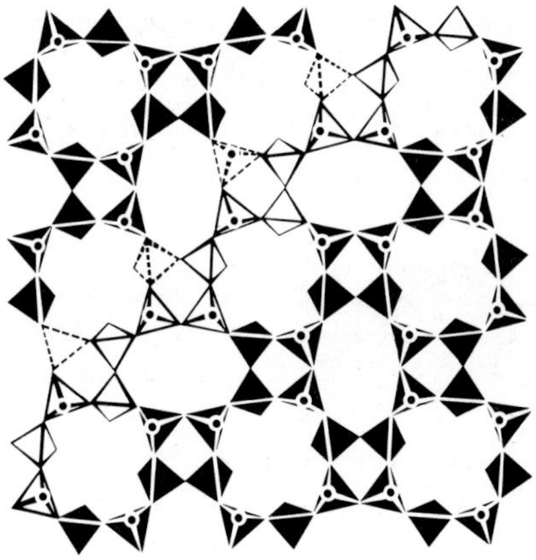

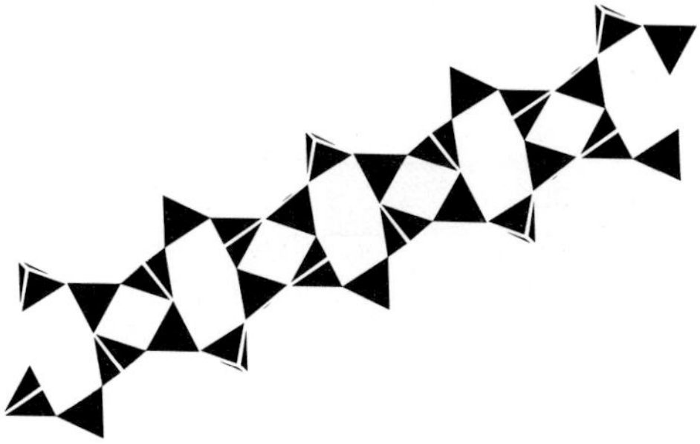

e

f

g

Fig. 10.29 a – g. Several complex silicate layers in **a** stilpnomelane, $KFe_{24}\{\boldsymbol{uB}, 2^2_\infty\}[^9(AlSi_{35})O_{84}]O_6(OH)_{24} \cdot nH_2O$. (The side view shows only the tetrahedra within the slab that is indicated by the *double arrow* in the vertical projection. The scale is reduced by a factor of 2 compared to **b** to **g**) **b** Zussmanite, $KFe^{+2}_{13}\{\boldsymbol{lB}, 1^2_\infty\}[^5(Al, Si)_9O_{21}]_2(OH)_{14}$; **c** reyerite, $(Na, K)Ca_7\{\boldsymbol{uB}, 1^2_\infty\}[^4Si_4O_{10}]\{\boldsymbol{uB}, 2^2_\infty\}[^4(AlSi_7)O_{19}](OH)_4 \cdot 3H_2O$; **d** rhodesite, $HKCa_2\{\boldsymbol{lB}, 2^2_\infty\}[^3Si_8O_{19}] \cdot (6\text{-}n)H_2O$, and delhayelite, $Na_3K_7Ca_5\{\boldsymbol{lB}, 2^2_\infty\}[^3AlSi_7O_{19}]_2F_4Cl_2$; **e** naujakasite, $Na_6Fe\{\boldsymbol{uB}, 2^2_\infty\}[^4(Al_4Si_8)O_{26}]$; **f** carletonite, $Na_4KCa_4\{\boldsymbol{lB}, 2^2_\infty\}[^6Si_8O_{18}][CO_3]_4(OH, F) \cdot H_2O$, and **g** synthetic $M\{\boldsymbol{uB}, 2^2_\infty\}[^2(AlSi)O_4]_2(hT)$, M = Ca, Sr, Ba. Two such sublayers are linked via the apical oxygen atoms marked by *dots* and lying on a mirror plane

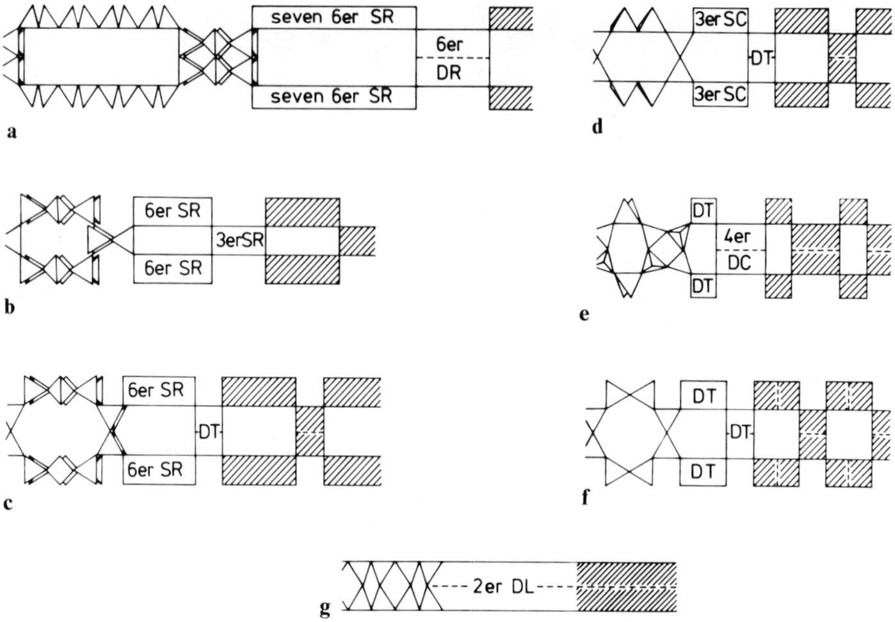

Fig. 10.30. Schematical representation of the curled nature of composite silicate layers in **a** stilpnomelane; **b** zussmanite; **c** reyerite; **d** rhodesite and delhayelite; **e** naujakasite; **f** carletonite; and **g** synthetic $M\{\boldsymbol{uB}, 2^2_\infty\}[^2(AlSi)O_4]_2(hT)$, M = Ca, Sr, Ba.
SR single ring; *DR* double ring; *SC* single chain; *DC* double chain; *DT* double tetrahedron; *DL* double layer (cf. with Fig. 10.29)

one-dimensional chains, zero-dimensional groups of various size (between 2 and 24 tetrahedra), or they do not exist if the two sublayers are linked via all tetrahedra.

In Fig. 10.31 the structural characteristics of the known double layer silicates and zussmanite are plotted in such a way that the correlation between the sizes of the linking and nonlinking units on the one hand, and the mean cation radius and the ratio $n_M : n_T$ on the other, becomes visible. The diagrams indicate that for the cation-poor phases ($n_M : n_T \lesssim 0.5$) with large cations (with the exception of leucosphenite), the linking units are infinitely extended in two dimensions, whereas for the cation-richer phases with smaller cations both the linking and nonlinking units have finite extension. In the central region of the diagrams either the linking or the nonlinking units are one-dimensionally extended chains.

A crystallochemical interpretation of this diagram can be given along the same lines as for the hydrous single layer silicates (see Sec. 10.5.3). Attempts to understand the formation of double layer silicates of cations of various sizes from silica-rich aqueous solutions begin with a consideration of the cations having $\langle r^{[6]}_M \rangle \simeq 0.72$ Å $= r^{[6]}_{Mg}$. Due to the good fit between the mesh sizes of tetrahedral and octahedral layers, the kaolinite-like layer arrangement will be formed from less silica-rich starting materials, and the mica-like layer arrangement from more silica-rich materials. In these arrangements all tetrahedra of a tetrahedral layer have the same directedness and, therefore, no tetrahedra are available for linking to tetrahedra of another tetrahedral layer to form a double layer. As cation size increases

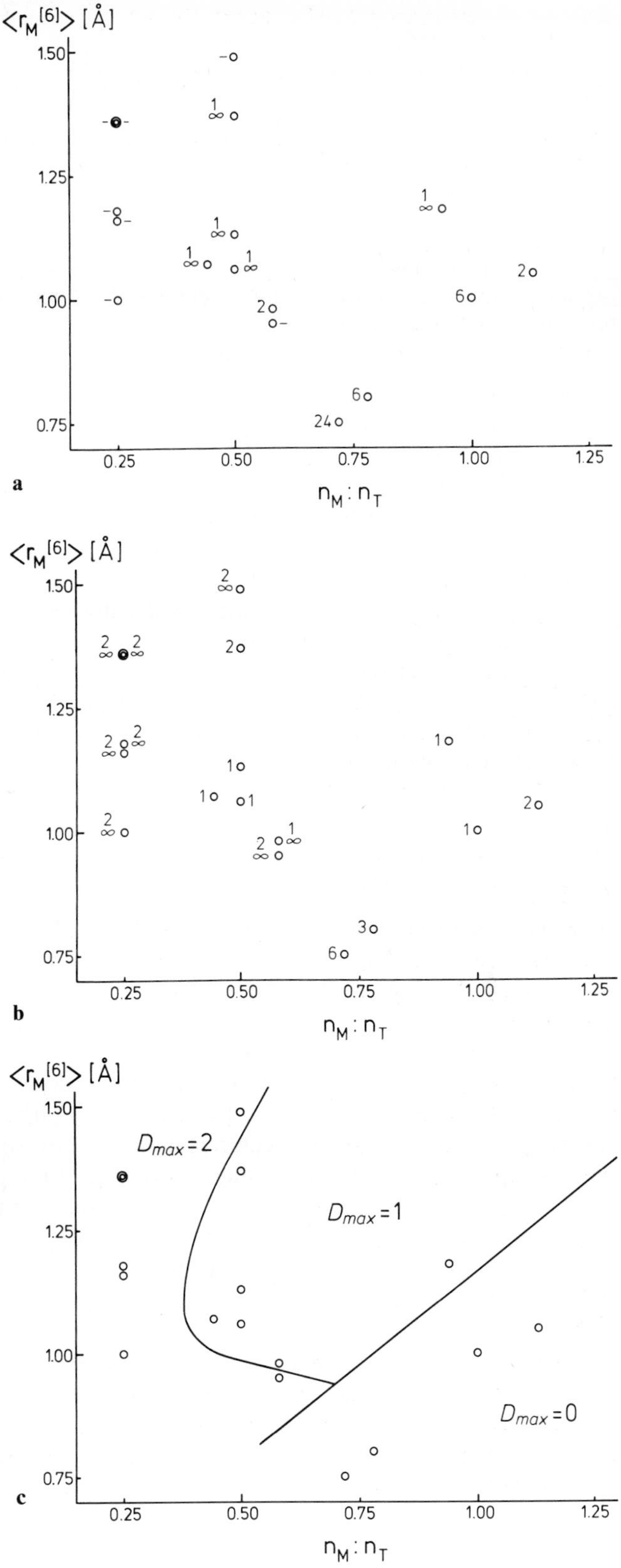

Fig. 10.31 a–c. Correlation between mean cation radius $\langle r_{M}^{[6]} \rangle$ and atomic ratio $n_M : n_T$ of double layer silicates and zussmanite on the one hand, and the size of **a** the nonlinking units and **b** the linking units of the composite layers [in number of tetrahedra (see Table 10.19)]. **c** Map indicating the fields of existence of the maximum of the dimensionalities, D_{max}, of linking and nonlinking units

in the cation-rich phases, the layer misfit causes inversion of some of the tetrahedra giving rise to curled layers with zero-dimensional regions of equal directedness. However, in more siliceous materials, the increase in cation size leads to the formation of layers with one-dimensional regions of equal directedness (Tables 10.14 and 10.15). These inverted regions become available for linkage to another tetrahedral layer if the starting material is very silica-rich.

In single layer silicates the size of the regions of equal directedness decreases as the misfit between tetrahedral and "octahedral" layer increases. However, since both sides of such single layers are equivalent, the regions of equal directedness on both sides of a layer are equal in size and number.

In double layer silicates of corresponding cations, the apical oxygen atoms of the nonlinking units are linked to cation–oxygen polyhedra, while those of the linking units are bonded to tetrahedra of the other sublayer. Since the sides of a sublayer are no longer equivalent, the sizes of the nonlinking and linking units do not have to be equal. Moreover, the number of tetrahedra per unit cell of the nonlinking units does not have to be equal to the number of tetrahedra in the linking units of a double layer. Of the two kinds of units only the sizes of the nonlinking units are controlled by the misfit between the "octahedral" and tetrahedral parts of the structure. As a result, in the silicates of Table 10.19 with small cation radii and relatively high $n_M : n_T$ ratios (stilpnomelane, zussmanite, reyerite, and carletonite), the size of the nonlinking units decreases with increasing cation radius from 24 tetrahedra arranged in seven sechser rings, to 6 tetrahedra in one sechser ring, to 2 tetrahedra in a double tetrahedron (Figs. 10.30, 10.31 a). With mean cation radii $\langle r_M^{[6]} \rangle \geq 1.0$ Å, the misfit is so large that double layers are formed only from very silica-rich starting materials. The linking units are then tetrahedral chains only one single chain wide so that the misfit can be relieved within a very short distance perpendicular to the direction of the nonlinking chains.

Within a tetrahedral double layer, ignoring its adjustment to the "octahedral" part of the structure, strain will accumulate mainly within the linking units. The more extended the linking units are, the larger the strain. This explains why linking units with more than two tetrahedra exist only in the silicates with rather small cations (stilpnomelane $\langle r \rangle = 0.75$ Å, zussmanite $\langle r \rangle \simeq 0.8$ Å, leucosphenite $\langle r \rangle = 0.95$ Å, and naujakasite $\langle r \rangle = 0.98$ Å), and in the double layer silicates with an atomic ratio T:O = 1:2 ($M[Al_2Si_2O_8]$ and $Rb[AlSi_3O_8]$(hT)). In the latter silicates, the strain is reduced by partial replacement of Si^{+4} by Al^{+3}, thereby reducing not only the repulsive forces by decreasing the valence of the tetrahedral ions, but also the stiffness of the T–O bond by replacing some of the more covalent Si–O bonds by more ionic Al–O bonds. However, even under these favorable conditions, double layers with all their tetrahedra belonging to the linking regions form only from starting materials very high in ($SiO_2 + Al_2O_3$) content, such as melts.

The stability of the double layers of leucosphenite may also be explained by the partial replacement of tetravalent silicon by trivalent boron. Since the misfit between the mesh sizes of the "octahedral" and tetrahedral layers is reduced by replacement of Si^{+4} by the smaller B^{+3}, such highly linked double layers in leucosphenite are formed at higher $n_M : n_T$ values than the double layer silicates in which Si^{+4} is partially replaced by the larger Al^{+3} ions.

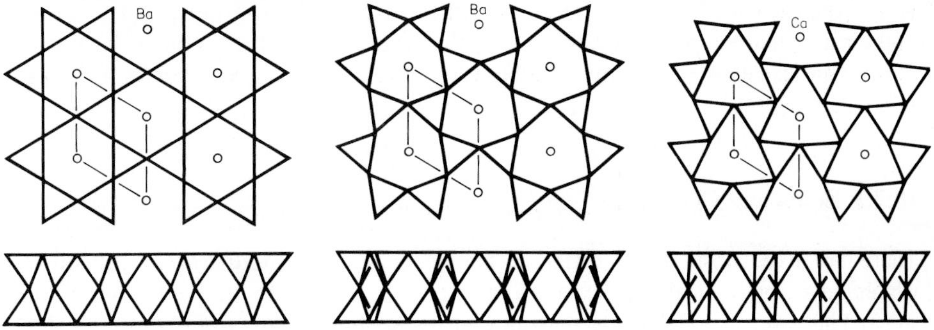

Fig. 10.32 a–c. Influence of cation size on the zweier double layers in **a** $Ba[Al_2Si_2O_8]$(hT) above 570 K; **b** $Ba[Al_2Si_2O_8]$(hT) below 570 K; **c** $Ca[Al_2Si_2O_8]$(hT) (Takéuchi and Donnay 1959)

Since the number of double layer silicates is as yet rather small and since these silicates differ widely in chemical composition and structure, the correlations between cation properties and the shape of the double layers are barely observable. However, the small group of isostructural double layer silicates of composition $M[AlSiO_4]_2$ with M = Ca, Sr, Ba, and Pb allow finer details of the influence of cation size on a particular type of double layer to be studied.

This structure type (Fig. 7.25a) is formed over a considerable range of cation size ($r^{[6]}$ = 1.00 Å to 1.36 Å). This may be due to the fact that in these silicates all oxygen atoms are bonded to two tetrahedral cations – Si or Al – so that the electrostatic charge of −1 per two tetrahedra is rather evenly distributed over the entire double layer. The interlayer cations, therefore, play the same role as the potassium or calcium ions located between the tetrahedral–octahedral–tetrahedral composite layers of, for example, the micas; that is, they compensate for the weak charges of the composite layers and are very weakly bonded. Their influence on the shape of the silicate layers is, therefore, rather small but, of course, stronger for the smaller calcium ion than for the larger barium ion. As a result, the tetrahedral double layer in the barium silicate assumes its ideal hexagonal symmetry at high temperature and is only slightly distorted below 570 K, while the smaller calcium ions produce a much larger trigonal distortion of the double layers (Fig. 10.32) by attracting three of the oxygen atoms in each sechser ring more than the three others.

10.7 Influence of Cation Properties on the Shape of Silicate Frameworks

With increasing degree of condensation of $[SiO_4]$ tetrahedra, there is a steady decrease in the $n_M : n_{Si}$ ratio. As a consequence, the influence of the cations M on the structure of the silicates and, in particular, on the topology and the conformation of their tetrahedral anions, decreases in the sequence from the monosilicates, to the ring-, chain-, and phyllosilicates, to the tectosilicates. It eventually becomes zero at the extreme end of the condensation process, i.e., in the polymorphs of silica.

10.7.1 Silica and Clathrasils

There are five substantially different known crystalline modifications of silica which have topologically different frameworks and can be derived from pure silica. They are, in order of decreasing density, stishovite containing octahedrally coordinated silicon, and coesite, quartz, keatite, and cristobalite containing corner-shared [SiO_4] tetrahedra. Of these, quartz and cristobalite both exist in at least two varieties with the same topology, but differing mainly in the Si–O–Si bond angles. Which one of these phases is formed depends on thermodynamic properties, such as pressure and temperature and on kinetic conditions, such as structure of the starting material or p_H.

A sixth phase of pure SiO_2 with chains of edge-shared tetrahedra (Fig. 7.2) is known as fibrous silica and can be synthesized by heating a mixture of SiO_2 and Si in the presence of oxygen (Weiss and Weiss 1954). Another three topologically different SiO_2 frameworks of corner-shared [SiO_4] tetrahedra have been obtained by removing the guest molecules from the clathrasils defined in Sec. 8.4.5 by thermal treatment. Although the guest molecules are neutral species and not cations, these phases deserve at least some consideration here.

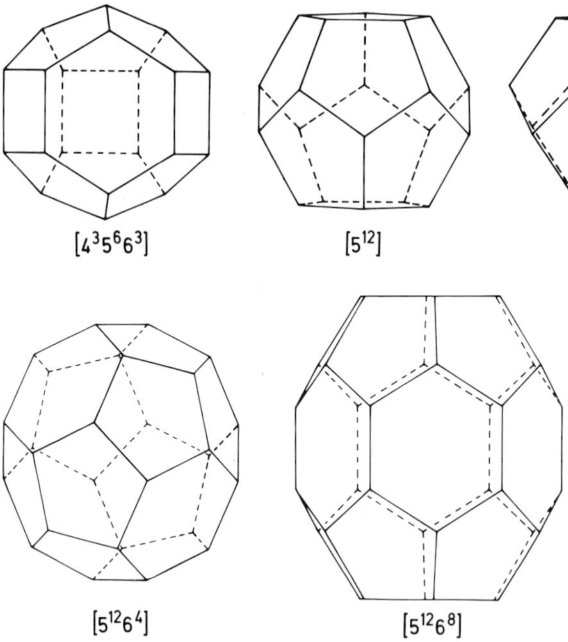

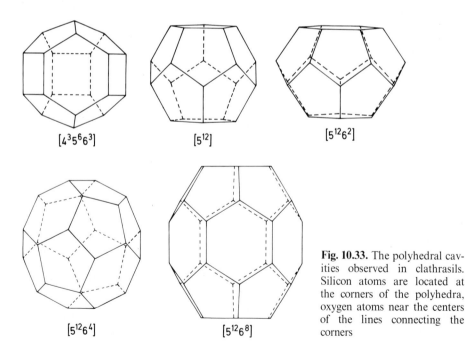

Fig. 10.33. The polyhedral cavities observed in clathrasils. Silicon atoms are located at the corners of the polyhedra, oxygen atoms near the centers of the lines connecting the corners

Table 10.20 Geometric shape and approximate free diameters of the cages present in the silica frameworks of clathrasils (Gerke et al. 1984)

Cage	$[4^35^66^3]$	$[5^{12}]$	$[5^{12}6^2]$	$[5^{12}6^4]$	$[5^{12}6^8]$
Geometric shape	Sphere	Sphere	Rotation ellipsoid	Sphere	Rotation ellipsoid
Approximate free diameters [Å]	$d \cong 5.7$	$d \cong 5.7$	$d_1 \cong 5.8$ $d_2 \cong 7.7$	$d \cong 7.5$	$d_1 \cong 7.7$ $d_2 \cong 11.2$

In the presence of air, silica crystallizes from aqueous solutions under hydrothermal conditions as quartz, cristobalite, keatite, or coesite. However, in the presence of a number of suitable inorganic and organic molecules, clathrasils can form under corresponding conditions by the condensation of low molecular weight silicic acid around these molecules. The neutral molecules in the crystalline phases formed are enclosed in polyhedral cavities (cages) of the tetrahedral silica frameworks. Since the molecules act as templates during the condensation process, the kind of cage formed depends on the size and shape of the guest molecule present. The cages observed in the three known clathrasils are shown in Fig. 10.33, their dimensions are given in Table 10.20, and the numbers of cages per unit cell are listed in Table 10.21 (Gerke et al. 1984).

As expected from the formation mechanism proposed above, the type of clathrasil formed is controlled by the size and shape of the guest molecules enclosed in the cavities. This is clearly demonstrated in Table 10.22, which lists the guest molecules which have been used successfully in the formation of the different clathrasils. Melanophlogite, $46\,SiO_2 \cdot 2\,M^{12} \cdot 6\,M^{14}$, with M^{12} and M^{14} guest molecules in the 12-sided and 14-sided cages, respectively, forms in the presence of small molecules which just fit into the small $[5^{12}]$ and $[5^{12}6^2]$ cages. Dodecasil 3C can be described by the chemical formula $136\,SiO_2 \cdot 16\,M^{12} \cdot 8\,M^{16}$, where M^{12} are again the guest molecules in the small $[5^{12}]$ pentagondodecahedra and M^{16} are somewhat larger molecules in the 16-hedron $[5^{12}6^4]$. Dodecasil 1H, $34\,SiO_2 \cdot 3\,M^{12} \cdot 2\,M^{12\prime} \cdot M^{20}$, contains the large 20-hedron $[5^{12}6^8]$ in addition to the dodecahedra $[5^{12}]$ and $[4^35^66^3]$ (Fig. 10.34) and is formed in the presence of even larger molecules, such as adamantylamine $C_{10}H_{15} \cdot NH_2$.

The presence and orientation of the guest molecules in the clathrasil cages are clearly visible in electron density distribution maps obtained from accurate single crystal X-ray diffraction data (Gies 1983, 1984; Gerke and Gies 1984).

Table 10.21 Cages present in the silica frameworks of clathrasils (Gerke et al. 1984)

Clathrasil	Number of cages per unit cell					SiO_2 per unit cell
	$[4^35^66^3]$	$[5^{12}]$	$[5^{12}6^2]$	$[5^{12}6^4]$	$[5^{12}6^8]$	
Melanophlogite	–	2	6	–	–	46
Dodecasil 1H	2	3	–	–	1	34
Dodecasil 3C	–	16	–	8	–	136

Table 10.22 Guest molecules M^n which have successfully been used in the synthesis of various clathrasils (Gerke et al. 1984)

Clathrasil	M^n	Guest molecules
Melanophlogite	M^{12} M^{14}	N_2, Kr, Xe; CH_4 N_2, N_2O, CO_2, Kr, Xe; CH_3NH_2
Dodecasil 3C	M^{12} M^{16}	N_2 SF_6, Kr, Xe; CH_3NH_2, $C_2H_5NH_2$, $(CH_3)_2CHNH_2$, $(CH_3)_3CNH_2$, $(CH_3)_3CCH_2NH_2$, $(CH_3)_2NH$, $(C_2H_5)_2NH$, $(CH_3)_3N$, $(CH_3)_2NC_2H_5$ 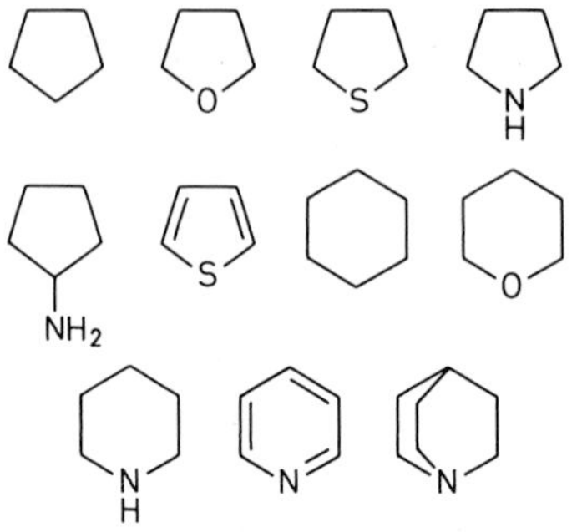
Dodecasil 1H	M^{12}, $M^{12'}$ M^{20}	N_2 $(C_2H_5)_2NH$ 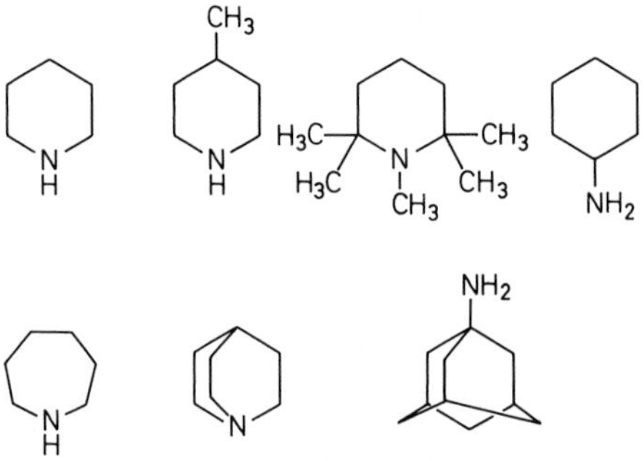

A suitable size for the molecule is not the only condition it has to fulfill to be occluded as a guest species in the synthesis of a clathrasil. Gerke et al. (1984) quote the following parameters in order of importance:

1) The molecule must have sufficient room within the particular cage of the clathrasil (molecular size).
2) The guest compound must be stable in the presence of water under the conditions of synthesis (ca. 475 K, 15 MPa) (stability).
3) The guest molecule should fit to the inner surface of the cage with as many van der Waals contacts as possible, but with the least deformation of the molecule and/or the cage (molecular shape).

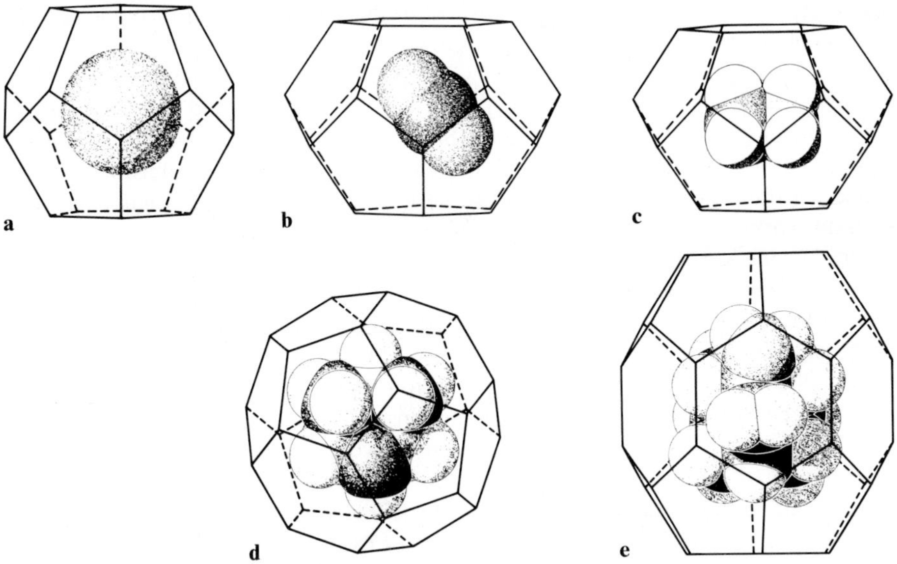

Fig. 10.34 a–e. Orientation of various guest molecules in cages of the host frameworks of clathrasils (Gerke et al. 1984).
a Xenon in $[5^{12}]$ of melanophlogite; **b** CO_2 in $[5^{12}6^2]$ of melanophlogite; **c** $CH_3 \cdot NH_2$ in $[5^{12}6^2]$ of melanophlogite; **d** piperidine 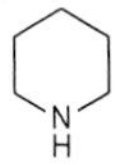 in $[5^{12}6^4]$ of dodecasil 3C; **e** adamantylamine 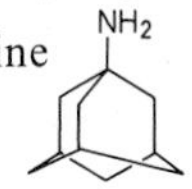 in $[5^{12}6^8]$ of dodecasil 1H

4) The tendency of the guest substance to form complexes with the solvent should be weak (solvatization tendency).
5) If the other parameters are equally favorable, those molecules with a higher stiffness will have a higher tendency to form a clathrasil (molecule stiffness).
6) If all other relevant parameters are equal, the tendency to form a clathrasil will increase as the basicity of the guest compound or the polarizability of its molecules increases (basicity).

The three clathrasils, which are the silica analogues of the well-known clathrate hydrates (gas hydrates) (Jeffrey 1984), can be considered as intermediates between the pure silica polymorphs and the usual framework silicates. The tetrahedral frameworks of the silica polymorphs coesite, quartz, keatite, and cristobalite, and the clathrasils melanophlogite, dodecasil 3C, and dodecasil 1H are all electrically neutral. Because the framework densities of the silica polymorphs are high (Table 8.6), these phases have their own fields of thermodynamic stability. However, as the framework density decreases, the frameworks become less stable. The tetrahedral frameworks of the three clathrasils are, therefore, only thermodynamically stable when at least a considerable fraction of their cages are occupied by suitable neutral guest molecules. Nevertheless, their host frameworks show a remarkable metastability even after their guest molecules are removed by heating to temperatures between 1075 and 1275 K (Gies et al. 1982).

If the term silicate is used in the strict sense of the word to describe a salt of a silicic acid, then the synthetic phases $K_2Ce\{\boldsymbol{oB},{}^3_\infty\}[Si_6O_{15}]$, $Rb_6\{\boldsymbol{oB},{}^3_\infty\}[Si_{10}O_{23}]$, and $Cs_6\{\boldsymbol{oB},{}^3_\infty\}[Si_{10}O_{23}]$ (see Table 7.8) are the only tectosilicates observed so far. They contain interrupted tetrahedral frameworks or, in other words, not all their $[SiO_4]$ tetrahedra are quaternary: $K_2Ce[Si_6O_{15}]$ has only tertiary tetrahedra, whereas $Rb_6[Si_{10}O_{23}]$ and $Cs_6[Si_{10}O_{23}]$ have both quaternary and tertiary ones. In the terminology of Wells (1975), $K_2Ce[Si_6O_{15}]$ has a 3-dimensional 3-connected framework of $[SiO_4]$ tetrahedra, the two alkali silicates have 3-dimensional (3, 4)-connected frameworks, and quartz, cristobalite, etc. have 3-dimensional 4-connected frameworks.

In 4-connected frameworks the incorporation of cations is made possible by partial replacement of tetravalent silicon by other tetrahedrally coordinated cations of lower valence, such as B^{+3}, Fe^{+3}, Ga^{+3}, Be^{+2}, Zn^{+2}, Mg^{+2}, and of course Al^{+3}. In such compounds, which are, in general, also called tectosilicates, the number of non-tetrahedral M cations per tetrahedral T ions depends on the amount of silicon replaced and on the valence of the replacing ions.

The amount of silicon replacement can be very small. For example, a pure silica phase with the tridymite-type tetrahedral framework seems to be thermodynamically unstable due to its relatively low framework density of 22.6 T ions per 1000 $Å^3$ compared to 26.6 T/1000 $Å^3$ for quartz (Table 8.6). It is widely agreed that tridymite requires a certain number of cations to fill some of the cavities in the framework in order to become a stable phase. This may be concluded not only from the fact that all natural and synthetic tridymites have been shown to contain at least a few tenths of a percent of ionic impurities, but has been demonstrated conclusively in the laboratory by Flörke (1956).

He applied a potential difference of 250 V to a pellet of tridymite at temperatures between 1475 and 1625 K and observed transformation of tridymite to cristobalite near the anode, i.e., in that part of the pellet from which the impurities had been extracted by solid state electrolysis. This, together with phase studies in the systems Li_2O-SiO_2, Na_2O-SiO_2, and K_2O-SiO_2 (Holmquist 1961), led to the conclusion that tridymite is not a phase in the pure SiO_2 system, but instead requires alkali or hydrogen ions to balance the charges resulting from minor replacement of Si^{+4} by Al^{+3}. It should, however, be mentioned that no reliable data are available on the minimum amounts of the different impurities which are required to stabilize the tridymite framework (e.g., Hirota and Ono 1977; Schneider et al. 1980). Consequently, there is still some debate about whether tridymite should or should not be considered to be a phase of silica.

The amount of impurities that can enter the tridymite structure can reach 50 mole percent. Corresponding phases, such as $Na[AlSiO_4]$, $Na_3K[AlSiO_4]_4$ (nepheline), $K[AlSiO_4]$ (kalsilite), $Ba[MgSiO_4]$, and others are typical examples of *stuffed derivatives* of silica (using the terminology of Buerger, 1954) in which non-tetrahedral cations are "stuffed" into voids of the three-dimensional tetrahedral framework.

10.7.2 Zeolites

The vast majority of zeolites have 3-dimensional 4-connected nets of $[SiO_4]$ and $[AlO_4]$ tetrahedra in which the atomic ratio of these tetrahedra varies considerably. For example, the ideal composition of mordenite is given as $Na[AlSi_5O_{12}] \cdot 3H_2O$ and that of gismondine as $Ca[Al_2Si_2O_8] \cdot 4H_2O$ (Barrer 1978). However, the Al : Si ratio of zeolites can be much smaller. Values as low as 1 : 450 (Nakamoto and Takahashi 1982) and 1 : 4000 (Olson et al. 1980) have been reported for the synthetic zeolite ZSM-5. The name silicalite-1 has been proposed for this zeolite in order to indicate that it is a silica phase and that the very small alumina content is a nonessential impurity (Flanigen et al. 1978).

The negative charge per framework tetrahedron increases with increasing degree of replacement of Si^{+4} by Al^{+3}. The non-framework cations supply the positive charge necessary to balance the negative charge of the framework. Due to the large variation in the Al : Si ratio of the zeolite frameworks, the number of cations per tetrahedron required for charge balance varies considerably from one cation per two tetrahedra in, for example, sodalite hydrate $Na_6[Al_6Si_6O_{24}] \cdot 8H_2O$ and synthetic zeolite A $Na_{12}[Al_{12}Si_{12}O_{48}] \cdot 27H_2O$, to virtually zero in the precursors of the synthetic zeolites silicalite-1 and silicalite-2. It is clear that the influence of the cations on the structure of the tetrahedral frameworks decreases as the ratio $n_M : n_T$ decreases.

As the framework density of the zeolites decreases, the frameworks become progressively less stable and there is an increasing tendency to stuff at least part of the channel volume with "fillers". In general, these fillers are cations, such as Na^+, K^+, Ca^{+2}, Ba^{+2}, etc., usually in a more or less hydrated form. With decreasing aluminum content of the zeolite framework, the filling function is taken over by neutral species, such as H_2O or other inorganic or organic molecules of suitable size and shape.

It is, therefore, mainly the "size" (or more appropriately, the dimensions) and valence (which would be zero for neutral molecules) of the fillers which influence the structure of the zeolite frameworks. The electronegativity of the fillers, which has been shown to be very important in silicates with low degrees of condensation (i.e., chain and ring silicates), seems to have a much weaker influence in framework silicates.

The structure-directing role of cations in zeolite synthesis has recently been reviewed by Barrer (1981). According to this work:

1) Na^+ in the reaction mixture favors the formation of sodalite and cancrinite hydrates, gismondine types, gmelinites, faujasites, and zeolite A;
2) K^+ favors the formation of chabazite and zeolite L;
3) Solutions containing Ca^{+2} favor the formation of thomsonite and epistilbite;
4) Sr^{+2} favors zeolites of the heulandite and ferrierite types;
5) Ba^{+2} and Li^+ ions favor different kinds of synthetic zeolite frameworks.

Sometimes a mixture of a small and a large cation has the same effect as a cation of intermediate size. This indicates that the average cation radius $\langle r_M \rangle$ is one of the structure-directing cation properties.

For example, zeolite A can be synthesized from solutions containing Na^+ ions and also from solutions containing a mixture of Li^+, Cs^+, and $[N(CH_3)_4]^+$ ions (Barrer 1981).

Indeed, some zeolites are more easily synthesized from solutions of two or more non-framework cations than from solutions containing only one kind of cation. For example, offretite and mazzite-type zeolites and the so-called zeolite EAB are formed from solutions containing a mixture of Na^+ and $[N(CH_3)_4]^+$ cations.

The structure-directing influence of the cations is exerted in at least two ways:

1) by changing the kinetics of formation of the crystalline zeolite from the low molecular species in the aqueous solution by a condensation process and
2) by stabilizing the crystalline zeolite framework.

It is only the latter type of effect which has been considered in the earlier part of this treatment.

As yet, no clear picture has arisen from the literature about the extent to which the structure-directing influence of the cations is due to one or other of the above two causes, despite the fact that such knowledge is of substantial importance in manufacturing and using zeolites.

10.7.3 Aluminosilicates $M[(Al,Si)_4O_8]$: Feldspars and Their Polymorphs

The feldspars are by far the most abundant minerals in the earth's crust, constituting some 65 vol% (Table 1.3). Like the zeolites they rarely have an atomic ratio Al : Si > 1 : 1 but, in contrast to the zeolites, their Al : Si ratio is very seldom < 1 : 3. In this section the discussion will be restricted to the pure end members of the alkali feldspars $M^+[AlSi_3O_8]$ and the alkaline earth feldspars $M^{+2}[Al_2Si_2O_8]$, including comparisons with their normal pressure polymorphs.

The feldspars are constructed from a 3-dimensional 4-connected net of $[AlO_4]$ and $[SiO_4]$ tetrahedra (Fig. 4.9b) which can be described as a loop-branched dreier framework. In the alkaline earth feldspars, each $[AlO_4]$ tetrahedron is corner-linked to four $[SiO_4]$ tetrahedra and vice versa; in other words, they have an ordered Al/Si distribution among the available T sites. This is in agreement with Loewenstein's aluminum avoidance rule (Loewenstein 1954) which indicates that, in general, two $Si^{[4]}-O-Al^{[4]}$ groups have a lower energy content than one $Al^{[4]}-O-Al^{[4]}$ plus one $Si^{[4]}-O-Si^{[4]}$ group. The Al/Si distribution in alkali feldspars is more or less disordered depending on the thermal history.

The non-tetrahedral M ions occupy asymmetrical voids in the tetrahedral framework.

10.7.3.1 $M[AlSi_3O_8]$ Phases with M = Li, Na, K, Rb, Cs

At room temperature the Na^+ ions in low-albite are irregularly surrounded by seven oxygen atoms with a mean Na–O distance of 2.63 Å (Winter et al. 1977; Harlow and Brown 1980), while in high-albite they are surrounded by eight oxygen neighbors with ${}^8\langle d(\text{Na}-\text{O})\rangle = 2.74$ Å and a ninth oxygen atom slightly further away at 3.37 Å (Prewitt et al. 1976) (Table 10.24 and Fig. 10.35a). Similarly, in microcline, the low temperature form of potassium feldspar $K[AlSi_3O_8]$, there are eight short K–O distances with an average of 2.92 Å and an additional oxygen atom at 3.34 Å (Brown and Bailey 1964). In sanidine, the high temperature form of potassium feldspar, and in synthetic $Rb[AlSi_3O_8]$, the larger monovalent cations have nine oxygen atoms at mean distances ${}^9\langle d(\text{M}-\text{O})\rangle$ of 2.97 Å and 3.08 Å, respectively (Weitz 1972; Gasperin 1971).

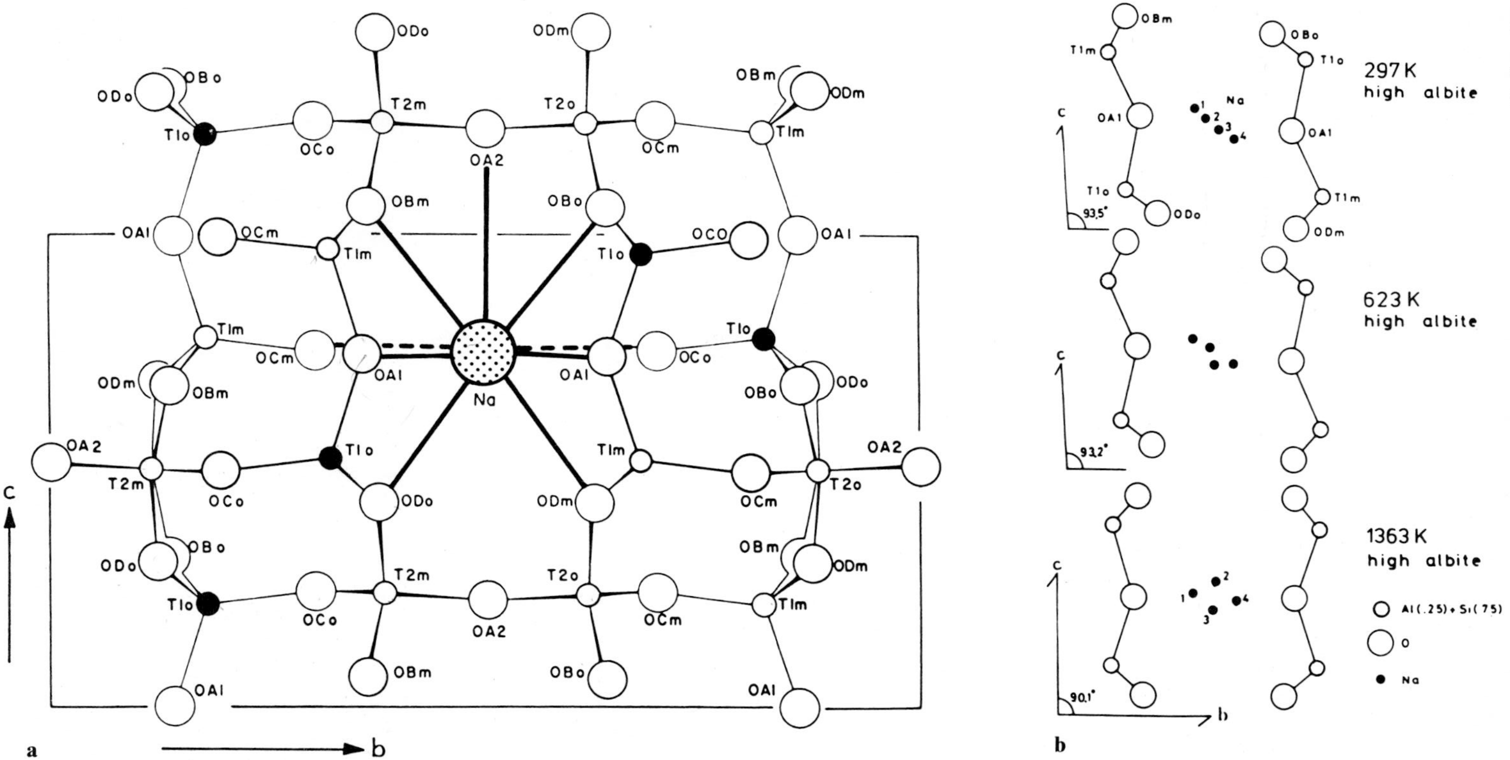

Fig. 10.35 a, b. Projection of the structure of albite, Na[AlSi$_3$O$_8$], on (100) (after Prewitt et al. 1976).

a The average position of the sodium ion in the void of the tetrahedral framework of monoclinic albite. The sodium–oxygen distances listed in Table 10.24 are indicated by *heavy lines.* ● aluminum; ○ silicon; ◯ oxygen. **b** The split positions that a sodium atom can occupy in the void at different temperatures

Table 10.23 Normal pressure aluminosilicates of composition $M[AlSi_3O_8]$

Composition	Name	Structure type	$r_M^{[6]}$ [Å] [a]	B [Å^2] [b]	V_{ox} [Å^3] [c]	Mean T–O–T bond angle [°]	Reference
$Li[AlSi_3O_8]$	Virgilite	Quartz	0.74	n. d.[d]	~ 20.7	n. d.[d]	French et al. 1978
	Synthetic	Keatite		n. d.	n. d.	152.1 [e]	Ostertag et al. 1968
$Na[AlSi_3O_8]$	Low-albite	Feldspar	1.02	2.5	20.8	140.5	Harlow and Brown 1980
	High-albite			7.6	20.8	140.4	Prewitt et al. 1976
$K[AlSi_3O_8]$	Low-microcline	Feldspar	1.38	1.4	22.5	141.9	Brown and Bailey 1964
	Sanidine				22.5	142.0	Weitz 1972
$Rb[AlSi_3O_8]$	Synthetic	Feldspar	1.49	1.9	23.3	142.8	Gasperin 1971
$Cs[AlSi_3O_8]$	Unknown	–	1.70	–	–	–	–

[a] $r_M^{[6]}$ = Cation radius for coordination number six;
[b] B = Isotropic temperature factor;
[c] V_{ox} = Volume per oxygen atom;
[d] n. d. = Not determined;
[e] Extrapolated from values given by Li and Shlichta (1974)

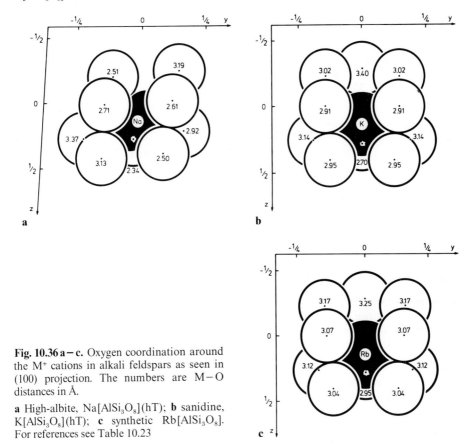

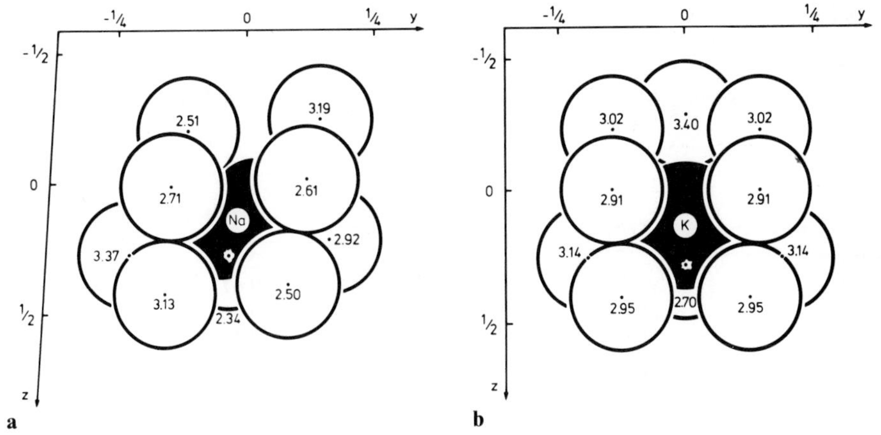

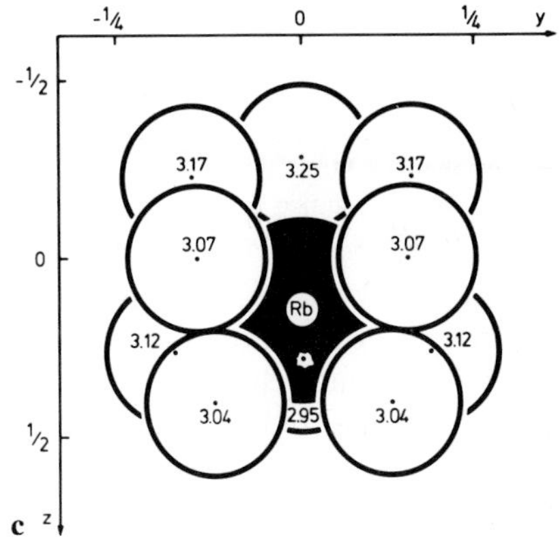

Fig. 10.36 a – c. Oxygen coordination around the M$^+$ cations in alkali feldspars as seen in (100) projection. The numbers are M–O distances in Å.

a High-albite, Na[AlSi$_3$O$_8$](hT); **b** sanidine, K[AlSi$_3$O$_8$](hT); **c** synthetic Rb[AlSi$_3$O$_8$]. For references see Table 10.23

The tetrahedral framework of the feldspars is rather rigid. Indeed, it is much less flexible in regard to rotation of its tetrahedra than the leucite-type framework discussed in the next section (10.7.4). Its rather oblate void (Fig. 10.35 a) is unable to accommodate a relatively small sodium ion in a central position in such a way that it is in close contact with all the oxygen atoms of the walls of the void. As a consequence, an Na$^+$ ion occupies one or other of four off-center split positions (Fig. 10.35 b). In time-averaged crystal structure refinements, this disorder of the Na$^+$ ion expresses itself as an unusually high isotropic temperature factor, B (see Sec. 3.1.2.4), on the sodium ion if it is assumed to be located at the center of the site. For example, whereas the B value of a cation, such as Na$^+$, is expected to be around 0.8 Å^2 in other structures, this value rises to 2.5 Å^2 in low-albite and 7.6 Å^2 in high-albite (Table 10.23).

A larger cation, such as K$^+$ or Rb$^+$, causes a widening of the void by slight rotation of the tetrahedra in order to increase the rather short cation–oxygen distances in albite. At the same time, two longer M–O distances in the void decrease, thereby increasing the coordination number of the cations from seven in low-albite to nine in potassium and rubidium feldspar (Fig. 10.36). From the M–O distances and distortion indices listed in Table 10.24, it is apparent that the

Table 10.24 Cation–oxygen distances (Å) and distortion indices of the $[MO_9]$ clusters in alkali feldspars. (For references see Table 10.23)

Oxygen \ Feldspar	Low-albite	High-albite	Microcline	Sanidine	Rb feldspar
OA1	2.573	2.606	2.877	2.901	3.068
	2.671	2.704	2.881	2.901	3.068
OA2	2.372	2.340	2.750	2.686	2.95
OBo	2.461	2.512	2.961	3.017	3.169
OBm	(3.465)	3.186	3.136	3.017	3.169
OCo	2.961	2.917	2.907	3.147	3.124
OCm	(3.266)	(3.371)	(3.335)	3.147	3.124
ODo	2.437	2.497	2.892	2.940	3.041
ODm	2.996	3.133	2.993	2.940	3.041
$^{CN}\langle d(M-O)\rangle$ [a]	2.634	2.737	2.925	2.966	3.084
$^{9}\langle d(M-O)\rangle$ [b]	2.797	2.807	2.971	2.966	3.084
Distortion index [c]	0.391	0.367	0.197	0.155	0.07

[a] Mean M–O distance averaged over the oxygen atoms considered to belong to the coordination sphere of M;
[b] Mean M–O distance averaged over the nine nearest oxygen atoms;
[c] Distortion index $[d(M-O)_{max} - d(M-O)_{min}] : {}^{9}\langle d(M-O)\rangle$

cation – oxygen coordination becomes more regular in the sequence low-albite – high-albite – microcline – sanidine – rubidium feldspar. In $Rb[AlSi_3O_8]$ the large Rb^+ ion fills the void so tightly that there is very little variation in the nine Rb–O distances. Moreover, the isotropic temperature factor decreases, indicating that the K^+ and Rb^+ ions vibrate about one position near the center of the void instead of occupying several split positions statistically.

The average T–O–T bond angle of 140.5° in albite is almost identical to the value 140° assumed to represent an unstrained Si–O–Si bond (see Sec. 3.1.2.5). This indicates that the tetrahedral framework in albite has very little strain, in agreement with calculations of electrostatic energies which show that $Na[AlSi_3O_8]$ is more "stable" than $K[AlSi_3O_8]$ (Brown and Fenn 1979). However, because the feldspar framework is inflexible to tetrahedral rotation, replacement of Na^+ by the large cations K^+ and Rb^+ causes only a slight increase in the mean T–O–T angles. Concurrently, the framework is expanded only a little by the cation replacement as seen from the changes in the volume per oxygen atom, V_{ox}, in Table 10.23.

Replacement of the tightly fitting Rb^+ ion by the considerably larger Cs^+ ion would require an increase in the M–O distances by approximately 0.2 Å. This cannot be achieved merely by rotation of the tetrahedra, but would require substantial angular distortions of the tetrahedra and/or expansion of the T–O bonds. It is, therefore, understandable why Borutskaya's (1975) and Henderson's (1976) efforts failed to synthesize cesium feldspar $Cs[AlSi_3O_8]$.

The nonexistence of a lithium feldspar can be explained along similar lines. Lithium ions tend to be tetrahedrally or octahedrally coordinated by oxygen, and coordination numbers higher than six are unknown. In order to occupy the large irregular void in the feldspar framework, the small Li^+ ion would have to attract

Table 10.25 Normal pressure aluminosilicates of composition $M[Al_2Si_2O_8]$

Composition	Name	Structure type	$r_M^{[6]}$ [Å] [a]	B [Å^2] [b]	V_{ox} [Å^3] [c]	Mean T−O−T bond angle [°]	Reference
$Mg[Al_2Si_2O_8]$	Synthetic	Quartz	0.72	n. d. [d]	~ 21.1	n. d.	Schreyer and Schairer 1961
$Ca[Al_2Si_2O_8]$	Anorthite	Feldspar	1.00	0.8 to 1.4	20.9	142.9	Wainwright and Starkey 1971
	Synthetic	Hexacelsian		n. d.	20.7	134	Takéuchi and Donnay 1959
	Synthetic	Immm		0.5, 1.4	21.4	133.6	Takéuchi et al. 1973
$Sr[Al_2Si_2O_8]$	Slawsonite	Paracelsian	1.16	0.7	21.7	128.7	Griffen et al. 1977
	Synthetic	Feldspar		1.1	21.9	136.9	Chiari et al. 1975
	Synthetic	Hexacelsian		n. d.	22.2	n. d.	Pentinghaus 1980
$Ba[Al_2Si_2O_8]$	Paracelsian	Paracelsian	1.36	n. d.	23.3	132.0	Bakakin and Belov 1961
	Celsian	Feldspar		1.0	23.0	138.4	Griffen and Ribbe 1976
	Hexacelsian (synthetic)	Hexacelsian		n. d.	23.6	132	Takéuchi 1958

[a] $r_M^{[6]}$ = Cation radius for coordination number six;
[b] B = Isotropic temperature factor;
[c] V_{ox} = Volume per oxygen atom;
[d] n. d. = Not determined

some oxygen atoms even more and repel others even further than the Na^+ ion does in albite. This would probably require that the lithium ion be located almost completely on one side of the void leaving a considerable part of the void empty – a rather unfavorable situation. As a consequence, $Li[AlSi_3O_8]$ crystallizes as stuffed derivatives of quartz and keatite, i.e., with the lithium ions stuffed into voids of quartz-type and keatite-type tetrahedral frameworks.

10.7.3.2 $M[Al_2Si_2O_8]$ Phases with M = Mg, Ca, Sr, Ba

Whereas only three types of tetrahedral frameworks in the $M^+[AlSi_3O_8]$ series are observed at normal pressure – the keatite- and quartz-types for M = Li and the feldspar-type for M = Na, K, and Rb – the series $M^{+2}[Al_2Si_2O_8]$ is more complicated.

By analogy with the alkali system, the small Mg ion forms an aluminosilicate with the quartz-type framework and the feldspar structure is taken up by the larger cations M = Ca, Sr, and Ba. However, each of the latter three divalent cations is also known to form two additional normal pressure structures as shown in Table 10.25. The schematic diagram in Fig. 10.37 illustrates that for all three cations, the feldspar phase transforms into a hexacelsian-type phase at higher temperatures. This is the double layer structure type displayed in Fig. 10.32. An even less stable crystalline $Ca[Al_2Si_2O_8]$ phase can be synthesized from a glass of anorthite composition. It has the Immm-type framework which will be described in Sec. 10.7.5. On the other hand, the stable room temperature phases of $Sr[Al_2Si_2O_8]$ and $Ba[Al_2Si_2O_8]$ have the paracelsian-type framework which transforms to the corresponding feldspar phases during heating.

The larger number of structure types with composition $M^{+2}[Al_2Si_2O_8]$ compared to the three $M^+[AlSi_3O_8]$ types is mainly due to the fact that for M ions of equal size, the $M^{+2}-O$ bond is stronger than the M^+-O bond and, therefore, has a stronger influence on the tetrahedral part of the structure. In addition, since the Al–O bond is weaker than the Si–O bond, and since the $M^{+2}[Al_2Si_2O_8]$ series has a higher Al:Si ratio than the $M^+[AlSi_3O_8]$ series, the aluminosilicate anions of the former series are more susceptible to the influence of the cations than are the anions of the latter series.

The stronger influence of the divalent cations on the aluminosilicate anions is also indicated by the quite normal values of the isotropic temperature factors B of

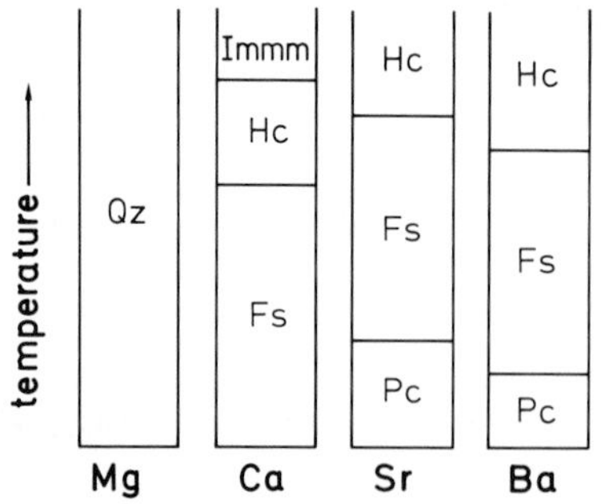

Fig. 10.37. The stability relations of the $M[Al_2Si_2O_8]$ phases as a function of temperature.
Qz quartz-type; *Pc* paracelsian-type; *Fs* feldspar-type; *Hc* hexacelsian-type; *Immm* Immm-type

these cations (Table 10.25) compared to the larger (and in the case of Na^+, much larger) values for the monovalent cations (Table 10.24).

10.7.4 Aluminosilicates $M[AlSi_2O_6]$

Considering, once again, only the normal pressure tectosilicate phases, there are three different structure types with the composition $M[AlSi_2O_6]$:

1) the quartz-type in the high temperature phase $Li[AlSi_2O_6]$-III;
2) the keatite-type in another high temperature phase $Li[AlSi_2O_6]$-II;
3) the leucite-type in the aluminosilicates of Na, K, Rb, and Cs.

In addition to these tectosilicates, there are the clinopyroxenes spodumene, $LiAl^{[6]}[Si_2O_6]$, and jadeite, $NaAl^{[6]}[Si_2O_6]$, in which the $[SiO_4]$ tetrahedra are corner-linked to form zweier single chains, but the aluminum ions are octahedrally coordinated (Table 10.26).

Of the above three framework types, the leucite-type seems to be energetically most favorable since it exists over a wide range of cation size, namely, from $r^{[6]}_{Na} = 1.02$ Å to $r^{[6]}_{Cs} = 1.70$ Å. The highest possible symmetry this framework can possess, the "aristotype" in the nomenclature of Megaw (1973), is cubic with space group Ia3d and lattice constant $a_0 \simeq 13.75$ Å, corresponding to an average volume per framework oxygen $V_{ox} \simeq 27.1$ Å^3. This is the maximum state of expansion of the leucite-type framework. The tetrahedral framework contains large voids with point symmetry 32 defined by twelve oxygen atoms all equidistant at about 3.50 Å from the center of the void. In addition to these large voids, there are smaller ones of point symmetry 222 which form the "windows" between pairs of the large voids.

The largest alkali ion, Cs^+, is not big enough to fill the large voids in such a way that it is in contact with all twelve surrounding oxygen atoms in the walls. Consequently, in pollucite, six of the twelve oxygen atoms are slightly closer to the Cs^+ ion (3.394 Å), while the other six Cs–O bonds are weaker (3.561 Å) (Beger 1969). This configuration is achieved by a slight rotation of the $[AlO_4]$ and $[SiO_4]$ tetrahedra.

As the cation radius decreases, the tetrahedral framework is increasingly distorted by further rotation of the $[TO_4]$ tetrahedra away from the most symmetrical orientation of the aristotype in order to provide better contacts between the cations and oxygen atoms. As a consequence, the space group symmetry is lowered to $I4_1/a$, a tetragonal subgroup of the cubic space group Ia3d of the aristotype. In leucite, $K[AlSi_2O_6]$, six of the K–O distances lie between 2.956 and 3.138 Å (average 3.014 Å) with the other six lying between 3.506 and 3.759 Å (average 3.659 Å), leading to an overall mean value of 3.337 Å (Mazzi et al. 1976).

Sodium ions do not occupy the large voids of the leucite-type framework since the rotation angles of the $[TO_4]$ tetrahedra which would be required to bring the small sodium into contact with six oxygen atoms are too large. Consequently, in analcime, the Na^+ ions occupy the smaller voids in the windows between the large voids (Mazzi and Galli 1978). The large voids are then filled with water molecules to give analcime the chemical composition $Na[AlSi_2O_6] \cdot H_2O$.

From $Cs[AlSi_2O_6]$ to $K[AlSi_2O_6]$ there is a steady decrease in the volume per framework oxygen, V_{ox}. The incorporation of water into the voids in analcime causes an increase in its value of V_{ox} (Table 10.26).

Table 10.26 Normal pressure silicates of composition $M[AlSi_2O_6]$

Composition	Name	Structure type	$r_M^{[6]}$ [Å] [a]	V_{ox} [Å^3] [b]	Mean T–O–T bond angle [°]	Reference
$Li[AlSi_2O_6]$-III	γ-Spodumene	Quartz	0.74	21.5	151.6	Li 1968
$Li[AlSi_2O_6]$-II	β-Spodumene	Keatite	0.74	21.7	149.4	Li and Peacor 1968
$Na[AlSi_2O_6] \cdot H_2O$	Analcime	Leucite	1.02	26.9	144.4	Mazzi and Galli 1978
$K[AlSi_2O_6]$	Leucite (lT)	Leucite	1.38	24.5	138.4	Mazzi et al. 1976
$Rb[AlSi_2O_6]$	Synthetic	Leucite	1.49	25.3	n. d. [c]	Martin and Lagache 1975
$Cs[AlSi_2O_6]$	Pollucite	Leucite	1.70	26.7	144.5	Martin and Lagache 1975; Beger 1969
$LiAl^{[6]}[Si_2O_6]$	Spodumene	Clinopyroxene	0.74	16.2	139.0	Clark et al. 1969
$NaAl^{[6]}[Si_2O_6]$	Jadeite	Clinopyroxene	1.02	16.7	139.1	Cameron et al. 1973

[a] $r_M^{[6]}$ = Cation radius for coordination number six;
[b] V_{ox} = Volume per oxygen atom;
[c] n. d. = Not determined.

Of the six tetrahedral frameworks listed in the upper part of Table 10.26, the structure of leucite, K[$AlSi_2O_6$], seems to be the most stable at ambient temperatures and pressures. This is indicated by its mean T–O–T bond angle of 138.4° which is very near to the value 140° corresponding to an unstrained Si–O–Si angle. As the cation radius increases from K^+ to Cs^+, the mean T–O–T angle of the leucite-type framework increases to 144.5° indicating a slight destabilization of the framework. The concurrent replacement of K^+ by the smaller Na^+ ion and the incorporation of water also increases the mean bond angle T–O–T.

The even smaller Li^+ ion would lead to a further destabilization of the leucite-type framework and, infact, no lithium aluminosilicate with such a framework has been reported. There are, instead, two known high temperature phases of Li[$AlSi_2O_6$] which have the denser frameworks of keatite and quartz (Li 1968; Li and Peacor 1968). However, both these phases are metastable, probably because their frameworks are rather strained due to the presence of large T–O–T angles of 149.4 and 151.6°.

The phase which is stable at ambient temperature and pressure is, therefore, the chain silicate spodumene, $LiAl^{[6]}[Si_2O_6]$, which has a T–O–T angle of 139.0° (Clark et al. 1969).

The favorable T–O–T angle of 139.1° in jadeite, $NaAl^{[6]}[Si_2O_6]$ (Cameron et al. 1973), the corresponding sodium clinopyroxene, explains why jadeite is the stable phase in the system $Na_2O-Al_2O_3-SiO_2$ rather than an anhydrous "analcime" Na[$AlSi_2O_6$].

10.7.5 Aluminosilicates M[$AlSiO_4$]

The influence of the cations on the tetrahedral framework is perhaps most obvious from the crystal structures of the series $LiAlSiO_4$, $NaAlSiO_4$, $KAlSiO_4$, $RbAlSiO_4$, and $CsAlSiO_4$. In all these phases, [AlO_4] and [SiO_4] tetrahedra share corners to build three-dimensional frameworks.

$LiAlSiO_4$ is known in two topologically different modifications: the low temperature form, eucryptite, is isostructural with phenakite, $Be_2^{[4]}[SiO_4]$, and willemite, $Zn_2^{[4]}[SiO_4]$ (Winkler 1953), wherein the beryllium and zinc atoms, respectively, are tetrahedrally coordinated along with the silicon atoms.

A structure refinement in progress (Hesse 1984a) indicates a statistical distribution of silicon and aluminum atoms, although Al/Si ordering within microdomains has not been ruled out completely. At 1245 K eucryptite transforms slowly to a high temperature phase, often called β-eucryptite. In this phase aluminum and silicon atoms occupy the Si positions of the quartz structure either in an ordered way or in a disordered way, depending on the conditions prevailing during crystallization of the material (Behruzi and Hahn 1971; Schulz and Tscherry 1972; Guth 1979).

$NaAlSiO_4$, $KAlSiO_4$, $RbAlSiO_4$, and $CsAlSiO_4$ all have structures containing zweier layers of [AlO_4] and [SiO_4] tetrahedra. The free corners of one-half of the tetrahedra point to one side of the layer, the other half to the other side, and the layers are linked to form three-dimensional frameworks via these free corners. The structures of these $MAlSiO_4$ phases differ in their directedness Δ, i.e., in the way in

Table 10.27 Normal pressure aluminosilicates of composition $M[AlSiO_4]$ [a]

Composition	Name	Structure type	V_{ox} [Å^3] [b]	Mean T−O−T bond angle [°]	Reference
$Li[AlSiO_4]$ (lT)	α-Eucryptite	Phenakite	19.9	130.8	Hesse 1984a
$Li[AlSiO_4]$ (hT)	β-Eucryptite	Quartz	22.3	148	Guth 1979
$Na[AlSiO_4]$	Synthetic	Beryllonite	21.9	130	Klaska 1974
$Na[AlSiO_4]$	Synthetic (Na-nepheline)	Tridymite	22.5	n.d. [c]	Donnay et al. 1959; Klaska 1974
$(Na_{1-x}K_x)[AlSiO_4]$	Nepheline	Tridymite	22.6−22.8	140	e.g. Dollase 1970
$Na[AlSiO_4]$	Carnegieite	Cristobalite	23.6	n.d.	Klingenberg et al. 1981
$K[AlSiO_4]$	Tetrakalsilite	?	24.3	n.d.	Benedetti et al. 1977
$K[AlSiO_4]$	Kalsilite	Tridymite	25.1	146	Perrotta and Smith 1965
$K[AlSiO_4]$	Kaliophilite-01	Kaliophilite	25.3	143	Gregorkiewitz 1980
$Rb[AlSiO_4]$	Synthetic	Icmm	26.9	149	Klaska and Jarchow 1975
$Cs[AlSiO_4]$	Synthetic	Icmm	28.5	157	Jarchow 1976, personal commun.

[a] Several additional polymorphs of $K[AlSiO_4]$ have been described, some of which may have new structure types;
[b] V_{ox} = Volume per oxygen atom;
[c] n.d. = Not determined

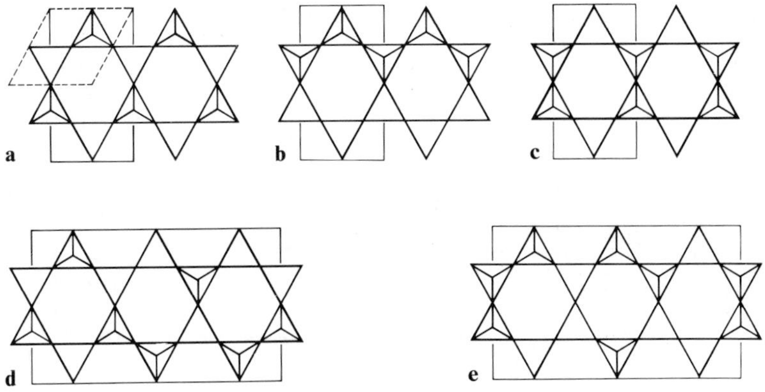

Fig. 10.38 a – e. The undistorted layers found in M[$AlSiO_4$] phases of the **a** tridymite-type; **b** Icmm-type; **c** Immm-type; **d** beryllonite-type; **e** kaliophilite-01.
For each layer one unit cell is indicated.

which the free corners of the tetrahedra point up and down. The five ways observed to date are called the beryllonite-type (after the mineral beryllonite $NaBePO_4$), the tridymite-type, the cristobalite-type, the kaliophilite-type, and the Icmm-type (Table 10.27). The layers which form these types of framework are represented somewhat schematically in Fig. 10.38, together with a layer that leads to a framework of symmetry Immm found in a metastable phase of anorthite composition, Ca[$Al_2Si_2O_8$] (see 10.7.3.2). Therefore, proceeding from $LiAlSiO_4$ to $CsAlSiO_4$ there is a sequence of at least seven different structure types: phenakite – quartz – beryllonite – tridymite – cristobalite – kaliophilite – Icmm.

This variation in structure type, which is in fact caused by the increase in cation size, does not, however, reflect the influence of cation size on the silicate anion in the same way as demonstrated for the chain silicates in Secs. 10.1 to 10.3 and the layer silicates in Secs. 10.5 and 10.6. In other words, in the $MAlSiO_4$ phases, the relatively small number of monovalent cations are no longer able to determine the chain type present in the respective framework. Otherwise we would expect zweier frameworks for $LiAlSiO_4$, instead of the observed phenakite-type structure of α-eucryptite (Hesse 1984a) and the dreier framework of β-eucryptite (Guth 1979), and we would expect dreier frameworks for $NaAlSiO_4$ and $KAlSiO_4$ instead of the observed zweier frameworks (Klaska 1974; Klingenberg et al. 1981; Perrotta and Smith 1965; Gregorkiewitz and Schäfer 1980). Furthermore, we would predict frameworks containing shortened zweier chains for $RbAlSiO_4$ and $CsAlSiO_4$ instead of the observed frameworks containing straight zweier chains (b_0 = 5.337 Å and 5.435 Å, respectively) (Klaska and Jarchow 1975; Jarchow 1976, personal communication). However, the [AlO_4] and [SiO_4] tetrahedra do form a framework in accordance with the rules given in Chap. 9, and the various cations can only impose limited alterations on the framework in order to satisfy their own "needs". These needs include contact with as many oxygen atoms as possible for their size and, for the small lithium ion, an oxygen coordination in accord with the small size of the cation and the valence angles required by the considerably covalent character of the Li – O bond.

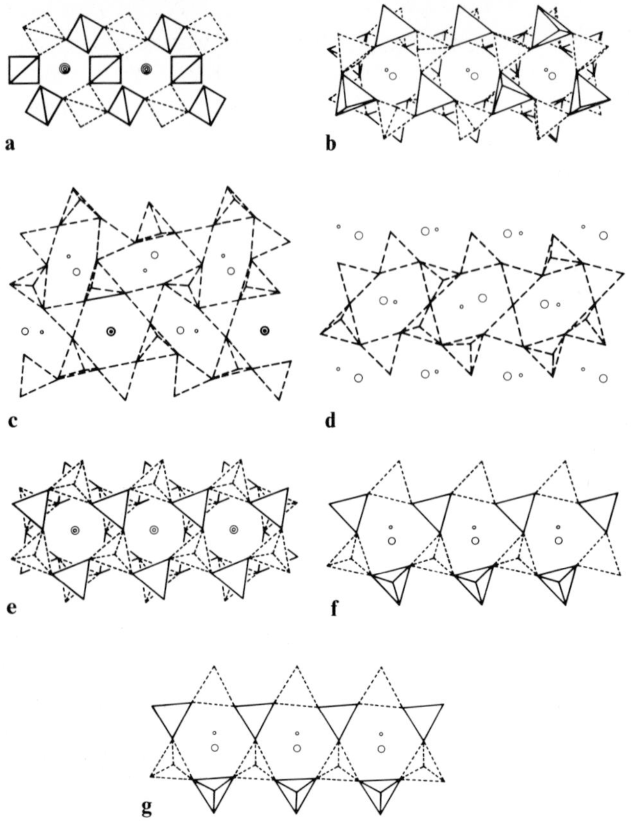

Fig. 10.39 a – g. Projections of the structures of several $M[AlSiO_4]$ phases showing the distorted layers. —— $[SiO_4]$ tetrahedra; - - - $[(Al,Si)O_4]$ tetrahedra; ···· $[AlO_4]$ tetrahedra. **a** β-eucryptite, $Li[AlSiO_4]$(hT) (projection of a slab only half a unit cell thick is shown); **b** synthetic beryllonite-type $Na[AlSiO_4]$; **c** nepheline, $Na_{1-x}K_x[AlSiO_4]$, $0 \leq x < 0.25$; **d** kaliophilite-01, $K[AlSiO_4]$; **e** kalsilite, $K[AlSiO_4]$; **f** synthetic $Rb[AlSiO_4]$; **g** synthetic $Cs[AlSiO_4]$

In this series of structures kalsilite, $K[AlSiO_4]$, with a tridymite-type framework, seems to possess the least strain since its average T–O–T angle (Table 10.27) has the smallest deviation from the value 142° suggested for an "unstrained" Si–O–Al angle[4]. The tetrahedral layers contain rings of trigonal symmetry with the bond angle (Al, Si)–O–(Al, Si) at the oxygen atoms linking adjacent layers having a higher value (163°) than the average of the six corresponding Si–O–Si angles (155.2°) in monoclinic low-tridymite, SiO_2 (Kato and Nukui 1976). The potassium ions fill the open spaces in the tridymite-type framework leading to a *stuffed derivative* of tridymite. These potassium ions fit neatly into the holes between two sechser rings and are symmetrically surrounded by nine oxygen atoms: three basal oxygen atoms each from the upper and lower sechser ring and three oxygen atoms which connect the two rings at the same level as the potassium ion (Fig. 10.39 e). The potassium–oxygen distances vary between 2.77 and 2.99 Å with a mean value of 2.90 Å. The larger rubidium ions in synthetic $Rb[AlSiO_4]$

4 The "unstrained" $Si^{[4]}$–O–$Al^{[4]}$ bond angle is slightly larger – perhaps by about 2° – than the "unstrained" $Si^{[4]}$–O–$Si^{[4]}$ angle, assumed to be equal to the grand mean value of 140° for the Si–O–Si angles in all silicates (see Sec. 3.1.2.5).

require more space than the smaller potassium ions. This additional space cannot be achieved merely by widening the tridymite-like tetrahedral framework, but instead requires a change in topology. This change consists only of a change of directedness of half the [TO_4] tetrahedra in the layer. As a result, the zweier chains with all tetrahedra pointing up alternate with zweier chains with all tetrahedra pointing down, reducing the symmetry of the framework from hexagonal to orthorhombic (Fig. 10.39f).

In addition, the tetrahedra which are superimposed in the projection in Fig. 10.39 have a staggered orientation in kalsilite, but an eclipsed one in Rb[$AlSiO_4$]. For these reasons, together with the fact that the two symmetrically nonequivalent O···O···O angles between the basal oxygen atoms of a sechser ring are slightly more similar in the rubidium than in the potassium and sodium silicates, there is an increase in the size of the sites available for the cations. Therefore, in this Icmm-type framework the Rb^+ ions have eleven neighbors in the range 2.91 to 3.54 Å, with a mean value of 3.27 Å compared to the nine neighbors of K^+ in kalsilite. However, the accommodation of the larger Rb^+ ion is achieved at the expense of an increase in the T−O−T bond angle between the layers from 163° in K[$AlSiO_4$] to 179.3° in Rb[$AlSiO_4$], a value that is energetically very unfavorable.

The framework with Icmm topology is able to accommodate even the very large cesium ion by widening the sechser rings to nearly the maximum value, at which point the O···O···O angles between the basal oxygen atoms of a sechser ring become equal (Fig. 10.39g). This brings the Cs^+ cations into contact with 13 oxygen atoms at distances between 3.12 Å and 3.76 Å, with a mean value of 3.46 Å. In addition to the formation of a large bond angle of 174° at the oxygen atoms which link adjacent zweier layers, the resultant adjustment of the framework increases the bond angle at the oxygen atoms that link adjacent zweier chains within one layer from 151° in the rubidium compound to 177° in the cesium compound. This is a clear indication of the high degree of tension which the cesium cation generates in the framework as a consequence of its large size and electropositivity.

In contrast, smaller cations than potassium have difficulty in achieving proper contacts with oxygen atoms on all sides in the tridymite-type or Icmm-type framework. Although it is possible to synthesize Na[$AlSiO_4$] with the tridymite structure (Donnay et al. 1959), natural nephelines always show considerable replacement of sodium by potassium in accordance with the formula $Na_{1-x}K_x$[$AlSiO_4$]. From the structure of a plutonic nepheline with $x \simeq 0.25$, shown in Fig. 10.39c, it can be seen that there are two kinds of sechser rings in the structure instead of one in kalsilite. The larger potassium ions are accommodated in the more regular rings, and are more or less symmetrically coordinated by nine oxygen atoms at distances ranging from 2.96 Å to 3.06 Å. The sodium ions, on the other hand, are accommodated in highly distorted rings, and are irregularly coordinated by seven or eight oxygen atoms with bond distances in the much larger range of 2.51 Å to 2.86 Å (average 2.62 Å). Unfortunately, no crystal structure analysis of pure sodium nepheline has yet been completed, so it is not known how the tridymite-type framework responds to the small ions in the larger cavities of this structure.

In synthetic Na[$AlSiO_4$] with the beryllonite-type framework (Fig. 10.39b), the adjustment between the cations and the tetrahedral framework again requires an inversion of the orientation of part of the $[TO_4]$ tetrahedra relative to the tridymite-type. This produces two kinds of sechser rings: one with the apices of adjacent tetrahedra alternately pointing up and down, that is, with the same sequence of directedness UDUDUD as in tridymite, and the other type with a sequence UUDUDD. While the more symmetrical channels of the first kind accommodate sodium ions which have three oxygen neighbors at 2.44 Å to 2.50 Å and five more distant ones between 2.88 Å and 3.10 Å (average 2.84 Å), the sodium ions in the channels of the second kind are coordinated by six oxygen atoms between 2.34 Å and 2.88 Å (average 2.53 Å). Not only is the coordination of the sodium ions far from normal, as these values show, but there is also a considerable degree of strain within the tetrahedral framework. This is evident from the fact that the mean value (129.9°) of the T–O–T angle in this structure is considerably lower than the grand mean value of 140° reported for Si–O–Si bonds in Sec. 3.1.2.5.

The small lithium ions are either octahedrally or tetrahedrally coordinated by six or four oxygen ions, respectively, the tetrahedral coordination being more common. In fact, both known polymorphs of $LiAlSiO_4$ contain four-coordinated lithium. Such close cation–oxygen contact cannot be achieved by further distortion of channels in a tetrahedral framework built from zweier layers as in the case of the $MAlSiO_4$ phases with larger M^+ ions. Thus, the phenakite- and quartz-type structures of α- and β-eucryptite have no pronounced layer-like arrangement of tetrahedra. The mean T–O–T angles of 130.8° and 148.3°, respectively (Table 10.27), indicate that both frameworks are considerably strained.

The contraction of the tetrahedral framework by small cations and the dilation by large ones is demonstrated in Table 10.27 for this series of silicates, along with the average volume V_{ox} available to an oxygen atom in the structure.

10.7.6 Comparison of Framework Silicates

The topologically distinct normal pressure phases[5] in the systems SiO_2-MAlO_2 (M = Li, Na, K, Rb, Cs) are surveyed in Table 10.28 together with their mean T–O–T angles and volume per framework oxygen atom, V_{ox}.

Perhaps the most striking feature of this Table is that it contains four polymorphs of SiO_2, with three of these structure types (quartz, cristobalite, and tridymite), occurring in several topologically identical phases.

This indicates that for SiO_2 it is impossible for one structure type to fulfill all the structural requirements of:

1) ideal bond lengths $d(Si-O) = 1.605$ Å;
2) ideal O–Si–O bond angles of 109.5°;
3) strain-free Si–O–Si bond angles of 140°;
4) good space filling with as many unstrained bonds per unit volume as possible (in order to achieve as much free energy per unit volume as possible): in an ionic oxide this would mean $V_{ox} \simeq 15$ Å^3.

5 Topologically identical phases are those which have the same topology of bonds between neighboring atoms. In this sense, high- and low-quartz are topologically identical phases, whereas quartz and cristobalite are topologically different phases.

Table 10.28 Structure types in the system SiO_2-MAlO_2 (M = Li, Na, K, Rb, Cs) together with their volumes per oxygen atom, V_{ox}, and mean bond angles ⟨∢ T–O–T⟩

	–	Li (0.74 Å)	Na (1.02 Å)	K (1.38 Å)	Rb (1.49 Å)	Cs (1.70 Å)	Molecules
SiO_2	Qz 18.8 143.6 Kt 20.0 153 Cr 21.4 146.4 Tr 22.1 150						Mel 26.4 168.8 D3C 26.9 174.8 D1H 27.1 170.6
$M[AlSi_3O_8]$		Qz 20.7 n.d. Kt n.d. n.d.	Fs 20.8 140.5	Fs 22.5 141.9	Fs 23.3 142.8	–	
$M[AlSi_2O_6]$		Px 16.2 139.0 Qz 21.5 151.6 Kt 21.7 149.4	Px 16.7 139.1 Leu 26.9 144.4	Leu 24.5 138.4	Leu 25.3 n.d.	Leu 26.7 144.5	
$M[AlSiO_4]$		Ph 19.9 130.8 Qz 22.3 148	Ber 21.9 130 Cr 23.3 n.d. Tr 22.5 ~140	Tr ~ 25 146 Kp 25.3 143	Icmm 26.9 149	Icmm 28.5 159	

n.d. = Not determined; Qz = quartz-type; Kt = keatite-type; Cr = cristobalite-type; Tr = tridymite-type; Px = pyroxene-type; Fs = feldspar-type; Leu = leucite-type; Ph = phenakite-type; Ber = beryllonite-type; Kp = kaliophilite-type; Icmm = Icmm-type; Mel = melanophlogite (Gies 1983; Gies et al. 1982); D3C = dodecasil 3C (Gies 1984); D1H = dodecasil 1H (Gerke and Gies 1984). For further references see Tables 10.23, 10.26, and 10.27

Key:

Cation ($r^{[6]}$ [Å])
Structure type V_{ox} [$Å^3$] ⟨∢ T–O–T⟩ [°]

The high covalency of the Si–O bond does not allow large deviations from the ideal values of the bond lengths and bond angles, in particular the rigid O–Si–O angle. The structures adopted by silica are, therefore, compromises between the above various structural requirements. Of the four requirements mentioned, the first two are so strong in SiO_2 that only small deviations from the ideal values are possible (see Sec. 3.1). Somewhat larger deviations from the ideal are permitted in the case of the Si–O–Si angle and, in particular, V_{ox}.

From a comparison in the various silica frameworks of the deviations $\Delta Si{-}O{-}Si = |\langle \sphericalangle Si{-}O{-}Si \rangle_{obs} - 140°|$ of the Si–O–Si angles from 140°, and $\Delta V_{ox} = (V_{ox})_{obs} - 15$ Å³ of the V_{ox} values from the value for a densely packed oxygen ion arrangement, it is clear that quartz is expected to be the energetically most favorable polymorph of SiO_2 because the values $\Delta Si{-}O{-}Si = 3.6°$ and $\Delta V_{ox} = 3.8$ Å³ are the lowest for this phase (Table 10.28). Cristobalite and tridymite, respectively, are increasingly less stable, but the position of keatite in this series is uncertain because the observed value of its average Si–O–Si angle is rather inaccurate.

The silica frameworks of the three clathrasils (see Sec. 10.7.1) are very unstable due to very large values of $\Delta Si{-}O{-}Si$ and ΔV_{ox}. However, when the large cages in their frameworks are filled with neutral molecules, they become thermodynamically stable due to the additional energy gained by the van der Waals bonds between these molecules and the framework atoms.

Coupled replacement of Si^{+4} by $Li^{+} + Al^{+3}$ leads to an increase in $\Delta T{-}O{-}T$ to 11.6° and 8° in the quartz-type phases $Li[AlSi_2O_6]$-III and $Li[AlSiO_4]$ indicating that the tetrahedral framework becomes considerably destabilized. On the other hand, in the keatite-type, such replacement produces a decrease in $\Delta T{-}O{-}T$ which is not compensated by a considerable increase of V_{ox}. This suggests that in contrast to the quartz structure, the keatite-type framework is stabilized by Si/Li+Al substitution. However, for both frameworks the mean T–O–T angles approach 150° and, as a result, spodumene with a single chain, rather than a framework structure, is formed. This arises because in the spodumene single chain, the $[SiO_4]$ tetrahedra have such a high degree of rotational freedom that essentially unstrained Si–O–Si angles ($\Delta T{-}O{-}T = 1°$) can form. This angular contribution to the stabilization of the pyroxene-type $LiAl[Si_2O_6]$ is augmented by a density increase ($\Delta V_{ox} = 1.2$ Å³ compared with $\Delta V_{ox} = 3.8$ Å³ and 5.0 Å³ for quartz and keatite, respectively).

The frameworks of cristobalite and tridymite have voids large enough to enclose one Na^{+}, i.e., one cation per two $[TO_4]$ tetrahedra, without substantial expansion. In the case of the tridymite framework a 50% replacement of Si by Na+Al instead of Li+Al leads to a considerable stabilization due to a decrease in $\Delta T{-}O{-}T$ from 10° to 0°. Even in $K[AlSiO_4]$, which has the tridymite framework, the $\Delta T{-}O{-}T$ value is 4° lower than in tridymite itself. However, the corresponding angular contribution to the stabilization is roughly balanced by an increase in ΔV_{ox} from 7.1 to 10 Å³.

The cations Rb^{+} and Cs^{+} are too large to fit into the cavities of the frameworks of any of the silica polymorphs. However, a 50% substitution of Si by the slightly larger Al cation leads to the formation of a new framework type of Icmm symmetry containing cavities just large enough to accommodate these large monovalent cations. The high values of ΔT−O−T (9° for Rb[$AlSiO_4$] and 19° for Cs[$AlSiO_4$]), and ΔV_{ox} (11.9 Å^3 and 13.5 Å^3, respectively), suggest that these compounds are not particularly stable.

Alkali cations larger than Li^+ are too large to fit into the small voids of the quartz and keatite frameworks. It has already been mentioned that Na^+ and K^+ stabilize the tridymite and/or the cristobalite framework, provided that there is one alkali ion per two tetrahedra (50% Si/M+Al replacement) to fill all the suitable voids available. For smaller replacement – 33% in M[$AlSi_2O_6$] and 25% in M[$AlSi_3O_8$] – too many of the voids remain empty for the structure to be stable. Therefore, two new framework types are formed, one with a large cavity per three tetrahedra – the leucite-type – and another with a large cavity per four tetrahedra – the feldspar-type. There are at least two reasons why these two framework types have such a remarkable stability:

1) they both have T−O−T angles very near to 140°, and
2) their cavities can adjust to the size of the cations from Na^+ to Rb^+ (feldspar) and Cs^+ (leucite) without a substantial increase in V_{ox}.

ΔT−O−T and ΔV_{ox} do not represent the only contributions to the stability of a particular framework type for a given composition; corresponding deviations $\Delta d(\mathrm{T-O}) = |\langle d(\mathrm{T-O})\rangle_{obs} - d(\mathrm{T-O})_{ideal}|$ and $\Delta \mathrm{O-T-O} = |\langle \sphericalangle \mathrm{O-T-O}\rangle_{obs} - 109.5°|$ supply similar contributions and would also have to be considered in a more detailed discussion. However, much more energy is required to produce, let's say, a 10% deviation from the ideal values of d(T−O) and ∢O−T−O than a 10% deviation from the ideal values of ∢T−O−T and V_{ox}. The relatively small Δd(T−O) and ΔO−T−O values necessary to produce a given level of destabilization are, therefore, of the same order of magnitude as the accuracy of the observed d(T−O) and ∢O−T−O values in most of our present crystal structure determinations. It is the much larger relative deviations of ∢T−O−T and V_{ox} from their ideal values, both of which are well outside the limits of error, that make ΔT−O−T and ΔV_{ox} suitable indicators for the study of the relative stabilities of tectosilicates.

Table 10.28 clearly shows the influence of monovalent cations on the TO_2 frameworks. With the exception of Li[$AlSi_2O_6$] and Na[$AlSiO_4$], only one or two topologically distinct normal pressure polymorphs are observed. From Table 10.29 it is evident that the situation is quite different for M[$Al_2Si_2O_8$] phases with divalent cations. Mg[$Al_2Si_2O_8$] exists in the high-quartz structure type only, but the corresponding aluminosilicates of Ca, Sr, and Ba each have three topologically different phases at normal pressure. In each case the feldspar-type phase has the least strained silicate anion (smallest ΔT−O−T) of the three polymorphs. In Ca[$Al_2Si_2O_8$] the low ΔT−O−T value of 2.9° for the feldspar anorthite relative to values of 6° and 6.4° for the hexacelsian-type and the Immm-type, respectively, probably explains why anorthite is the stable Ca[$Al_2Si_2O_8$] polymorph, despite the fact that it has a slightly lower density than the other two polymorphs.

The correlation between ΔV_{ox}, ΔT−O−T, and the relative stability of the tetrahedral frameworks in SiO_2 and alkali and akaline earth aluminosilicates is indicated in Fig. 10.40 which is a summary of the data given in Tables 10.28 and 10.29. The pyroxene phases of $LiAlSi_2O_6$ and $NaAlSi_2O_6$ have the most stable

Table 10.29 Comparison of the structure types of SiO_2 and the aluminosilicates $M^{+2}[Al_2Si_2O_8]$, (M = Be, Mg, Ca, Sr, Ba), and $M^{+}[AlSi_3O_8]$, (M = Li, Na, K, Rb, Cs)

<table>
<tr><td colspan="4">–</td><td>Be
(~ 0.42 Å)</td><td colspan="2">Mg
(0.72 Å)</td><td colspan="3">Ca
(1.00 Å)</td><td colspan="3">Sr
(1.16 Å)</td><td colspan="3">Ba
(1.36 Å)</td><td></td><td></td></tr>
<tr><td rowspan="2">Qz
18.8
143.6</td><td rowspan="2">Kt
20.0
153</td><td rowspan="2">Cr
21.4
146.4</td><td rowspan="2">Tr
22.1
150</td><td>–</td><td colspan="2">Qz
21.1
n. d.</td><td>Fs
20.9
142.9</td><td>Hc
20.7
134</td><td>Immm
21.4
133.6</td><td>Pc
21.7
128.7</td><td>Fs
21.9
136.9</td><td>Hc
22.2
n. d.</td><td>Pc
23.3
133.2</td><td>Fs
23.0
138.4</td><td>Hc
23.6
132</td><td></td><td></td></tr>
<tr><td></td><td>Qz
20.7
n. d.</td><td>Kt
n. d.
n. d.</td><td colspan="3">Fs
20.8
140.5</td><td colspan="3"></td><td colspan="3">Fs
22.5
141.9</td><td>Fs
23.3
142.8</td><td>–</td></tr>
<tr><td colspan="4">–</td><td></td><td colspan="2">Li
(0.74 Å)</td><td colspan="3">Na
(1.02 Å)</td><td colspan="3"></td><td colspan="3">K
(1.38 Å)</td><td>Rb
(1.49 Å)</td><td>Cs
(1.70 Å)</td></tr>
</table>

V_{ox} = Volume per oxygen atom; n. d. = not determined; Hc = hexacelsian-type; Pc = paracelsian-type; Immm = Immm-type. The other types are explained in Table 10.28. For references see Tables 10.23 and 10.25

Key:

Cation ($r^{[6]}$ [Å])
Structure type V_{ox} [$Å^3$] ⟨∢ T–O–T⟩ [°]

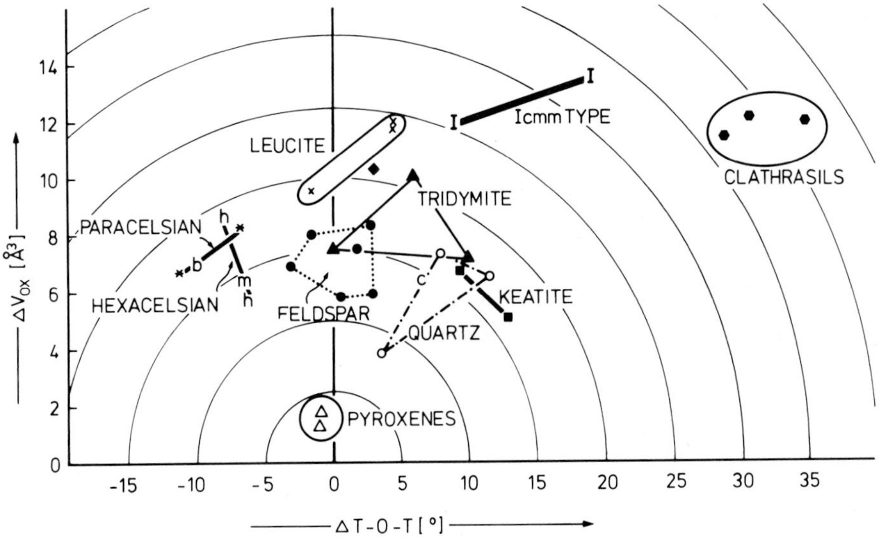

Fig. 10.40. Correlation between the oxygen density given as $\Delta V_{ox} = V_{ox} - 15$ Å^3, the angular strain given as $\Delta T-O-T = \langle \sphericalangle T-O-T \rangle_{obs} - 140°$, and the relative stability of the tetrahedral frameworks and chains of silica and Al containing alkali and alkaline earth silicates. The frameworks and chains are the less stable, the larger the distance from the point $\Delta V_{ox} = 0$, $\Delta T-O-T = 0$ is

△	pyroxenes	●	feldspars	◆	kaliophilite
⬣	clathrasils	■	keatite-type	▲	tridymite-type
○	quartz-type	c	cristobalite	b	beryllonite-type
×	leucite-type	*	paracelsian-type	h	hexacelsian-type
m	Immm-type	I	Icmm-type		

silicate anions since they have the lowest ΔV_{ox} and $\Delta T-O-T$ values. The stability of the silicate anions decreases as the distance of their data points from the point (0, 0), that is, from $V_{ox} = 15$ Å^3 and $\sphericalangle T-O-T = 140°$, increases. Figure 10.40 shows that the stability of the more common anions decreases in the order: pyroxenes – feldspar, quartz – keatite, tridymite, hexacelsian, paracelsian – leucite – Icmm-type – clathrasils.

The larger the cations M (or the guest molecules of the clathrasils), the more the tetrahedral frameworks expand, i.e., the larger the value of ΔV_{ox} becomes. This expansion is of course higher for two monovalent cations M^+ than for one divalent M^{+2}. For divalent cations the expansion is generally accompanied by a narrowing of the average T–O–T angle (Sr and Ba feldspars, paracelsians, hexacelsians, Immm-type). For monovalent cations these angles are in general somewhat larger than 140° (alkali feldspars, leucites, derivatives of the quartz, tridymite, keatite, and Icmm-type), whereas the mean T–O–T angles are much larger in clathrasils with neutral guest molecules.

11 Conclusion

11.1 Correlation Between Classification Parameters and Cation Properties

In the main part of this book it has been shown that the silicates, more than any other group of chemical compounds with the exception of the organic carbon compounds, exhibit an enormous chemical and structural variability. In order to demonstrate this variability, a large number of silicates with different structures were described. Using these silicates as examples, it has been shown that the atomic structure of a silicate is controlled mainly by its chemical composition and to a lesser extent, by temperature and pressure. To obtain a better understanding of the large number of observed silicates, a classification has been described and used which is based on crystal chemical principles and, therefore, reflects structural relations between the silicates.

Nevertheless, the large number of examples treated has probably concealed some of the correlations between structure and chemical composition developed in the main part of the book. It may, therefore, be advisable to give a more condensed summary of these correlations. The parameters used for the crystal chemical classification of silicates provide a suitable basis for such a summary.

(1) Coordination Number of Silicon, CN. For silicon compounds of general formula $M_rSi_sA_t$ containing Si–A–M bonds, the tendency of silicon to have a coordination number $\boldsymbol{CN} > 4$ is higher the more electronegative the nearest neighbors A are. For very high values of χ_A ($\chi_F = 4.10$) octahedral coordination $[SiA_6]$ is favored, while for lower χ_A (e.g., $\chi_C = 2.50$) tetrahedral $[SiA_4]$ groups are formed. Oxygen, with its intermediate value of electronegativity ($\chi_O = 3.50$), can have $[SiO_6]$ as well as $[SiO_4]$ polyhedra, although it has a preference for the latter.

For a given ligand A of silicon, such as oxygen, the influence of the next-nearest neighbors M of silicon is such that highly electronegative M atoms bring about a slight increase in silicon's tendency for $\boldsymbol{CN} > 4$, whereas strongly electropositive cations M favor $\boldsymbol{CN} = 4$.

Restricting the future discussion to compounds with oxygen as nearest neighbor of silicon, the covalency of the Si–O bond increases as the value of χ_M decreases. As a consequence, the $Si^{[6]}$–O bond in $[SiO_6]$ groups is less covalent than the $Si^{[4]}$–O bond in $[SiO_4]$ tetrahedra, and the $Si^{[4]}$–O bond is less covalent if the next-nearest silicon neighbors M are highly electronegative than if they are electropositive cations.

For an isolated $[SiO_6]$ group, the average dimensions are $\langle d(Si^{[6]}-O)\rangle = 1.77$ Å, $\langle d(O\cdots O)\rangle = 2.50$ Å, $\langle \sphericalangle O-Si^{[6]}-O\rangle = 90°$ and $180°$. The average dimensions of an isolated $[SiO_4]$ group are $\langle d(Si^{[4]}-O)\rangle = 1.62$ Å, $\langle d(O\cdots O)\rangle = 2.64$ Å, $\langle \measuredangle O-Si^{[4]}-O\rangle = 109.47°$, and $\langle \measuredangle Si^{[4]}-O-Si^{[4]}\rangle = 140°$.

These values represent the dimensions of unstrained $[SiO_n]$ groups. The energy E required for say, a 5% deviation from these values decreases in the order: $E_{d(Si-O)} > E_{d(O\cdots O)} \approx E_{\measuredangle O-Si-O} > E_{\measuredangle Si-O-Si}$ and is higher for tetrahedral coordination of silicon than for octahedral Si due to the higher covalency of the $Si^{[4]}-O$ bond.

Silicate anions tend to come as near as possible to these "unstrained" dimensions, but there are a number of *stress factors* which cause deviations from the ideal values:

1) increasing degree of condensation of the $[SiO_n]$ polyhedra;
2) cation properties, in particular:
   a) the cation electronegativity χ_M causes slight changes in the covalency of the Si–O bond as well as changes in the effective charge per $[SiO_n]$ polyhedron and, consequently, changes in the repulsive forces between the polyhedra.
   b) The cation valence v_M has a direct influence on the atomic ratio $n_M : n_{Si}$ between cations and silicon, and an indirect influence on the ratio $n_M : n_O$. Because each terminal oxygen atom tends to be in contact with one or more cations M, and since each cation tends to be regularly and densely surrounded by oxygen atoms, the valence of the cations has a strong influence on the structure of the silicate anions.
   c) The cation radius r_M is primarily responsible for determining the size of the coordination polyhedron of the cation and exerts a strong influence on its coordination number. If there are no oxygen atoms available in a silicate other than those which are bonded to silicon, the oxygen atoms are both part of the silicate anion and part of the cation–oxygen polyhedra. Competition between silicon and the cations M for the available bonding space around the oxygen atoms means that compromises are necessary between the steric requirements of the silicate anions and the cation–oxygen polyhedra.
3) An increase in temperature causes an increase in the atomic thermal vibration amplitudes and, therefore, an increase in the effective cation radii. This increase is higher the larger the cation–oxygen polyhedra are and, in particular, is higher for $[MO_n]$ than for $[SiO_n]$ polyhedra. The resulting differential thermal expansion of $[MO_n]$ and $[SiO_n]$ polyhedra means that the mutual adjustment between the cationic part of the structure and the silicate anions varies with temperature.
4) An increase in pressure leads to a differential compression of $[MO_n]$ and $[SiO_n]$ polyhedra and, therefore, structural changes which are inverse to the changes brought about by an increase in temperature.

In silicates containing $[SiO_4]$ groups the structural changes are:

a) rotations of the $[SiO_4]$ tetrahedra and, to a lesser extent, the $[MO_n]$ polyhedra;
b) angular distortions of the $[MO_n]$ polyhedra and, to a lesser extent, the $[SiO_4]$ tetrahedra;
c) bond length distortions of the $[MO_n]$ polyhedra and, to lesser extent, the $[SiO_4]$ tetrahedra.

The amount of stress exerted on a particular silicate anion by the "stress factors" 1) to 4) can be roughly estimated from the magnitude of the deviations of the bond lengths $d(Si-O)$ and bond angles $\sphericalangle O-Si-O$ and $\sphericalangle Si-O-Si$ from the values for unstressed tetrahedra.

(2) Linkedness of [SiO_n] Groups, L. An increase in the linkedness of an [SiO_n] group from $L=0$ to $L=3$, i.e., from an isolated [SiO_n] polyhedron containing no oxygen atom in common with others to one which shares three oxygen atoms with another [SiO_n] polyhedron (face sharing), corresponds to a large decrease in the Si···Si distance between silicon atoms of neighboring [SiO_n] polyhedra. This decrease is even more drastic for [SiO_4] tetrahedra than for [SiO_6] octahedra. As a consequence, the relative stability of silicate anions decreases in the order:

isolated – corner-sharing – edge-sharing – face-sharing polyhedra,

and the decrease is larger for silicates with tetrahedrally than for silicates with octahedrally coordinated silicon.

Since highly electronegative cations lead to a lower effective charge per [SiO_n] polyhedron, an increase in χ_M favors a higher degree of linkedness of the [SiO_n] groups.

(3) Connectedness of [SiO_n] Polyhedra, s. As the connectedness s increases, i.e., as the number of [SiO_n] polyhedra linked to one [SiO_n] polyhedron increases, the average Si···Si distance decreases. The corresponding increase in the repulsive forces between the silicon atoms causes a destabilization of silicate anions containing [SiO_n] groups of higher connectedness. This destabilization is partly balanced by a decrease in the formal charge per [SiO_n] polyhedron. For a given atomic ratio Si : O of the silicate anion, highly electronegative cations will favor anions with higher values of s.

(4) Terminated and Cyclic Anions, t and r. Cyclic silicate anions, especially those with low values of the ring periodicity P^r, in general have lower average Si···Si distances than terminated anions (oligosilicate anions). Consequently, terminated silicate anions are energetically more favorable than cyclic ones, particularly when they occur in silicates of highly electropositive cations.

(5) Branchedness of Silicate Anions, B. Branched silicate anions, especially loop-branched ones, in general have shorter average Si···Si distances and are, therefore, less stable than unbranched ones. However, cations with high χ_M values reduce the repulsive forces between the [SiO_n] polyhedra and, therefore, have a stabilizing influence.

This stabilizing influence of highly electronegative cations is less pronounced in silicates with high Si : O ratios, i.e., in those with higher dimensionality of the silicate anions, because of the resultant lower effective charges per [SiO_n] polyhedron. The stress exerted on branched silicate anions by the relatively strong repulsive forces can also be reduced by the presence of non-silicate anions, such as O^{-2}, OH^-, F^-, etc. and/or water molecules in the cation–oxygen polyhedra, and by the so-called soft cations Na^+, K^+, Ca^{+2}, and Ba^{+2} which make weaker demands on the silicate anion than the less polarizable cations, such as Li^+, Mg^{+2}, Al^{+3}, etc.

(6) Number of Different Silicate Anions, N_{an}. The general principle of parsimony for a given composition implies that a system tends to keep the number of geometrically and chemically distinct groups as small as possible. This means that silicates with only one kind of silicate anion ($N_{an} = 1$) are, in general, energetically more favorable than those with two or more different silicate anions. Crystalline mixed-anion silicates with $N_{an} = 2$ are, therefore, rather rare, and those with $N_{an} \geq 3$ have not been observed as yet. By analogy with the branched silicates, mixed-anion silicates are favored in the presence of strongly electronegative cations, soft cations, and additional non-silicate anions and/or water. This is evident from the survey of crystalline mixed-anion silicates given in Table 11.1.

(7) Multiplicity of Silicate Anions, M. The multiplicity of a silicate anion is strongly correlated with its Si : O ratio and is, therefore, mainly controlled by the $SiO_2 : M_rO$ ratio of the material from which the silicate is formed. From the data compiled in Table 7.10, it is clear that the number of silicates decreases rapidly as the multiplicity of their anions increases. In fact, this decrease is evident within each class of silicates having different values of dimensionality, i.e., for the oligosilicates, cyclosilicates, and the chain and layer silicates, and it clearly expresses the periodic character of the crystal chemical system of silicates. The oligosilicates may be used to attempt a crystallochemical explanation of this observation.

For this explanation it is assumed that the number of oligosilicates observed within a group with constant $\boldsymbol{M}$ is positively correlated with the relative stability of the corresponding type of multiple tetrahedra. This assumption is supported by the observation that the absolute numbers of natural oligosilicates and their abundance also decrease as $\boldsymbol{M}$ increases.

Figure 11.1 gives a schematic representation of a number of unbranched multiple tetrahedra and an unbranched single chain. For oligosilicates (see Sec. 7.2.2.1.1) the number of chemically nonequivalent $[SiO_4]$ tetrahedra increases from 1 in the monosilicates ($\boldsymbol{M} = 1$) and the double tetrahedra $[Si_2O_7]$ of the disilicates

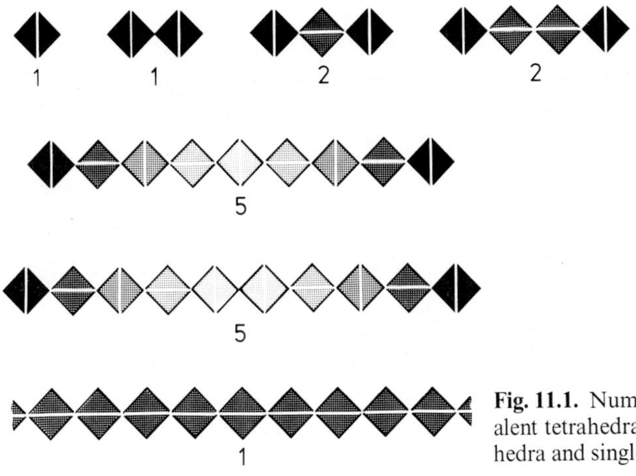

Fig. 11.1. Number of chemically nonequivalent tetrahedra in unbranched multiple tetrahedra and single chains

Table 11.1 Survey of mixed-anion silicates

Name	Formula						Ref.
	"Soft" cations	Electronegative cations	"Hard" cations	Silicate anions		Additional ligands	
				A_1	A_2		
$\boldsymbol{D}(A_1)=0,\ \boldsymbol{D}(A_2)=0$							
Ganomalite	Ca_2	Pb_3		$[SiO_4]$	$[Si_2O_7]$		[1]
Synthetic		Pb_3Cd_2		$[SiO_4]$	$[Si_2O_7]$		[2]
Synthetic	K_2	Be_2Zn_2		$[SiO_4]$	$[Si_2O_7]$		[3]
Synthetic		Pb_{11}		$[SiO_4]$	$[Si_2O_7]$	O_6	[4]
Zoisite, clinozoisite, piemontite, epidote	Ca_2	$(Al, Fe, Mn)_3$		$[SiO_4]$	$[Si_2O_7]$	$O(OH)$	[5]
Dellaite $(Y-C_6S_3H)$	Ca_6			$[SiO_4]$	$[Si_2O_7]$	$(OH)_2$	[6]
Macfallite	Ca_2	$(Mn, Al)_3$		$[SiO_4]$	$[Si_2O_7]$	$(OH)_3$	[7]
Sursassite	Mn_2	Al_3		$[SiO_4]$	$[Si_2O_7]$	$(OH)_3$	[8]
Rustumite	Ca_{10}			$[SiO_4]$	$[Si_2O_7]_2$	$(OH)_2Cl_2$	[9]
Queitite		Pb_4Zn_2		$[SiO_4]$	$[Si_2O_7]$	$[SO_4]$	[10]
Synthetic	K_2	Mn_2Zn_4		$[SiO_4]_2$	$[Si_2O_7]$		[11]
Synthetic	Ba_6	H	Li_2	$[SiO_4]_2$	$[Si_2O_7]$	Cl	[12]
Vesuvianite	Ca_{10}	$Al_4(Fe,$	$Mg)_2$	$[SiO_4]_5$	$[Si_2O_7]_2$	$(OH)_4$	[13]
Kilchoanite	Ca_6			$[SiO_4]$	$[Si_3O_{10}]$		[14]
Synthetic	Ho_4			$[SiO_4]$	$[Si_3O_{10}]$		[15]
Ardennite		$Mn_4(Al,$	$Mg)_6$	$[SiO_4]$	$[Si_3O_{10}]$	$[(As, V)O_4](OH)_6$	[16]
Synthetic	BaY_4			$[Si_2O_7]$	$[Si_3O_{10}]$		[17]
Synthetic	$BaGd_4$			$[Si_2O_7]$	$[Si_3O_{10}]$		[18]
Synthetic	Er_4	Pb		$[Si_2O_7]$	$[Si_3O_{10}]$		[19]
Traskite	$(Ca, Sr)Ba_{24}$	$(Mn, Fe, Ti)_4(Ti, Fe, Al,$	$Mg)_{12}$	$[Si_2O_7]_6$	$\{\boldsymbol{r}\}[Si_{12}O_{36}]$	$(O, OH)_{30}Cl_6 \cdot 14H_2O$	[20]
Eudialyte	$Na_{12}(Ca, RE)_6$	$(Zr, Nb)_{3+x}(Fe, Mn,$	$Mg)_3$	$\{\boldsymbol{r}\}[Si_3O_9]_2$	$\{\boldsymbol{r}\}[Si_9(O, OH)_{27}]_2$	Cl_y	[21]

$\boldsymbol{D}(A_1) = 0$, $\boldsymbol{D}(A_2) = 1$							
Synthetic	Ca_4		Li_2	$[SiO_4]$	$\{1^{1}_{\infty}\}[^{3}Si_3O_9]$		[22]
Joesmithite	$Ca_2(Ca,$	$Pb)Be_2(Fe,$	$Mg)_5$	$[SiO_4]_2$	$\{1^{1}_{\infty}\}[^{2}Si_2O_6]_2$	$O_2(OH)_2$	[23]
Miserite	$K_2Ca_{10}(Y,RE)_2$			$[Si_2O_7]_2$	$\{4^{1}_{\infty}\}[^{3}Si_{12}O_{30}]$	$(OH)_2F_2$	[24]
Bavenite	Ca_4	Be_2Al_2		$[Si_3O_{10}]$	$\{2^{1}_{\infty}\}[^{2}Si_6O_{16}]$	$(OH)_2$	[25]
$\boldsymbol{D}(A_1) = 0$, $\boldsymbol{D}(A_2) = 2$							
Meliphanite	$(Na,Ca)_4Ca_4$	Be_4		$[SiO_4]$	$\{1^{2}_{\infty}\}[^{4}Si_7O_{20}]$	F_4	[26]
$\boldsymbol{D}(A_1) = 1$, $\boldsymbol{D}(A_2) = 1$							
Chesterite		$(Fe,$	$Mg)_{17}$	$\{2^{1}_{\infty}\}[^{2}Si_4O_{11}]_2$	$\{3^{1}_{\infty}\}[^{2}Si_6O_{16}]_2$	$(OH)_6$	[27]
$\boldsymbol{D}(A_1) = 1$, $\boldsymbol{D}(A_2) = 2$							
Okenite	Ca_{10}			$\{2^{1}_{\infty}\}[^{3}Si_6O_{16}]$	$\{1^{2}_{\infty}\}[^{3}Si_6O_{15}]_2$	$18\,H_2O$	[28]
$\boldsymbol{D}(A_1) = 2$, $\boldsymbol{D}(A_2) = 2$							
Reyerite	$(Na,K)Ca_7$			$\{1^{2}_{\infty}\}[^{4}Si_4O_{10}]$	$\{2^{2}_{\infty}\}[^{4}(AlSi_7)O_{19}]$	$(OH)_4 \cdot 3\,H_2O$	[29]

$\boldsymbol{D}(A_1)$ and $\boldsymbol{D}(A_2)$ represent the dimensionalities of the two silicate anions.

References: [1] Engel 1972; [2] Engel 1972; [3] Balko et al. 1979; [4] Kato 1982; [5] Dollase 1968, 1969; Carbonin and Molin 1980; [6] Safronov et al. 1981; [7] Moore et al. 1979; [8] Mellini et al. 1984; [9] Nevskii et al. 1979; [10] Hess and Keller 1980; [11] Simonov and Belov 1976; [12] Il'inets et al. 1982; [13] Coda et al. 1970; [14] Taylor 1971; [15] Felsche 1972; [16] Donnay and Allmann 1968; [17] de Hair 1980; [18] de Hair 1980; [19] Ansell and Wanklyn 1975; [20] Malinovskii et al. 1976; [21] Giuseppetti et al. 1971; Golyshev et al. 1971; [22] Castrejón et al. 1983; [23] Moore 1969; [24] Scott 1976; [25] Cannillo et al. 1966b; [26] Dal Negro et al. 1967; [27] Veblen and Burnham 1978; [28] Merlino 1983; [29] Merlino 1972a

($M = 2$), to 2 in triple tetrahedra $[Si_3O_{10}]$ ($M = 3$) and quadruple tetrahedra $[Si_4O_{13}]$ ($M = 4$), to 3 in fivefold tetrahedra $[Si_5O_{16}]$ ($M = 5$), to 4 in the eightfold tetrahedra $[Si_8O_{25}]$ ($M = 8$), and to 5 in the ninefold $[Si_9O_{28}]$ ($M = 9$) and tenfold $[Si_{10}O_{31}]$ ($M = 10$) tetrahedra.

According to the principle of parsimony, a substance should be more stable the smaller its number of nonequivalent building units is. Consequently, the "stability" of unbranched multiple tetrahedra is expected to decrease monotonously as their multiplicity M increases. This expectation is in good agreement with the number of known oligosilicates which decreases drastically from $M = 1$ to $M = 6$. It does not, however, explain the apparent slight increase in stability at higher multiplicities indicated by the existence of oligosilicates with $M = 8$, 9, and 10.

To explain this stability increase, a comparison has to be made between the energy contents of the various chemically nonequivalent tetrahedra of the oligosilicate anions. In a monosilicate all tetrahedra are chemically equivalent, and even those which are crystallographically nonequivalent will have almost the same energy content. In contrast, the two kinds of $[SiO_4]$ tetrahedra in a triple or a quadrupole tetrahedron, Q^1 and Q^2, respectively, are energetically considerably different, as indicated by the degree of shading of the corresponding tetrahedra in Fig. 11.1. The larger the distance of a tetrahedron from the end of a multiple tetrahedron, the smaller the energy difference between adjacent tetrahedra will be.

Moreover, the smaller the average of the energy differences between adjacent tetrahedra, the smaller the destabilizing effect will be, so that the stability of the multiple tetrahedron should converge to a limiting value $\sigma_{\Delta E \to 0}$, provided that this is the only operative influence (curve B in Fig. 11.2).

In an unbranched silicate chain all tetrahedra are chemically equivalent; any energy differences between them are due to crystallographic nonequivalence and are, therefore, small. According to the principle of parsimony, such silicate chains should be more stable than multiple tetrahedra with multiplicities $M \gtrsim 6$. This is expected to be particularly true for single chains of low periodicity P because in such chains not only are all the tetrahedra chemically equivalent, but, in addition, the number of crystallographically nonequivalent tetrahedra is small, namely $\leq P$.

A silicate with unbranched multiple tetrahedra of high multiplicity M may be considered to be a single chain silicate that contains one lattice fault every M tetrahedra. The fault energy causes a decrease in the stability which is larger the smaller M is (curve C in Fig. 11.2).

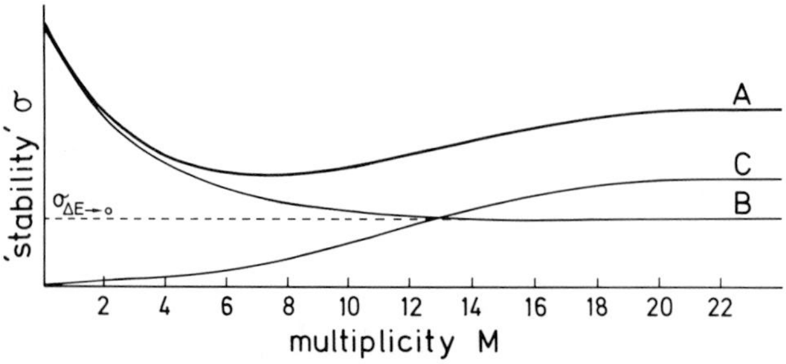

Fig. 11.2. The relative "stability" of multiple tetrahedra (*curve A*) as a sum of the two effects described in the text (*curves B* and *C*)

The two effects complement each other so that the resulting stability of a multiple tetrahedron goes through a minimum as the multiplicity increases (curve A in Fig. 11.2). From the oligosilicates known to exist, it can be assumed that this minimum of stability is near $\boldsymbol{M} = 7$.

(8) Dimensionality of Silicate Anions, D. Silicate anions with tetrahedrally coordinated silicon have a tendency to possess the highest dimensionality possible for any given Si : O ratio. This rule has a number of exceptions, but can in part be explained by the observation that anions with higher dimensionality generally achieve charge balance within shorter distances than anions with the same Si : O ratio and a lower dimensionality. Another reason is that for a given Si : O value, the number of loops, especially those of the small vierer rings, increases as the dimensionality of the anion decreases, causing an increase in the repulsive forces between the silicon atoms.

(9) Stretching Factor of Tetrahedral Chains, f_s. Within an unbranched single chain each $[SiO_4]$ tetrahedron carries a formal charge -2. Highly electropositive cations transfer their valence electrons almost completely to the silicate anions leading to an effective charge per $[SiO_4]$ which is almost as high as the formal charge. Due to the strong repulsive forces between the tetrahedra, such chains are highly stretched and, therefore, have a stretching factor $f_s \simeq 1.0$. As the electronegativity χ_M increases, the repulsive forces become weaker and the tetrahedral chains become less stretched, i.e., their f_s values become considerably smaller than 1.0.

As the mean valence $\langle v_M \rangle$ of the cations increases, the stretching factor decreases in order to involve all terminal oxygen atoms of the chain in cation–oxygen bonds.

The influence of cation size on the stretching degree of a tetrahedral chain is much weaker than the corresponding influences of cation electronegativity and valence. It can, therefore, only be detected in isotypic or homeotypic silicates where the mutual adjustment between the cation–oxygen polyhedra and the tetrahedral chain requires a slightly more stretched chain for a slightly larger cation.

(10) Periodicity of Tetrahedral Chains, P. The chain periodicity $\boldsymbol{P}$, in general, increases with a decrease in the stretching factor of the chain. For a maximum stretched chain with $f_s = 1.0$, all bridging oxygen atoms lie on a straight line so that the lowest possible value of $\boldsymbol{P}$ is 1. For a chain with $f_s < 1.0$, the periodicity is necessarily larger than 1, i.e., at least $\boldsymbol{P} = 2$. The more folded or coiled a tetrahedral chain is ($f_s \ll 1.0$), the higher its periodicity has to be because of the steric hindrance of bridging oxygen atoms (Fig. 10.1).

Because of the specific geometry of the structure of the pyroxenoids and pyroxenes, they do not follow this general trend.

In this homologous series of structures the periodicity of the tetrahedral chains increases as the stretching factor increases.

11.2 Influence of Cation Radius on the Structure of Silicates

The strong influence of the electronegativity and valence of the cations on the structure of silicates has been explained in the preceding section. In contrast, the influence of cation radius on silicate structures is less pronounced. Within a morphotropic series (i.e., a series of silicates $M_rSi_sO_t$ which differs only in the nature of M), the influence of the cation radius r_M alone is rarely sufficient to bring about a change in ***CN***, linkedness ***L***, branchedness ***B***, or multiplicity ***M***. However, within the limits of commensurability between the cation–oxygen polyhedra and the silicate anions, considerable changes of cation radius can bring about changes in chain periodicity (as in the pyroxenoids and pyroxenes) and dimensionality ***D*** (for example, from single rings to single chains, Table 10.1). In general, the influence of r_M is restricted to the production of distortions in a given structure type. This is particularly true for silica-rich silicates, as discussed in detail in Sec. 10.7 in the case of several series of framework silicates. At this point the discussion will be limited to the silica polymorphs which contain 3-dimensional 4-connected nets, and to the stuffed derivatives of silica with the general formula $M[AlSiO_4]$, since these phases display a crystallochemical principle of far-reaching importance.

Figure 10.39 demonstrates that the tetrahedral frameworks of these phases become more expanded with an increase in the size of the non-tetrahedral cation M. Starting from a maximum expanded framework of aristotype symmetry in which the voids between the tetrahedra are very regular, the incorporation of cations with decreasing radius r_M leads to a progressive distortion of these voids in order to achieve reasonably strong M–O bonds.

The aluminosilicates $M[AlSiO_4]$ of the largest monovalent alkali ions Cs^+ and Rb^+ crystallize with the Icmm-type structure. For this structure type both the aristotype and the slightly distorted frameworks have eight $[TO_4]$ tetrahedra per unit cell ($n_T = 8$, Table 11.2). However, with the incorporation of the smaller K^+ ion a change to the tridymite structure type takes place. The unit cell of the aristotype tridymite framework contains four tetrahedra per unit cell, as does the unit cell of the rather disordered kalsilite polymorph of $K[AlSiO_4]$. In addition to this polymorph several other polymorphs of $K[AlSiO_4]$ are known for which reliable lattice constants have been reported: one has $n_T = 24$, the other $n_T = 64$. Nepheline, with the composition $Na_3K[AlSiO_4]_4$, has 16 tetrahedra in its unit cell, whereas for the three polymorphs of $Na[AlSiO_4]$ $n_T = 8$, 24, and 72.

The tetrahedral frameworks of all of these phases can be considered to be composed of flat layers with equal numbers of tetrahedra pointing up and down. The same is true for the silica polymorph cristobalite, and for a number of varieties of tridymite which have different regions of relative thermal stability and, possibly, slightly different amounts and kinds of stabilizing impurities. In these phases n_T varies between 4 in high-tridymite above 693 K, and 320 in volcanic low-tridymite and a high pressure tridymite.

This comparison indicates that the number n_T of $[TO_4]$ tetrahedra in the unit cell of these phases depends on two different kinds of parameters:

1) for a given composition there is an increase in n_T as the temperature decreases or the pressure increases;

Table 11.2 Number n_T of $[TO_4]$ tetrahedra per unit cell in the alkali aluminosilicates $M[AlSiO_4]$ and the crystalline silica polymorphs which contain tetrahedral frameworks (Structures marked with a star do not contain frameworks composed of flat tetrahedral layers)

Formula	Name	Structure type	n_T	Reference
SiO_2	Quartz	Quartz*	3	Le Page et al. 1980
	High-tridymite (> 693 K)	Tridymite	4	Kihara 1978
	High-tridymite (~ 490 K)	Tridymite	8	Dollase 1967
	Cristobalite	Cristobalite	8	Dollase 1965; Peacor 1973
	Keatite	Keatite*	12	Shropshire et al. 1959
	Coesite	Coesite*	16	Levien and Prewitt 1981
	Intermediate-tridymite (378 – 453 K)	Tridymite	24	Kihara 1977
	Low-tridymite (meteoritic)	Tridymite	48	Dollase and Baur 1976
	Tridymite PO-5 (natural)	Tridymite	160	Nukui et al. 1980
	Low-tridymite (volcanic)	Tridymite	320	Konnert and Appleman 1978
	Tridymite – MC (> 0.5 GPa)	Tridymite	320	Nukui et al. 1980
$LiAlSiO_4$	β-Eucryptite (hT) (> 753 K)	Quartz*	6	Guth and Heger 1979
	α-Eucryptite (lT)	Phenakite*	12	Hesse 1984a
	β-Eucryptite (lT) (< 753 K)	Quartz*	24	Guth and Heger 1979
$NaAlSiO_4$	Carnegieite	Cristobalite	8	Klingenberg et al. 1981
	Synthetic	Beryllonite	24	Klaska 1974
	Trinepheline	Tridymite	72	Klaska 1974
$Na_3K(AlSiO_4)_4$	Nepheline	Tridymite	16	Dollase 1970
$KAlSiO_4$	Kalsilite	Tridymite	4	Perrotta and Smith 1965
	Kaliophilite-01	Kaliophilite	24	Gregorkiewitz 1980
	Tetrakalsilite	Tridymite (?)	64	Benedetti et al. 1977
$RbAlSiO_4$	Synthetic	Icmm	8	Klaska and Jarchow 1975
$CsAlSiO_4$	Synthetic	Icmm	8	Klaska and Jarchow 1973; Gallagher et al. 1977

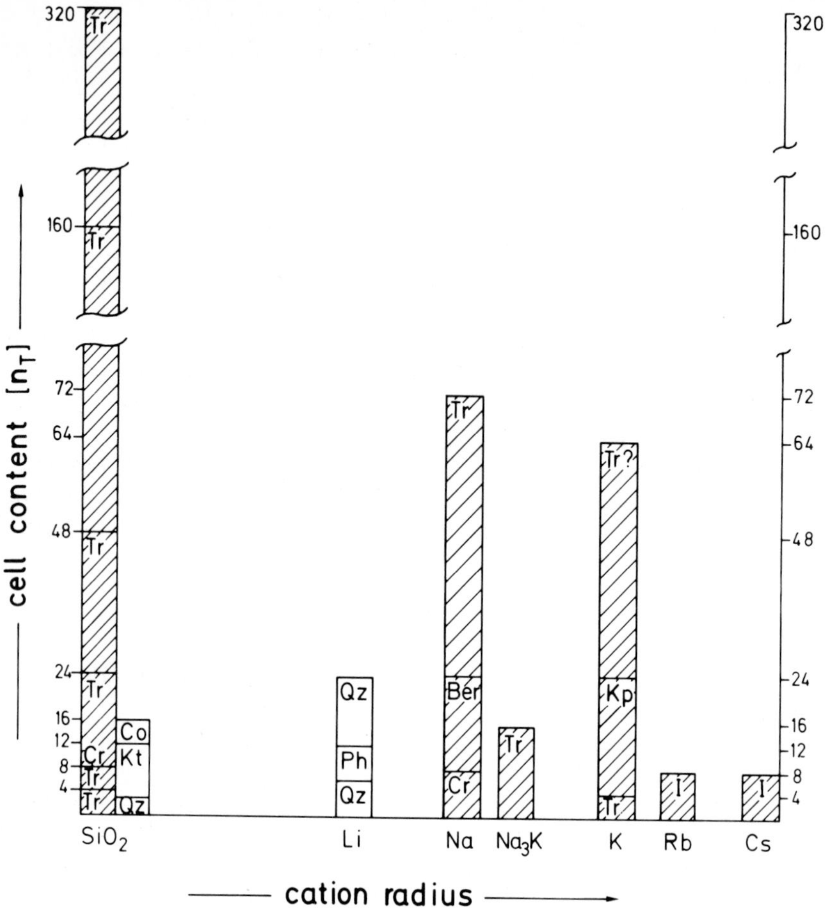

Fig. 11.3. Correlation between cation radius and number of tetrahedra per unit cell, n_T, in silica and the framework silicates $M[AlSiO_4]$.

Ber beryllonite-type; *Co* coesite; *Cr* cristobalite-type; *I* Icmm-type; *Kt* keatite; *Ph* phenakite; *Qz* quartz-type; *Tr* tridymite-type; *Kp* kaliophilite-01. *Hatched columns:* layered frameworks; *white columns:* nonlayered frameworks

2) at comparable temperature and pressure n_T increases with a decrease in the size of the non-tetrahedral cations M, the highest values being observed in the complete (cristobalite), or almost complete, absence of such cations.

The increase in n_T with decreasing temperature and decreasing size of the framework-stuffing, non-tetrahedral cations is probably more clearly deduced from the histogram in Fig. 11.3. It is mainly brought about by three different types of structural change:

a) a rotation of the $[TO_4]$ tetrahedra away from the orientation present in the aristotype structure in order to accomplish good contact between the M cations and oxygen atoms of the framework. Such a change causes a considerable increase in the framework density;

b) an increase in the degree of ordering of the Al/Si distribution causes only slight changes in the framework density, but alters the shape of the voids available for the cations;
c) non-tetrahedral cations in large, regular voids generally occupy the center of the voids and have relatively large amplitudes of thermal vibration. As the M cations become smaller and/or the voids become progressively smaller and more distorted, the cations occupy one or other of two or more positions off the center of the voids (split positions).

The first two of these changes are usually accompanied by a multiplication of the unit cell volume, but only rarely in the latter case.

The increase in unit cell content n_T with decreasing cation radius (and/or temperature) is not restricted to framework silicates, but is rather a general trend which becomes more pronounced, the higher the degree of condensation of the polyhedra in the crystal structure.

The trend which is particularly evident in the $MAlSiO_4$ series just described is somewhat obscured by the comparatively small unit cells of quartz and the three polymorphs of $LiAlSiO_4$, since these phases cannot form frameworks composed of tetrahedral layers due to the small cation size.

11.3 Characteristics of the Crystal Chemical Classification of Silicates

It has been shown that the various parameters used in the crystal chemical classification of silicates are strongly controlled by the electronegativity, valence, and size of the cations. The electronegativity and valence of an element are clearly chemical properties of general importance. The size of an atom or ion also has an influence on the character of its chemical bonds to other atoms or ions, but its main structural influence is exerted through the radius ratio $r_{\text{cation}} : r_{\text{oxygen}}$ which, in large part, controls the coordination number of the cations by oxygen. The cation size, therefore, plays an important role in determining the details of structures especially in the crystalline state.

Consequently, all the parameters used in the silicate classification described in this book are based on general chemical and crystallochemical properties. It is to be expected, therefore, that a classification based on these parameters reflects crystal chemical relationships between silicate structures which are in the same or neighboring categories of the classification. Moreover, such a classification may be expected to provide a better understanding of known silicate crystal structures and to enable sensible predictions to be made about unknown structures from their chemical compositions.

Since electronegativity, valence, and size of cations are properties of universal relevance, they permit the detection not only of correlations between similar substances, but also of the relationships between chemically and structurally distinct silicates, such as the various single chain and single ring silicates or the large and diverse class of hydrous and anhydrous layer silicates. In this respect such concepts, as simple as they are, have a much wider applicability than many of the modern, highly sophisticated qualitative theories which are, at the present time, only suitable for the comparison of isotypic or at least very similar substances. Therefore, when crystal chemical correlations between members of a large group of structurally distinct silicates are required, it may be useful to attempt this task using the procedure described in Chap. 10. Moreover, filing the silicates into the proper category of the crystal chemical classification can indicate which silicate species can sensibly be structurally and chemically correlated.

Finally, we may alter Samuel Butler's statement (see p. 89) by a more optimistic one, concluding:

The hardness of men's hearts makes an idol of classification, but a knowing heart will use it as an aid.

References

Abraham K, Gebert W, Medenbach O, Schreyer W, Hentschel G (1983) Eifelite, $KNa_3Mg_4Si_{12}O_{30}$, a new mineral of the osumilite group with octahedral sodium. Contrib Miner Petrol 82:252–258

Ahrens LH (1952) The use of ionization potentials. I. Ionic radii of the elements. Geochim Cosmochim Acta 2:155–169

Akai J (1982) Polymerization process of biopyribole in metasomatism at the Akatani ore deposit, Japan. Contrib Miner Petrol 80:117–131

Alberti A, Galli E (1980) The structure of nekoite, $Ca_3Si_6O_{15} \cdot 7\,H_2O$, a new type of sheet silicate. Am Miner 65:1270–1276

Alberti A, Vezzalini G, Tazzoli V (1981) Thomsonite: a detailed refinement with cross checking by crystal energy calculations. Zeolites 1:91–97

Allred AL, Rochow EG (1958) A scale of electronegativity based on electrostatic force. J Inorg Nucl Chem 5:264–268

Anderson CS, Bailey SW (1981) A new cation ordering pattern in amesite – $2H_2$. Am Miner 66:185–195

Annehed H, Fälth L, Lincoln FJ (1982) Crystal structure of synthetic makatite $Na_2Si_4O_8(OH)_2 \cdot 4\,H_2O$. Z Kristallogr 159:203–210

Ansell GB, Wanklyn B (1975) The crystal structure of the new rare-earth silicate $Er_4PbSi_5O_{17}$. J Chem Soc Chem Commun 1975:794

Appelo CAJ (1978) Layer deformation and crystal energy of micas and related minerals. I. Structural models for 1M and $2M_1$ polytypes. Am Miner 63:782–792

Araki T, Zoltai T (1972) The crystal structure of babingtonite. Z Kristallogr 135:355–373

Ashworth JR, Mallinson LG, Hutchison R, Biggar GM (1984) Chondrite thermal histories constrained by experimental annealing of Quenggouk orthopyroxene. Nature (Lond) 308:259–261

Baerlocher CH, Meier WM (1972) The crystal structure of synthetic zeolite Na-P1, an isotype of gismondine. Z Kristallogr 135:339–354

Bailey SW (1966) The status of clay mineral structures. Clays and Clay Minerals. Proceedings of the 14th national conference on clays and clay minerals, Berkeley, California. Pergamon, Oxford

Bakakin VV, Belov NV (1961) Crystal structure of paracelsian. Sov Phys Crystallogr 5:826–829

Bakakin VV, Solov'eva LP (1971) Crystal structure of $Fe_3BeSi_3O_9(F,OH)_2$, an example of a wollastonite-like silicon-oxygen chain based on Fe. Sov Phys Crystallogr 15:999–1005

Bakakin VV, Belov NV, Borisov SV, Solovyeva LP (1970) The crystal structure of nordite and its relationship to melilite and datolite-gadolinite. Am Miner 55:1167–1181

Balko VP, Bakakin VV, Gatilov YuV, Pavlyuchenko VS (1979) The crystal structure of potassium-zinc beryllosilicate $K_2Zn_2Be_2(SiO_4)(Si_2O_7)$. Sov Phys Dokl 24:875–876

Balko VP, Bakakin VV, Gatilov YuV (1980) Crystal structure of potassium beryllosilicate $K_2BeSi_4O_{10}$. Sov Phys Dokl 25:878–880

Baronnet A (1980) Polytypism in micas: a survey with emphasis on the crystal growth aspect. Chapter 5. In: Kaldis E (ed) Current topics in materials science, vol 5. North-Holland Publishing Company

Barrer RM (1978) Zeolites and clay minerals as sorbents and molecular sieves. Academic, London

Barrer RM (1981) Zeolites and their synthesis. Zeolites 1:130–140

Barrer RM, Villiger H (1969) The crystal structure of the synthetic zeolite L. Z Kristallogr 128:352–370

Barrer RM, Cole JF, Villiger H (1970) Chemistry of soil minerals. Part VII. Synthesis, properties and crystal structures of salt-filled cancrinites. J Chem Soc A 1970:1523–1531

Bartl H, Fischer K (1967) Untersuchung der Kristallstruktur des Zeolithes Laumontit. Neues Jahrb Miner Mh 1967:33–42

Bartl H, Pfeifer G (1976) Neutronenbeugungsanalyse des Apophyllit $KCa_4(Si_4O_{10})_2(F/OH) \cdot 8\,H_2O$. Neues Jahrb Miner Mh 1976:58–65

Basso R, Della Giusta A (1980) The crystal structure of a new manganese silicate. Neues Jahrb Miner Abh 138:333–342

Basso R, Dal Negro A, Della Giusta A, Ungaretti L (1975) The crystal structure of naujakasite, a double sheet silicate. Bull Grønlands Geol Unders 116:11–24

Baumgartner O, Völlenkle H (1977) Die Kristallstruktur der Verbindung $K_4SrGe_3O_9$, ein Cyclogermanat mit Zwölferringen. Z Kristallogr 146:261–268

Baur WH (1964) On the cation and water positions in faujasite. Am Miner 49:697–704

Baur WH (1970) Bond length variation and distorted coordination polyhedra in inorganic crystals. Trans Am Cryst Assoc 6:129–155

Baur WH (1971) The prediction of bond length variations in silicon-oxygen bonds. Am Miner 56:1573–1599

Baur WH (1977a) Variation of mean Si–O bond lengths in silicon-oxygen octahedra. J Solid State Chem 22:445–446

Baur WH (1977b) Silicon-oxygen bond lengths, bridging angles Si–O–Si and synthetic low tridymite. Acta Crystallogr B33:2615–2619

Baur WH (1978) Crystal structure refinement of lawsonite. Am Miner 63:311–315

Baur WH (1980) Straight Si–O–Si bridging bonds do exist in silicates and silicon dioxide polymorphs. Acta Crystallogr B36:2198–2202

Baur WH, Ohta T (1982) The Si_5O_{16} pentamer in zunyite refined and empirical relations for individual silicon-oxygen bonds. Acta Crystallogr B38:390–401

Beger RM (1969) The crystal structure and chemical composition of pollucite. Z Kristallogr 129:280–302

Behruzi M, Hahn T (1971) Hoch-$LiAlSiO_4$ und verwandte Phasen im System $LiAlSiO_4$–$LiGaSiO_4$–$LiAlGeO_4$–$LiGaGeO_4$. Z Kristallogr 133:405–421

Belokoneva EL, Egorov-Tismenko YuK, Simonov MA, Belov NV (1970) Crystal structure of zinc chkalovite $Na_2Zn[Si_2O_6]$. Sov Phys Crystallogr 14:918–919

Belokoneva EL, Sandomirskii PA, Simonov MA, Belov NV (1974) Crystal structure of cadmium pectolite $NaHCd_2[Si_3O_9]$. Sov Phys Dokl 18:629–630

Belov NV (1959) Crystal chemistry of large cation silicates. Consultants Bureau, New York

Belov NV, Maksimov BA, Nozik YuZ, Muradyan LA (1978) Refining the crystal structure of dioptase $Cu_6[Si_6O_{18}] \cdot 6\,H_2O$ by X-ray and neutron-diffraction methods. Sov Phys Dokl 23:215–217

Benedetti E, de Gennaro M, Franco E (1977) First occurrence in nature of tetrakalsilite. Atti Accad Naz Lincei, Cl Sci Fi Mat Nat 62:835–838 (ref from Am Miner 64:658, 1979)

Beran A, Bittner H (1974) Untersuchungen zur Kristallchemie des Ilvaits. Tschermaks Miner Petrogr Mitt 21:11–29

Berman H (1937) Constitution and classification of the natural silicates. Am Miner 22: 342–408

Bibber JW, Barnes CL, van der Helm D, Zuckerman JJ (1983) The crystal and molecular structure of bis(1,8-naphthalenedioxy)-silane, a contribution to the controversy over planar, four-coordinated silicon in orthosilicic acid esters. Angew Chem Int Ed Engl 22: 501 Angew Chem Suppl 1983:668–674

Bibby DM, Milestone NB, Aldridge LP (1979) Silicalite-2, a silica analogue of the alumino-silicate zeolite ZSM-11. Nature (Lond) 280:664–665

Binks JH, Duffy JA (1980) A molecular orbital treatment of the basicity of oxyanion units. J Noncryst Solids 37:387–400

Bissert G (1980) Verfeinerung der Struktur von Tinaksit, $Ca_2K_2NaTiO[Si_7O_{18}(OH)]$. Acta Crystallogr B36:259–263

Bissert G (1985) The crystal structure of a bikitaite, $Li[AlSi_2O_6] \cdot H_2O$, with an ordered Al/Si distribution. Neues Jahrb Miner Mh submitted

Bissert G, Liebau F (1970) Die Kristallstruktur von monoklinem Siliziumphosphat SiP_2O_7 AIII: Eine Phase mit $[SiO_6]$-Oktaedern. Acta Crystallogr B26:233–240

Bissert G, Liebau F (1984) Cubic $[N(C_4H_9)_4]H_7[Si_8O_{20}] \cdot 5.33\,H_2O$ – A double ring silicate with a zeolite like structure and $(H_2O)_{16}$ water clusters. Acta Crystallogr A40 Supplement, C 232

Blinov VA, Shumayatskaya NG, Voronkov AA, Ilyukhin VV, Belov NV (1977) Refinement of the crystal structure of wadeite $K_2Zr[Si_3O_9]$ and its relationship to kindred structural types. Sov Phys Crystallogr 22:31–35

Boer FP, van Remoortere FP (1970) Structural studies of pentacoordinate silicon. VI. Cyclobis (benzamidodimethylsilane). J Am Chem Soc 92:801–807

Boer FP, Turley JW (1969) Structural studies of pentacoordinate silicon. V. Methyl (2,2′, 3-nitrilodiethoxypropyl)silane. J Am Chem Soc 91:4134–4139

Boer FP, Turley JW, Flynn JJ (1968 a) Structural studies of pentacoordinate silicon. II. Phenyl (2,2′,2″-nitrilotriphenoxy)silane. J Am Chem Soc 90:5102–5105

Boer FP, Flynn JJ, Turley JW (1968 b) Structural studies of pentacoordinate silicon. III. Tetramethylammonium bis(o-phenylenedioxy)phenylsiliconate. J Am Chem Soc 90: 6973–6977

Borutskaya VL (1975) Synthesis of various rubidium and cesium feldspars. Dokl Akad Nauk SSSR 222:924–927

Boucher ML, Peacor DR (1968) The crystal structure of alamosite, $PbSiO_3$. Z Kristallogr 126:98–111

Bradley WF (1940) The structural scheme of attapulgite. Am Miner 25:405–410

Bradley JP, Brownlee DE, Veblen DR (1983) Pyroxene whiskers and platelets in interplanetary dust: evidence of vapour phase growth. Nature (Lond) 301:473–477

Bragg WL (1930) The structure of silicates. Z Kristallogr 74:237–305

Brauner K, Preisinger A (1956) Struktur und Entstehung des Sepioliths. Tschermaks Miner Petrogr Mitt III Folge 6:120–140

Brown BE, Bailey SW (1964) The structure of maximum microcline. Acta Crystallogr 17: 1391–1400

Brown GE, Fenn PM (1979) Structure energies of the alkali feldspars. Phys Chem Minerals 4: 83–100

Brown GE, Gibbs GV (1969) Oxygen coordination and the Si–O bond. Am Miner 54: 1528–1539

Brown GE, Gibbs GV (1970) Stereochemistry and ordering in the tetrahedral portion of silicates. Am Miner 55:1587–1607

Brown ID, Shannon RD (1973) Empirical bond-strength – bond-length curves for oxides. Acta Crystallogr A29:266–282

Brückner R, Chun H-U, Goretzki H, Sammet M (1980) XPS measurements and structural aspects of silicate and phosphate glasses. J Noncryst Solids 42:49–60

Bruno E, Chiari G, Facchinelli A (1976) Anorthite quenched from 1530 °C. I. Structure refinement. Acta Crystallogr B32:3270–3280

Bucchi R, Pesenti G (1982) Present energy problems of the Italian cement industry. Il Cimento 79:3–16

Buerger MJ (1954) The stuffed derivatives of the silica structures. Am Miner 39:600–614

Burnham CW (1971) The crystal structure of pyroxferroite from Mare Tranquillitatis. Geochim Cosmochim Acta Suppl 2 1:47–57

Buttler S in Keynes G, Hill B (1951)

Cameron M, Sueno S, Prewitt CT, Papike JJ (1973) High-temperature crystal chemistry of acmite, diopside, hedenbergite, jadeite, spodumene, ureite. Am Miner 58:594–618

Cannillo E, Mazzi F, Rossi G (1966 a) The crystal structure of neptunite. Acta Crystallogr 21:200–208

Cannillo E, Coda A, Fagnani G (1966 b) The crystal structure of bavenite. Acta Crystallogr 20:301–309

Cannillo E, Giuseppetti G, Tazzoli V (1967) The crystal structure of leucophanite. Acta Crystallogr 23:255–259

Cannillo E, Rossi G, Ungaretti L (1968) The crystal structure of macdonaldite. Accad Naz Lincei, Rend Classe Sci Fis Mat Nat, serie VIII 45:399–414

Cannillo E, Rossi G, Ungaretti L (1969) The crystal structure of delhayelite. Rend Soc Ital Mineral Petrol 26:3–15

Cannillo E, Mazzi F, Fang JH, Robinson PD, Ohya Y (1971) The crystal structure of aenigmatite. Am Miner 56:427–446

Cannillo E, Mazzi F, Rossi G (1972) The structure type of joaquinite. Tschermaks Miner Petrogr Mitt 17:233–246

Cannillo E, Dal Negro A, Rossi G (1973a) The crystal structure of latiumite, a new type of sheet silicate. Am Miner 58:466–470

Cannillo E, Rossi G, Ungaretti L (1973b) The crystal structure of elpidite. Am Miner 58:106–109

Carbonin S, Molin G (1980) Crystal-chemical considerations on eight metamorphic epidotes. Neues Jahrb Miner Abh 139:205–215

Carlisle EM (1970) Silicon: A possible factor in bone calcification. Sciences (NY) 167: 279–280

Castrejón MEV, Diazmirón LER, Campana C, West AR (1983) Silicate anion constitution of $Li_2Ca_4Si_4O_{13}$ and $Li_2Ca_2Si_5O_{13}$. J Mater Sci Lett 1:355–356

Černý P, Hawthorne FC, Jarosewich E (1980) Crystal chemistry of milarite. Can Mineralogist 18:41–57

Cervantes-Lee FJ, Dent Glasser LS, Glasser FP, Howie RA (1982) The structure of potassium barium silicate $K_2Ba_7Si_{16}O_{40}$. Acta Crystallogr B38:2099–2102

Chao GY (1971) The refinement of the crystal structure of apophyllite. II. Determination of the hydrogen positions by X-ray diffraction. Am Miner 56:1235–1242

Chao GY (1972) The crystal structure of carletonite $KNa_4Ca_4Si_8O_{18}(CO_3)_4(F,OH) \cdot H_2O$, a double-sheet silicate. Am Miner 57:765–778

Chao GY (1973) The crystal structure of gaidonnayite, orthorhombic $Na_2ZrSi_3O_9 \cdot 2\,H_2O$. Can Mineralogist 12:143–144

Chao GY (1978) Monteregianite, a new hydrous sodium potassium yttrium silicate mineral from Mont St Hilaire, Québec. Can Mineralogist 16:561–565

Chao GY, Watkinson DH (1974) Gaidonnayite, $Na_2ZrSi_3O_9 \cdot 2\,H_2O$, a new mineral from Mont St Hilaire, Québec. Can Mineralogist 12:316–319

Chernitsova NM, Pudovkina ZV, Voronkov AA, Ilyukhin VV, Pyatenko YuA (1980) Imandrite $Na_{12}Ca_3Fe_2[Si_6O_{18}]_2$ – a representative of a new branch in the lovozerite structural family. Sov Phys Dokl 25:337–339

Chiari G, Calleri M, Bruno E, Ribbe PH (1975) The structure of partially disordered, synthetic strontium feldspar. Am Miner 60:111–119

Ch'in-Hang, Simonov MA, Belov NV (1969) The crystalline structure of ramsayite $Na_2Ti_2Si_2O_9 = Na_2Ti_2O_3(Si_2O_6)$. Sov Phys Dokl 14:516–519

Chiragov MI, Mamedov KS, Belov NV (1969) Crystal structure of canasite $Ca_5Na_4K_2[Si_{12}O_{30}](OH, F)_4$. Dokl Akad Nauk SSSR 185:672–674

Clark JR, Appleman DE, Papike JJ (1969) Crystal-chemical characterization of clinopyroxenes based on eight new structure refinements. Miner Soc Am Spec Pap 2:31–50

Coda A, Rossi G, Ungaretti L (1967a) The crystal structure of aminoffite. Accad Naz Lincei, Rend Classe Sci Fis Mat Nat, serie VIII 43:225–232

Coda A, Dal Negro A, Rossi G (1967b) The crystal structure of krauskopfite. Accad Naz Lincei, Rend Classe Sci Fis Mat Nat, serie VIII 42:859–873

Coda A, Della Giusta A, Isetti G, Mazzi F (1970) On the crystal structure of vesuvianite. Atti Accad Scienze Torino 105:1–22

Coda A, Ungaretti L, Della Giusta A (1974) The crystal structure of leifite, $Na_6[Si_{16}Al_2(BeOH)_2O_{39}] \cdot 1.5\,H_2O$. Acta Crystallogr B30:396–401

Collins GAD, Cruickshank DWJ, Breeze A (1972) Ab initio calculations on the silicate ion, orthosilicic acid and their $L_{2,3}$ X-ray spectra. J chem Soc, Faraday Trans II 68:1189–1195

Colville AA, Anderson CP, Black PM (1971) Refinement of the crystal structure of apophyllite. I. X-ray diffraction and physical properties. Am Miner 56:1222–1233

Cosmovici CB (1980) Moleküle im interstellaren Raum. Sterne und Weltraum 1980:239–244

Cottrell T (1958) The strength of the chemical bonds, 2nd edn. Butterworths, London

Cradwick ME, Taylor HFW (1972) The crystal structure of $Na_2Mg_2Si_6O_{15}$. Acta Crystallogr B28:3583–3587

Crawford ES, Jefferson DA, Thomas JM (1977) Electron-microscope and diffraction studies of polytypism in stilpnomelane. Acta Crystallogr A33:548–553

Cruickshank DWJ (1961) The rôle of $3d$-orbitals in π-bonds between (a) silicon, phosphorus, sulphur, or chlorine and (b) oxygen or nitrogen. J chem Soc 1961:5486–5504

Czank M, Buseck P (1980) Crystal chemistry of silica-rich barium silicates. II. Electron microscopy of barium silicates containing multiple chains. Z Kristallogr 153:19–32

Czank M, Bissert G (1985) The crystal structure of $Li_2Mg_2[Si_4O_{11}]$. In preparation for Z Kristallogr.

Czank M, Liebau F (1980) Periodicity faults in chain silicates: A new type of planar lattice fault observed with high resolution electron microscopy. Phys Chem Minerals 6:85–93

Czank M, Liebau F (1983) Chain periodicity faults in pyroxferroite from lunar basalt 12021. Lunar and planetary science XIV. Part 1 (abstracts). Lunar and Planetary Institute, Houston, Texas, pp 144–145

Dago AM, Pushcharovskii DYu, Strelkova EE, Pobedimskaya EA, Belov NV (1980) Hydrothermal synthesis and crystal structure of $La_2Si_2O_7$. Sov Phys Dokl 25:417–420

Dal Negro A, Rossi G, Ungaretti L (1967) The crystal structure of meliphanite. Acta Crystallogr 23:260–264

Dent Glasser LS, Glasser FP (1968) The crystal structure of walstromite. Am Miner 53:9–13

Dent Glasser LS, Howie RA, Smart RM (1981) The structure of lead 'orthosilicate', $2\,PbO \cdot SiO_2$. Acta Crystallogr B37:303–306

Dikov YuP, Debolsky EI, Romashenko YuN, Dolin SP, Levin AA (1977) Molecular orbitals of $Si_2O_7^{6-}$, $Si_3O_{10}^{8-}$, etc. and mixed $(B, Al, P, Si)_m$ applied to clusters and X-ray spectroscopy data of silicates. Phys Chem Minerals 1:27–41

Dollase WA (1965) Reinvestigation of the structure of low cristobalite. Z Kristallogr 121: 369–377

Dollase WA (1967) The crystal structure at 220 °C of orthorhombic high tridymite from the Steinbach meteorite. Acta Crystallogr 23:617–623

Dollase WA (1968) Refinement and comparison of the structures of zoisite and clinozoisite. Am Miner 53:1882–1898

Dollase WA (1969) Crystal structure and cation ordering of piemontite. Am Miner 54:710–717

Dollase WA (1970) Least-squares refinement of the structure of a plutonic nepheline. Z Kristallogr 132:27–44

Dollase WA, Baur WH (1976) The superstructure of meteoritic low tridymite solved by computer simulation. Am Miner 61:971–978

Donnay G, Allmann R (1968) Si_3O_{10} groups in the crystal structure of ardennite. Acta Crystallogr B24:845–855

Donnay G, Barton R Jr (1972) Refinement of the crystal structure of elbaite and the mechanism of tourmaline solid solution. Tschermaks Miner Petrogr Mitt 18:273–286

Donnay G, Schairer JF, Donnay JDH (1959) Nepheline solid solutions. Miner Mag 32: 93–109

Donnay G, Morimoto N, Takeda H, Donnay JDH (1964) Trioctahedral one-layer micas. I. Crystal structure of a synthetic iron mica. Acta Crystallogr 17:1369–1373

Dorfman MD, Chiragov MI (1979) Hydrodelhayelite, a product of supergene alteration of delhaylite. New Data Minerals USSR 28:172–175

Drits VA, Goncharov YuI, Aleksandrova VA, Khadzhi VE, Dmitrik AL (1975a) New type of strip silicate. Sov Phys Crystallogr 19:737–741

Drits VA, Kashaev AA, Sokolova GV (1975b) Crystal structure of cymrite. Sov Phys Cristallogr 20:171–175

Dunitz JD (1979) X-Ray analysis and structure of organic molecules. Chapter 7. Cornell University Press, Ithaka

Dunitz JD (1980) Planar tetrakoordiniertes Silicium? – Eine Kritik. Angew Chem 92: 1070–1071 Angew Chem Int Ed Engl 19:1034

Durif PA, Averbuch-Pouchot MT, Guitel JC (1976) Structure cristalline de $(NH_4)_2SiP_4O_{13}$: un nouvel exemple de silicium hexacoordiné. Acta Crystallogr B32:2957–2960

Effenberger H (1980) Petalit, $LiAlSi_4O_{10}$: Verfeinerung der Kristallstruktur, Diskussion der Raumgruppe und Infrarot-Messung. Tschermaks Miner Petrogr Mitt 27:129–142

Effenberger H, Kirfel A, Will G, Zobetz E (1983) A further refinement of the crystal structure of thaumasite, $Ca_3Si(OH)_6CO_3SO_4 \cdot 12\,H_2O$. Neues Jahrb Miner Mh 1983:60–68

Eggleton RA (1972) The crystal structure of stilpnomelane. Part II. The full cell. Miner Mag 38:693–711

Engel G (1972) Ganomalite, an intermediate between the nasonite and apatite types. Naturwissenschaften 59:121–122

Eugster HP, Wones DR (1962) Stability relations of the ferruginous biotite, annite. J Petrol 3:82–125

Evans HT Jr (1973) The crystal structures of cavansite and pentagonite. Am Miner 58: 412–424

Evans HT Jr, Mrose ME (1977) The crystal chemistry of the hydrous copper silicates, shattuckite and planchéite. Am Miner 62:491–502

Evans BW, Johannes W, Oterdoom H, Trommsdorff V (1976) Stability of chrysotile and antigorite in the serpentinite multisystem. Schweiz Miner Petrogr Mitt 56:79–93

Fang JH, Robinson PD, Ohya Y (1972) Redetermination of the crystal structure of eudidymite and its dimorphic relationship to epididymite. Am Miner 57:1345–1354

Felsche J (1972) A new silicate structure containing linear $[Si_3O_{10}]$ groups. Naturwissenschaften 59:35–36

Ferraris G, Jones DW, Yerkess J (1972) A neutron-diffraction study of crystal structure of analcime, $NaAlSi_2O_6 \cdot H_2O$. Z Kristallogr 135:240–252

Fischer K (1963) The crystal structure determination of the zeolite gismondite $CaAl_2Si_2O_8 \cdot 4\,H_2O$. Am Miner 48:664–672

Fischer K (1966) Untersuchung der Kristallstruktur von Gmelinit. Neues Jahrb Miner Mh 1:1–13

Fischer K (1969) Verfeinerung der Kristallstruktur von Benitoit $BaTi[Si_3O_9]$. Z Kristallogr 129:222–243

Fischer RX, Tillmanns E (1983) The crystal structures of natural $Na_2Ca_2Si_3O_9$ from Mt. Shaheru (Zaire) and from the Mayener Feld (Eifel). Neues Jahrb Miner Mh 1983:49–59

Fischer RX, Tillmanns E (1984) Die Kristallstruktur von $Na_4CaSi_3O_9$ und dessen strukturelle Beziehungen zu $K_4SrGe_3O_9$ und $Ca_3Al_2O_6$. Z Kristallogr 166:245–256

Flanigen EM, Bennett JM, Grose RW, Cohen JP, Patton RL, Kirchner RM, Smith JV (1978) Silicalite, a new hydrophobic crystalline silica molecular sieve. Nature (Lond) 271: 512–516

Fleet ME (1977) The crystal structure of deerite. Am Miner 62:990–998

Fleet SG (1965) The crystal structure of dalyite. Z Kristallogr 121:349–368

Flörke OW (1956) Über das Einstoffsystem SiO_2. Naturwissenschaften 43:419–420

Flynn JJ, Boer FP (1969) Structural studies of hexacoordinate silicon. Tris(o-phenylenedioxy) siliconate. J Am Chem Soc 91:5756–5761

Foris CM, Zumsteg FC, Shannon RD (1979) Crystal data for $Na_4Mg_2Si_3O_{10}$. J Appl Crystallogr 12:405–406

Fortier S, Donnay G (1975) Schorl refinement showing composition dependence of the tourmaline structure. Can Mineralogist 13:173–177

Franzini M (1969) The A and B mica layers and the crystal structure of sheet silicates. Contrib Miner Petrol 21:203–224

Freed RL, Peacor DR (1967) Refinement of the crystal structure of johannsenite. Am Miner 52:709–720

Freed RL, Peacor DR (1969) Determination and refinement of the crystal structure of margarosanite. Z Kristallogr 128:213–228

French BM, Jezek PA, Appleman DE (1978) Virgilite: a new lithium aluminum silicate mineral from the Macusani glass, Peru. Am Miner 63:461–465

Fröhlich R (1984) $Pb_8[O_2(SO_4)(Si_4O_{13})]$, a new tetrasilicate. Acta Crystallogr A40, Supplement, C224

Gallagher SA, MacCarthy GJ, Smith DK (1977) Preparation and X-ray characterization of $CsAlSiO_4$. Mater Res Bull 12:1183–1190

Galli E (1975) Crystal structure refinement of mazzite. Soc Italiana di Mineralogia e Petrologia – Rendiconti 31:599–612

Galli E (1980) The crystal structure of roggianite, a zeolite-like silicate. In: Rees LV (ed) Proceedings of the 5th international conference on zeolites, Heyden, London, pp 205–213

Galli E, Alberti A (1971) The crystal structure of rinkite. Acta Crystallogr B27:1277–1284

Galli E, Passaglia E (1978) Vertumnite: its crystal structure and relationship with neutral and synthetic phases. Tschermaks Miner Petrogr Mitt 25:33–46

Galli E, Gottardi G, Pongiluppi D (1979) The crystal structure of the zeolite merlinoite. Neues Jahrb Miner Mh 1979:1–9

Gard JA, Tait JM (1972) The crystal structure of the zeolite offretite, $K_{1.1}Ca_{1.1}Mg_{0.7}[Si_{12.8}Al_{5.2}O_{36}] \cdot 15.2\ H_2O$. Acta Crystallogr B28:825–834

Gard JA, Taylor HFW (1960) The crystal structure of foshagite. Acta Crystallogr 13:785–793

Gard JA, Luke K, Taylor HFW (1981) Crystal structure of K-phase, $Ca_7Si_{16}O_{40}H_2$. Sov Phys Crystallogr 26:691–695

Gasperin M (1971) Structure cristalline de $RbAlSi_3O_8$. Acta Crystallogr B27:854–855

Gebert W, Medenbach O, Flörke OW (1983) Darstellung und Kristallographie von $K_2TiSi_6O_{15}$ – isotyp mit Dalyit $K_2ZrSi_6O_{15}$. Tschermaks Miner Petrogr Mitt 31:69–79

Geisser C, Hübner K (1984) Effective oxygen charge in SiO_2 and its dependence on the bond angle of oxygen. Crystal Res Technol 19:1149–1166

Gerke H, Gies H (1984) Studies on clathrasils. IV. Crystal structure of dodecasil 1H, a synthetic clathrate compound of silica. Z Kristallogr 166:11–22

Gerke H, Gies H, Liebau F (1984) Untersuchungen über Clathrasile. V. Synthese von Clathrasilen, den Clathratverbindungen von SiO_2. Z Anorg Allg Chem submitted

Ghose S, Thakur P (1985) Georgechaoite, $NaKZrSi_3O_9 \cdot 2\ H_2O$: a zirconosilicate with six-tetrahedral-repeat single chains and ordered Na-K distribution. Can Mineralogist submitted

Ghose S, Wan C (1978) Zektzerite, $NaLiZrSi_6O_{15}$: a silicate with six-tetrahedral-repeat double chains. Am Miner 63:304–310

Ghose S, Wan C (1979) Agrellite, $Na(Ca,RE)_2Si_4O_{10}F$: a layer structure with silicate tubes. Am Miner 64:563–572

Ghose S, Wan C, Chao GY (1980) Petarasite, $Na_5Zr_2Si_6O_{18}(Cl,OH) \cdot 2\ H_2O$, a zeolite-type zirconosilicate. Can Mineralogist 18:503–509

Gibbs GV (1982) Molecules as models for bonding in silicates. Am Miner 67:421–450

Gibbs GV, Hamil MM, Bartell LS, Yow H (1972) Correlations between Si–O bond length, Si–O–Si angle and bond overlap populations calculated using extended Hückel molecular orbital theory. Am Miner 57:1578–1613

Gibbs GV, Prewitt CT, Baldwin KJ (1977) A study of the structural chemistry of coesite. Z Kristallogr 145:108–123

Gies H (1983) Studies on clathrasils. III. Crystal structure of melanophlogite, a natural clathrate compound of silica. Z Kristallogr 164:247–257

Gies H (1984) Studies on clathrasils. VI. Crystal structure of dodecasil 3C, another synthetic clathrate compound of silica. Z Kristallogr 167:73–82

Gies H, Gerke H, Liebau F (1982) Chemical composition and synthesis of melanophlogite, a clathrate compound of silica. Neues Jahrb Miner Mh 1982:119–124

Ginderow D, Cesbron F, Sichére M-C (1982) Structure du silicate de béryllium et de sodium $Na_2Be_2Si_3O_9$. Acta Crystallogr B38:62–66

Giuseppetti G, Mazzi F, Tadini C (1971) The crystal structure of eudialyte. Tschermaks Miner Petrogr Mitt 16:105–127

Glidewell C (1975) Some chemical and structural consequences of non-bonded interactions. Inorg Chim Acta 12:219–227

Glidewell C (1977) Intramolecular non-bonded radii: Application to synthetic and naturally-occurring beryllates, aluminates, silicates, germanates, phosphates and arsenates. Inorg Chim Acta 25:77–90

Glidewell C (1978) Anomalous geometry in μ-oxo and μ-nitrido compounds of silicon, phosphorus, sulphur, and their heavier congeners: Sterochemical inactivity of lone pairs in the presence of ligands of low electronegativity. Inorg Chim Acta 29:L283–L284

Glidewell C, Liles DC (1978) The crystal and molecular structure of oxobis [triphenylsilicon (IV)]. Acta Crystallogr B34:124–128

Golovachev VP, Drozdov YuN, Kuz'min EA, Belov NV (1971) The crystal structure of phenaxite FeNaK(Si_4O_{10}) (KNaFe$[Si_4O_{10}]$). Sov Phys Dokl 15:902–904

Golyshev VM, Simonov VI, Belov NV (1971) Crystal structure of eudialyte. Sov Phys Crystallogr 16:70–74

Gordon EK, Samson S, Kamb WB (1966) Crystal structure of the zeolite paulingite. Science (Wash DC) 154:1004–1006

Gottardi G, Alberti A (1974) Domain structure in garronite: a hypothesis. Miner Mag 39:898–899

Gramaccioli CM, Pilati T, Liborio G (1979) Structure of a manganese (II) arsenatotrisilicate, $Mn_4[AsSi_3O_{12}(OH)]$: The presence of a new tetrapolyphosphate-like anion. Acta Crystallogr B35:2287–2291

Gramaccioli CM, Griffin WL, Mottana A (1980) Tiragalloite, $Mn_4[AsSi_3O_{12}(OH)]$, a new mineral and the first example of arsenatotrisilicate. Am Miner 65:947–952

Gramaccioli CM, Liborio G, Pilati T (1981) Structure of medaite, $Mn_6[VSi_5O_{18}(OH)]$: the presence of a new kind of heteropolysilicate anion. Acta Crystallogr B37:1972–1978

Gramlich V, Meier WM (1971) The crystal structure of hydrated NaA: a detailed refinement of a pseudosymmetric zeolite structure. Z Kristallogr 133:134–149

Gramlich-Meier R, Meier WM (1982) Constituent units and framework conformations in zeolite networks. J Solid State Chem 44:41–49

Gregorkiewitz M (1980) Synthese und Charakterisierung poröser Silicate. Dissertation, Technische Hochschule Darmstadt

Gregorkiewitz M, Schäfer H (1980) The structure of $KAlSiO_4$–Kaliophilite-01: application of the subgroup–supergroup relations to the quantitative space group determination of pseudosymmetric crystals. 6th European crystallogr meeting, Barcelona 1980 (abstracts), p 155

Griffen DT, Ribbe PH (1976) Refinement of the crystal structure of celsian. Am Miner 61:414–418

Griffen DT, Ribbe PH, Gibbs GV (1977) The structure of slawsonite, a strontium analog of paracelsian. Am Miner 62:31–35

Griscom DL (1977) The electronic structure of SiO_2: a review of recent spectroscopic and theoretical advances. J Noncryst Solids 24:155–234

Gross EB, Wainwright JEN, Evans BW (1965) Pabstite, the tin analogue of benitoite. Am Miner 50:1164–1169

Grosse H-P, Tillmanns E (1974) Bariummetasilicate, $BaSiO_3$(h). Cryst Struct Commun 3:603–605

Grove TL (1982) Use of exsolution lamellae in lunar clinopyroxenes as cooling rate speedometers: an experimental calibration. Am Miner 67:251–268

Guggenheim S, Bailey SW, Eggleton RA, Wilkes P (1982) Structural aspects of greenalite and related minerals. Can Mineralogist 20:1–18

Guillebert C, Le Bihan M-TH (1965) Contribution à l'étude structurale des silicates de cuivre: structure atomique de la papagoite. Bull Soc Fr Minér Cristallogr 88:119–121

Gunawardane RP, Cradwick ME, Dent Glasser LS (1973) Crystal structure of $Na_2BaSi_2O_6$. J chem Soc (Daltons transact) 1973:2397–2400

Gunawardane RP, Howie RA, Glasser FP (1982) Structure of lithium sodium yttrium silicate $Na_2LiYSi_6O_{15}$. Acta Crystallogr B38:1405–1408

Guth H (1979) Strukturuntersuchungen an dem eindimensionalen Li-Ionenleiter β-Eukryptit ($LiAlSiO_4$) mit Hilfe von Neutronenbeugung. Berichte KFK 2851, p 80

Guth H, Heger G (1979) Temperature dependence of the crystal structure of the one-dimensional Li^+-conductor β-eucryptite ($LiAlSiO_4$). In: Vashishta, Mundy, Shenoy (eds) Fast ion transport in solids. Elsevier, New York Amsterdam

Guth J-L, Hubert Y, Jordan D, Kalt A, Perati B, Wey R (1977) Un nouveau type de silice hydratée cristallisée $H_2Si_3O_7$. Compt Rend Acad Sci Paris, série D 285:1367–1370

Haga N (1973) The crystal structure of banalsite, $BaNa_2Al_4Si_4O_{16}$, and its relation to the feldspar structure. Miner J Sapporo 7:262–281

Hair de JTW (1980) The luminescence of Pr^{3+} in $BaY_4Si_5O_{17}$. J Solid State Chem 33:33–36

Harlow GE, Brown GE Jr (1980) Low albite: an X-ray and neutron diffraction study. Am Miner 65:986–995

Hassan I, Grundy HD (1983) Structure of basic sodalite, $Na_8Al_6Si_6O_{24}(OH)_2 \cdot 2H_2O$. Acta Crystallogr C39:3–5
Hawthorne FC, Grundy HD (1973) Refinement of the crystal structure of $NaScSi_2O_6$. Acta Crystallogr B29:2615–2616
Hawthorne FC, Grundy HD (1974) Refinement of the crystal structure of $NaInSi_2O_6$. Acta Crystallogr B30:1882–1884
Hawthorne FC, Grundy HD (1976) The crystal chemistry of the amphiboles. IV. X-ray and neutron refinement of the crystal structure of tremolite. Can Mineralogist 14:334–345
Hawthorne FC, Grundy HD (1977) Refinement of the crystal structure of $LiScSi_2O_6$ and structural variations in alkali pyroxenes. Can Mineralogist 15:50–58
Hawthorne FC, Ito J (1977) Synthesis and crystal-structure refinement of transition-metal orthopyroxenes. I. Orthoenstatite and (Mg, Mn, Co) orthopyroxene. Can Mineralogist 15:321–338
Hazen RM (1976) Effects of temperature and pressure on the crystal structure of forsterite. Am Miner 61:1280–1293
Hazen RM, Finger LW (1983) High-pressure and high-temperature crystallographic study of the gillespite I–II phase transition. Am Miner 68:595–603
Heinen W (1967a) Ion accumulation in bacterial systems. II. Properties of silicate-metabolizing cell-free extracts and particulate fractions from Proteus mirabilis. Arch Biochem Biophys 120:93–100
Heinen W (1967b) Ion accumulation in bacterial systems. III. Respiration-dependent accumulation of silicate by a particulate fraction from Proteus mirabilis cell-free extracts. Arch Biochem Biophys 120:101–107
Heinrich A, Gramlich V (1982) $Cs_2Cu_2Si_8O_{19}$, a new double layer silicate structure. Naturwissenschaften 69:142–143
Henderson CMB (1976) Substitution of Rb, Ti and Cs in K-feldspars. Prog Exp Petrol 3:53–56
Henmi C, Kusachi I, Kawahara A, Henmi K (1978) 7T wollastonite from Fuka, Okayama Prefecture. Miner J Sapporo 9:169–181
Henmi C, Kawahara A, Henmi K, Kusachi I, Takéuchi Y (1983) The 3T, 4T and 5T polytypes of wollastonite from Kushiro, Hiroshima Prefecture, Japan. Am Miner 68:156–163
Hess H, Keller P (1980) Die Kristallstruktur von Queitit, $Pb_4Zn_2SO_4SiO_4Si_2O_7$. Z Kristallogr 151:287–299
Hesse K-F (1979a) Refinement of the crystal structure of silicon diphosphate, SiP_2O_7 AIV – a phase with six-coordinated silicon. Acta Crystallogr B35:724–725
Hesse K-F (1979b) Die Kristallstruktur von Rhodesit, $H_2K_2Ca_4Si_{16}O_{38} \cdot 10H_2O$ – ein Silikat mit verzweigten Doppelschichten. Z Kristallogr 149:155–156
Hesse K-F (1984a) The crystal structure of α-eucryptite, $LiAlSiO_4$. Acta Crystallogr A40 Supplement, C224
Hesse K-F (1984b) Refinement of the crystal structure of wollastonite-2M (parawollastonite). Z Kristallogr 168:93–98
Hesse K-F, Liebau F (1980a) Crystal chemistry of silica-rich barium silicates. I. Refinement of the crystal structures of $Ba_4[Si_6O_{16}]$, $Ba_5[Si_8O_{21}]$ and $Ba_6[Si_{10}O_{26}]$, silicates with triple, quadruple and quintuple chains. Z Kristallogr 153:3–17
Hesse K-F, Liebau F (1980b) Crystal chemistry of silica-rich barium silicates. III. Refinement of the crystal structures of the layer silicates $Ba_2[Si_4O_{10}]$ (l) (sanbornite) and $Ba_2[Si_4O_{10}]$ (h). Z Kristallogr 153:33–41
Hesse K-F, Seifert F (1982) Site occupancy refinement of osumilite. Z Kristallogr 160: 179–186
Hesse K-F, Liebau F, Böhm H, Ribbe PH, Phillips MW (1977) Disodium zincosilicate, $Na_2ZnSi_3O_8$. Acta Crystallogr B33:1333–1337
Hesse K-F, Narita H, Liebau F (1982) Parabustamit $(Ca, Mn)_3[Si_3O_9]$-2M: Synthese und Kristallstruktur. Z Kristallogr 159:58–59
Higgins JB, Ribbe PH (1977) The classification of vertex-sharing octahedral chain structures. Neues Jahrb Miner Mh 1977:310–319
Hill RJ, Gibbs GV (1979) Variation in d(T–O), d(T$\cdots$T) and $\sphericalangle$TOT in silica and silicate minerals, phosphates and aluminates. Acta Crystallogr B35:25–30

Hill RJ, Gibbs GV, Craig JR, Ross FK, Williams JM (1977) A neutron diffraction study of hemimorphite. Z Kristallogr 146:241–259

Hill RJ, Newton MD, Gibbs GV (1983) A crystal chemical study of stishovite. J Solid State Chem 47:185–200

Hilmer W (1964) Die Kristallstruktur des sauren Kaliummetasilikates $K_4(HSiO_3)_4$. Acta Crystallogr 17:1063–1066

Hinze J (1967/68) Elektronegativität der Valenzzustände. Fortschr Chem Forsch 9:448–485

Hinze J, Whitehead MA, Jaffé HH (1963) Electronegativity. II. Bond and orbital electronegativities. J Am Chem Soc 85:148–154

Hirota K, Ono A (1977) On the stability of tridymite. Naturwissenschaften 64:39–40

Hochella MF Jr, Brown GE Jr, Ross FK, Gibbs GV (1979) High-temperature crystal chemistry of hydrous Mg- and Fe-cordierites. Am Miner 64:337–351

Hoebbel D, Wieker W, Franke P, Otto A (1975) Zur Konstitution des neuen Silicatanions $[Si_{10}O_{25}]^{10-}$. Z Anorg Allg Chem 418:35–44

Hoebbel D, Garzó G, Engelhardt G, Jancke H, Franke P, Wieker W (1976) Gaschromatographische und ^{29}Si-NMR-spektroskopische Untersuchungen an Kieselsäuretrimethylsilylestern. Z Anorg Allg Chem 424:115–127

Holmquist SB (1961) Conversion of quartz to tridymite. J Am Ceram Soc 44:82–86

Holweger H (1979) Abundances of the elements in the sun. 22nd Liége international symposium, institut d'astrophysique, Université de Liége, pp 117–138

Horiuchi H, Hirano M, Ito E, Matsui Y (1982) $MgSiO_3$ (ilmenite-type): single crystal X-ray diffraction study. Am Miner 67:788–793

Howie RA, West AR (1977) The crystal structure of $Rb_2Be_2Si_2O_7$. Acta Crystallogr B33: 381–385

Hübner K (1977) Chemical bond and related properties of SiO_2. I. Character of the chemical bond. Phys Stat Sol (a) 40:133–140

Huggins ML, Sun KH, Silverman A (1943) The vitreous state. J Am Ceram Soc 26:393–398

Ihara M, Odani K, Yoshida N, Fukunaga J, Setoguchi M, Higashi T (1984) The crystal structure of devitrite (disodium tricalcium hexasilicate), $Ca_3Na_2Si_6O_{16}$. Yogyo-Kyokai-Shi 92:373–378

Iijima S (1982) High-resolution electron microscopy of mcgillite. I. One-layer monoclinic structure. Acta Crystallogr A38:685–694

Il'inets AM, Ilyukhin VV, Belov NV, Nevskii NN (1982) Crystal structure of barium chlorosilicate. Sov Phys Dokl 27:996–997

Il'inets AM, Nevskii NN, Ilyukhin VV, Belov NV (1983) A new type of infinite silicate radical $[Si_{10}O_{25}]$ in the synthetic compound $LiBa_9[Si_{10}O_{25}]Cl_7(CO_3)$. Sov Phys Dokl 28:213–215

Ilyukhin VV, Belov NV (1962) The crystal structure of rubidium di(meta)fluoroberyllate, $RbBe_2F_5$. Sov Phys Dokl 6:851–854

Ilyushin GD, Voronkov AA, Nevskii NN, Ilyukhin VV, Belov NV (1981a) Crystal structure of hilairite $Na_2ZrSi_3O_9 \cdot 3\,H_2O$. Sov Phys Dokl 26:916–917

Ilyushin GD, Pudovkina ZV, Voronkov AA, Khomyakov AP, Ilyukhin VV, Pyatenko YuA (1981b) The crystal structure of a new natural modification of $K_2ZrSi_3O_9 \cdot H_2O$. Sov Phys Dokl 26:257–258

Ilyushin GD, Voronkov AA, Ilyukhin VV, Nevskii NN, Belov NV (1981c) Crystal structure of natural monoclinic catapleiite $Na_2ZrSi_3O_9 \cdot 2\,H_2O$. Sov Phys Dokl 26:808–810

Ito E, Matsui Y (1979) High-pressure transformations in silicates, germanates, and titanates with ABO_3 stoichiometry. Phys Chem Minerals 4:265–273

Jamieson PB (1967) Crystal Structure of $Na_2Si_3O_7$: a new type of silicate sheet. Nature (Lond) 214:794–796

Jansen M (1977) Silber (I)-disilicat. Acta Crystallogr. B33:3584–3586

Jansen M (1982) Zur Struktur von Kaliumdisilicat. Z Kristallogr 160:127–133

Jansen M, Keller H-L (1979) $Ag_{10}Si_4O_{13}$, das erste Tetrasilicat. Angew Chem 91:500

Jasmund K, Sylla HM, Freund F (1975) Solid solution in synthetic serpentine phases. Proceedings of the international clay conference 1975, Willmette, USA, pp 267–274

Jefferson DA (1978) The crystal structure of ganophyllite, a complex manganese aluminosilicate. I. Polytypism and structural variation. Acta Crystallogr A34:491–497

Jeffrey GA (1984) Hydrate inclusion compounds. In: Atwood JL, Davies JE and McNicol DD (eds) Inclusion compounds. Academic, London, pp 135–190

Johnsen O, Nielsen K, Søtofte I (1978) The crystal structure of emeleusite, a novel example of sechser-doppelkette. Z Kristallogr 147:279–296

Johnsen O, Leonardsen ES, Fälth L, Annehed H (1983) Crystal structure of kvanefjeldite: the introduction of $^{2}_{\infty}[Si_3O_7OH]$ layers with eight-membered rings. Neues Jahrb Miner Mh 1983:505–512

Jost KH (1962) Die Struktur eines sauren Natrium-Polyphosphates. Acta Crystallogr 15:951–955

Jost KH (1963) Zur Struktur des Maddrellschen Salzes. Acta Crystallogr 16:428

Joswig W, Takéuchi Y, Fuess H (1983) Neutron-diffraction study on the orientation of hydroxyl groups in margarite. Z Kristallogr 165:295–303

Kamb B (1965) Structure of ice VI. Science (Wash DC) 150:205–209

Kamb B, Davis BL (1964) Ice VII, the densest form of ice. Proc Natl Acad Sci USA 52:1433–1439

Kampf AR, Khan AA, Baur WH (1973) Barium chloride silicate with an open framework: verplanckite. Acta Crystallogr B29:2019–2021

Karpov OG, Pushcharovskii DYu, Pobedimskaya EA, Belov NV (1976) Crystal structures of the rare-earth silicates $NaPrSi_6O_{14}$ and $NaNdSi_6O_{14}$ a new $[Si_3O_7]_{\infty\infty}$ three level silicon-oxygen wall radical unbounded in two dimensions. Sov Phys Dokl 21:240–242

Karpov OG, Pobedimskaya EA, Belov NV (1977a) Crystal structure of a K, Ce silicate with a three-dimensional anion framework: $K_2CeSi_6O_{15}$. Sov Phys Crystallogr 22:215–217

Karpov OG, Pushcharovskii DYu, Pobedimskaya EA, Burshtein IF, Belov NV (1977b) The crystal structure of the rare-earth silicate $NaNdSi_6O_{13}(OH)_2 \cdot n\,H_2O$. Sov Phys Dokl 22:464–466

Kashaev AA, Sapozhnikov AN (1978) Crystal structure of armstrongite. Sov Phys Crystallogr 23:539–542

Kato K (1982) Die Kristallstruktur des Bleisilicats $Pb_{11}Si_3O_{17}$. Acta Crystallogr B38:57–62

Kato K, Nukui A (1976) Die Kristallstruktur des monoklinen Tief-Tridymits. Acta Crystallogr B32:2486–2491

Kato T, Takéuchi Y (1980) Crystal structures and submicroscopic textures of layered manganese silicates. J Miner Soc Japan 14:165–178

Kato T, Takéuchi Y (1983) The pyrosmalite group of minerals. I. Structure refinement of manganpyrosmalite. Can Mineralogist 21:1–6

Kawamura K, Iiyama JT (1981) Crystallochemistry and thermochemistry of sodipotassic copper silicate $Na_{2-2x}K_{2x}CuSi_4O_{10}$. Bull Minéral 104:387–395

Kawamura K, Kawahara A (1976) The crystal structure of synthetic copper sodium silicate: $Cu_3Na_2(Si_4O_{12})$. Acta Crystallogr B32:2419–2422

Kawamura K, Kawahara A, Iiyama JT (1978) The crystal structure of $Li_2Cu_5(Si_2O_7)_2$ and the proposal of new values for the effective ionic radii of Cu^{2+}. Acta Crystallogr B34: 3181–3185

Keller P (1972) Die Kristallchemie der Phosphat- und Arsenatminerale unter besonderer Berücksichtigung der Kationen-Koordinationspolyeder und des Kristallwassers. II. Die Kationen-Koordinationspolyeder und ihre Verknüpfung sowie eine neue Klassifikationsmöglichkeit. Neues Jahrb Miner Abh 117:217–252

Kerr IS (1974) Crystal structure of a synthetic lithium zeolite. Z Kristallogr 139:186–195

Keynes G, Hill B (1951) Samuel Butler's Notebooks. Cape, London, pp 202–203

Khan AA, Baur WH (1971) Eight-membered cyclosilicate rings in muirite. Sciences (NY) 173: 916–918

Khan AA, Baur WH, Forbes WC (1972) Synthetic magnesian merrihueite, dipotassium pentamagnesium dodecasilicate: a tetrahedral magnesiosilicate framework crystal structure. Acta Crystallogr B28:267–272

Khomyakov AP, Kurova TA, Shumyatskaya NG, Voronkov AA, Pyatenko YuA (1981) A new neutral zirconosilicate $Na_5Zr_2[Si_6O_{18}]Cl \cdot H_2O$ and its crystal structure. Sov Phys Dokl 26:265–267

Kihara K (1977) An orthorhombic superstructure of tridymite existing between about 105 and 180 °C. Z Kristallogr 146:185–203

Kihara K (1978) Thermal change in unit-cell dimensions, and a hexagonal structure of tridymite. Z Kristallogr 148:237–253

Kimata M, Ii N (1981) The crystal structure of synthetic åkermanite, $Ca_2MgSi_2O_7$. Neues Jahrb Miner Mh 1981:1–10

Kimata M, Ohashi H (1982) The crystal structure of synthetic gugiaite, $Ca_2BeSi_2O_7$. Neues Jahrb Miner Abh 143:210–222

Klaska K-H (1974) Strukturuntersuchungen an Tridymitabkömmlingen (Beryllonitreihe, Trinephelin). Dissertation, Universität Hamburg, Fachbereich Geowissenschaften

Klaska R, Jarchow O (1973) Die Kristallstruktur von $RbSiAlO_4$. Naturwissenschaften 60:299

Klaska R, Jarchow O (1975) Die Kristallstruktur und die Verzwillingung von $RbAlSiO_4$. Z Kristallogr 142:225–238

Klingenberg R, Felsche J, Miehe G (1981) Crystal data for the low-temperature form of carnegieite $NaAlSiO_4$. J Appl Crystallogr 14:66–68

Kocman V, Gait RI, Rucklidge J (1974) The crystal structure of bikitaite, $Li[AlSi_2O_6] \cdot H_2O$. Am Miner 59:71–78

Kokotailo GT, Chu P, Lawton SL, Meier WM (1978) Synthesis and structure of synthetic zeolite ZSM-11. Nature (Lond) 275:119–120

Konnert JH, Appleman DE (1978) The crystal structure of low tridymite. Acta Crystallogr B34:391–403

Kornev AN, Batalieva NG, Maksimov BA, Ilyukhin VV, Belov NV (1972) Crystal structure of the thalenite $Y_3[Si_3O_{10}](OH)$. Sov Phys Dokl 17:88–90

Kosoi AL (1976) The structure of babingtonite. Sov Phys Crystallogr 20:446–451

Kossiakoff AA, Leavens PB (1976) The crystal structure of eakerite, a calcium-tin silicate. Am Miner 61:956–962

Kostov I (1975) Crystal chemistry and classification of the silicate minerals. Bulgar Acad Sci, Geochem Mineral Petrol, Vol 1. Sofia, pp 5–41

Kovalenko VS, Mel'nikova EM, Tsinober LE (1977) Recrystallization of the "heavy phase" in the system $Na_2O-SiO_2-H_2O-FeO$. Sov Phys Dokl 22:61–62

Kudoh Y, Takéuchi Y (1979) Polytypism of xonotlite. I. Structure of an $A\bar{1}$ polytype. Miner J Sapporo 9:349–373

Kunze G (1956) Die gewellte Struktur des Antigorits. I. Z Kristallogr 108:82–107

Kunze G (1958) Die gewellte Struktur des Antigorits. II. Z Kristallogr 110:282–320

Kunze G (1959) Fehlordnungen des Antigorits. Z Kristallogr 111:190–212

Kuznetsova TP, Nevskii NN, Plyukhin VV, Belov NV (1980) Refinement of the crystal structure of Na, Ca triorthosilicate $Na_2Ca_3[Si_3O_{10}]$. Sov Phys Crystallogr 25:490–491

Kwasha LG (1976) List of meteoritic minerals. Meteoritika 35:136–138

Lasaga AC (1982) Optimization of CNDO for molecular orbital calculation on silicates. Phys Chem Minerals 8:36–46

Laughon RB (1971) The crystal structure of kinoite. Am Miner 56:193–200

Laves F (1932) Zur Klassifikation der Silikate. Geometrische Untersuchungen möglicher Silicium-Sauerstoff-Verbände als Verknüpfungsmöglichkeiten regulärer Tetraeder. Z Kristallogr 82:1–14

Le Bihan M-TH, Kalt A, Wey R (1971) Etude structurale de $KHSi_2O_5$ et $H_2Si_2O_5$. Bull Soc Fr Minér Cristallogr 94:15–23

Lee JH, Guggenheim S (1981) Single crystal X-ray refinement of pyrophyllite-1 Tc. Am Miner 66:350–357

Le Fur Y (1972) Structure du pentafluorodibéryllate $TlBe_2F_5$. Acta Crystallogr B28: 1159–1163

Le Fur Y, Aléonard S (1972) Structure du pentafluorodibéryllate $CsBe_2F_5$. Acta Crystallogr B28:2115–2118

Le Page Y, Perrault G (1976) Structure cristalline de la lemoynite $(Na, K)_2CaZr_2Si_{10}O_{26}$, 5–6 H_2O. Can Mineralogist 14:132–138

Le Page Y, Calvert LD, Gabe EJ (1980) Parameter variation in low-quartz between 94 and 298 K. J Phys Chem Solids 41:721–725

Levien L, Prewitt CT (1981) High-pressure crystal structure and compressibility of coesite. Am Miner 66:324–333

Li C-T (1968) The crystal structure of $LiAlSi_2O_6$ III (high-quartz solid solution). Z Kristallogr 127:327–348

Li C-T, Peacor DR (1968) The crystal structure of $LiAlSi_2O_6$-II ("β spodumene"). Z Kristallogr 126:46–65

Li C-T, Shlichta PJ (1974) Relation between the angle T–O–T and the stability of $LiAlSi_2O_6$ III and other high-quartz solid solutions. Z Kristallogr 140:100–113

Liebau F (1956) Bemerkungen zur Systematik der Kristallstrukturen von Silikaten mit hochkondensierten Anionen. Physikal Chem 206:73–92

Liebau F (1961a) Untersuchungen über die Größe des Si–O–Si – Valenzwinkels. Acta Crystallogr 14:1103–1109

Liebau F (1961b) Untersuchungen an Schichtsilikaten des Formeltyps $A_m(Si_2O_5)_n$. I. Die Kristallstruktur der Zimmertemperaturform des $Li_2Si_2O_5$. Acta Crystallogr 14:389–395

Liebau F (1961c) Untersuchungen an Schichtsilikaten des Formeltyps $A_m(Si_2O_5)_n$. IV. Über die Kristallstruktur des $Ag_2Si_2O_5$. Acta Crystallogr 14:537–538

Liebau F (1962a) Die Systematik der Silikate. Naturwissenschaften 49:481–491

Liebau F (1962b) Ein Stabilitätskriterium für Silikatstrukturen. Ber Dtsch Keram Ges 39: 72–74

Liebau F (1964) Über Kristallstrukturen zweier Phyllokieselsäuren, $H_2Si_2O_5$. Z Kristallogr 120:427–449

Liebau F (1968) Ein Beitrag zur Kristallchemie der Schichtsilikate. Acta Crystallogr B24: 690–699

Liebau F (1972) Crystal chemistry of silicon. Chapter 14-A. In: Wedepohl (ed) Handbook of Geochemistry, vol II/3. Springer, Berlin Heidelberg New York

Liebau F (1978) Silicates with branched anions: a crystallochemically distinct class. Am Miner 63:918–923

Liebau F (1980a) Classification of silicates. Rev Mineralogy 5:1–24

Liebau F (1980b) The role of cationic hydrogen in pyroxenoid crystal chemistry. Am Miner 65:981–985

Liebau F (1981) The influence of cation properties on the conformation of silicate and phosphate anions. In: O'Keeffe M, Navrotsky A (eds) Structure and bonding in crystals, vol II. Academic, London, pp 197–232

Liebau F (1982) Principles of silicate classification based on crystal chemistry. Sov Phys Crystallogr 27:66–73

Liebau F (1983) Zeolites and clathrasils – two distinct classes of framework silicates. Zeolites 3:191–193

Liebau F (1984a) Correlations between Si–O bond length, Si–O–Si angle and static and dynamic disorder in silica – an empirical method to correct Si–O bond lengths. Acta Crystallogr A40 Supplement, C254

Liebau F (1984b) Pentacoordinate silicon intermediate states during silicate condensation and decondensation – Crystallographic support. Inorg Chim Acta 89:1–7

Liebau F, Pallas I (1981) The influence of cation properties on the shape of silicate chains. Z Kristallogr 155:139–153

Liebau F, Bissert G, Köppen N (1968) Synthese und kristallographische Eigenschaften einiger Phasen im System $SiO_2-P_2O_5$. Z Anorg Allg Chem 359:113–134

Lindemann W, Wögerbauer R, Berger P (1979) Die Kristallstruktur von Karpholith $(Mn_{0,97}Mg_{0,08}\overset{II}{Fe}_{0,07})(Al_{1,90}\overset{III}{Fe}_{0,01})Si_2O_6(OH)_4$. Neues Jahrb Miner Mh 1979:282–287

Lindqvist B (1966) Hydrothermal synthesis studies of potash-bearing sesquioxide-silica systems. Geol För Stockh Förh 88:133–178

Little EJ Jr, Jones MM (1960) A complete table of electronegativities. J Chem Educ 37: 231–233

Liu L (1979) High pressure phase transformations in the joins $Mg_2SiO_4-Ca_2SiO_4$ and $MgO-CaSiO_3$. Contrib Mineral Petrol 69:245–247

Livingstone A, Atkin D, Hutchison D, Al-Hermez HM (1976) Iraqite, a new rare-earth mineral of the ekanite group. Miner Mag 40:441–445

Loewenstein W (1954) The distribution of aluminum in the tetrahedra of silicates and aluminates. Am Miner 39:92–96

Löns J, Schulz H (1967) Strukturverfeinerung von Sodalith, $Na_8Si_6Al_6O_{24}Cl_2$. Acta Crystallogr 23:434–436

Lopes-Vieira A, Zussman J (1969) Further detail on the crystal structure of zussmanite. Miner Mag 37:49–60

Louisnathan SJ (1970) The crystal structure of synthetic soda melilite, $CaNaAlSi_2O_7$. Z Kristallogr 131:314–321

Louisnathan SJ, Gibbs GV (1972) Aluminum–silicon distribution in zunyite. Am Miner 57:1089–1108

Louisnathan SJ, Smith JV (1970) Crystal structure of tilleyite: Refinement and coordination. Z Kristallogr 132:288–306

Lucchetti G, Penco AM, Rinaldi R (1981) Saneroite, a new natural hydrated Mn-silicate. Neues Jahrb Miner Mh 1981:161–168

Lüschen H (1968) Die Namen der Steine. Ott, Thun und München

MacGillavry CH, Korst WL, Weichel Moore EJ, van der Plas HJ (1956) The crystal structure of ferrocarpholite. Acta Crystallogr 9:773–776

Machatschki F (1928) Zur Frage der Struktur und Konstitution der Feldspate. Zentralbl Miner Abt A 1928:97–104

Machida K, Adachi G, Shiokawa J, Shimada M, Koizumi M (1982) Structure and high-pressure polymorphism of strontium metasilicate. Acta Crystallogr B38:386–389

Majling J, Hanic F (1980) Phase chemistry of condensed phosphates. In: Grayson M and Griffith EJ (eds) Topics in phosphorus chemistry, vol 10. Wiley, New York, pp 341–502

Maksimov BA, Belov NV (1981) High-temperature X-ray studies of $Na_5YSi_4O_{12}$ single crystals. Sov Phys Dokl 26:1026–1029

Maksimov BA, Merinov BV, Borovkov VS, Ivanov-Shits AK, Kharitonov YuA, Belov NV (1979) Structural characteristics of the family of rare earth silicates $Na_5TRSi_4O_{12}$ – a new class of solid electrolytes. Sov Phys Crystallogr 24:151–154

Maksimov BA, Mel'nikov OK, Zhdanova TA, Ilyukhin VV, Belov NV (1980a) Crystal structure of $Na_4Sc_2Si_4O_{13}$. Sov Phys Dokl 25:143–145

Maksimov BA, Kalinin VR, Merinov BV, Ilyukhin VV, Belov NV (1980b) Crystal structure of the rare-earth Na, Y metasilicate $\{Na_3YSi_3O_9\} \times 4 = Na_{12}Y_4[Si_{12}O_{36}]_\infty$. Sov Phys Dokl 25:415–417

Malinovskii YuA (1983) Crystal structure of $Na_2BaNd_2[Si_4O_{12}][CO_3]$. Sov Phys Dokl 28: 829–831

Malinovskii YuA, Pobedimskaya EA, Belov NV (1976) Crystal structure of traskite. Sov Phys Dokl 21:426–428

Malinovskii YuA, Pobedimskaya EA, Belov NV (1977) Crystal structure of tienshanite. Sov Phys Dokl 22:544–545

Malinovskii YuA, Yamnova NA, Belov NV (1981) Refined crystal structure of leucosphenite. Sov Phys Dokl 26:372–375

Malinovskii YuA, Baturin SV, Bondareva OS (1983) A new island silicate radical $[Si_4O_{13}]$ in the structure of $NaBa_3Nd_3[Si_2O_7][Si_4O_{13}]$. Sov Phys Dokl 28:809–812

Malinovskii YuA, Bondareva OS, Baturin SV (1984) Crystal structure of $NaBaNd[Si_3O_9]$. Dokl Akad Nauk SSSR 275:372–375

Marr JM, Glasser FP (1979) Synthesis and properties of zektzerite, $LiNaZrSi_6O_{15}$, and its isotypes. Miner Mag 43:171–173

Martin RF, Lagache M (1975) Cell edges and infrared spectra of synthetic leucites and pollucites in the system $KAlSi_2O_6$–$RbAlSi_2O_6$–$CsAlSi_2O_6$. Can Mineralogist 13:275–281

Martin C, Tordjman I, Durif A (1975) Structure cristalline du polyphosphate de baryumpotassium, $Ba_2K(PO_3)_5$. Z Kristallogr 141:403–411

Matsubara S (1980a) The crystal structure of orthoericssonite. Miner J Sapporo 10:107–121

Matsubara S (1980b) The crystal structure of nagashimalite, $Ba_4(V^{3+}, Ti)_4[(O, OH)_2|Cl|Si_8B_2O_{27}]$. Miner J Sapporo 10:131–142

Matsubara S (1981) Taneyamalite, $(Na, Ca)(Mn^{2+}, Mg, Fe^{3+}, Al)_{12}Si_{12}(O, OH)_{44}$, a new mineral from the Iwaizawa mine, Saitama Prefecture, Japan. Miner Mag 44:51–53

Matsubara S, Kato A, Yui S (1982) Suzukiite, $Ba_2V_2^{4+}[O_2|Si_4O_{12}]$, a new mineral from the Mogurazawa mine, Gumma Prefecture, Japan. Miner J Sapporo 11:15–20

Mayer H (1974) Die Kristallstruktur von $Si_5O[PO_4]_6$. Monatsh Chem 105:46–54

Mazzi F, Galli E (1978) Is each analcime different? Am Miner 63:448–460

Mazzi F, Galli E (1983) The tetrahedral framework of chabazite. Neues Jahrb Miner Mh 1983:461–480

Mazzi F, Rossi G (1980) The crystal structure of taramellite. Am Miner 65:123–128

Mazzi F, Galli E, Gottardi G (1976) The crystal structure of tetragonal leucite. Am Miner 61: 108–115

Mazzi F, Ungaretti L, Dal Negro A, Petersen OV, Rönsbo JG (1979) The crystal structure of semenovite. Am Miner 64:202–210

McCauley JW, Newnham RE (1973) Structure refinement of a barium mica. Z Kristallogr 137:360–367

McDonald WS, Cruickshank DWJ (1967) A reinvestigation of the structure of sodium metasilicate, Na_2SiO_3. Acta Crystallogr 22:37–43

Meagher EP (1976) The atomic arrangement of pellyite: $Ba_2Ca(Fe, Mg)_2[Si_6O_{17}]$. Am Miner 61:67–73

Meagher EP (1980) Stereochemistry and energies of single two-repeat silicate chains. Am Miner 65:746–755

Meagher EP, Tossell JA, Gibbs GV (1979) A CNDO/2 molecular orbital study of the silica polymorphs quartz, cristobalite and coesite. Phys Chem Minerals 4:11–21

Megaw HD (1973) Crystal structures: a working approach. Chapter 14, 15. Saunders, Philadelphia

Meier WM (1961) The crystal structure of mordenite (ptilolite). Z Kristallogr 115:439–450

Meier WM (1968) SCI monograph on "molecular sieves". Society of Chem Industry, London, p 10

Meier R, Ha T-K (1980) A theoretical study of electronic structure of disiloxane $[(SiH_3)_2O]$ and its relation to silicates. Phys Chem Minerals 6:37–46

Mellini M (1981) Refinement of the crystal structure of låvenite. Tschermaks Miner Petrogr Mitt 28:99–112

Mellini M (1982) The crystal structure of lizardite 1T: hydrogen bonds and polytypism. Am Miner 67:587–598

Mellini M, Merlino S (1978) Caysichite: a double crankshaft chain structure. Can Mineralogist 16:81–88

Mellini M, Merlino S (1979) Refinement of the crystal structure of wöhlerite. Tschermaks Miner Petrogr Mitt 26:109 – 123

Mellini M, Merlino S (1982) The crystal structure of cascandite, $CaScSi_3O_8(OH)$. Am Miner 67:604–609

Mellini M, Merlino S, Rossi G (1977) The crystal structure of tuscanite. Am Miner 62:1114–1120

Mellini M, Merlino S, Pasero M (1984) X-ray and HRTEM study of sursassite: crystal structure, stacking disorder and sursassite-pumpellyite intergrowth. Phys Chem Minerals 10:99–105

Menchetti S, Sabelli C (1979) The crystal structure of baratovite. Am Miner 64:383–398

Merinov BV, Maksimov BA, Kharitonov YuA, Belov NV (1978) Crystal structure of the rare-earth silicate $Na_5LuSi_4O_{12}$. Sov Phys Dokl 23:291–293

Merinov BV, Maksimov BA, Belov NV (1980) Crystal structure of sodium-scandium silicate $Na_5ScSi_4O_{12}$. Sov Phys Dokl 25:885–888

Merlino S (1969) Tuhualite crystal structure. Sciences (NY) 166:1399–1401

Merlino S (1972a) New tetrahedral sheets in reyerite. Nature Phys Sci 238:124–125

Merlino S (1972b) The crystal structure of zeophyllite. Acta Crystallogr B28:2726–2732

Merlino S (1974) The crystal structure of wenkite. Acta Crystallogr B30:1262–1266

Merlino S (1983) Okenite, $Ca_{10}Si_{18}O_{46} \cdot 18\,H_2O$: the first example of a chain and sheet silicate. Am Miner 68:614–622

Metcalf-Johansen J, Hazell RG (1976) The crystal structure of sorensenite, $Na_4SnBe_2(Si_3O_9)_2 \cdot 2\,H_2O$. Acta Crystallogr B32:2553–2556

Meyer H, Nagorsen G (1979) Struktur und Reaktivität des Orthokohlensäure- und Orthokieselsäureesters des Brenzcatechins. Angew Chem 91:587–588 Angew Chem Int Ed Engl 18:551–553

Mielke JE (1979) Composition of the earth's crust and distribution of the elements. In: Siegel FR (ed) Review of research on modern problems in geochemistry, vol 16. Earth sciences. Int Assoc Geochem Cosmochem, UNESCO, pp 13–37

Mitchell RS (1979) Mineral names: what do they mean. Van Norstrand, New York

Miura Y, Kato T, Rucklidge J, Matsueda H (1981) Natroapophyllite, a new orthorhombic sodium analog of apophyllite. II. Crystal structure. Am Miner 66:416–423

Mizota T, Komatsu M, Chihara K (1983) A refinement of the crystal structure of ohmilite, $Sr_3(Ti, Fe^{3+})(O, OH)(Si_2O_6)_2 \cdot 2\text{–}3\ H_2O$. Am Miner 68:811–817

Moore PB (1969) Joesmithite: a novel amphibole crystal chemistry. Mineral Soc Am Spec Pap 2:111–115

Moore PB (1970a) Edge-sharing silicate tetrahedra in the crystal structure of leucophoenicite. Am Miner 55:1146–1166

Moore PB (1970b) Structural hierarchies among minerals containing octahedrally coordinating oxygen. I. Stereoisomerism among corner-sharing octahedral and tetrahedral chains. Neues Jahrb Miner Mh 1970:163–173

Moore PB (1974) Structural hierarchies among minerals containing octahedrally coordinating oxygen. II. Systematic retrival and classification of octahedral edge-sharing clusters: an epistemological approach. Neues Jahrb Miner Abh 120:205–227

Moore PB, Araki T (1979) Crystal structure of synthetic $Ca_3Mn_2^{3+}O_2[Si_4O_{12}]$. Z Kristallogr 150:287–297

Moore PB, Ito J, Steele IM (1979) Macfallite and orientite: calcium manganese(III) silicates from upper Michigan. Miner Mag 43:325–331

Moore PB, Araki T, Ghose S (1982) Hyalotekite, a complex lead borosilicate: its crystal structure and the lone-pair effect of Pb(II). Am Miner 67:1012–1020

Morimoto N, Koto K (1969) The crystal structure of orthoenstatite. Z Kristallogr 129:65–83

Morimoto N, Nakajima Y, Syono Y, Akimoto S, Matsui Y (1975) Crystal structures of pyroxene-type $ZnSiO_3$ and $ZnMgSi_2O_6$. Acta Crystallogr B31:1041–1049

Morosin B (1972) Structure and thermal expansion of beryl. Acta Crystallogr B28:1899–1903

Mortier WJ, Pluth JJ, Smith JV (1976) Crystal structure of natural zeolite offretite after carbon monoxide adsorption. Z Kristallogr 144:32–41

Müller WF, Wlotzka F (1982) Mineralogical study of the Leoville meteorite (CV3): macroscopic texture and transmission electron microscopic observations. Lunar and Planetary Science XIII, pp 558–559

Muradyan LA, Simonov VI (1977) Full-matrix and block-diagonal refinement of the atomic structures of crystals by the method of least squares. Sov Phys Crystallogr 22:277–281

Murakami T, Takéuchi Y (1979) Structure of synthetic rhodonite, $Mn_{0.685}Mg_{0.315}SiO_3$, and compositional transformations in pyroxenoids. Miner J Sapporo 9:286–304

Murakami T, Takéuchi Y, Tagai T, Koto K (1977) Lithium-hydrorhodonite. Acta Crystallogr B33:919–921

Murakami N, Kato T, Hirowatari F (1983) Katayamalite, a new Ca-Li-Ti silicate mineral from Iwagi Islet, southwest Japan. Miner J Sapporo 11:261–268

Nagy B (1954) Multiplicity and disorder in the lattice of ekmanite. Am Miner 39:946–956

Nakamoto H, Takahashi H (1982) Hydrophobic natures of zeolite ZSM-5. Zeolites 2:67–68

Náray-Szabó ST (1930) Ein auf der Kristallstruktur basierendes Silikatsystem. Z Phys Chem Abt B9:356–377

Narita H, Koto K, Morimoto N (1975) The crystal structure of nambulite $(Li, Na)Mn_4Si_5O_{14}(OH)$. Acta Crystallogr B31:2422–2426

Narita H, Koto K, Morimoto N (1977) The crystal structures of $MnSiO_3$ polymorphs (rhodonite- and pyroxmangite-type). Miner J Sapporo 8:329–342

Naumova IS, Pobedimskaya EA, Pushcharovskii DYu, Belov NV, Altukhova YuN (1976) Crystal structure of a synthetic K-beryllosilicate of the epididymite group. Sov Phys Dokl 21:422–424

Nevskii NN, Ilyukhin VV, Belov NV (1979) Determination of the crystal structure of rustumite. Sov Phys Dokl 24:598–600

Newton MD (1981) Theoretical probes of bonding in the disiloxy group. In: O'Keeffe M and Navrotsky A (eds) Structure and bonding in crystals, vol I. Academic, New York, pp 175–193

Newton MD, Gibbs GV (1980) Ab initio calculated geometries and charge distributions for H_4SiO_4 and $H_6Si_2O_7$ compared with experimental values for silicates and siloxanes. Phys Chem Minerals 6:305–312

Nguyen N, Choisnet J, Raveau B (1980) Silicates synthétiques à structure milarite. J Solid State Chem 34:1–9

Noll W, Kircher H, Sybertz W (1960) Über synthetischen Kobaltchrysotil und seine Beziehungen zu anderen Solenosilikaten. Beitr Miner Petrogr 7:232–241

Novak GA, Gibbs GV (1971) The crystal chemistry of the silicate garnets. Am Miner 56:791–825

Nuber B, Schmetzer K (1981) Strukturverfeinerung von Liddicoatit. Neues Jahrb Miner Mh 1981:215–219

Nukui A, Yamaoka S, Nakazawa H (1980) Pressure-induced phase transitions in tridymite. Am Miner 65:1283–1286

Oberhammer H, Boggs JE (1980) Importance of (p–d) π bonding in the siloxane bond. J Am Chem Soc 102:7241–7244

Oehlschlegel G (1971) Das binäre Teilsystem $BaO \cdot 2\,SiO_2 - 2\,BaO \cdot 3\,SiO_2$. Glastechn Ber 44: 194–204

Ohashi H (1981) Studies on the Si–O distances in $NaM^{3+}Si_2O_6$ pyroxenes. J Jpn Assoc Miner Petrol Econ Geol 76:308–311

Ohashi Y, Finger LW (1975) Pyroxenoids: a comparison of refined structures of rhodonite and pyroxmangite. Carnegie Inst Wash Year Book 74:564–569

Ohashi Y, Finger LW (1978) The role of octahedral cations in pyroxenoid crystal chemistry. I. Bustamite, wollastonite and the pectolite-schizolite-serandite series. Am Miner 63:274–288

Ohashi Y, Finger LW (1981) The crystal structure of santaclaraite, $CaMn_4[Si_5O_{14}(OH)](OH) \cdot H_2O$: the role of hydrogen atoms in the pyroxenoid structure. Am Miner 66:154–168

Ohashi H, Taketoshi F, Toshikazu O (1981) Structure of $Co_3Al_2Si_3O_{12}$ garnet. J Jpn Assoc Miner Petrol Econ Geol 76:58–60

O'Keeffe M, Hyde BG (1976) Cristobalites and topologically related structures. Acta Crystallogr B32:2923–2936

O'Keeffe M, Hyde BG (1978) On Si–O–Si configurations in silicates. Acta Crystallogr B34: 27–32

Okkerse C (1970) Porous silica. In: Linsen BG (ed) Physical and chemical aspects of adsorbents and catalysts. Chapter 5. Academic, London, pp 213–264

Olson DH, Hagg WO, Lago RM (1980) Chemical and physical properties of the ZSM-5 substitutional series. J Catal 61:390–396

Olson DH, Kokotailo GT, Lawton SL, Meier WM (1981) Crystal structure and structure-related properties of ZSM-5. J Phys Chem 85:2236–2243

Ostertag W, Fischer GR, Williams JP (1968) Thermal expansion of synthetic β-spodumene and β-spodumene-silica solid solutions. J Am Ceram Soc 51:651–654

Ozawa T, Takéuchi Y, Mori H (1979) 145 Å superstructure in $Me_{\sim(x+2)/3}T_{\sim(x-2)/3}O_x$ (Me = Mg, Sc; T = Si, Mg). In: Modulated structures. pp 324–326. American Institute of Physics

Ozawa T, Takéuchi Y, Takahata T, Donnay G, Donnay JDH (1983) The pyrosmalite group of minerals. II. The layer structure of mcgillite and friedelite. Can Mineralogist 21:7–17

Pabst A (1959) Structures of some tetragonal sheet silicates. Acta Crystallogr 12:733–739

Pant AK (1968) A reconsideration of the crystal structure of β-$Na_2Si_2O_5$. Acta Crystallogr B24:1077–1083

Pant AK, Cruickshank DWJ (1968) The crystal structure of α-$Na_2Si_2O_5$. Acta Crystallogr B24:13–19

Papike JJ, Zoltai T (1967) Ordering of tetrahedral aluminum in prehnite, $Ca(Al, Fe^{3+})[Si_3AlO_{10}](OH)_2$. Am Miner 52:974–984

Pauling L (1939) The nature of the chemical bond, 1st edn. Cornell University Press, Ithaca, NY

Pauling L (1967) The nature of the chemical bond, 3rd edn, Chapter 13–16. Cornell University Press, Ithaka, NY, pp 543 ff, rule 1: Radius ratio rule pp 544–547, rule 2: Electrostatic valence rule pp 547–559, rule 3: Sharing of elements rule pp 559–562

Pauling L (1980) The nature of silicon-oxygen bonds. Am Miner 65:321–323

Peacor DR (1973) High-temperature single-crystal study of the cristobalite inversion. Z Kristallogr 138:274–298

Peacor DR, Buerger MJ (1962) The determination and refinement of the structure of narsarsukite $Na_2TiOSi_4O_{10}$. Am Miner 47:539–556

Peacor DR, Niizeki N (1963) The redetermination and refinement of the crystal structure of rhodonite, $(Mn, Ca)SiO_3$. Z Kristallogr 119:98–116

Peacor DR, Dunn PJ, Sturman BD (1978) Marsturite, $Mn_3CaNaHSi_5O_{15}$, a new mineral of the nambulite group from Franklin, New Jersey. Am Miner 63:1187–1189

Pentinghaus H (1975) Hexacelsiane. Fortschr Miner 53 Suppl 1:65

Pentinghaus H (1980) Polymorphie in den feldspatbildenden Systemen $A^{+}T^{3+}T_3^{4+}O_8$ und $A^{2+}T_2^{3+}T_2^{4+}O_8$. Alkali- und Erdalkali-, Bor-, Aluminium-, Gallium-, Eisensilikate und -Germanate. Habilitationsschrift, Westfälische Wilhelms-Universität, Münster

Perdikatsis B, Burzlaff H (1981) Strukturverfeinerung am Talk $Mg_3[(OH)_2Si_4O_{10}]$. Z Kristallogr 156:177–186

Perrault G, Szymański JT (1982) Steacyite, a new name, and a re-evalution of the nomenclature of "ekanite"-group minerals. Can Mineralogist 20:59–63

Perrault G, Harvey Y, Pertsowsky R (1975) La yofortierite, un nouveau silicate hydrate de manganese de St-Hilaire, P.Q. Can Mineralogist 13:68–74

Perrotta AJ, Smith JV (1965) The crystal structure of kalsilite, $KAlSiO_4$. Miner Mag 35: 588–595

Petersen OV, Rönsbo JG (1972) Semenovite – a new mineral from the Ilímaussaq alkaline intrusion, south Greenland. Lithos 5:163–173

Phillips MW, Ribbe PH (1973) The structures of monoclinic potassium-rich feldspars. Am Miner 58:263–270

Phillips MW, Gibbs GV, Ribbe PH (1974) The crystal structure of danburite: A comparison with anorthite, albite, and reedmergnerite. Am Miner 59:79–85

Pinckney LR, Finger LW, Hazen RM, Burnham CW (1981) Crystal structure of pyroxmangite at high temperature. Carnegie Institution Wash Year Book 80:380–384

Pluth JJ, Smith JV (1973) The crystal structure of scawtite, $Ca_7(Si_6O_{18})(CO_3)\cdot 2\,H_2O$. Acta Crystallogr B29:73–80

Pozas IMM, Rossi G, Tazzoli V (1975) Re-examination and crystal structure analysis of litidionite. Am Miner 60:471–474

Prewitt CT (1967) Refinement of the structure of pectolite, $Ca_2NaHSi_3O_9$. Z Kristallogr 125:298–316

Prewitt CT, Sueno S, Papike JJ (1976) The crystal structures of high albite and monalbite at high temperatures. Am Miner 61:1213–1225

Price GD, Pluth JJ, Smith JV, Bennett JM, Patton RL (1982) Crystal structure of tetrapropylammonium fluoride containing precursor to fluoride silicalite. J Am Chem Soc 104: 5971–5977

Price GD, Putnis A, Agrell SO, Smith DGW (1983) Wadsleyite, natural β-$(Mg, Fe)_2SiO_4$ from the Peace River meteorite. Can Mineralogist 21:29–35

Prince E (1971) Refinement of the crystal structure of apophyllite. III. Determination of the hydrogen positions by neutron diffraction. Am Miner 56:1243–1251

Pudovkina ZV, Chernitsova NM, Voronkov AA, Pyatenko YuA (1980) Crystal structure of zirsinalite $Na_6Ca\{Zr[Si_6O_{18}]\}$ Sov Phys Dokl 25:69–70

Pushcharovskii DYu, Burshtein IV, Naumova IS, Pobedimskaya EA, Belov NV (1976) Crystal structure of $Na_2Mn(SiO_3)_2$. Sov Phys Dokl 21:365–367

Pushcharovskii DYu, Karpov OG, Pobedimskaya EA, Belov NV (1977) The crystal structure of $K_3NdSi_6O_{15}$. Sov Phys Dokl 22:292–293

Pushcharovskii DYu, Dago AM, Pobedimskaya EA (1981) Interaction of cation and anion motifs in the structures of TR-silicates, germanates and phosphates. Vestn Mosk Un – Ta 5:1–8 (translated from Vestnik Moskovskogo Universiteta Geologiya 36:10–17)

Pyatenko YuA, Zhdanova TA, Voronkov AA (1979) Crystal structure of $K_4Sc_2(OH)_2Si_4O_{12}$. Sov Phys Dokl 24:794–796

Quintana P, West AR (1981) Synthesis of $Li_2ZrSi_6O_{15}$, a zektzerite-related phase. Miner Mag 44:361–362

Radoslovich EW, Norrish K (1962) The cell dimensions and symmetry of layer-lattice silicates. I. Some structural considerations. Am Miner 47:599–616

Rapoport PA, Burnham CW (1973) Ferrobustamite: the crystal structures of two Ca, Fe bustamite-type pyroxenoids. Z Kristallogr 138:419–438

Rayner JH, Brown G (1973) The crystal structure of talc. Clays and Clay Minerals 21:103–114

Reid AF, Li C, Ringwood AE (1977) High-pressure silicate pyrochlores, $Sc_2Si_2O_7$ and $In_2Si_2O_7$. J Solid State Chem 20:219–226

Ribbe PH, Prunier AR Jr (1977) Stereochemical systematics of ordered C2/c silicate pyroxenes. Am Miner 62:710–720

Ribbe PH, Gibbs GV, Hamil MM (1977) A refinement of the structure of dioptase, $Cu_6[Si_6O_{18}] \cdot 6\,H_2O$. Am Miner 62:807–811

Richard P, Perrault G (1972) Structure cristalline de l'ekanite de St-Hilaire, P.Q. Acta Crystallogr B28:1994–1999

Rinaldi R, Pluth JJ, Smith JV (1974) Zeolites of the phillipsite family. Refinement of the crystal structure of phillipsite and harmotome. Acta Crystallogr B30:2426–2433

Rinaldi R, Pluth JJ, Smith JV (1975) Crystal structure of mazzite dehydrated at 600 °C. Acta Crystallogr B31:1603–1608

Ringwood AE, Reid AF, Wadsley AD (1967) High-pressure $KAlSi_3O_8$, an aluminosilicate with sixfold coordination. Acta Crystallogr 23:1093–1095

Robinson EA (1963) Characteristic vibrational frequencies of oxygen compounds of silicon, phosphorus and chlorine: correlation of stretching frequencies and force constants with bond lengths and bond orders. Can J Chem 41:3021–3033

Robinson PD, Fang JH (1970) The crystal structure of epididymite. Am Miner 55:1541–1549

Robson HE, Shoemaker DP, Ogilvie RA, Manor PC (1973) Synthesis and crystal structure of zeolite rho – a new zeolite related to Linde type A. In: Meier WM and Uytterhoeven JB (eds) Molecular sieves. Advances in chemistry, series 121. Am Chem Soc, p 106

Ronov AB, Yaroshevsky AA (1972) Earth's crust geochemistry. In: Fairbridge RW (ed) The encyclopedia of geochemistry and environmental sciences. Van Nostrand, New York, pp 234–254

Ross FK (1980) Experimental charge density studies in minerals and other inorganic compounds. Trans Am Cryst Assoc 16:79–95

Rossi G, Tazzoli V, Ungaretti L (1974) The crystal structure of ussingite. Am Miner 59:335–340

Rothbauer R (1971) Untersuchung eines $2M_1$-Muskovits mit Neutronenstrahlen. Neues Jahrb Miner Mh 1971:143–154

Roy DM, Mumpton FA (1956) Stability of minerals in the system $ZnO-SiO_2-H_2O$. Econ Geol 51:432–443

Rucklidge JC, Kocman V, Whitlow SH, Gabe EJ (1975) The crystal structures of three canadian vesuvianites. Can Mineralogist 13:15–21

Saburi S, Kusachi I, Henmi C, Kawahara A, Henmi K, Kawada I (1976) Refinement of the structure of rankinite. Miner J Sapporo 8:240–246

Saburi S, Kawahara A, Henmi C, Kusachi I, Kihara K (1977) The refinement of the crystal structure of cuspidine. Miner J Sapporo 8:286–298

Sackerer D, Nagorsen G (1977) Die Kristallstruktur von Bis(äthylendiamin)kupfer-(II)-tris (brenzcatechino)silicat. Z Anorg Allg Chem 437:188–192

Sadanaga R, Marumo F, Takéuchi Y (1961) The crystal structure of harmotome, $Ba_2Al_4Si_{12}O_{32} \cdot 12\,H_2O$. Acta Crystallogr 14:1153–1163

Safronov AN, Nevskii NN, Ilyukhin VV, Belov NV (1980) Crystal structure of the tin silicate $Na_8\{Sn[Si_6O_{18}]\}$. Sov Phys Dokl 25:962–964

Safronov AN, Nevskii NN, Ilyukhin VV, Belov NV (1981) Refinement of the crystal structure of the cementite phase Y-C_6S_3H. Sov Phys Dokl 26:129–130

Safronov AN, Nevskii NN, Ilyukhin VV, Belov NV (1983) New type of silicate radical Si_5O_{16} in the structure of $Na_4Sn_2[Si_5O_{16}] \cdot H_2O$. Sov Phys Dokl 28:304–305

Saf'yanov YuN, Vasil'eva NO, Golovachev VP, Kuz'min EA, Belov NV (1983) Crystal structure of lamprophyllite. Sov Phys Dokl 28:207–209

Sandomirskii PA, Simonov MA, Belov NV (1975) Crystal structure of $Na_2LiFe[Si_2O_5]_3$. Sov Phys Dokl 20:239–241

Sandomirskii PA, Simonov MA, Belov NV (1977) Crystal structure of synthetic Mn-milarite $K_2Mn_5[Si_{12}O_{30}] \cdot H_2O$. Sov Phys Dokl 22: 181–183

Sasaki S, Fujino K, Takéuchi Y, Sadanaga R (1980) On the estimation of atomic charges by the X-ray method for some oxides and silicates. Acta Crystallogr A36: 904–915

Sasaki S, Takéuchi Y, Fujino K, Akimoto S-I (1982) Electron-density distributions of three orthopyroxenes, $Mg_2Si_2O_6$, $Co_2Si_2O_6$, and $Fe_2Si_2O_6$. Z Kristallogr 158: 279–297

Schichl H, Völlenkle H, Wittmann A (1973) Die Kristallstruktur von $Rb_6Si_{10}O_{23}$. Monatsh Chem 104: 854–863

Schmahl W (1981) Struktur und Kristallchemie einer Batisit-Varietät aus der Westeifel (Ba, K)(K, Na)(Na, Ca)(Ti, Fe, Nb, Zr)$_2$ Si_4O_{14}. Diplomarbeit, Universität Mainz

Schneider H, Wohlleben K, Majdic A (1980) Incorporation of impurities in tridymites from a used silica brick. Miner Mag 43: 879–883

Schnering HGv, Hoppe R (1961) Die Kristallstruktur des $SrZnO_2$. Z Anorg Allg Chem 312: 87–98

Schomaker V, Stevenson DP (1941) Some revisions of the covalent radii and the additivity rule for the lengths of partially ionic single covalent bonds. J Am Chem Soc 63: 37–40

Schreyer W, Schairer JF (1961) Metastable solid solutions with quartz-type structures on the join $SiO_2-MgAl_2O_4$. Z Kristallogr 116: 60–82

Schulz H, Tscherry V (1972) Structural relations between the low- and high-temperature forms of β-eucryptite ($LiAlSiO_4$) and low and high quartz. I. Low-temperature form of β-eucryptite and low quartz. II. High-temperature form of β-eucryptite and high quartz. Acta Crystallogr B28: 2168–2173, 2174–2177

Schweinsberg H, Liebau F (1974) Die Kristallstruktur des $K_4[Si_8O_{18}]$: Ein neuer Silikat-Schichttyp. Acta Crystallogr B30: 2202–2213

Scott JD (1976) Crystal structure of miserite, a Zoltai type 5 structure. Can Mineralogist 14: 515–528

Seifert F, Schreyer W (1969) Stability relations of $K_2Mg_5Si_{12}O_{30}$, an end member of the merrihueite-roedderite group of meteoritic minerals. Contrib Mineral Petrol 22: 190–207

Shannon RD (1976) Revised effective ionic radii and systematic studies of interatomic distances in halides and chalcogenides. Acta Crystallogr A32: 751–767

Shannon J, Katz L (1970) The structure of barium silicon niobium oxide, $Ba_3Si_4Nb_6O_{26}$: a compound with linear silicon–oxygen–silicon groups. Acta Crystallogr B26: 105–109

Shannon RD, Prewitt CT (1969) Effective ionic radii in oxides and fluorides. Acta Crystallogr B25: 925–946

Shannon RD, Prewitt CT (1970) Revised values of effective ionic radii. Acta Crystallogr B26: 1046–1048

Shannon RD, Chen H-Y, Berzins T (1977) Ionic conductivity in $Na_5GdSi_4O_{12}$. Mater Res Bull 12: 969–973

Shropshire J, Keat PP, Vaughan PA (1959) The crystal structure of keatite, a new form of silica. Z Kristallogr 112: 409–413

Shumyatskaya NG, Voronkov AA, Belov NV (1971) X-ray diffraction study of leucosphenite. Sov Phys Crystallogr 16: 416–422

Shumyatskaya NG, Blinov VA, Voronkov AA, Ilyukhin VV (1973) Hydrothermal synthesis and crystal structure of Ti wadeite. Sov Phys Dokl 18: 17–19

Shumyatskaya NG, Voronkov AA, Pyatenko YaA (1980) Sazhinite, $Na_2Ce[Si_6O_{14}(OH)] \cdot n\,H_2O$: a new representative of the dalyite family in crystal chemistry. Sov Phys Crystallogr 25: 419–423

Siegrist T, Petter W, Hulliger F (1982) Samarium pyrosilicate sulfide, $Sm_4S_3Si_2O_7$. Acta Crystallogr B38: 2872–2874

Simonov MA, Belov NV (1976) Some structural features of zinc silicates: synthetic $K_2Mn_2Zn_2[SiO_4]_2[Si_2O_7]$ and natural $Zn_4[Si_2O_7](OH)_2 \cdot H_2O$ (hemimorphite). Sov Phys Dokl 21: 476–478

Simonov MA, Egorov-Tismenko YuK, Belov NV (1968) Crystal structure of Na, Cd-silicate $Na_2CdSi_2O_6 = Na_6Cd_3[Si_6O_{18}]$. Sov Phys Dokl 12: 662–664

Simonov MA, Egorov-Tismenko YuK, Belov NV (1975/76) Refined crystal structure of chkalovite $Na_2Be[Si_2O_6]$. Sov Phys Dokl 20: 805–807

Simonov MA, Egorov-Tismenko YuK, Belov NV (1977) Improved crystal structure of Na, Cd-triortho-silicate $Na_2Cd_3[Si_3O_{10}]$. Sov Phys Dokl 22:545–547

Simonov MA, Egorov-Tismenko YuK, Belov NV (1978a) Refinement of crystal structure of Na, Cd-triorthosilicate $Na_4Cd_2[Si_3O_{10}]$. Sov Phys Dokl 23:6–8

Simonov MA, Belokoneva EL, Egorov-Tismenko YuK, Belov NV (1978b) The crystal structures of natural rhodonite $CaMn_4[Si_5O_{15}]$ and synthetic $NaHCd_4[Ge_5O_{15}]$ and $LiHCd_4[Ge_5O_{15}]$. Sov Phys Dokl 22:241–242

Sinclair W, Ringwood AE (1978) Single crystal analysis of the structure of stishovite. Nature (Lond) 272:714–715

Slater JC (1964) Atomic radii in crystals. J Chem Phys 41:3199–3204

Slater JC (1972) Symmetry and energy bands in crystals. Chapters 3 and 4. Dover, New York

Smith JV (1977) Enumeration of 4-connected 3-dimensional nets and classification of framework silicates. I. Perpendicular linkage from simple hexagonal net. Am Miner 62:703–709

Smith JV (1978) Enumeration of 4-connected 3-dimensional nets and classification of framework silicates. II. Perpendicular and near-perpendicular linkages from 4.8^2, 3.12^2 and 4.6.12 nets. Am Miner 63:960–969

Smith JV (1979) Enumeration of 4-connected 3-dimensional nets and classification of framework silicates. III. Combination of helix and zigzag, crankshaft and saw chains with simple 2D nets. Am Miner 64:551–562

Smith JV (1983) Enumeration of 4-connected 3-dimensional nets and classification of framework silicates: Combination of 4-1 chain and 2D nets. Z Kristallogr 165:191–198

Smith JV, Bailey SW (1963) Second review of Al–O and Si–O tetrahedral distances. Acta Crystallogr 16:801–811

Smith JV, Bennett JM (1981) Enumeration of 4-connected 3-dimensional nets and classification of framework silicates: the infinite set of ABC-6 nets; the Archimedean and δ-related nets. Am Miner 66:777–788

Smith JV, Rinaldi F (1962) Framework structures formed from parallel four- and eight-membered rings. Miner Mag 33:202–212

Smith JV, Rinaldi F, Dent Glasser LS (1963) Crystal structures with chabazite framework. II. Hydrated Ca-chabazite at room temperature. Acta Crystallogr 16:45–53

Smolin YuI (1970) New silicon-oxygen radical. Ternary two-level Si_6O_{15} ring in the structure of $[Ni(en)_3]Si_2O_5 \cdot 8.7\,H_2O$. Sov Phys Crystallogr 15:23–27

Smolin YuI (1975) In: Hoebbel D, Wieker W, Franke P, Otto A: Zur Konstitution des neuen Silicatanions $[Si_{10}O_{25}]^{10-}$. Z Anorg Allg Chem 418:35–44

Smolin YuI, Shepelev YuF (1970) The crystal structures of the rare earth pyrosilicates. Acta Crystallogr B26:484–492

Smolin YuI, Shepelev YuF, Butikova IK (1970) The crystal structure of the low-temperature form of samarium pyrosilicate, $Sm_2Si_2O_7$. Sov Phys Crystallogr 15:214–219

Smolin YuI, Shepelev YuF, Butikova IK (1972) Crystal structure of $4[Cu(NH_2CH_2CH_2NH_2)_2] \cdot Si_8O_{20} \cdot 38\,H_2O$. Sov Phys Crystallogr 17:10–15

Smolin YuI, Shepelev YuF, Titov AP (1973) Refinement of the crystal structure of thortveitite $Sc_2Si_2O_7$. Sov Phys Crystallogr 17:749–750

Smolin YuI, Shepelev YuF, Pomes R, Hoebbel D, Wieker W (1976) The silicon-oxygen radical $Si_8O_{18}(OH)_2$ in the $2[Co(NH_2CH_2CH_2NH_2)_3]Si_8O_{18}(OH)_2 \cdot 16.4\,H_2O$ crystal. Sov Phys Crystallogr 20:567–570

Smolin YuI, Shepelev YuF, Pomes R, Hoebbel D, Wieker W (1979) Determination of the crystal structure of tetramethylammonium silicate $8[N(CH_3)_4]Si_8O_{20} \cdot 64.8\,H_2O$ at $t = -100\,°C$. Sov Phys Crystallogr 24:19–23

Sokolova GV, Kashaev AA, Drits VA, Ilyukhin VV (1983) The crystal structure of fedorite. Sov Phys Crystallogr 28:95–96

Solov'eva LP, Borisov SV, Bakakin VV (1972) New skeletal structure in the crystal structure of barium chloroaluminosilicate. Sov Phys Crystallogr 16:1035–1038

Sorrell CA, Negas T (1963) Metastable rubidium aluminum silicate with a hexagonal sheet structure. Science (Wash DC) 141:917

Stewart RF, Whitehead MA, Donnay G (1980) The ionicity of the Si-O bond in low-quartz. Am Miner 65:324–326

Stout GH, Jensen LH (1968) X-ray structure determination. Macmillan, London

Strelkova EE, Karpov OG, Litvin BN, Pobedimskaya EA, Belov NV (1977) A K–Ce silicate with a three-dimensional silicon-oxygen radical [Si_6O_{15}]. Synthesis and X-ray analysis. Sov Phys Crystallogr 22:98–99

Strunz H (1938) Systematik und Struktur der Silikate. Z Kristallogr 98:60–83

Strunz H, Tennyson C (1970) Mineralogische Tabellen, 5. Aufl. Akademische Verlagsgesellschaft, Leipzig

Sueno S, Cameron M, Prewitt CT (1976) Orthoferrosilite: High-temperature crystal chemistry. Am Miner 61:38–53

Swanson DK, Prewitt CT (1983) The crystal structure of $K_2Si^{VI}Si_3^{IV}O_9$. Am Miner 68:581–585

Szymański JT, Owens DR, Roberts AC, Ansell HG, Chao GY (1982) A mineralogical study and crystal-structure determination of nonmetamict ekanite, $ThCa_2Si_8O_{20}$. Can Mineralogist 20:65–75

Tacke R, Wannagat U (1979) Syntheses and properties of bioactive organo-silicon compounds. Top Curr Chem 84:1–76

Tacke R, Linoh H, Attar-Bashi MT, Sheldrick WS, Ernst L, Niedner R, Frohnecke J (1982) Sila-pharmaca, 26th communication [1]. Preparation and properties of silicon compounds with potential curare-like activity. III. Z Naturforsch 37b:1461–1471

Tagai T, Ried H, Joswig W, Korekawa M (1982) Kristallographische Untersuchungen eines Petalits mittels Neutronenbeugung und Transmissionselektronenmikroskopie. Z Kristallogr 160:159–170

Takéuchi Y (1958) A detailed investigation of the structure of hexagonal $BaAl_2Si_2O_8$ with reference to its α-β inversion. Miner J Sapporo 2:311–332

Takéuchi Y (1965) Structures of brittle micas. Clays and Clay Minerals. Proceedings of the 13th national conference, pp 1–25

Takéuchi Y, Donnay G (1959) The crystal structure of hexagonal $CaAl_2Si_2O_8$. Acta Crystallogr 12:465–470

Takéuchi Y, Joswig W (1967) The structure of haradaite and a note on the Si-O bond lengths in silicates. Miner J Sapporo 5:98–123

Takéuchi Y, Kudoh Y (1977) Hydrogen bonding and cation ordering in Magnet Cove pectolite. Z Kristallogr 146:281–292

Takéuchi Y, Sadanaga R (1966) Structural studies of brittle micas. I. The crystal structure of xanthophyllite refined. Miner J Sapporo 4:424–437

Takéuchi Y, Haga N, Ito J (1973) The crystal structure of monoclinic $CaAl_2Si_2O_8$: a case of monoclinic structure closely simulating orthorhombic symmetry. Z Kristallogr 137:380–398

Takéuchi Y, Kudoh Y, Yamanaka T (1976) Crystal chemistry of the serandite-pectolite series and related minerals. Am Miner 61:229–237

Takéuchi Y, Kudoh Y, Ito J (1977) High-temperature derivative structure of pyroxene. Proc Japan Acad Ser B53:60–63

Takéuchi Y, Kudoh Y, Ito J (1984a) New series of superstructures based on a clinopyroxene. I. The structure of the 'enstatite-IV' series, $[Mg_{(x-12)/3}Sc_4][Li_{4/3}Si_{(x-4)/3}]O_x$, with x = 100, 112 or 124. Acta Crystallogr B40:115–125

Takéuchi Y, Mori H, Kudoh Y, Ito J (1984b) New series of superstructures based on a clinopyroxene. II. The structure of the Sc Series of enstatite-IV, $[Mg_{\sim(x-7.5)/3}Sc_{\sim3}][Mg_{2/3}Si_{(x-4)/3}]O_x$, with x = 100, 112 or 124. Acta Crystallogr B40:126–132

Taylor WH (1933) The structure of sanidine and other feldspars. Z Kristallogr 85:425–442

Taylor HFW (1971) The crystal structure of kilchoanite, $Ca_6(SiO_4)(Si_3O_{10})$, with some comments on related phases. Miner Mag 38:26–31

Taylor D (1972) The thermal expansion behaviour of the framework silicates. Miner Mag 38:593–604

Tennyson Ch (1960) Berylliummineralien und ihre pegmatitische Paragenese in den Graniten von Tittling/Bayerischer Wald. Neues Jahrb Miner Abh 94:1253–1265

Thilo E, Wodtcke F (1958) Über die „neutralen" Silbersilikate: $Ag_4SiO_4(Ag_2O_3)_n$, $(Ag_2Si_2O_5)_n$ und $AgAl_2[AlSi_3O_{10}](OH)_2$. Z Anorg Allg Chem 295:247–261

Thompson JB Jr (1978) Biopyriboles and polysomatic series. Am Miner 63:239–249

Tillmanns E, Gebert W, Baur WH (1973) Computer simulation of crystal structures applied to the solution of the superstructure of cubic silicondiphosphate. J Solid State Chem 7:69–84

Tokonami M, Horiuchi H, Nakano A, Akimoto S, Morimoto N (1979) The crystal structure of the pyroxene-type $MnSiO_3$. Miner J Sapporo 9:424–426

Toraya H, Iwai S, Marumo F, Hirao M (1977) The crystal structure of taeniolite, $KLiMg_2Si_4O_{10}F_2$. Z Kristallogr 146:73–83

Tossell JA (1975a) The electronic structures of silicon, aluminum and magnesium in tetrahedral coordination with oxygen from SCF-X_α MO calculations. J Am Chem Soc 97:4840–4844

Tossell JA (1975b) The electronic structures of Mg, Al and Si in octahedral coordination with oxygen from SCF $X\alpha$ MO calculations. J Phys Chem Solids 36:1273–1280

Tossell JA (1977) A comparison of silicon-oxygen bonding in quartz and magnesian olivine from X-ray spectra and molecular orbital calculations. Am Miner 62:136–141

Tossell JA, Gibbs GV (1976) Molecular orbital studies of angular distortions resulting from tetrahedral edge sharing in silicon oxides, sulfides and hydrides. J Mol Struct 35:273–287

Tossell JA, Gibbs GV (1977) Molecular orbital studies of geometries and spectra of minerals and inorganic compounds. Phys Chem Minerals 2:21–57

Tossell JA, Gibbs GV (1978) The use of molecular-orbital calculations on model systems for the prediction of bridging-bond-angle variations in siloxanes, silicates, silicon nitrides and silicon sulfides. Acta Crystallogr A34:463–472

Trojer FJ (1968) The crystal structure of parawollastonite. Z Kristallogr 127:291–308

Trojer FJ (1969) The crystal structure of a high-pressure polymorph of $CaSiO_3$. Z Kristallogr 130:185–206

Turley JW, Boer FP (1968) Structural studies of pentacoordinate silicon. I. Phenyl-(2,2′,2″-nitrilotriethoxy)silane. J Am Chem Soc 90:4026–4030

Turley JW, Boer FP (1969) Structural studies of pentacoordinate silicon. IV. m-Nitrophenyl(2,2′,2″-nitrilotriethoxy)silane. J Am Chem Soc 91:4129–4134

Ullmanns Encyklopädie der technischen Chemie (1976) 4. Aufl, Band 12. Chemie, Weinheim

Urch DS (1969) Direct evidence for *3d–2pπ*-bonding in oxy-anions. J Chem Soc (A): 3026–3028

Urusov VS (1970) Relation of effective charge for Si and O to structure and composition for silicates. Geochem Int 1970:143–147

Vaughan PA (1966) The crystal structure of the zeolite ferrierite. Acta Crystallogr 21:983–990

Veblen DR, Burnham CW (1978) New biopyriboles from Chester, Vermont. II. The crystal chemistry of jimthompsonite, clinojimthompsonite, and chesterite and the amphibole-mica reaction. Am Miner 63:1053–1073

Veblen DR, Buseck PR (1979) Chain-width order and disorder in biopyriboles. Am Miner 64:687–700

Veblen DR, Buseck PR (1981) Hydrous pyriboles and sheet silicates in pyroxenes and uralites: intergrowth microstructures and reaction mechanism. Am Miner 66:1107–1134

Vezzalini G (1984) A refinement of Elba dachiardite: opposite acentric domains simulating a centric structure. Z Kristallogr 166:63–71

Viswanathan K (1981) The crystal structure of a Mg-rich-carpholite. Am Miner 66:1080–1085

Völlenkle H (1981) Verfeinerung der Kristallstrukturen von Li_2SiO_3 und Li_2GeO_3. Z Kristallogr 154:77–81

Völlenkle H, Wittmann A, Nowotny H (1967) Zur Kristallstruktur von $CuGeO_3$. Monatsh Chem 98:1352–1357

Völlenkle H, Wittmann A, Nowotny H (1968) Die Kristallstruktur von $Li_2(Si_{0.25}Ge_{0.75})_2O_5$. Z Kristallogr 126:37–45

Volodina GF, Rumanova IM, Belov NV (1963) Crystal structure of kainosite, $Ca_2(Y, RE)_2[Si_4O_{12}]CO_3 \cdot H_2O$. Dokl Akad Nauk SSSR 149:173–175

Vorma A (1963) Crystal structure of stokesite, $CaSnSi_3O_9 \cdot 2\,H_2O$. Miner Mag 33:615–617

Voronkov MG (1973) Bio-organosilicon chemistry. Chem Brit 9:411–415

Voronkov MG (1979) Biological activity of silatranes. Top Curr Chem 84:77–136

Voronkov AA, Zhdanova TA, Pyatenko YuA (1974) Refinement of the structure of vlasovite $Na_2ZrSi_4O_{11}$ and some characteristics of the composition and structure of the zirconosilicates. Sov Phys Crystallogr 19:152–156

Voronkov AA, Pudovkina ZV, Blinov VA, Ilyukhin VV, Pyatenko YuA (1979a) Crystal structure of kazakovite $Na_6Mn\{Ti[Si_6O_{18}]\}$. Sov Phys Dokl 24:132–134

Voronkov MG, Kashaev AA, Zel'bst EA, Frolov YuL, D'yakov VM, Gubanova LI (1979b) New organic pentacoordinated silicon compounds: (aroyloxymethyl)-trifluorosilanes. Crystal structure of (4-bromobenzoyloxymethyl)-trifluorosilane. Sov Phys Dokl 24: 594–596

Vukcevich MR (1972) A new interpretation of the anomalous properties of vitreous silica. J Noncryst Solids 11:25–63

Wada M, Inoue A, Okutani S et al. (1983) I.P.C.R. air-borne experiments in 1981. Rep Inst Phys Chem Res 59:1–19

Wainwright J, Starkey J (1971) A refinement of the structure of anorthite. Z Kristallogr 133:75–84

Wan C, Ghose S (1978) Inesite, a hydrated calcium manganese silicate with five-tetrahedral-repeat double chains. Am Miner 63:563–571

Wan C, Ghose S, Gibbs GV (1977) Rosenhahnite, $Ca_3Si_3O_8(OH)_2$: crystal structure and the stereochemical configuration of the hydroxylated trisilicate group, $Si_3O_8(OH)_2$. Am Miner 62:503–512

Weber HP (1983) Ferrosilite III, the high-temperature polymorph of $FeSiO_3$. Acta Crystallogr C39:1–3; Ferrosilite III, the high-temperature polymorph of $FeSiO_3$: erratum. Acta Crystallogr. C39:508

Wedepohl KH (1971) Geochemistry. Holt, Reinhart and Winston, New York

Weiss A, Weiss A (1954) Zur Kenntnis der faserigen Siliciumdioxyd-Modifikation. Z Anorg Allg Chem 276:95–112

Weitz G (1972) Die Struktur des Sanidins bei verschiedenen Ordnungsgraden. Z Kristallogr 136:418–426

Wells AF (1975) Structural inorganic chemistry, 4th edn. Clarendon, Oxford

Wenk H-R (1973) The structure of wenkite. Z Kristallogr 137:113–126

Wenk HR (1974) Howieite, a new type of chain silicate. Am Miner 59:86–97

Werner D (1966) Die Kieselsäure im Stoffwechsel von Cyclotella cryptica Reimann, Lewin und Guillard. Arch Mikrobiol. 55:278–308

White JS Jr, Arem JE, Nelen JA, Leavens PB, Thomssen RW (1973) Brannockite a new tin mineral. Mineral Res 4:73–76 (In: Am Miner 58:1111)

Whittaker EJW (1956) The structure of chrysotile. II. Clino-chrysotile. Acta Crystallogr 9:855–862

Whittaker EJW, Wicks FJ (1970) Chemical differences among the serpentine "polymorphs": a discussion. Am Miner 55:1025–1047

Wicks FJ, Whittaker EJW (1975) A reappraisal of the structures of the serpentine minerals. Can Mineralogist 13:227–243

Wiech G, Zöpf E, Chun H-U, Brückner R (1976) X-ray spectroscopic investigation of the structure of silica, silicates and oxides in the crystalline and vitreous state. J Noncryst Solids 21:251–261

Winkler HGF (1953) Tief-$LiAlSiO_4$ (Eukryptit). Acta Crystallogr 6:99

Winnewisser G, Mezger PG, Breuer H-D (1974) Interstellar molecules. Top Curr Chem 44: 1–81. Springer, Berlin Heidelberg New York

Winter JK, Ghose S (1979) Thermal expansion and high-temperature crystal chemistry of the Al_2SiO_5 polymorphs. Am Miner 64:573–586

Winter JK, Ghose S, Okamura FP (1977) A high-temperature study of the thermal expansion and the anisotropy of the sodium atom in low albite. Am Miner 62:921–931

Wojnowski W, Peters K, Böhm MC, Schnering HGv (1984) Die Struktur des Spiro-bis(ethylendithia)silans. Z Anorg Allg Chem in press

Woodrow PJ (1967) The crystal structure of astrophyllite. Acta Crystallogr 22:673–678

Wright AF, Lehmann MS (1981) The structure of quartz at 25 and 590 °C determined by neutron diffraction. J Solid State Chem 36:371–380

Würthwein E-U, Schleyer P von R (1979) Planar tetrakoordiniertes Silicium. Angew Chem 91:588–589 Angew Chem Int Ed Engl 18:553–554

Yada K (1971) Study of microstructure of chrysotile asbestos by high resolution electron microscopy. Acta Crystallogr A27:659–664

Yagi T, Mao H-K, Bell PM (1978) Structure and crystal chemistry of perovskite-type $MgSiO_3$. Phys Chem Minerals 3:97–110

Yamada H, Matsui Y, Ito E (1983) Crystal-chemical characterization of $NaAlSiO_4$ with the $CaFe_2O_4$ structure. Miner Mag 47:177–181

Yamanaka T, Mori H (1981) The structure and polytypes of α-$CaSiO_3$ (pseudowollastonite). Acta Crystallogr B37:1010–1017

Yip KL, Fowler WB (1974) Electronic structure of SiO_2. II. Calculations and results. Phys Rev B10:1400–1408

Young BR, Hawkes JR, Merriman RJ, Sayles MT (1978) Bazirite, $BaZrSi_3O_9$, a new mineral from Rockall Island, Inverness-shire, Scotland. Miner Mag 42:35–40

Zayakina NV, Rozhdestvenskaya IV, Nekrasov IYa, Dadze TP (1980) Crystal structure of Sn, Na-silicate $Na_8SnSi_6O_{18}$. Sov Phys Dokl 25:669–671

Zoltai T (1960) Classification of silicates and other minerals with tetrahedral structures. Am Miner 45:960–973

Zuckerman B (1977) Interstellar molecules. Nature (Lond) 268:491–495

Zvyagin VV (1960) Electron-diffraction determination of the structure of kaolinite. Sov Phys Crystallogr 5:32–42

Appendix I: Periodic Table of the Elements

I A	II A	III B	IV B	V B	VI B	VII B	VIII B			I B	II B	III A	IV A	V A	VI A	VII A	VIII A
H 1																	**He** 2
Li 3	**Be** 4											**B** 5	**C** 6	**N** 7	**O** 8	**F** 9	**Ne** 10
Na 11	**Mg** 12											**Al** 13	**Si** 14	**P** 15	**S** 16	**Cl** 17	**Ar** 18
K 19	**Ca** 20	**Sc** 21	**Ti** 22	**V** 23	**Cr** 24	**Mn** 25	**Fe** 26	**Co** 27	**Ni** 28	**Cu** 29	**Zn** 30	**Ga** 31	**Ge** 32	**As** 33	**Se** 34	**Br** 35	**Kr** 36
Rb 37	**Sr** 38	**Y** 39	**Zr** 40	**Nb** 41	**Mo** 42	**Tc** 43	**Ru** 44	**Rh** 45	**Pd** 46	**Ag** 47	**Cd** 48	**In** 49	**Sn** 50	**Sb** 51	**Te** 52	**I** 53	**Xe** 54
Cs 55	**Ba** 56	°	**Hf** 72	**Ta** 73	**W** 74	**Re** 75	**Os** 76	**Ir** 77	**Pt** 78	**Au** 79	**Hg** 80	**Tl** 81	**Pb** 82	**Bi** 83	**Po** 84	**At** 85	**Rn** 86
Fr 87	**Ra** 88	†	**Ku** 104	**Ha** 105													

° **Lanthanides**	**La** 57	**Ce** 58	**Pr** 59	**Nd** 60	**Pm** 61	**Sm** 62	**Eu** 63	**Gd** 64	**Tb** 65	**Dy** 66	**Ho** 67	**Er** 68	**Tm** 69	**Yb** 70	**Lu** 71
† **Actinides**	**Ac** 89	**Th** 90	**Pa** 91	**U** 92	**Np** 93	**Pu** 94	**Am** 95	**Cm** 96	**Bk** 97	**Cf** 98	**Es** 99	**Fm** 100	**Md** 101	**No** 102	**Lr** 103

Appendix II: Electronegativities, Atomic Radii, and Nonbonding Radii of the Elements

Element	Symbol	Atomic number	Electronegativity		Atomic radii[c]	Nonbonded radii[d]
			A & R[a]	P[b]	[Å]	[Å]
Actinium	Ac	89	1.00	1.1	1.95	
Aluminum	Al	13	1.47	1.61	1.25	1.85
Americium	Am	95	1.2	1.3	1.75	
Antimony	Sb	51	1.82	2.05	1.45	1.88
Argon	Ar	18				
Arsenic	As	33	2.20	2.18	1.15	1.58
Astatine	At	85	1.90	2.2		
Barium	Ba	56	0.97	0.89	2.15	
Berkelium	Bk	97	1.2	1.3		
Beryllium	Be	4	1.47	1.57	1.05	1.39
Bismuth	Bi	83	1.67	2.02	1.60	
Boron	B	5	2.01	2.04	0.85	1.33
Bromine	Br	35	2.74	2.96	1.15	1.56
Cadmium	Cd	48	1.46	1.69	1.55	
Calcium	Ca	20	1.04	1.00	1.80	
Californium	Cf	98	1.2	1.3		
Carbon	C	6	2.50	2.55	0.70	1.25
Cerium	Ce	58	1.08	1.12	1.85	
Cesium	Cs	55	0.86	0.79	2.60	
Chlorine	Cl	17	2.83	3.16	1.00	1.44
Chromium	Cr	24	1.56	1.66	1.40	
Cobalt	Co	27	1.70	1.88	1.35	
Copper	Cu	29	1.75	1.90	1.35	
Curium	Cm	96	1.2	1.3		
Dysprosium	Dy	66	1.10	1.22	1.75	
Einsteinium	Es	99	1.2	1.3		
Erbium	Er	68	1.11	1.24	1.75	
Europium	Eu	63	1.01		1.85	
Fermium	Fm	100	1.2	1.3		
Fluorine	F	9	4.10	3.98	0.50	1.08
Francium	Fr	87	0.86	0.7		
Gadolinium	Gd	64	1.11	1.20	1.80	
Gallium	Ga	31	1.82	1.81	1.30	
Germanium	Ge	32	2.02	2.01	1.25	1.58
Gold	Au	79	1.42	2.54	1.35	

[a] Electronegativity values from Allred and Rochow (1958) supplemented by values from Little and Jones (1960)
[b] Electronegativity values from Pauling (1967)
[c] Atomic radii from Slater (1964, 1972)
[d] Nonbonding radii from Glidewell (1975)

Appendix II (continued)

Element	Symbol	Atomic number	Electronegativity		Atomic radii [c]	Nonbonded radii [d]
			A & R [a]	P [b]	[Å]	[Å]
Hafnium	Hf	72	1.23	1.3	1.55	
Hahnium	Ha	105				
Helium	He	2				
Holmium	Ho	67	1.10	1.23	1.75	
Hydrogen	H	1	2.1	2.20	0.25	0.92
Indium	In	49	1.49	1.78	1.55	
Iodine	I	53	2.21	2.66	1.40	1.85
Iridium	Ir	77	1.55	2.20	1.35	
Iron	Fe	26	1.64	1.83	1.40	
Krypton	Kr	36				
Kurchatovium	Ku	104				
Lanthanum	La	57	1.08	1.10	1.95	
Lawrencium	Lr	103		1.3		
Lead	Pb	82	1.55	2.33	1.80	
Lithium	Li	3	0.97	0.98	1.45	
Lutetium	Lu	71	1.14	1.27	1.75	
Magnesium	Mg	12	1.23	1.31	1.50	
Manganese	Mn	25	1.60	1.55	1.40	
Mendelevium	Md	101	1.2	1.3		
Mercury	Hg	80	1.44	2.00	1.50	
Molybdenum	Mo	42	1.30	2.16	1.45	
Neodymium	Nd	60	1.07	1.14	1.85	
Neon	Ne	10				
Neptunium	Np	93	1.22	1.36	1.75	
Nickel	Ni	28	1.75	1.91	1.35	
Niobium	Nb	41	1.23	1.6	1.45	
Nitrogen	N	7	3.07	3.04	0.65	1.14
Nobelium	No	102	1.2	1.3		
Osmium	Os	76	1.52	2.2	1.30	
Oxygen	O	8	3.50	3.44	0.60	1.13
Palladium	Pd	46	1.35	2.20	1.40	
Phosphorus	P	15	2.06	2.19	1.00	1.46
Platinum	Pt	78	1.44	2.28	1.35	
Plutonium	Pu	94	1.22	1.28	1.75	
Polonium	Po	84	1.76	2.0	1.90	
Potassium	K	19	0.91	0.82	2.20	
Praseodymium	Pr	59	1.07	1.13	1.85	
Promethium	Pm	61	1.07		1.85	
Protactinium	Pa	91	1.14	1.5	1.80	
Radium	Ra	88	0.97	0.9	2.15	
Radon	Rn	86				
Rhenium	Re	75	1.46	1.9	1.35	
Rhodium	Rh	45	1.45	2.28	1.35	
Rubidium	Rb	37	0.89	0.82	2.35	
Ruthenium	Ru	44	1.42	2.2	1.30	
Samarium	Sm	62	1.07	1.17	1.85	
Scandium	Sc	21	1.20	1.36	1.60	
Selenium	Se	34	2.48	2.55	1.15	1.58
Silicon	Si	14	1.74	1.90	1.10	1.55
Silver	Ag	47	1.42	1.93	1.60	
Sodium	Na	11	1.01	0.93	1.80	

Appendix II (continued)

Element	Symbol	Atomic number	Electronegativity		Atomic radii[c]	Nonbonded radii[d]
			A & R[a]	P[b]	[Å]	[Å]
Strontium	Sr	38	0.99	0.95	2.00	
Sulfur	S	16	2.44	2.58	1.00	1.45
Tantalum	Ta	73	1.33	1.5	1.45	
Technetium	Tc	43	1.36	1.9	1.35	
Tellurium	Te	52	2.01	2.1	1.40	1.87
Terbium	Tb	65	1.10		1.75	
Thallium	Tl	81	1.44	2.04	1.90	
Thorium	Th	90	1.11	1.3	1.80	
Thulium	Tm	69	1.11	1.25	1.75	
Tin	Sn	50	1.72	1.96	1.45	1.88
Titanium	Ti	22	1.32	1.54	1.40	
Tungsten	W	74	1.40	2.36	1.35	
Uranium	U	92	1.22	1.38	1.75	
Vanadium	V	23	1.45	1.63	1.35	
Xenon	Xe	54				
Ytterbium	Yb	70	1.06		1.75	
Yttrium	Y	39	1.11	1.22	1.80	
Zinc	Zn	30	1.66	1.65	1.35	
Zirconium	Zr	40	1.22	1.33	1.55	

Appendix III: Ionic Radii

Element	Atomic number	Symbol	Valence	Electron configuration	Spin	Coordination number	Ionic radius [Å] Ahrens[a]	S & P[b]
Actinium	89	Ac	+3	$6p^6$		6	1.18	1.12
Aluminum	13	Al	+3	$2p^6$		4		0.39 *
						5		0.48
						6	0.51	0.535 *
Americium	95	Am	+2	$5f^7$		7		1.21
						8		1.26
						9		1.31
			+3	$5f^6$		6	1.07	0.975
						8		1.09
			+4	$5f^5$		6	0.92	0.85
						8		0.95
Ammonium		NH_4	+1			6	1.48 P	
Antimony	51	Sb	+3	$5s^2$		4 PY		0.76
						5		0.80
						6	0.76	0.76
			+5	$4d^{10}$		6	0.62	0.60 *
Arsenic	33	As	+3	$4s^2$		6	0.58	0.58
			+5	$3d^{10}$		4		0.335 *
						6	0.46	0.46 *
Astatine	85	At	+7	$5d^{10}$		6	0.62	0.62
Barium	56	Ba	+2	$5p^6$		6	1.34	1.35
						7		1.38
						8		1.42
						9		1.47
						10		1.52
						11		1.57
						12		1.61
Berkelium	97	Bk	+3	$5f^8$		6		0.96
			+4	$5f^7$		6		0.83
						8		0.93
Beryllium	4	Be	+2	$1s^2$		3		0.16
						4		0.27 *
						6	0.35	0.45

LS: low spin configuration;
HS: high spin configuration;
PY: pyramidal coordination;
SQ: square planar coordination;
* most reliable values;
** estimated values.
[a] Values from Ahrens (1952), values marked with P from Pauling (1967)
[b] Effective ionic radii from Shannon and Prewitt (1969, 1970) and Shannon (1976)

Appendix III (continued)

Element	Atomic number	Symbol	Valence	Electron configuration	Spin	Coordination number	Ionic radius [Å] Ahrens[a]	S & P[b]
Bismuth	83	Bi	+3	$6s^2$		5		0.96
						6	0.96	1.03 *
						8		1.17
			+5	$5d^{10}$		6	0.74	0.76 **
Boron	5	B	+3	$1s^2$		3		0.01 *
						4		0.11 *
						6	0.23	0.27
Bromine	35	Br	−1	$4p^6$		6	1.95 P	1.96
			+3	$4p^2$		4 SQ		0.59
			+5	$4s^2$		3 PY		0.31
			+7	$3d^{10}$		4		0.25
						6	0.39	0.39
Cadmium	48	Cd	+2	$4d^{10}$		4		0.78
						5		0.87
						6	0.97	0.95
						7		1.03
						8		1.10
						12		1.31
Calcium	20	Ca	+2	$3p^6$		6	0.99	1.00
						7		1.06 *
						8		1.12 *
						9		1.18
						10		1.23
						12		1.34
Californium	98	Cf	+3	$6d^1$		6		0.95
			+4	$5f^8$		6		0.821
						8		0.92
Carbon	6	C	+4	$1s^2$		3		−0.08
						4	0.15 P	0.15
						6	0.16	0.16
Cerium	58	Ce	+3	$6s^1$		6	1.07	1.01
						7		1.07 **
						8		1.143
						9		1.196
						10		1.25
						12		1.34
			+4	$5p^6$		6	0.94	0.87
						8		0.97
						10		1.07
						12		1.14
Cesium	55	Cs	+1	$5p^6$		6	1.67	1.67
						8		1.74
						9		1.78
						10		1.81
						11		1.85
						12		1.88
Chlorine	17	Cl	−1	$3p^6$		6	1.81 P	1.81
			+5	$3s^2$		3 PY		0.12
			+7	$2p^6$		4		0.08 *
						6	0.27	0.27

Appendix III (continued)

Element	Atomic number	Symbol	Valence	Electron configuration	Spin	Coordination number	Ionic radius [Å] Ahrens[a]	S & P[b]
Chromium	24	Cr	+2	$3d^4$	LS	6		0.73**
					HS	6		0.80
			+3	$3d^3$		6	0.63	0.615*
			+4	$3d^2$		4		0.41
						6		0.55
			+5	$3d^1$		4		0.345
						6		0.49**
						8		0.57
			+6	$3p^6$		4		0.26
						6	0.52	0.44
Cobalt	27	Co	+2	$3d^7$	HS	4		0.58
						5		0.67
					LS	6		0.65
					HS	6	0.72	0.745*
						8		0.90
			+3	$3d^6$	LS	6		0.545*
					HS	6	0.63	0.61
			+4	$3d^5$		4		0.40
					HS	6		0.53
Copper	29	Cu	+1	$3d^{10}$		2		0.46
						4		0.60**
						6	0.96	0.77**
			+2	$3d^9$		4		0.57
						4SQ		0.57*
						5		0.65*
						6	0.72	0.73
			+3	$3d^8$	LS	6		0.54
Curium	96	Cm	+3	$5f^7$		6		0.97
			+4	$5f^6$		6		0.85
						8		0.95
Deuterium	1	D	+1	$1s^0$		2		−0.10
Dysprosium	66	Dy	+2	$4f^{10}$		6		1.07
						7		1.13
						8		1.19
			+3	$4f^9$		6	0.92	0.912
						7		0.97
						8		1.027
						9		1.083
Erbium	68	Er	+3	$4f^{11}$		6	0.89	0.890
						7		0.945
						8		1.004
						9		1.062
Europium	63	Eu	+2	$4f^7$		6		1.17
						7		1.20
						8		1.25
						9		1.30
						10		1.35
			+3	$4f^6$		6	0.98	0.947
						7		1.01
						8		1.066
						9		1.120

Appendix III (continued)

Element	Atomic number	Symbol	Valence	Electron configuration	Spin	Coordination number	Ionic radius [Å] Ahrens[a]	S & P[b]
Fluorine	9	F	−1	$2p^6$		2		1.285
						3		1.30
						4		1.31
						6	1.36 P	1.33
			+7	$1s^2$		6	0.08	0.08
Francium	87	Fr	+1	$6p^6$		6	1.80	1.80
Gadolinium	64	Gd	+3	$4f^7$		6	0.97	0.938
						7		1.00
						8		1.053
						9		1.107
Gallium	31	Ga	+3	$3d^{10}$		4		0.47 *
						5		0.55
						6	0.62	0.620 *
Germanium	32	Ge	+2	$4s^2$		6	0.73	0.73
			+4	$3d^{10}$		4		0.390 *
						6	0.53	0.530 *
Gold	79	Au	+1	$5d^{10}$		6	1.37	1.37
			+3	$5d^8$		4 SQ		0.68
						6	0.85	0.85
			+5	$5d^6$		6		0.57
Hafnium	72	Hf	+4	$4f^{14}$		4		0.58
						6	0.78	0.71
						7		0.76
						8		0.83
Holmium	67	Ho	+3	$4f^{10}$		6	0.91	0.901
						8		1.015
						9		1.072
						10		1.12
Hydrogen	1	H	−1				2.08 P	
			+1	$1s^0$		1		−0.38
						2		−0.18
Hydroxyl		OH	−1			2		1.32
						3		1.34
						4		1.35 **
						6		1.37 **
Indium	49	In	+3	$4d^{10}$		4		0.62
						6	0.81	0.800 *
						8		0.92
Iodine	53	I	−1	$5p^6$		6	2.16 P	2.20
			+5	$5s^2$		3 PY		0.44 *
						6	0.62	0.95
			+7	$4d^{10}$		4		0.42
						6	0.50	0.53
Iridium	77	Ir	+3	$5d^6$		6		0.68 **
			+4	$5d^5$		6	0.68	0.625
			+5	$5d^4$		6		0.57 **
Iron	26	Fe	+2	$3d^6$	HS	4		0.63
					HS	4 SQ		0.64
					LS	6		0.61 **
					HS	6	0.74	0.780 *
					HS	8		0.92

Appendix III (continued)

Element	Atomic number	Symbol	Valence	Electron configuration	Spin	Coordination number	Ionic radius [Å]	
							Ahrens[a]	S & P[b]
(Iron)			+3	$3d^5$	HS	4		0.49*
						5		0.58
					LS	6		0.55
					HS	6	0.64	0.645*
					HS	8		0.78
			+4	$3d^4$		6		0.585
			+6	$3d^2$		4		0.25
Lanthanum	57	La	+3	$4d^{10}$		6	1.14	1.032
						7		1.10
						8		1.160
						9		1.216
						10		1.27
						12		1.36
Lead	82	Pb	+2	$6s^2$		4 PY		0.98
						6	1.20	1.19
						7		1.23
						8		1.29
						9		1.35
						10		1.40
						11		1.45
						12		1.49
			+4	$5d^{10}$		4		0.65**
						5		0.73**
						6	0.84	0.775
						8		0.94
Lithium	3	Li	+1	$1s^2$		4		0.590*
						6	0.68	0.76*
						8		0.92
Lutetium	71	Lu	+3	$4f^{14}$		6	0.85	0.861
						8		0.977
						9		1.032
Magnesium	12	Mg	+2	$2p^6$		4		0.57
						5		0.66
						6	0.66	0.720*
						8		0.89
Manganese	25	Mn	+2	$3d^5$	HS	4		0.66
					HS	5		0.75
					LS	6		0.67**
					HS	6	0.80	0.830*
					HS	7		0.90
						8		0.96
			+3	$3d^4$		5		0.58
					LS	6		0.58
					HS	6	0.66	0.645*
			+4	$3d^3$		4		0.39
						6	0.60	0.530*
			+5	$3d^2$		4		0.33
			+6	$3d^1$		4		0.255
			+7	$3p^6$		4		0.25
						6	0.46	0.46

Appendix III (continued)

Element	Atomic number	Symbol	Valence	Electron configuration	Spin	Coordination number	Ionic radius [Å] Ahrens[a]	S & P[b]
Mercury	80	Hg	+1	$6s^1$		3		0.97
						6		1.19
			+2	$5d^{10}$		2		0.69
						4		0.96
						6	1.10	1.02
						8		1.14
Molybdenum	42	Mo	+3	$4d^3$		6		0.69 **
			+4	$4d^2$		6	0.70	0.650
			+5	$4d^1$		4		0.46
						6		0.61
			+6	$4p^6$		4		0.41 *
						5		0.50
						6	0.62	0.59 *
						7		0.73
Neodymium	60	Nd	+2	$4f^4$		8		1.29
						9		1.35
			+3	$4f^3$		6	1.04	0.983
						8		1.109 *
						9		1.163
						12		1.27 **
Neptunium	93	Np	+2	$5f^5$		6		1.10
			+3	$5f^4$		6	1.10	1.01
			+4	$5f^3$		6	0.95	0.87
						8		0.98
			+5	$5f^2$		6		0.75
			+6	$5f^1$		6		0.72
			+7	$6p^6$		6	0.71	0.71
Nickel	28	Ni	+2	$3d^8$		4		0.55
						4 SQ		0.49
						5		0.63 **
						6	0.69	0.690 *
			+3	$3d^7$	LS	6		0.56 *
					HS	6		0.60 **
			+4	$3d^6$	LS	6		0.48
			+3	$4d^2$		6		0.72
Niobium	41	Nb	+4	$4d^1$		6	0.74	0.68 **
						8		0.79
			+5	$4p^6$		4		0.48
						6	0.69	0.64
						7		0.69
						8		0.74
Nitrogen	7	N	−3	$2p^6$		4		1.46
			+3	$2s^2$		6	0.16	0.16
			+5	$1s^2$		3		−0.104
						6	0.13	0.13
Nobelium	102	No	+2	$5f^{14}$		6		1.1 **
Osmium	76	Os	+4	$5d^4$		6	0.69	0.630
			+5	$5d^3$		6		0.575 **
			+6	$5d^2$		5		0.49
						6		0.545 **

Appendix III (continued)

Element	Atomic number	Symbol	Valence	Electron configuration	Spin	Coordination number	Ionic radius [Å]	
							Ahrens[a]	S & P[b]
(Osmium)			+7	$5d^1$		6		0.525 **
			+8	$5p^6$		4		0.39
Oxygen	8	O	−2	$2p^6$		2		1.35
						3		1.36
						4		1.38
						6	1.40 P	1.40
						8		1.42
Palladium	46	Pd	+1	$4d^9$		2		0.59
			+2	$4d^8$		4 SQ		0.64
						6	0.80	0.86
			+3	$4d^7$		6		0.76
			+4	$4d^6$		6	0.65	0.615
Phosphorus	15	P	+3	$3s^2$		6	0.44	0.44
			+5	$2p^6$		4		0.17 *
						5		0.29
						6	0.35	0.38
Platinum	78	Pt	+2	$5d^8$		4 SQ		0.60
						6	0.80	0.80
			+4	$5d^6$		6	0.65	0.625
			+5	$5d^5$		6		0.57 **
Plutonium	94	Pu	+3	$5f^5$		6	1.08	1.00
			+4	$5f^4$		6	0.93	0.86
						8		0.96
			+5	$5f^3$		6		0.74 **
			+6	$5f^2$		6		0.71
Polonium	84	Po	+4	$6s^2$		6		0.94
						8		1.08
			+6	$5d^{10}$		6	0.67	0.67
Potassium	19	K	+1	$3p^6$		4		1.37
						6	1.33	1.38
						7		1.46
						8		1.51
						9		1.55
						10		1.59
						12		1.64
Praseodymium	59	Pr	+3	$4f^2$		6	1.06	0.99
						8		1.126
						9		1.179
			+4	$4f^1$		6	0.92	0.85
						8		0.96
Promethium	61	Pm	+3	$4f^4$		6	1.06	0.97
						8		1.093
						9		1.144
Protactinium	91	Pa	+3	$5f^2$		6	1.13	1.04 **
			+4	$6d^1$		6	0.98	0.90
						8		1.01
			+5	$6p^6$		6	0.89	0.78
						8		0.91
						9		0.95

Appendix III (continued)

Element	Atomic number	Symbol	Valence	Electron configuration	Spin	Coordination number	Ionic radius [Å] Ahrens[a]	S & P[b]
Radium	88	Ra	+2	$6p^6$		6	1.43	
						8		1.48
						12		1.70
Rhenium	75	Re	+4	$5d^3$		6	0.72	0.63
			+5	$5d^2$		6		0.58 **
			+6	$5d^1$		6		0.55 **
			+7	$5p^6$		4		0.38
						6	0.56	0.53
Rhodium	45	Rh	+3	$4d^6$		6	0.68	0.665
			+4	$4d^5$		6		0.60
			+5	$4d^4$		6		0.55
Rubidium	37	Rb	+1	$4p^6$		6	1.47	1.52
						7		1.56
						8		1.61
						9		1.63 **
						10		1.66
						11		1.69
						12		1.72
						14		1.83
Ruthenium	44	Ru	+3	$4d^5$		6		0.68
			+4	$4d^4$		6	0.67	0.620
			+5	$4d^3$		6		0.565 **
			+7	$4d^1$		4		0.38
			+8	$4p^6$		4		0.36
Samarium	62	Sm	+2	$4f^6$		7		1.22
						8		1.27
						9		1.32
			+3	$4f^5$		6	1.00	0.958
						7		1.02 **
						8		1.079
						9		1.132
						12		1.24
Scandium	21	Sc	+3	$3p^6$		6	0.81	0.745 *
						8		0.870 *
Selenium	34	Se	−2	$4p^6$		6	1.98 P	1.98
			+4	$4s^2$		6	0.50	0.50
			+6	$3d^{10}$		4		0.28 *
						6	0.42	0.42
Silicon	14	Si	+4	$2p^6$		4		0.26 *
						6	0.42	0.400 *
Silver	47	Ag	+1	$4d^{10}$		2		0.67
						4		1.00
						4 SQ		1.02
						5		1.09
						6	1.26	1.15
						7		1.22
						8		1.28
			+2	$4d^9$		4 SQ		0.79
						6	0.89	0.94
			+3	$4d^8$		4 SQ		0.67
						6		0.75

Appendix III (continued)

Element	Atomic number	Symbol	Valence	Electron configuration	Spin	Coordination number	Ionic radius [Å] Ahrens[a]	S & P[b]
Sodium	11	Na	+1	$2p^6$		4		0.99
						5		1.00
						6	0.97	1.02
						7		1.12
						8		1.18
						9		1.24
						12		1.39
Strontium	38	Sr	+2	$4p^6$		6	1.12	1.18
						7		1.21
						8		1.26
						9		1.31
						10		1.36
						12		1.44
Sulfur	16	S	−2	$3p^6$		6	1.84 P	1.84
			+4	$3s^2$		6	0.37	0.37
			+6	$2p^6$		4		0.12*
						6	0.30	0.29
Tantalum	73	Ta	+3	$5d^2$		6		0.72**
			+4	$5d^1$		6		0.68**
			+5	$5p^6$		6	0.68	0.64
						7		0.69
						8		0.74
Technetium	43	Tc	+4	$4d^3$		6		0.645
			+5	$4d^2$		6		0.60**
			+7	$4p^6$		4		0.37
						6	0.56	0.56
Tellurium	52	Te	−2	$5p^6$		6	2.21 P	2.21
			+4	$5s^2$		3		0.52
						4		0.66
						6	0.70	0.97
			+6	$4d^{10}$		4		0.43
						6	0.56	0.56*
Terbium	65	Tb	+3	$4f^8$		6	0.93	0.923
						7		0.98**
						8		1.040
						9		1.095
			+4	$4f^7$		6	0.81	0.76
						8		0.88
Thallium	81	Tl	+1	$6s^2$		6	1.47	1.50
						8		1.59
						12		1.70**
			+3	$5d^{10}$		4		0.75
						6	0.95	0.885
						8		0.98
Thorium	90	Th	+4	$6p^6$		6	1.02	0.94
						8		1.05
						9		1.09*
						10		1.13**
						11		1.18
						12		1.21

Appendix III (continued)

Element	Atomic number	Symbol	Valence	Electron configuration	Spin	Coordination number	Ionic radius [Å] Ahrens[a]	S & P[b]
Thulium	69	Tm	+2	$4f^{13}$		6		1.03
						7		1.09
			+3	$4f^{12}$		6	0.87	0.880
						8		0.994
						9		1.052
Tin	50	Sn	+2	$5s^2$		6	0.93	
						8		1.22
			+4	$4d^{10}$		4		0.55
						5		0.62
						6	0.71	0.690*
						7		0.75
						8		0.81
Titanium	22	Ti	+2	$3d^2$		6		0.86**
			+3	$3d^1$		6	0.76	0.670*
			+4	$3p^6$		4		0.42
						5		0.51
						6	0.68	0.605*
						8		0.74
Tungsten	74	W	+4	$5d^2$		6	0.70	0.66
			+5	$5d^1$		6		0.62
			+6	$5p^6$		4		0.42*
						5		0.51
						6	0.62	0.60*
Uranium	92	U	+3	$5f^3$		6		1.025
			+4	$5f^2$		6	0.97	0.89
						7		0.95**
						8		1.00*
						9		1.05
						12		1.17**
			+5	$5f^1$		6		0.76
						7		0.84**
			+6	$6p^6$		2		0.45
						4		0.52
						6	0.80	0.73*
						7		0.81**
						8		0.86
Vanadium	23	V	+2	$3d^3$		6	0.88	0.79
			+3	$3d^2$		6	0.74	0.640*
			+4	$3d^1$		5		0.53
						6	0.63	0.58*
						8		0.72**
			+5	$3p^6$		4		0.355*
						5		0.46*
						6	0.59	0.54
Xenon	54	Xe	+8	$4d^{10}$		4		0.40
						6		0.48
Ytterbium	70	Yb	+2	$4f^{14}$		6		1.02
						7		1.08**
						8		1.14
			+3	$4f^{13}$		6	0.86	0.868*

Appendix III (continued)

Element	Atomic number	Symbol	Valence	Electron configuration	Spin	Coordination number	Ionic radius [Å]	
							Ahrens[a]	S & P[b]
(Ytterbium)						7		0.925 **
						8		0.985
						9		1.042
Yttrium	39	Y	+3	$4p^6$		6	0.92	0.900 *
						7		0.96
						8		1.019 *
						9		1.075
Zinc	30	Zn	+2	$3d^{10}$		4		0.60 *
						5		0.68 *
						6	0.74	0.740 *
						8		0.90
Zirconium	40	Zr	+4	$4p^6$		4		0.59
						5		0.66
						6	0.79	0.72 *
						7		0.78 *
						8		0.84 *
						9		0.89

Subject Index

abundances of elements 1, 3
abundance of Si 1, 3
acid pyroxenoids 179, 180
additional ligands 187–194, 268–270
adjustment, mutual 175, 182, 184, 195, 199, 208, 217, 223
Al content of feldspars 246
– – of zeolites 244
Al/Si distribution 12, 88, 108, 246, 277
Al/Si replacement 12, 85, 214, 238
aluminosilicates, definition 13
aluminum avoidance rule 88, 108, 246
aluminum silicates, definition 13
amphoteric character of Si 14
angles O–Si–O 15, 18–20, 31, 36, 161, 168
– Si–O–Si 15, 18, 21, 24–30, 36, 44, 161, 168, 209, 262
– T–O–T 248, 250, 251, 254–256, 258, 260–265
angular distortion 30, 267
anhydrous single layer silicates 198–205, 208, 212, 229–231
aristotype 253, 274
atomic distances 22

b_{oct} 213, 218, 220
b_{tetr} 213, 218, 220
B/Si distribution 88
ball and stick model 8
beryllonite-type 256–258, 261, 265, 275, 276
bond angles O–Si–O 15, 18–20, 31, 36, 161, 168
– – Si–O–Si 15, 18, 21, 24–30, 36, 44, 161, 168, 209, 262
– – T–O–T 248, 250, 251, 254–256, 258, 260–265
bond character, covalent 34–44, 46, 266
– –, ionic 33, 46, 266
– –, π 36, 38
– –, σ 42
– –, Si–O 32, 46
bond energies 5
bond length distortion 30, 267
bond lengths $d(Si^{[4]}–O)$ 14–24, 161, 168
– – $d(Si^{[6]}–O)$ 15, 30–32
bond strength 17, 21, 163
Bragg classification 136
branch 36
branched anions, complementation 88
– –, definition 54, 57
branched silicates 88, 94, 96, 97, 100, 104, 113, 119–122, 124–128, 187–191, 268
branchedness 54, 57, 268
building units, chain-like 146–151
– –, ladder-like 146
– –, layer-like 151–155
– –, polyhedral 147, 155–159
– –, secondary 145
– –, tubular 147, 150, 151

cages 117, 159, 240–243
cage, fundamental 157
carbon compounds 5
cation-poor single layer silicates 213–223, 236
cation-rich single layer silicates 213, 214, 220, 221, 223–229, 236
chain conformation 170–187
chain flatness 170
chain-like building units 146–151
chain multiplicity faults 110
chain periodicity 57, 101–104, 170–186, 273
chain periods 80, 180, 185, 197, 209
chain silicates 78–81, 83, 170–187, 206–212
channels 157, 159
charge balance 164–168, 273
charge density distribution 38, 48
charge distribution 165, 168
class 63
classification, Bragg's 136
–, Kostov's 137
–, Zoltai's 13, 138–143
classification based on $[MO_n]$ polyhedra 159
– categories 63
– of tectosilicates 143–159
– parameters 52–63, 76, 266–273
– procedure 82–84
columnar chains 109–111
complementation of branched anions 88

complex anion 55–56
condensation reactions 41
conformation of chains 170–187
connectedness 51, 144, 161, 268
connectedness difference 161, 163–166
convolution of layers 199–203
coordination number of oxygen 16, 21, 30, 37
– – – Si 14, 41, 52, 266
correlation $d\,(Si^{[4]}-O)/B(O)$ 23
– $d\,(Si^{[4]}-O)/CN(O)$ 17, 21
– $d\,(Si^{[6]}-O)/CN(O)$ 30
– $d\,(Si^{[4]}-O)/O:Si$ 17
– $d\,(Si^{[4]}-O)/p_O$ 20, 21
– $d\,(Si^{[4]}-O)/\boldsymbol{s}$ 20
– $d\,(Si^{[4]}-O/\sphericalangle\,Si-O-Si$ 21
– f_s/r_M 177, 182
– f_s/v_M 177, 182
– f_s/χ_M 177–179
– $\boldsymbol{P}/f_s$ 177, 183
– $r_M/\boldsymbol{P}$ 173, 184
– $v_M/\boldsymbol{P}$ 175, 184
– $\chi_M/\boldsymbol{P}$ 175, 184
– $\sphericalangle\,Si-O-Si/B(O)$ 29
– $\sphericalangle\,Si-O-Si/CN(O)$ 25
corrugated layers 199–203, 215, 224
cosmic abundance of Si 1
cosmic dust 1, 2
coupled replacement of Si 262
covalency of Si–O bond 34–44, 46, 266
cristobalite-type 256, 261, 265, 275
crustal abundance 3, 4
curled layers 227–229, 238
curling units 227
cyclic anions 57, 62, 268
cyclosilicates 70, 72, 83, 96–102, 172, 188, 191–195, 202–205, 268
–, branched 97, 100, 188
–, unbranched 96, 98, 191–195, 202–205

(d-p) π bond 36, 37, 45
$d\,(Si-O_{br})$ 17, 18
$d\,(Si-O_{term})$ 17, 18
$d\,(Si^{[4]}-O)$ 14–24, 161, 168
$d\,(Si^{[6]}-O)$ 15, 30–32
decasilicate, unbranched 94, 95
decondensation reactions 41
deformation electron density 38, 48
destabilization of anions 191, 268
dicyclosilicates 70, 97, 99, 168, 194, 202
Dimensionalität 62
dimensionality 57, 161, 166–168, 273
dioctahedral layer silicates 213, 223
diphyllosilicates 70, 121–126, 231–239
–, anhydrous 123, 231–239
–, branched 124–125, 232
–, hybrid 124, 125
–, hydrous 122–126, 231–237
–, unbranched 121–124, 232
dipolysilicates 70, 80, 105, 108–112, 140, 184–187, 202–205
–, branched 112, 113, 188–190
–, even-periodic 80, 105, 109–112
–, hybrid 112, 113
–, unbranched 80, 105, 108, 109, 111, 184–187, 202–205
directedness 117, 126, 198, 216, 255, 260
–, regions of equal 216–223, 227–229, 233, 238
disilicates 18, 25, 70, 93
disiloxanes 51
disorder 22, 26, 27
distance $d\,(Si^{[4]}-O)$ 14–24, 161, 168
– $d\,(Si^{[6]}-O)$ 15, 30–32
distortion, angular 30, 267
–, bond length 30, 267
– of $[SiO_6]$ 30
distribution, Al/Si 12, 88, 108, 246, 277
diversity of compounds 5
double bond character, Si–O 36
double chain silicates 70, 80, 105, 108–112, 140, 184–187, 202–205
– – –, branched 112, 113, 188–190
– – –, even periodic 80, 105, 109–112
– – –, hybrid 112, 113
– – –, unbranched 80, 105, 108, 109, 111, 131, 184–187, 202–205
double crankshaft chain 148, 149
double layers 122–125, 321
– –, branched 124, 125
– –, hybrid 124, 125
– –, unbranched 124, 125
double layer silicates 121–126, 231–239
– – –, anhydrous 123, 231–239
– – –, branched 124–125, 232
– – –, hybrid 124–125
– – –, hydrous 122–126, 231–237
– – –, unbranched 121–124, 232
double oxides 69
double rings 97
double ring silicates 70, 97, 99, 168, 194, 202
dreier chain silicates 78–81, 102–113, 172–180
dreier double chains 78, 81, 105–111
dreier single layers 114, 118, 217
dynamic disorder 22, 26

edge-sharing polyhedra 93, 132, 161, 268
eightfold tetrahedron 94, 95
einer chains 104, 108
electron density, residual 37, 48
electronegativity, influence on $\boldsymbol{CN}$ 41, 266
–, influence on $d\,(Si-O)$ 48–51
–, influence on structure 174–181, 184, 266–273

electrostatic bond strength 17, 163
even-periodic chain silicates 79–81, 102–112, 174–178, 181, 185
extraterrestrial material 1, 2

face-sharing polyhedra 161, 268
family 63
feldspars, cation positions 246–252
–, framework rigidity 249
feldspar-type 248, 251, 252, 261, 264, 265
five-coordinated Si 39–43
fivefold chain silicate 109–113, 207–212
fivefold tetrahedron 94, 165
flat layers 218
flatness of chains 170
folded layers 199–202
formulae, structural 72, 105, 106
four-coordinated Si 14, 35–38, 266
fourfold chain silicates 109–111, 168, 184–186, 206–212
fourfold tetrahedron 94
framework density 155, 156, 243, 245, 276
framework, interpenetrating 128
frameworks 61, 127, 128
framework silicates 70, 72, 83, 126–129, 143–159, 239–265, 274–277
frequency distribution of silicates 130–135
fundamental anion 55, 62
Fundamental-Anion 62
fundamental cage 157
fundamental chains 57, 59
Fundamental-Kette 62
fünfer chain silicates 79, 81, 103–109

gemischt-verzweigt 62
geschlossen-verzweigt 62
glass, sodium-silicate 49
glossary 62
grand mean value $\langle d\,(Si^{[4]}-O)\rangle$ 14, 16
– – – $\langle d\,(Si^{[6]}-O)\rangle$ 15
– – – $\langle \measuredangle\, Si-O-Si\rangle$ 24, 30
group 63
group electronegativity 39
guest molecules in clathrasils 241–243

"hard" cations 188, 192, 270
hexacelsian-type 124, 251, 252, 264, 265
hexahydroxosilicates 69
hexaoxosilicates 11, 15, 30, 41, 42, 51, 69, 90–93, 130, 132, 162, 266
hybrid anion 55–57
hybrid silicates 112, 113, 124, 128
hydrous layer silicates 212–238
hydrous phyllosilicates 212–238

Icmm-type 256, 257, 259, 261, 262, 265, 275, 276

ilmenite-type 31, 69
Immm-type 251, 252, 257, 264, 265
inosilicates 72, 78–81, 83, 170–187, 206–212
inter-chain angle Si–O–Si 209
intergrowth talc-pyroxene 113
inter-layer cations 223, 239
interpenetrating silicates 128, 145
interrupted frameworks 126, 144, 166, 201, 203, 205, 244
interstellar material 1, 2
intra-chain angle Si–O–Si 209
inversion of tetrahedra 216–229, 238, 260
ionic bond 33, 46, 266
ionicity of Si–O bond 33, 46, 266
ionic radii 308–318
ionic radius of Si 33
ionization energies 35

kaliophilite-type 256, 261, 265, 275
kaolinite-like layers 213, 214, 220, 221, 223–229, 236
keatite-type 248, 254, 261, 262, 264, 265
Kettenperiodizität 62
komplexe Anionen 62
Kostov classification 137

ladder-like chain units 146
lattice fault 110–113
layer convolution 199–203
layer misfit 214, 238
layer-like building units 151–155
layers, corrugated 199–203, 215, 224
–, curled 227–229, 238
–, flat 212
–, folded 199–202
–, rolled 224
–, tubular 224–227
–, warped 199, 202
layer silicates 70, 72, 83, 113–126, 195–239
– –, anhydrous 123, 198–205, 208, 212, 229–239
– –, hydrous 212–238
leucite-type 254, 261, 263, 265
linkedness 51, 162, 268
linking unit 233–238
Loewenstein's rule 88, 108, 246
loop-branched anion 54–62

M–O bond 14, 48–51
Mehrfachpolyeder 62
mesh-size 213, 218–221
metasilicates 70
meteorites 1, 2
mica-like layers 213–223, 236
micas 212–216, 218–221

misfit of layers 214, 238
mixed-anion silicates 76, 88, 92, 94, 96, 99, 110, 115, 118, 122, 126, 129, 135, 142, 269, 270
mixed-branched anion 54
mixed coordination number anions 63, 64
mixed coordination number phase 91–93
mixed linkedness 63
model substance 201
models 7
monocyclosilicates 70, 96, 98, 192
monophyllosilicates 70, 113–121, 195–231
–, anhydrous 198–205, 208, 212, 229–231
–, branched 119–121, 188–189
–, hydrous 229–231
–, unbranched 115, 196
monopolysilicates 70, 101–107, 170–191
–, branched 104–105, 187–191
–, unbranched 101–104, 170–187
monosilicates 70
multiple anions 57
multiple chains 57, 78
multiple chain silicates 80–81, 105–113, 184–188, 203–212
– – –, branched 112, 113, 187–188
– – –, hybrid 112, 113, 188
– – –, unbranched 105–111, 184–187
multiple polyhedra 57, 64
multiple rings 97
multiple tetrahedra, branched 94, 96
– –, unbranched 93–95, 269–272
multiplicity 57, 269
Multiplizität 62
mutual adjustment 175, 182, 184, 195, 199, 208, 217, 223

nesosilicates 72
nesosubsilicates 72
neuner chain silicate 81, 103
ninefold tetrahedron 94, 95
nomenclature, chemical 69–70
–, mineralogical 71–72
nonasilicate 94, 95
nonbonded interactions 44–46, 51
nonbonded radii 45, 305
nonlinking unit 233–238
number of different silicate anions 76–78, 269–271

occurrence of Si 1, 4
octahedral layers 212–214
octahedral layer splitting 216, 223
octasilicate 94, 95
odd-periodic chain silicates 173, 177–181, 184, 186
offen-verzweigt 62
oligopolysilicates, branched 112, 113, 187–188
–, hybrid 112, 113, 188
–, unbranched 105–111, 184–187
oligosilicate anions 94–96, 269–272
oligosilicates 70, 83, 94, 269–273
–, branched 94, 96
–, unbranched 93–95, 269–273
orbital participation 36, 39, 42, 44
order 63
oriented intergrowth 113
open-branched anion 54
O–Si–O angles 15, 18–20, 31, 36, 161, 168

π-bond character 36, 38
paracelsian-type 251, 252, 264, 265
parsimony principle 269, 272
Pauling's rules 17, 33, 162, 163
pentacoordinate silicon 39–43
pentapolysilicates 106–107, 109–110, 206–212
pentasilicate 94, 95
periodicity 57–61, 101–109, 132–134, 170, 178, 183, 273
periodic system of silicate anions 65–68
perovskite-type 31, 91
phase transformations 41
phenakite-type 256, 261, 275
phyllosilicates 70, 72, 83, 113–126, 195–239
–, anhydrous 123, 198–205, 208, 212, 229–239
–, hydrous 212–238
planar fourfold coordination 38
polyhedral building units 147, 155–159
polyhedral models 8
polyhedral cavities 155–159, 240–243
polyphosphates 104, 178, 181–183, 195
polysilicates 70, 78–81, 83, 170–187, 206–212
pressure, influence on structure 267
primary polyhedron 54
principle of parsimony 269, 272
procedure, classification 82–84
pyknolites 156, 159
pyrochlore-type 31, 91
pyroxene-type 255, 261, 262, 265
pyroxenoids, acid 179, 180

Q^s unit 54
quadruple chain silicates 109–111, 168, 184–186, 206–212
quadruple tetrahedron 94, 165
quartz-type 248, 251, 254, 256, 261, 262, 264, 265, 275, 276
quaternary polyhedron 54

quinary polyhedron 54
quintuple chain silicates 109–113, 207–212
quintuple tetrahedron 94, 165

radii, Shannon/Prewitt 308–318
radius of curvature 220, 225
radius of cations, influence on structure 173, 177, 182, 184, 267, 273, 274, 276
reaction path concept 41
reduction of strain 214–229, 238
regions of equal directedness 216–223, 227–229, 233, 238
reliability index 15
replacement Al/Si 12, 85, 214, 238
– B/Si 238
– of cations M 227
– of Si 12, 214, 229, 238, 244, 262
residual charge 46
ring anions 57, 62, 268
ring periodicity 59, 61, 268
Ringperiodizität 62
ring silicates 70, 72, 83, 96–102, 172, 188, 191–195, 202–205, 268
– –, branched 97, 100, 188
– –, unbranched 96, 98, 191–195, 202–205
rolled layers 224
rotation angle of tetrahedra 215, 218–221, 223
rotation of tetrahedra 170, 184, 215–223, 249, 250, 253, 267, 276
rules for fundamental chains 59
rules for topology 161–169

SBU 145–147
σ-bond character 36, 42
Schläfli symbol 154
sechser chain silicates 81, 101, 103, 231
sechser single layers 115
secondary building units 145–147
secondary polyhedron 54
senary polyhedron 54
sharing coefficient 138–142
shrinkage of chains 80
Si/Al replacement 12, 85, 214, 238
[SiA_5] group 39–41
[SiF_6] group 42
Si–O bond 14–51
Si:O ratio 78, 119, 124, 126, 166, 167
Si–O–Si angles 15, 18, 21, 24–30, 36, 44, 161, 168, 209, 262
[SiO_4] tetrahedron, dimensions 14, 266
[SiO_6] octahedron, dimensions 15, 266
siebener chain silicates 81, 103
silicate anion, definition 12, 85
silicon salts 69
single chain phosphates 104, 178, 181–183, 195
single chain silicates 70, 101–107, 170–191
– – –, branched 104–105, 187–191
– – –, even-periodic 102–103, 174–177
– – –, odd-periodic 102–103, 173, 177–178
– – –, unbranched 80, 101, 102, 170–187
single chains, branched 104
– –, unbranched 79, 101, 102
single layers, curled 227–229
single layer silicates 70, 113–121, 195–231
– – –, 1:1 213, 214, 220, 221, 223–229, 236
– – –, 2:1 213–223, 236
– – –, anhydrous 198–205, 208, 212, 229–231
– – –, branched 119–121, 188, 189
– – –, cation-poor 213–223
– – –, cation-rich 213, 214, 220, 221, 223–229, 236
– – –, hydrous 229–231
– – –, unbranched 115, 196
single rings 62, 96–102
single ring silicates 96, 98, 192
– – –, branched 97
– – –, unbranched 96, 98
singular polyhedron 54
six-coordinated Si 11, 15, 30, 41, 42, 51, 69, 90–93, 130, 132, 162, 266
sixfold chains 110, 111
sodalite framework 146
"soft" cations 187–192, 268–270
solar abundance 3
sorosilicates 72, 93
sp^3 hybrid 35, 39, 43, 44
$sp^3 d$ hybrid 39, 43, 44
$sp^3 d^2$ hybrid 42, 44
specific volume of oxygen 33, 248–254, 256, 261–265
sphere packing model 7
split positions 247–249, 277
splitting of octahedral layer 216, 223
static disorder 22, 26
stereographic plot 10
straight Si–O–Si bonds 26, 28, 38
strain-free T–O–T angle 24, 30, 45, 250
strain reduction 214–229, 238
stress factors 267
stretching factor 80, 171, 176, 273
stretching factor frequency distribution 181
stretching of chains 79, 80
structural formula 72, 105, 106
structure models 7
stuffed derivatives 244, 252, 258, 274
subclass 63

subgroup 63
substitution of Si 12, 214, 229, 238, 244, 262
super class 76

T atom 12
talc-like layers 213–223, 236
tectosilicates 70, 72, 83, 126–129, 143–159, 239–265, 274–277
temperature factor 23, 30, 249, 251
temperature, influence on structure 267, 274
tenfold tetrahedron 85, 94, 95
terminated anions 57, 64, 268
terrestrial abundance of Si 3
tertiary polyhedron 51
tetrahedral rotation angle 215, 218–221, 223
tetrahedron inversion 216–229, 238, 260
tetrahedron rotation 170, 184, 215–223, 249, 250, 253, 267, 276
tetrahedron tilting 215, 223–224
tetraoxosilicate 11, 14, 93–129, 131
tetrapolysilicates 70, 109, 206–212
tetrasilicates 70, 94, 95
thermal ellipsoid 11, 27
thermal motions 22, 26, 27, 267
tilting of tetrahedra 215, 223, 224
topology rules 161–169
T–O–T angles 248, 250, 251, 254–256, 258, 260–265
transition state 41
tricyclosilicates 70
tridymite-type 256, 257, 261, 265, 275, 276
trigonal bipyramid 39–43
trioctahedral layer silicates 213, 223
triple chain silicates 109, 110, 186, 206–212
triple tetrahedra 93, 165
tripolysilicates 70, 109, 110, 186, 206–212
trisilicates, branched 94, 96
–, unbranched 70, 93, 95
tubular building units 147, 150, 151
tubular chain silicates 109–111, 168, 203
tubular layers 224–227

unbranched anion 54
uniform-anion silicate 76, 130, 131, 269
unverzweigt 62

valence electron distribution 38, 48
valence, influence on structure 175, 177, 182, 184, 267, 273
Verknüpfungstyp 62
Verknüpfungszahl 62
verzweigt 62
Verzweigtheit 62
vierer chain silicates 79, 81, 101, 103–111
vierer layer silicates 114, 115, 120, 123
vierundzwanziger chain silicate 81, 103
vitreous silica 49
V_{ox} 33, 248–254, 256, 261–265

wadeite-type 31
warped layers 199, 202

zeolites 128, 146–153, 155–159, 244–246
zeolite synthesis 245
Zoltai classification 13, 138–143
zweier chain silicates 80, 101–102, 104–113, 205–212
zweier double layer silicates 122–123, 231–239
zweier single layers 60, 86
zweier single layer silicates 86, 114, 116, 119, 195–197, 199–203, 208–212
zwölfer chain silicates 81, 101, 103

Substance Index

acmite 102
actinolite 71
aenigmatite 104, 164, 188
afghanite 147
agrellite 187, 188
åkermanite 18
alamosite 81, 101, 103, 140, 174, 175
albite 2, 34, 71, 246–249, 261
alkali feldspars 2, 4, 246–252, 260–265
almandine 2
aluminosilicates 13, 246–265
aluminum silicates 13
amesite 196, 220, 223
aminoffite 94, 95
amphiboles 4, 49, 71, 105, 111, 113, 186
analcime 147, 156, 158, 253, 254
andalusite 71
andradite 2, 34
annite 214
anorthite 2, 88, 251, 263
anthophyllite 111, 140
antigorite 2, 196, 197, 221, 222, 224, 226
apatite 4
apophyllite 73, 115, 141, 196–198, 221, 225, 227
ardennite 140, 164, 270
arfvedsonite 80
armenite 204
armstrongite 118, 196, 217, 219
astrophyllite 88, 104, 164, 187, 188, 190
augite 2
axinite 71

babingtonite 103, 180
bafertisite 71
banalsite 153
bannisterite 229
baotite 98, 140, 192
baratovite 98
barium phyllosilicates 117, 167, 205, 207–212
barium polysilicates 110, 205–212
barrerite 71
barylite 142
barysilite 94
batisite 81, 103, 140, 174, 185
bavenite 112, 113, 164, 188, 271
bazirite 192
bementite 196, 221, 227
benitoite 96, 98, 140, 192
bertrandite 142
beryl 96, 98, 142
beryllonite 257
bikitaite 11, 147, 155, 156
biopyriboles 184
biotite 4
brannockite 99, 194, 204
brewsterite 147
bronzite 2
bustamite 81, 103, 180

Ca-bustamite 180
canasite 184, 186
cancrinite 146, 147, 150, 245
carletonite 124, 141, 189, 197, 231–238
carnegieite 256, 275
carpholite 102, 174
caryopilite 229
cascandite 103, 180
catapleiite 98, 192
cavansite 196, 217, 219
caysichite 141, 185
celsian 34, 251, 252
chabazite 128, 142, 147, 156, 245
chamosite 2
chelates 41
chesterite 129, 141, 271
chkalovite 81, 103, 174, 185, 231
chlorite 71
chrysotile 71, 196, 221, 224–226
clathrasils 23, 27, 29, 156, 159, 240, 262, 265
clathrate hydrates 243
clay minerals 4, 195, 198, 212–229
clinobronzite 2
clinochrysotile 2
clinoenstatite 2
clinohypersthene 2
clinojimthompsonite 186
clinozoisite 129, 270
cobalt chrysotile 221, 225–227
coesite 28, 30, 34, 38, 47, 128, 142, 146, 155, 156, 240, 275, 276
combeite 98, 192

copper polygermanate 104
cordierite 2, 128
cristobalite 2, 71, 128, 129, 142, 156, 225, 240, 260, 261, 264, 265, 275, 276
cummingtonite 50
cuprorivaite 205
cuspidine 19
cymrite 141, 232, 233

dachiardite 147, 149
dalyite 115, 118, 141, 196, 201, 205
danburite 18, 87
datolite 141
deerite 73, 104, 188
delhayelite 124, 141, 189, 197, 231–236
dellaite 71, 270
devitrite 184, 186
diopside 2, 80, 102, 111
dioptase 99, 140, 193
disilicates 18, 20
dodecasil 1H 147, 155, 156, 241, 243, 261
dodecasil 3C (ZSM-39) 147, 241, 243, 261

eakerite 73, 97, 164, 188
earth alkali feldspars 2, 251–253, 260–265
edingtonite 147
eifelite 194
ekanite 196, 203, 205
ekmanite 229
elbaite 193
elpidite 81, 141, 184, 186
emeleusite 185, 205
enstatite 2, 47, 79, 80, 111, 140, 172, 175, 186, 190, 197, 208
epididymite 78, 105, 110, 111, 146, 184, 186, 205
epidote 129, 140, 270
epistilbite 147, 245
erionite 147
esters 100, 188
α-eucryptite 255, 256, 260, 275
β-eucryptite 255, 256, 258, 260, 275
eudialyte 96, 98, 129, 141, 193, 270
eudidymite 189

fassaite 2
faujasite 147, 155, 156, 245
fayalite 2, 47
fedorite 232
feldspar polymorphs 251, 252
feldspars 2, 12, 59, 61, 73, 88, 127, 128, 146, 148, 151, 152, 155, 189, 246, 261, 265
feldspars, alkali 2, 4, 246–252, 260–265
feldspars, earth alkali 2, 251–253, 260–265
feldspar, potassium 2, 12, 61, 246, 261
fenaksite 71, 188
ferrierite 147, 150, 245
ferrobustamite 103
ferrocarpholite 80, 174
ferroferritschermakite 72
ferrosilite I 172
ferrosilite III 81, 101, 103, 140, 172, 173
fibrous silica 93, 132, 161, 162, 240
fluoroberyllates 199, 201
forsterite 2, 8, 34, 39, 43, 47
foshagite 81, 103

gaidonnayite 81, 103, 174, 231
ganomalite 270
garnets 2, 49, 51
garronite 152
gas hydrates 243
gehlenite 141
georgechaoite 103, 174
gillespite 28, 196, 201, 203, 205
gismondine 128, 147, 152, 245
glaucophane 50, 80
gmelinite 71, 146, 147, 150, 245
greenalite 2, 229
grossular 2
grunerite 50, 80
gugiaite 18

haradaite 81, 103, 171, 174, 187, 190
harmotome 128, 148, 151, 152, 156
hedenbergite 102
hemimorphite 18
heulandite 147, 245
hexacelsian 122, 124, 251
hexaoxosilicates 11, 15, 30, 41, 42, 51, 69, 90–93, 130, 132, 162, 266
hilairite 103, 176, 192
holdenite 142
hortonolite 2
howieite 188
hyalotekite 100, 188
hydrodelhayelite 232
hypersthene 2

ice 129
ilmenite 4
ilvaite 18
imandrite 98, 193
inesite 81, 105, 140, 184, 186
iraqite 99, 194
iron chrysotile 221, 225

jadeite 102, 253, 254
jimthompsonite 74, 111, 141, 186
joaquinite 192
joesmithite 129, 271
johannsenite 80, 102

kainosite 192
kaliophilite-01 256–258, 265, 275, 276
kalsilite 244, 256, 258, 275
kaolinite 141, 214, 220, 222, 223
katayamalite 193
kazakovite 98, 193
keatite 127–129, 156, 240, 262, 275, 276
kilchoanite 129, 140, 270
kinoite 95
kosmochlor 2
K phase 232
krauskopfite 79, 81, 103, 171, 174
krinovite 2
kvanefjeldite 188
kyanite 71

lamprophyllite 18
latiumite 86, 124, 141, 231
laumontite 128, 147
låvenite 19
lawsonite 18
leifite 28, 30, 126–128, 142, 144, 189
lemoynite 189
leucite 71, 254
leucophanite 81, 103, 171, 174
leucophoenicite 93
leucosphenite 189, 232, 233, 238
levyne 147
liddicoatite 193
Li-hydrorhodonite 180
liotite 147
lithium silicates 34, 80, 102, 174, 185, 196, 197, 200–205
litidionite 112, 140, 188
lizardite 196, 220, 223
lorenzenite 102

macdonaldite 124, 189, 231–233
macfallite 270
Maddrell's salt 180
magnetite 4
majorite 2
makatite 196
manganoan wollastonite 179
manganpyrosmalite 115, 196, 221, 225, 227
margarite 216, 218, 223
margarosanite 192
marsturite 103, 180
mazzite 147, 150, 246
mcgillite 196, 221, 229
medaite 94, 95, 165
melanophlogite 71, 128, 142, 147, 155, 156, 241, 261
melilite 2, 18
meliphanite 88, 119, 121, 129, 164, 188, 271
merlinoite 147, 150, 152
merrihueite 2, 205
metahalloysite 225, 227
Mg-rhodonite 103
micas 115–117, 195, 198, 212, 214
microcline 12, 246
milarite 97, 99, 194, 204
milarite-type phases 99
miserite 109, 110, 129, 142, 164, 168, 184, 186, 271
Mn-bustamite 180
monteregianite 232, 233
monticellite 2
montmorillonite 2
mordenite 147, 149, 150, 156, 158, 245
muirite 96, 98, 140, 193
muscovite 115, 116, 196, 223

nagashimalite 98, 192
nambulite 81, 103, 180
Na-nepheline 256–259
narsarsukite 81, 105, 109, 149, 151, 185, 205
natroapophyllite 196, 221
natrolite 142, 147
naujakasite 122, 124, 141, 231–238
nekoite 118, 195, 196, 198, 217, 219
nepheline 2, 244, 256, 258, 259, 275
neptunite 128, 144
nordite 88, 104, 188

offretite 146, 147, 150, 246
ohmilite 81, 103, 174
okenite 78, 105, 118, 129, 142, 184, 186, 271
oligoclase 71
olivine 2, 4, 39, 49, 71, 94
organosilicon compounds 39–41, 90
orthoclase 2, 71, 127, 142, 156
orthoenstatite 34, 37, 48, 101, 102, 171
orthoericssonite 18
orthoferrosilite 102, 197
osumilite 194

pabstite 192
palygorskite 2, 116, 117, 196, 216, 217, 219, 222
papagoite 192
paracelsian 87, 148, 151, 152, 251
parawollastonite 180
parsettensite 229
paulingite 147, 158
pecoraite 225, 227
pectolite 81, 103, 180, 195, 197
pellyite 28. 104, 188
pentagonite 28, 196, 217, 219
petalite 12, 28, 196, 201–203, 205
petarasite 98, 193
phenakite 34, 142, 255, 276
phillipsite 147, 152

phosphates 6
phyllosilicic acid 196, 200
piemontite 270
pigeonite 2
plagioclase 2, 4, 71, 88
pollucite 253, 254
potassium feldspar 2, 12, 61, 246, 261
prehnite 85, 121, 164, 188
pseudowollastonite 98, 172
pyriboles 111
pyrope 34
pyrophyllite 213, 214, 218
pyrosmalite 141
pyroxenes 4, 49, 101, 111–113, 181, 183, 186, 195, 265, 273
pyroxenoids 173, 183, 273
pyroxenoids, acid 179, 180
pyroxferroite 79, 81, 101, 103, 172
pyroxmangite 81, 103, 140

quartz 2, 4, 34, 47, 128, 142, 155, 156, 240, 260–265, 275
queitite 270

ramsayite 102, 174
rankinite 18
reyerite 115, 122, 124, 129, 141, 232–238, 271
rhipidolite 2
rhodesite 124, 189, 231–236
rhodonite 101, 103, 140, 172, 180, 186
rhönite 2
ringwoodite 2
rinkite 19
roedderite 1, 2, 194, 204
roggianite 126, 189
rosenhahnite 73, 95, 140
rustumite 71, 140, 270

sanbornite 34, 116, 168, 196, 197, 201, 202, 205, 212
saneroite 104, 164, 180, 188
sanidine 12, 246, 249
santaclaraite 103, 180
sauconite 214
sazhinite 118, 196, 219
scawtite 99, 193
schizolite 180
semenovite 119, 122, 189
sepiolite 2, 116, 117, 196, 216, 217, 219, 222
serandite 79, 81, 103, 179, 180, 195, 197, 198
shattuckite 80, 102, 174
silica 21–30, 38, 93, 128, 129, 132, 162, 240, 260, 261, 264, 274, 275
silica glass 49
silicalite-1 (ZSM-5) 128, 147, 150, 155, 156, 245
silicalite-2 (ZSM-11) 150, 245
silicon diphosphate 69
silicon monophosphate 28, 31, 91, 92, 129
silicon monosulfide 2
silicon monoxide 2
silicon-organic compounds 39–41, 90
sillimanite 105, 108, 141
sinoite 2
slawsonite 251
sodalite 2, 71, 128, 142, 146, 147, 156, 189, 245
sodalite hydrate 245
sorensenite 79, 81, 103, 174, 176
spessartine 71
α-spodumene 12, 102, 253, 254, 262
β-spodumene 254
γ-spodumene 254
steacyite 97, 99, 141, 194, 205
stilbite 147
stilpnomelane 122, 124, 232–235, 238
stishovite 31, 34, 47, 91, 92
stokesite 81, 101, 103, 140, 171, 174, 231
Strahlstein 71
sulfates 6
sursassite 270
suzukiite 103, 174

taeniolite 215
talc 113, 196, 197, 213, 214, 217, 218, 222
taneyamalite 72, 188
taramellite 96, 98, 192
tetrakalsilite 256, 275
thalenite 95
thaumasite 31, 69, 91
thomsonite 128, 147, 189, 245
thortveitite 28, 34, 140
tienshanite 97, 164, 188
tilleyite 19
tinaksite 112, 113, 188
tiragalloite 95
tourmalines 98, 193
traskite 96, 98, 193, 270
tremolite 50, 71, 80, 105, 111, 186
tridymite 2, 28, 59, 61, 127–129, 155, 156, 244, 260, 261, 264, 265, 275
trinepheline 275
tuhualite 105, 142, 185, 205, 231
tuscanite 86, 124, 231

ureyite 2, 102
ussingite 126, 142, 144

vermiculite 71
verplanckite 98
vertumnite 232
vesuvianite 71, 129, 270
virgilite 248
vlasovite 104, 140, 187, 188, 190

wadeite 98, 192
wadsleyite 2
walstromite 172, 192
wenkite 126, 128, 189
willemite 255
wöhlerite 19
wollastonite 2, 71, 81, 101–103, 140, 172, 175, 179, 186, 195, 197
wollastonite 1T 102
wollastonite 2M 102
wollastonite 7T 103

xanthophyllite 214, 215
xonotlite 28, 78, 81, 105, 140, 184, 186, 195

yagiite 2
yofortierite 196, 216, 219
yugawaralite 147

zektzerite 81, 185, 205
zeolite A 128, 156, 245
zeolite L 146, 147, 150, 245
zeolite Li-A (BW) 147, 150
zeolite Losod 147
zeolite P1 152
zeolite rho 147, 156
zeolites 128, 146, 155, 156, 159
zeolites, fillers in 155
zeolite TMA-E (AB) 147
zeolite ZK-5 147
zeophyllite 121, 141, 188
zircon 2, 34, 140
zirsinalite 98, 192
zoisite 129, 270
ZSM-5 (silicalite-1) 128, 147, 150, 155, 156, 245
ZSM-11 (silicalite-2) 150, 245
ZSM-39 (dodecasil 3C) 147, 241, 243, 261
zunyite 27, 28, 73, 94, 96, 140, 164, 165, 188
zussmanite 59, 141, 188, 233, 234, 236, 238

Formula Index

Each phase referred to in the book has been indexed by its chemical formula. Within each formula the cations are, in general, arranged in the sequence of their valence and, in the second instance, in the sequence of their group number in the periodic system of the elements, main group elements preceding transition elements. In minerals with isomorphous replacement, instead, minor constituents replacing the main constituents in small amounts are preceded by the main constituents. The formulae are arranged alphabetically as sets under each of the cations and non-silicate anions. For example, apophyllite, $\mathrm{KCa_4[Si_4O_{10}]_2(F,OH)\cdot 8\,H_2O}$, appears under K, under Ca, under F, and under $\mathrm{O^{-2}}$, $\mathrm{(OH)^-}$. There are special sets for inorganic non-silicates, for silicon compounds containing organic groups, for phases with $\mathrm{SiO_2}$ frameworks, and for general formulae. Within each set the phases are listed in alphabetical order according to the cations in the formulae. This order is represented by the sequence of the following calcium silicates:
$\mathrm{Be_2Ca_3[Si_3O_{10}](OH)_2}$ – $\mathrm{Ca[Al_2Si_2O_8]}$ – $\mathrm{Ca_3[Si_3O_9]}$ – $\mathrm{Ca_2Al[AlSiO_7]}$ – $\mathrm{Ca_3Al_2[SiO_4]_3}$ – $\mathrm{Ca_2Ba[Si_3O_9]}$ – $\mathrm{Ca_2Ba_4(Fe,Mg)[Si_{12}O_{34}]}$ – $\mathrm{HNaCa_2[Si_3O_9]}$ – etc.

Ag

$\mathrm{Ag_2[Si_2O_5]}$ 197, 205
$\mathrm{Ag_4[Si_2O_6]}$ 102, 174, 197
$\mathrm{Ag_6[Si_2O_7]}$ 18
$\mathrm{Ag_{10}[Si_4O_{13}]}$ 94, 95

Al[4]

$\mathrm{Al[(AlSi)O_5]}$ 105, 108, 141
$\mathrm{Al_{13}[Si_5O_{16}](OH,F)_{18}O_4Cl}$ 28, 73, 94, 96, 140, 165, 188
$\mathrm{Ba[AlSiO_4]_2\cdot H_2O}$ 141, 232
$\mathrm{Ba[AlSi_2O_6](OH,Cl)}$ 152
$\mathrm{Ba[Al_2Si_2O_8]}$ 34, 87, 263
$\mathrm{Ba[Al_2Si_2O_8]}$ (feldspar-type) 34, 189, 251, 252, 264
$\mathrm{Ba[Al_2Si_2O_8]}$(hT) (hexacelsian-type) 122, 124, 232, 233, 235, 236, 238, 239, 251, 252, 264
$\mathrm{Ba[Al_2Si_2O_8]}$(1T) (paracelsian-type) 87, 148, 152, 251, 252, 264
$\mathrm{Ba[Al_2Si_6O_{16}]\cdot 6\,H_2O}$ 128, 148, 152, 156
$\mathrm{BaCa_2Al_3[Al_3Si_9O_{30}]\cdot 2\,H_2O}$ 204
$\mathrm{Be_2Ca_4Al_2[Si_3O_{10}][Si_6O_{16}](OH)_2}$ 112, 188, 271
$\sim\mathrm{Ca_4Al_4Si_4O_6(OH)_{24}\cdot 3\,H_2O}$ 232
$\mathrm{Ca[AlSi_2O_6]_2\cdot 4\,H_2O}$ 128
$\mathrm{Ca[Al_2Si_2O_8]}$ 88, 263
$\mathrm{Ca[Al_2Si_2O_8]}$ (feldspar-type) 88, 189, 251, 252, 264
$\mathrm{Ca[Al_2Si_2O_8]}$(hT) (hexacelsian-type) 122, 124, 232–239, 251, 252, 264
$\mathrm{Ca[Al_2Si_2O_8]}$ (Immm-type) 251, 252, 264
$\mathrm{Ca[Al_2Si_2O_8]\cdot 4\,H_2O}$ 128, 152, 245
$\mathrm{Ca[Al_2Si_4O_{12}]\cdot 6\,H_2O}$ 128, 142, 156
$\mathrm{CaAl_2[AlSiO_5]_2(OH)_2}$ 216, 218, 223
$\mathrm{Ca_2Al[AlSiO_7]}$ 2, 141
$\mathrm{Ca_2Al[AlSi_3O_{10}](OH)_2}$ 85, 121, 188
$\mathrm{Ca_2Al_2Sn[Si_6O_{18}](OH)_2\cdot 2\,H_2O}$ 73, 97, 188
$\mathrm{(Ca,K)_4[(Al,Si)_5O_{11}][SO_4,CO_3]}$ 141
$\mathrm{CaMg_2Al[Al_3SiO_{10}](OH)_2}$ 214, 215
$\mathrm{Ca_2(Mg,Al)[(Al,Si)_2O_7]}$ 2
$\sim\mathrm{CaMg_2Ti[Al_2SiO_{10}]}$ 2
$\mathrm{Ca_{1-x}Na_x[Al_{2-x}Si_{2+x}O_8]}$ 88
$\mathrm{Cs[AlSiO_4]}$ 256, 258, 261, 263, 275
$\mathrm{Cs[AlSi_2O_6]}$ 253, 261
$\mathrm{Cs[AlSi_3O_8]}$ 250
$\mathrm{(Fe,Mg)_6[AlSi_3O_{10}](O,OH)_8}$ 2
$\mathrm{(Fe^{+2},Mg,Mn,Fe^{+3})_{<9}[Si_9(Al,Si)_3O_{30}](OH)_6X_n\cdot 6\,H_2O}$ 229
$\mathrm{H_2(Ba,Ca)_{10}[(Al,Si)_{20}O_{43}][SO_4]_3\cdot H_2O}$ 126, 128, 189
$\mathrm{H_4Ca_4[(Al_4Si_8)O_{26}](OH)_4\cdot \sim 6.5\,H_2O}$ 126, 189
$\mathrm{H_2KCa_2[AlSi_7O_{19}]\cdot 6\,H_2O}$ 189, 232
$\mathrm{HNa_2[(AlSi_3)O_9]}$ 126, 142, 144
$\mathrm{H_2Na_6[Be_2Al_2Si_{16}O_{41}]\cdot 1.5\,H_2O}$ 142
$\mathrm{KAlSi_3O_8}$(hP) (hollandite-type) 90, 92
$\mathrm{K[AlSiO_4]}$ 244, 256, 258, 261, 262, 275
$\mathrm{K[AlSi_2O_6]}$ 253, 254, 261
$\mathrm{K[(AlSi_3)O_8]}$ (feldspar-type) 12, 61, 73, 127, 142, 156, 189, 246, 248, 249, 261, 264
$\mathrm{KAl_2[AlSi_3O_{10}](OH)_2}$ 115, 116, 196
$\mathrm{K_{1.7}Ca_6[(Al_{5.7}Si_{4.3})O_{22}][SO_4]_{1.4}[CO_3]_{0.6}}$ 86, 124, 231
$\mathrm{K_5Ca_2[Al_9Si_{23}O_{64}]\cdot 24\,H_2O}$ 150, 152
$\mathrm{KCa_2(Be_2Al)[Si_{12}O_{30}]\cdot 0.75\,H_2O}$ 97, 99, 194, 204
$\mathrm{(K,Ca,Na)_2[Al_3Si_{11}O_{28}]\cdot 12\,H_2O}$ 158
$\mathrm{KFe_3[AlSi_3O_{10}](OH)_2}$ 214
$\mathrm{KFe_{13}[(Al,Si)_9O_{21}]_2(OH)_{14}}$ 188, 235
$\mathrm{KFe_{24}[(AlSi_{35})O_{84}]O_6(OH)_{24}\cdot n\,H_2O}$ 122, 124, 232, 234

$K(Fe, Mg, Mn, Al)_{13}[(Al, Si)_9O_{21}]_2(OH)_{14}$ 141
$(K, H_2O)_3(Mn, Fe^{+3}, Mg, Al)_{<9}[Si_{12}O_{30}](OH)_6X_n \cdot 6\,H_2O$ 229
$KMg_2Al_3[(Al_2Si_{10})O_{30}]$ 99, 194
$(K, Sr, H_2O)_2(Ca, Na, Mg, Fe)_6[(Al_{3.66}Si_{6.34})O_{22}][SO_4]_{1.4}[CO_3]_{0.5}[O_4H_4]_{0.1}$ 86, 124, 231
$LiAlSiO_4$ 260, 261, 275, 277
$Li[AlSiO_4]$ (hT) (quartz-type) 255, 256, 261, 262
$Li[AlSi_2O_6]$-II (keatite-type) 253, 254, 261, 262
$Li[AlSi_2O_6]$-III (quartz-type) 253, 254, 261, 262
$Li[AlSi_2O_6] \cdot H_2O$ 11, 156
$Li[AlSi_3O_8]$ (quartz-type) 246, 248, 252, 261, 264
$Li[AlSi_3O_8]$ (keatite-type) 246, 248, 252, 261, 264
$Li[AlSi_3O_8]$ (feldspar-type) 250
$LiAl_2[AlSiO_5](OH)_4$ 220
$LiAl[SiO_4]$ (1T) (phenakite-type) 255–257, 261
$LiAl[Si_2O_5]_2$ 12, 28, 196, 201–205
$Mg[Al_2Si_2O_8]$ (quartz-type) 251, 252, 263, 264
$Mg_2[Al_4Si_5O_{18}]$ 2, 128
$MgAl_2[AlSiO_5]_2(OH)_2$ 218
$Mg_2Al[AlSiO_5](OH)_4$ 220, 223
$(Mg, Fe, Al)_6[(Al, Si)_4O_{10}](OH)_8$ 2
$Na[AlSiO_4]$ 2, 244, 256–261, 275
$Na[AlSi_3O_8]$ 34, 189, 246–249, 261, 264
$Na[AlSi_2O_6] \cdot H_2O$ 156, 158, 253, 254
$Na[AlSi_5O_{12}] \cdot 3\,H_2O$ 149, 156, 158, 245
$Na_2[Al_2Si_3O_{10}] \cdot 2\,H_2O$ 142
$Na_4[Al_3Si_3O_{12}]Cl$ 2, 128, 142, 146, 156, 189
$Na_5[Al_5Si_{19}O_{48}] \cdot 12\,H_2O$ 149
$Na_6[Al_6Si_6O_{24}] \cdot 8\,H_2O$ 245
$Na_8[Al_8Si_8O_{32}] \cdot 16\,H_2O$ 152
$Na_{12}[(Al_{12}Si_{12})O_{48}] \cdot 27\,H_2O$ 128, 156, 245
$Na_2Ba[Al_4Si_4O_{16}]$ 153
$Na_6Be_2[(Al_2Si_{16})O_{39}](OH)_2 \cdot 1.5\,H_2O$ 28, 126, 128, 144, 189
$NaCa_2[(Al, Si)_{10}O_{20}] \cdot 6\,H_2O$ 128, 189
$(Na_2, Ca)[Al_2Si_7O_{18}] \cdot 7\,H_2O$ 146
$Na_2Ca_5[Al_{12}Si_{20}O_{64}] \cdot 28\,H_2O$ 152
$(Na_2, Ca)_2[Al_4Si_8O_{24}] \cdot 12\,H_2O$ 146
$Na_6Ca[AlSiO_4]_6[CO_3] \cdot 2\,H_2O$ 146
$NaCaAl[Si_2O_7]$ 18
$(Na, Ca)_3(Fe, Al, Mg)_5[(Al, Si)_4O_{11}]_2(OH)_2$ 80
$(Na_2, Ca, Mg)_{29}[Al_{58}Si_{134}O_{384}] \cdot 240\,H_2O$ 156
$(Na, Cs)_3[Al_3Si_9O_{24}] \cdot 11\,H_2O$ 156
$Na_6Fe[(Al_4Si_8)O_{26}]$ 122, 124, 141, 232, 235
$Na_{1-x}K_x[AlSiO_4]$ 256, 259
$(Na, K)_5[Al_5Si_{11}O_{32}] \cdot 10\,H_2O$ 152
$NaK_2[Al_3Si_9O_{24}] \cdot 7\,H_2O$ 146, 150
$Na_3K[AlSiO_4]_4$ 244, 275
$(Na, K, Ca)_3Mg_2Ca[Al_{10}Si_{26}O_{72}] \cdot 28\,H_2O$ 150
$(Na, K)Ca_7[Si_4O_{10}][(AlSi_7)O_{19}](OH)_4 \cdot 3\,H_2O$ 115, 122, 124, 129, 141, 232, 234, 271
$Na_3K_7Ca_5[AlSi_7O_{19}]_2F_4Cl_2$ 124, 141, 189, 197, 232, 234
$Pb[Al_2Si_2O_8]$ (hexacelsian-type) 232, 233, 238, 239
$Rb[AlSiO_4]$ 256, 259, 261, 263, 275
$Rb[AlSi_2O_6]$ 254, 261
$Rb[AlSi_3O_8]$ (feldspar-type) 246, 248, 249, 250, 261, 264
$Rb[AlSi_3O_8]$ (hT) 232, 233, 238
$Sr[Al_2Si_2O_8]$ (feldspar-type) 251, 252, 264
$Sr[Al_2Si_2O_8]$ (hT) (hexacelsian-type) 124, 232–239, 251, 252, 264
$Sr[Al_2Si_2O_8]$ (paracelsian-type) 251, 252, 264

$Al^{[6]}$

$Al[(AlSi)O_5]$ 105, 108, 141
$Al_2[Si_2O_5](OH)_4$ 141, 214, 220, 222, 223, 225, 227
$Al_2[Si_2O_5]_2(OH)_2$ 213, 214, 218
$Al_{13}[Si_5O_{16}](OH, F)_{18}O_4Cl$ 28, 73, 94, 96, 140, 165, 188
$Be_3Al_2[Si_6O_{18}]$ 96, 98, 142
$\sim Ca_4Al_4Si_4O_6(OH)_{24} \cdot 3\,H_2O$ 232
$CaAl_2[AlSiO_5]_2(OH)_2$ 216, 218, 223
$CaAl_2[Si_2O_7](OH)_2 \cdot H_2O$ 18
$Ca_2Al[AlSi_3O_{10}](OH)_2$ 85, 121, 188
$Ca_2Al_3[SiO_4][Si_2O_7]O(OH)$ 129
$Ca_3Al_2[SiO_4]_3$ 51
$Ca_2Al_2Fe[SiO_4][Si_2O_7]O(OH)$ 140, 270
$Ca_{10}Al_4(Fe, Mg)_2[SiO_4]_5[Si_2O_7]_2(OH)_4$ 270
$Ca_2(Al, Fe, Mn)_3[SiO_4][Si_2O_7]O(OH)$ 270
$CaB_3Al_6(Al, Li)_3[Si_6O_{18}]O_9(O, OH, F)_4$ 193
$Ca_2Cu_2Al_2[Si_4O_{12}](OH)_6$ 192
$Ca_2(Fe, Al)_3[SiO_4][Si_2O_7]O(OH)$ 129
$CaMg_2Al[Al_3SiO_{10}](OH)_2$ 214, 215
$Ca_{10}(Mg, Fe)_2Al_4[SiO_4]_5[Si_2O_7]_2(OH)_4$ 129
$Ca_2(Mn, Al)_3[SiO_4][Si_2O_7](OH)_3$ 270
$(Ca, Sr)Ba_{24}(Mg, Mn, Fe, Al, Ti)_{16}[Si_2O_7]_6[Si_{12}O_{36}](O, OH)_{30}Cl_6 \cdot 14\,H_2O$ 96, 99, 193, 270
$Co_3Al_2[SiO_4]_3$ 51
$FeAl_2[Si_2O_6](OH)_4$ 80, 174
$Fe_3Al_2[SiO_4]_3$ 51
$KAl_2[AlSi_3O_{10}](OH)_2$ 115, 116, 196
$K(Fe, Mg, Mn, Al)_{13}[(Al, Si)_9O_{21}]_2(OH)_{14}$ 141
$KMg_2Al_3[(Al_2Si_{10})O_{30}]$ 99, 194
$(K, Na, Ca, H_2O)(Mn, Mg, Fe^{+2}, Zn, Al)_{9.6}[Si_{12}O_{30}](OH)_{10} \cdot 4\,H_2O$ 229

$LiAl_2[AlSiO_5](OH)_4$ 220
$LiAl[Si_2O_6]$ (pyroxene-type) 12, 102, 253, 254, 261
$MgAl_2[AlSiO_5]_2(OH)_2$ 218
$Mg_2Al[AlSiO_5](OH)_4$ 220, 223
$Mg_3Al_2[SiO_4]_3$ 34, 51
$(Mg, Fe, Al)_6[(Al, Si)_4O_{10}](OH)_8$ 2
$(Mg, Fe, Al)_4[Si_2O_5]_3(O, OH)_2 \cdot 4\,H_2O$ 196, 216, 217
$MnAl_2[Si_2O_6](OH)_4$ 102, 174
$Mn_2Al_3[SiO_4][Si_2O_7](OH)_3$ 270
$Mn_3Al_2[SiO_4]_3$ 51
$Mn_4Al_6[(As, V)O_4][SiO_4]_2[Si_3O_{10}](OH)_6$ 140, 270
$NaAlSiO_4$(hP) ($CaFe_2O_4$-type) 90, 91, 92
$NaAl[Si_2O_6]$ (pyroxene-type) 50, 102, 253, 254, 261
$NaB_3Al_6(Al, Li)_3[Si_6O_{18}]O_9(OH)_4$ 193
$(Na, Ca)_3(Fe, Al, Mg)_5[(Al, Si)_4O_{11}]_2(OH)_2$ 80
$(Na, Ca)(Mn, Fe, Al, Mg)_{12}[Si_6O_{17}]_2(O, OH)_{10}$ 188
$Na(Fe^{+2}, Fe^{+3}, Mn, Al, Mg)_{12}[Si_6O_{17}]_2(O, OH)_{10}$ 188
$Na_2Mg_3Al_2[Si_4O_{11}]_2(OH)_2$ 50, 80
$Na_X(Mg, Al)_2[Si_4O_{10}](OH)_2 \cdot 4\,H_2O$ 2

As

$HMn_4As[Si_3O_{10}]O_3$ 95
$Mn_4Al_6[(As, V)O_4][SiO_4]_2[Si_3O_{10}](OH)_6$ 140, 270
$Mn_6[Zn(OH)_4][Zn_2SiAs_2O_{12}(OH)_2](OH)_2$ 142

B

$Ba_4B_2(Fe, Ti)_4[Si_4O_{12}]_2O_5Cl_X$ 96, 98, 192
$Ba_4B_2(V, Ti)_4[Si_4O_{12}]_2O_3(O, OH)_2Cl$ 98, 192
$CaB_2[Si_2O_7]O$ 18, 87
$CaB_3Al_6(Al, Li)_3[Si_6O_{18}]O_9(O, OH, F)_4$ 193
$Ca_2Ba_2Pb_2B_2[(Si_{1.5}Be_{0.5})Si_8O_{28}]F$ 100, 188
$HCa[BSiO_5]$ 141
$NaB_3Al_6(Al, Li)_3[Si_6O_{18}]O_9(OH)_4$ 193
$Na_4BaTi_2[B_2Si_{10}O_{28}]O_2$ 189, 232
$Na_9KCa_2Ba_6(Mn, Fe)_6(Ti, Nb, Ta)_6B_{12}[Si_{18}O_{54}]_2O_{15}(OH)_2$ 97, 188

Ba

$Ba[AlSiO_4]_2 \cdot H_2O$ 141, 232
$Ba[AlSi_2O_6](OH, Cl)$ 152
$Ba[Al_2Si_2O_8]$ 34, 87, 263
$Ba[Al_2Si_2O_8]$ (feldspar-type) 34, 189, 251, 252, 264
$Ba[Al_2Si_2O_8]$(hT) (hexacelsian-type) 122, 124, 232–239, 251, 252, 264
$Ba[Al_2Si_2O_8]$(lT) (paracelsian-type) 87, 148, 152, 251, 252, 264
$Ba[Al_2Si_6O_{16}] \cdot 6\,H_2O$ 128, 148, 152, 156
$BaSiO_3$ 172, 175
$Ba[Si_2O_5]$ 117, 167, 205, 207–212
$Ba[Si_2O_5]$(lT) 34, 116, 117, 167, 168, 196, 197, 201, 202, 205, 207–212
$Ba[Si_2O_5]$(hT) 197, 201, 208–212
$Ba_2[Si_2O_6]$(hT) 34, 79, 80, 102, 174, 175, 207, 209–212
$Ba_3[Si_4O_{11}]$ 207, 210
$Ba_4[Si_6O_{16}]$ 73, 80, 110, 166, 206, 207, 209–211
$Ba_5[Si_8O_{21}]$ 80, 110, 141, 206, 207, 209–212
$Ba_6[Si_{10}O_{26}]$ 80, 110, 111, 141, 206, 207, 209–212
$Ba_7[Si_{12}O_{31}]$ 212
$Ba_{M+1}[Si_{2M}O_{5M+1}]$ 206
$Ba_4B_2(Fe, Ti)_4[Si_4O_{12}]_2O_5Cl_X$ 96, 98, 192
$Ba_4B_2(V, Ti)_4[Si_4O_{12}]_2O_3(O, OH)_2Cl$ 98, 192
$BaCa_2Al_3[Al_3Si_9O_{30}] \cdot 2\,H_2O$ 204
$Ba_{10}(Ca, Mn, Ti)_4[Si_8O_{24}](Cl, O, OH)_{12} \cdot 4\,H_2O$ 96, 99, 140, 193
$BaCu[Si_4O_{10}]$ 196, 205
$BaFe[Si_4O_{10}]$ 28, 196, 201, 203, 205
$BaGd_4[Si_2O_7][Si_3O_{10}]$ 270
$BaMgSiO_4$ 244
$Ba_{12}(Mn, Ti, Fe)_6[Si_4O_{12}]_3(O, OH)_2(OH, H_2O)_7Cl_9$ 98
$Ba_3Nb_6[Si_2O_7]_2O_{12}$ 19, 28
$Ba_2RE_2FeTi_2[Si_4O_{12}]_2O_2$ 192
$BaSn[Si_3O_9]$ 192
$(Ba, Sr)(Mn, Fe)_2(Fe, Ti)[Si_2O_7](O, OH)_2$ 18
$BaTi[Si_3O_9]$ 96, 98, 140, 192
$Ba_4(Ti, Nb)_8[Si_4O_{12}]O_{16}Cl$ 98, 140, 192
$Ba_2V_2[Si_4O_{12}]O_2$ 103, 174
$BaY_4[Si_2O_7][Si_3O_{10}]$ 270
$BaZr[Si_3O_9]$ 192
$Be_2Ba[Si_2O_7]$ 142
$Ca_2Ba[Si_3O_9]$ 98, 192
$Ca_2Ba_4(Fe, Mg)_4[Si_{12}O_{34}]$ 28, 104, 188
$Ca_2Ba_2Pb_2B_2[(Si_{1.5}Be_{0.5})Si_8O_{28}]F$ 100, 188
$(Ca, Sr)Ba_{24}(Mg, Mn, Fe, Ti)_{16}[Si_2O_7]_6[Si_{12}O_{36}](O, OH)_{30}Cl_6 \cdot 14\,H_2O$ 96, 99, 193, 270
$H_4Ba_2[Si_4O_{12}] \cdot 4\,H_2O$ 79, 81, 103, 174
$H_2(Ba, Ca)_{10}[(Al, Si)_{20}O_{43}][SO_4]_3 \cdot H_2O$ 126, 128, 189
$H_2Ca_4Ba[Si_8O_{19}]_2 \cdot (8+x)H_2O$ 124, 189, 232
$HLi_2Ba_6[SiO_4]_2[Si_2O_7]Cl$ 270
$K_2Ba_7[Si_4O_{10}]_4$ 28, 196, 205
$LiBa_9[Si_{10}O_{25}][CO_3]Cl_7$ 115, 196
$Na_2Ba[Al_4Si_4O_{16}]$ 153
$Na_2Ba[Si_2O_6]$ 102, 171, 174
$NaBaNd[Si_3O_9]$ 98
$NaBa_3Nd_3[Si_2O_7][Si_4O_{13}]$ 94, 96, 164, 188
$Na_2BaNd_2[Si_4O_{12}][CO_3]$ 98

$Na_2BaTi_2[Si_4O_{12}]O_2$ 81, 103, 140, 174, 185
$Na_4BaTi_2[B_2Si_{10}O_{28}]O_2$ 189, 232
$Na_9KCa_2Ba_6(Mn, Fe)_6(Ti, Nb, Ta)_6B_{12}[Si_{18}O_{54}]_2O_{15}(OH)_2$ 97, 188
$Na_2(Sr, Ba)_2Ti_3[Si_2O_7]_2(OH, F)_2O_2$ 18

Be

$BeSiO_3$ 175
$Be_2[SiO_4]$ 34, 142, 255
$Be_3Al_2[Si_6O_{18}]$ 96, 98, 142
$Be_2Ba[Si_2O_7]$ 142
$BeCa_2[Si_2O_7]$ 18
$Be_2Ca_3[Si_3O_{10}](OH)_2$ 94, 95
$Be_2Ca_4Al_2[Si_3O_{10}][Si_6O_{16}](OH)_2$ 112, 188, 271
$Be_2Ca_2(Ca, Pb)(Fe, Mg)_5[SiO_4]_2[Si_2O_6]_2O_2(OH)_2$ 129, 271
$BeFe_3[Si_3O_9](OH)_2$ 103
$Ca_2Ba_2Pb_2B_2[(Si_{1.5}Be_{0.5})Si_8O_{28}]F$ 100, 188
$H_2[Be_4Si_2O_9]$ 142
$H_2Na_6[Be_2Al_2Si_{16}O_{41}] \cdot 1.5\ H_2O$ 142
$H_xNa_{0-2}(Ca, Na)_8(Fe, Mn, Zn, Ti)RE_2[(Si, Be)_{10}(O, F)_{24}]_2$ 123, 189
$K_2Be[Si_4O_{10}]$ 196, 205
$K_2Be_2[Si_6O_{15}]$ 196, 203, 205
$K_2Be_2Zn_2[SiO_4][Si_2O_7]$ 270
$KCa_2(Be_2Al)[Si_{12}O_{30}] \cdot 0.75\ H_2O$ 97, 99, 194, 204
$Na_2Be_2[Si_3O_9]$ 98, 192
$Na_6Be_2[(Al_2Si_{16})O_{39}](OH)_2 \cdot 1.5\ H_2O$ 28, 126, 128, 144, 189
$Na_2Be_2[Si_6O_{15}] \cdot H_2O$ 105, 111, 146, 186, 189, 205
$Na_6Be_3[Si_6O_{18}]$ 81, 103, 174, 185
$Na_2Be_2Ca_2[Si_4O_{12}]F_2$ 81, 103, 174
$Na_4Be_2Sn[Si_3O_9]_2 \cdot 2\ H_2O$ 79, 81, 103, 174, 176
$(Na, Ca)_4Be_4Ca_4[SiO_4][Si_7O_{20}]F_4$ 121, 129, 188, 271
$Rb_2Be_2[Si_2O_7]$ 28

$[CO_3]^{-2}$

$Ca_3[Si(OH)_6][SO_4][CO_3] \cdot 12\ H_2O$ 31, 69
$Ca_5[Si_2O_7][CO_3]_2$ 19
$Ca_7[Si_6O_{18}][CO_3] \cdot 2\ H_2O$ 99, 193
$(Ca, K)_4[(Al, Si)_5O_{11}][SO_4, CO_3]$ 141
$(Ca_3RE)Y_4[Si_8O_{20}][CO_3]_6(OH) \cdot 7\ H_2O$ 141, 185
$Ca_2Y_2[Si_4O_{12}][CO_3] \cdot H_2O$ 192
$K_{1.7}Ca_6[(Al_{5.7}Si_{4.3})O_{22}][SO_4]_{1.4}[CO_3]_{0.6}$ 86, 124, 231
$(K, Sr, H_2O)_2(Ca, Na, Mg, Fe)_6[(Al_{3.66}Si_{6.34})O_{22}][SO_4]_{1.4}[CO_3]_{0.5}[O_4H_4]_{0.1}$ 86, 124, 231
$LiBa_9[Si_{10}O_{25}][CO_3]Cl_7$ 115, 196
$Na_2BaNd_2[Si_4O_{12}][CO_3]$ 98
$Na_6Ca[AlSiO_4]_6[CO_3] \cdot 2\ H_2O$ 146
$Na_4KCa_4[Si_8O_{18}][CO_3]_4(OH, F) \cdot H_2O$ 124, 141, 189, 197, 232, 235

Ca

$BaCa_2Al_3[Al_3Si_9O_{30}] \cdot 2\ H_2O$ 204
$Ba_{10}(Ca, Mn, Ti)_4[Si_8O_{24}](Cl, O, OH)_{12} \cdot 4\ H_2O$ 96, 99, 140, 193
$BeCa_2[Si_2O_7]$ 18
$Be_2Ca_3[Si_3O_{10}](OH)_2$ 94, 95
$Be_2Ca_4Al_2[Si_3O_{10}][Si_6O_{16}](OH)_2$ 112, 188, 271
$Be_2Ca_2(Ca, Pb)(Fe, Mg)_5[SiO_4]_2[Si_2O_6]_2O_2(OH)_2$ 129, 271
$CaSiO_3$ 98, 101, 140, 173
$CaSiO_3$ (hT) 98
$CaSiO_3$ (hP) (perovskite-type) 91, 92
$\sim Ca_4Al_4Si_4O_6(OH)_{24} \cdot 3\ H_2O$ 232
$Ca[Al_2Si_2O_8]$ 88, 189, 251, 263, 264
$Ca[Al_2Si_2O_8]$ (feldspar-type) 88, 189, 251, 252, 264
$Ca[Al_2Si_2O_8]$(hT) (hexacelsian-type) 122, 124, 232–239, 251, 252, 264
$Ca[Al_2Si_2O_8]$ (Immm-type) 251, 252, 264
$Ca[(AlSi_2)O_6]_2 \cdot 4\ H_2O$ 128
$Ca[(Al_2Si_2)O_8] \cdot 4\ H_2O$ 128, 152, 245
$Ca[(Al_2Si_4)O_{12}] \cdot 6\ H_2O$ 128, 142, 156
$Ca[Si_2O_5]$ 202
$Ca_3[Si_2O_7]$ 18
$Ca_3[Si_3O_9]$ 2, 81, 101–103, 140, 173, 175, 180, 186, 197
$Ca_3[Si_3O_9]$(hT) (α) 98, 192
$Ca_3[Si_3O_9]$(mP) (δ) 98, 192
$Ca_3[Si_6O_{15}] \cdot 7\ H_2O$ 118, 195, 196, 198, 217, 219
$Ca_3[Si(OH)_6][SO_4][CO_3] \cdot 12\ H_2O$ 31, 69
$Ca_4[Si_2O_7](OH, F)_2$ 19
$Ca_4[Si_3O_9](OH)_2$ 81, 103
$Ca_5[Si_2O_7][CO_3]_2$ 19
$Ca_6[SiO_4][Si_2O_7](OH)_2$ 71, 270
$Ca_6[SiO_4][Si_3O_{10}]$ 129, 140, 270
$Ca_6[Si_6O_{17}](OH)_2$ 28, 81, 105, 140, 186
$Ca_7[Si_8O_{19}]_2(OH)_2$ 232
$Ca_7[Si_6O_{18}][CO_3] \cdot 2\ H_2O$ 99, 193
$Ca_{10}[SiO_4][Si_2O_7]_2(OH)_2Cl_2$ 71, 140, 270
$Ca_{10}[Si_6O_{16}][Si_6O_{15}]_2 \cdot 18\ H_2O$ 105, 118, 129, 142, 186, 271
$Ca_{13}[Si_5O_{14}]_2F_8(OH)_2 \cdot 6\ H_2O$ 121, 141, 188
$CaAl_2[AlSiO_5]_2(OH)_2$ 216, 218, 223
$Ca_2Al[AlSiO_7]$ 2, 141
$Ca_2Al[AlSi_3O_{10}](OH)_2$ 85, 121, 188
$CaAl_2[Si_2O_7](OH)_2 \cdot H_2O$ 18
$Ca_2Al_3[SiO_4][Si_2O_7]O(OH)$ 129
$Ca_3Al_2[SiO_4]_3$ 51
$Ca_2Al_2Fe[SiO_4][Si_2O_7]O(OH)$ 140, 270
$Ca_{10}Al_4(Fe, Mg)_2[SiO_4]_5[Si_2O_7]_2(OH)_4$ 270
$Ca_2(Al, Fe, Mn)_3[SiO_4][Si_2O_7]O(OH)$ 270
$Ca_2Al_2Sn[Si_6O_{18}](OH)_2 \cdot 2\ H_2O$ 73, 97, 188

$CaB_2[Si_2O_7]O$ 18, 87
$CaB_3Al_6(Al, Li)_3[Si_6O_{18}]O_9(O, OH, F)_4$ 193
$Ca_2Ba[Si_3O_9]$ 98, 192
$Ca_2Ba_4(Fe, Mg)_4[Si_{12}O_{34}]$ 28, 104, 188
$Ca_2Ba_2Pb_2B_2[(Si_{1.5}Be_{0.5})Si_8O_{28}]F$ 100, 188
$CaCo[Si_2O_6]$ 102
$CaCu[Si_4O_{10}]$ 196, 203, 205
$Ca_2Cu_2[Si_3O_{10}] \cdot 2\,H_2O$ 95
$Ca_2Cu_2Al_2[Si_4O_{12}](OH)_6$ 192
$CaFe[Si_2O_6]$ 102
$(Ca, Fe)_3[Si_3O_9]$ 103
$(Ca, Fe)_7[Si_7O_{21}]$ 103
$Ca_3Fe_2[SiO_4]_3$ 34
$Ca_2(Fe, Al)_3[SiO_4][Si_2O_7]O(OH)$ 129
$CaFe_2(Fe, Mn)[Si_2O_7]O(OH)$ 18
$(Ca, K)_4[(Al, Si)_5O_{11}][SO_4, CO_3]$ 141
$Ca_4Li_2[SiO_4][Si_3O_9]$ 271
$Ca_4Li_4[Si_{10}O_{26}]$ 188
$CaMg[Si_2O_6]$ 80, 102, 111
$Ca_2Mg[Si_2O_7]$ 18
$CaMg_2Al[Al_3SiO_{10}](OH)_2$ 214, 215
$Ca_2(Mg, Al)[(Al, Si)_2O_7]$ 2
$Ca(Mg, Fe)[SiO_4]$ 2
$Ca_2(Mg, Fe)_5[Si_4O_{11}]_2(OH)_2$ 71
$Ca_{10}(Mg, Fe)_2Al_4[SiO_4]_5[Si_2O_7]_2(OH)_4$ 129
$\sim CaMg_2Ti[Al_2SiO_{10}]$ 2
$CaMn[Si_2O_6]$ 80, 102
$(Ca, Mn)_3[Si_3O_9]$ 81, 103, 179
$Ca_2Mn_7[Si_{10}O_{28}](OH)_2 \cdot 5\,H_2O$ 81, 105, 140, 186
$Ca_3Mn_2[Si_4O_{12}]O_2$ 103, 174
$Ca_2(Mn, Al)_3[SiO_4][Si_2O_7](OH)_3$ 270
$Ca_{0.95}(Mn, Fe, Mg)_{2.05}[Si_3O_9]$ 180
$Ca_{2.35}(Mn, Fe, Mg)_{0.65}[Si_3O_9]$ 180
$Ca_{1-x}Na_x[Al_{2-x}Si_{2+x}O_8]$ 88
$CaNi[Si_2O_6]$ 102
$Ca_2Pb[Si_3O_9]$ 98, 192
$Ca_2Pb_3[SiO_4][Si_2O_7]$ 270
$(Ca_3RE)Y_4[Si_8O_{20}][CO_3]_6(OH) \cdot 7\,H_2O$ 141, 185
$Ca_2Sn_2[Si_6O_{18}] \cdot 4\,H_2O$ 81, 101, 103, 140, 174
$(Ca, Sr)Ba_{24}(Mg, Mn, Fe, Al, Ti)_{16}[Si_2O_7]_6[Si_{12}O_{36}](O, OH)_{30}Cl_6 \cdot 14\,H_2O$ 96, 99, 193, 270
$Ca_2Th[Si_4O_{10}]_2$ 196, 203, 205
$CaV[Si_4O_{10}]O \cdot 4\,H_2O$ 28, 196, 217, 219
$Ca_2Y_2[Si_4O_{12}][CO_3] \cdot H_2O$ 192
$CaZr[Si_6O_{15}] \cdot 2.5\,H_2O$ 118, 196, 217, 219
$(Fe, Ca)_7[Si_7O_{21}]$ 79, 81, 101, 103
$(Fe, Mn, Ca)[SiO_3]$ 140
$H_2(Ba, Ca)_{10}[(Al, Si)_{20}O_{43}][SO_4]_3 \cdot H_2O$ 126, 128, 129
$H_4Ca_4[(Al_4Si_8)O_{26}](OH)_4 \cdot \sim 6.5\,H_2O$ 126, 189
$HCa[BSiO_5]$ 141
$H_2Ca_3[Si_3O_{10}]$ 73, 95, 140
$H_2Ca_4Ba[Si_8O_{19}]_2 \cdot (8+x)\,H_2O$ 124, 189, 232
$HCa_2Fe_2[Si_5O_{15}]$ 103, 180
$HCaMn_4[Si_5O_{15}](OH) \cdot H_2O$ 103, 180
$HCaSc[Si_3O_9]$ 103, 180
$H_2KCa_2[AlSi_7O_{19}] \cdot 6\,H_2O$ 189, 232
$HKCa_2[Si_8O_{19}] \cdot 5\,H_2O$ 124, 189, 232, 234
$HNaCa_2[Si_3O_9]$ 81, 103, 180, 197
$HNaCaMn_3[Si_5O_{15}]$ 103, 180
$HNa(Ca, Mn)_2[Si_3O_9]$ 180, 198
$H_2Na_4(Ca, Mn)[Si_6O_{16}]$ 189
$H_xNa_{0-2}(Ca, Na)_8(Fe, Mn, Zn, Ti)RE_2[(Si, Be)_{10}(O, F)_{24}]_2$ 123, 189
$HNaK_2Ca_2Ti[Si_7O_{19}]O$ 112, 188
$HNa(Mn, Ca)_2[Si_3O_9]$ 79, 81, 179, 180, 197, 198
$HNaMn_3Ca[Si_5O_{15}]$ 180
$K_{1.7}Ca_6[(Al_{5.7}Si_{4.3})O_{22}][SO_4]_{1.4}[CO_3]_{0.6}$ 86, 124, 231
$K_5Ca_2[Al_9Si_{23}O_{64}] \cdot 24\,H_2O$ 150, 152
$KCa_4[Si_4O_{10}]_2(F, OH) \cdot 8\,H_2O$ 73, 115, 141, 196–198, 221, 225–228
$K_{16}Ca_4[Si_{12}O_{36}]$ 193
$KCa_2(Be_2Al)[Si_{12}O_{30}] \cdot 0.75\,H_2O$ 97, 99, 194, 204
$(K, Ca, Na)_2[Al_3Si_{11}O_{28}] \cdot 12\,H_2O$ 158
$(K, \square)(Ca, RE, Na)_2(RE, Th)[Si_8O_{20}]$ 99, 194
$K_2Ca_{10}(Y, RE)_2[Si_2O_7]_2[Si_{12}O_{30}](OH)_2F_2$ 110, 111, 129, 142, 186, 271
$(K, Na, Ca, H_2O)(Mn, Mg, Fe^{+2}, Zn, Al)_{9.6}[Si_{12}O_{30}](OH)_{10} \cdot 4\,H_2O$ 229
$(K, Na)_{2.5}(Ca, Na)_7[Si_8O_{19}]_2(OH, F)_2 \cdot H_2O$ 232
$K_{1-x}(Na, Ca)_{2-y}Th_{1-z}[Si_8O_{20}]$ 97, 99, 141, 194, 204
$(K, Sr, H_2O)_2(Ca, Na, Mg, Fe)_6[(Al_{3.66}Si_{6.34})O_{22}][SO_4]_{1.4}[CO_3]_{0.5}[O_4H_4]_{0.1}$ 86, 124, 231
$Li_4Ca_4[Si_{10}O_{26}]$ 112, 188, 190
$Li_3KCa_7Ti_2[Si_6O_{16}]_2F_2$ 98
$Li_3(K, Na)Ca_7(Ti, Fe, Mn)_2[Si_6O_{18}]_2(OH, F)_2$ 193
$Mg_5Ca_2[Si_4O_{11}]_2(OH)_2$ 50, 80, 105, 111, 186
$(Mn, Ca)_5[Si_5O_{15}]$ 101, 103, 140, 180, 196
$(Mn, Fe, Ca)_7[Si_7O_{21}]$ 81
$Na_2Be_2Ca_2[Si_4O_{12}]F_2$ 81, 103, 174
$NaCa_2[(Al, Si)_{10}O_{20}] \cdot 6\,H_2O$ 128, 189
$(Na_2, Ca)[Al_2Si_7O_{18}] \cdot 7\,H_2O$ 146
$Na_2Ca_5[Al_{12}Si_{20}O_{64}] \cdot 28\,H_2O$ 152
$(Na_2, Ca)_2[Al_4Si_8O_{24}] \cdot 12\,H_2O$ 146
$Na_6Ca[AlSiO_4]_6[CO_3] \cdot 2\,H_2O$ 146
$NaCa_4[Si_4O_{10}]_2F \cdot 8\,H_2O$ 196, 221
$Na_2Ca_3[Si_3O_{10}]$ 95
$Na_2Ca_4[Si_8O_{20}]F_2$ 187, 188
$Na_4Ca_4[Si_6O_{18}]$ 98, 192
$Na_4Ca_6[Si_{12}O_{32}]$ 186

$Na_{16}Ca_4[Si_{12}O_{36}]$ 193
$NaCaAl[Si_2O_7]$ 18
$(Na, Ca)_4Be_4Ca_4[SiO_4][Si_7O_{20}]F_4$ 121, 129, 188, 271
$(Na, Ca)_3(Ca, Ce)_4(Ti, Nb)[Si_2O_7]_2(O, F)_4$ 19
$(Na, Ca)_2(Ca, Fe, Mn, Ti)(Zr, Nb)[Si_2O_7]OF$ 19
$Na_{12}Ca_3Fe_2[Si_6O_{18}]_2$ 98, 193
$(Na, Ca)_3(Fe, Al, Mg)_5[(Al, Si)_4O_{11}]_2(OH)_2$ 80
$(Na_2, Ca, Mg)_{29}[Al_{58}Si_{134}O_{384}] \cdot 240\ H_2O$ 156
$(Na, Ca)(Mn, Fe, Al, Mg)_{12}[Si_6O_{17}]_2(O, OH)$ 188
$Na_{12}(Ca, RE)_6(Fe, Mn, Mg)_3(Zr, Nb)_{3+x}[Si_3O_9]_2[Si_9(O, OH)_{27}]_2Cl_y$ 96, 99, 129, 141, 193, 270
$Na_6CaZr[Si_6O_{18}]$ 98, 192
$Na_2Ca_4Zr(Nb, Ti)[Si_2O_7]_2O_2F(O, F)$ 19
$(Na, K)Ca_7[Si_4O_{10}][(AlSi_7)O_{19}](OH)_4 \cdot 3\ H_2O$ 115, 122, 124, 129, 141, 232, 234, 271
$Na_3K_7Ca_5[AlSi_7O_{19}]_2F_4Cl_2$ 124, 141, 189, 197, 232, 234
$Na_4KCa_4[Si_8O_{18}][CO_3]_4(OH, F) \cdot H_2O$ 124, 141, 189, 197, 232, 235
$Na_4K_2Ca_5[Si_{12}O_{30}](OH, F)_4$ 186
$Na_9KCa_2Ba_6(Mn, Fe)_6(Ti, Nb, Ta)_6B_{12}[Si_{18}O_{54}]_2O_{15}(OH)_2$ 97, 188
$(Na, K, Ca)_3Mg_2Ca[Al_{10}Si_{26}O_{72}] \cdot 28\ H_2O$ 150
$(Na, K)_2CaZr_2[Si_5O_{13}]_2 \cdot 5-6\ H_2O$ 189
$Na_4(Na, Mn)_2(Sr, Ca)_2(Zn, Mg, Fe, Mn)RE_2[Si_{12}O_{34}]$ 104, 188

Cd

$HNaCd_2[Si_3O_9]$ 180
$Na_2Cd_3[Si_3O_{10}]$ 95
$Na_4Cd_2[Si_3O_{10}]$ 95
$Na_6Cd_3[Si_6O_{18}]$ 193
$Pb_3Cd_2[SiO_4][Si_2O_7]$ 270

Ce

$HNa_2Ce[Si_6O_{15}] \cdot n\ H_2O$ 118, 196, 219, 223
$K_2Ce[Si_6O_{15}]$ 126, 142, 144, 166, 167, 189, 199, 202, 205, 243
$(Na, Ca)_3(Ca, Ce)_4(Ti, Nb)[Si_2O_7]_2(O, F)_4$ 19

Cl

$Al_{13}[Si_5O_{16}](OH, F)_{18}O_4Cl$ 28, 73, 94, 96, 140, 165, 188
$Ba[AlSi_2O_6](OH, Cl)$ 152
$Ba_4B_2(Fe, Ti)_4[Si_4O_{12}]_2O_5Cl_x$ 96, 98, 192
$Ba_4B_2(V, Ti)_4[Si_4O_{12}]_2O_3(O, OH)_2Cl$ 98, 192
$Ba_{10}(Ca, Mn, Ti)_4[Si_8O_{24}](Cl, O, OH)_{12} \cdot 4\ H_2O$ 96, 99, 140, 193
$Ba_{12}(Mn, Ti, Fe)_6[Si_4O_{12}]_3(O, OH)_2(OH, H_2O)_7Cl_9$ 98
$Ba_4(Ti, Nb)_8[Si_4O_{12}]O_{16}Cl$ 98, 140, 192
$Ca_{10}[SiO_4][Si_2O_7]_2(OH)_2Cl_2$ 71, 140, 270
$(Ca, Sr)Ba_{24}(Mg, Mn, Fe, Al, Ti)_{16}[Si_2O_7]_6[Si_{12}O_{36}](O, OH)_{30}Cl_6 \cdot 14\ H_2O$ 96, 99, 193, 270
$HLi_2Ba_6[SiO_4]_2[Si_2O_7]Cl$ 270
$LiBa_9[Si_{10}O_{25}][CO_3]Cl_7$ 115, 196
$Mn_8[Si_6O_{15}](OH, Cl)_{10}$ 115, 196, 221, 225, 227
$(Mn, Fe)_8[Si_6O_{15}](OH, Cl)_{10}$ 141, 196, 221, 229
$Na_4[Al_3Si_3O_{12}]Cl$ 2, 128, 142, 146, 156, 189
$Na_{12}(Ca, RE)_6(Fe, Mn, Mg)_3(Zr, Nb)_{3+x}[Si_3O_9]_2[Si_9(O, OH)_{27}]_2Cl_y$ 96, 99, 129, 141, 193, 270
$Na_3K_7Ca_5[AlSi_7O_{19}]_2F_4Cl_2$ 124, 141, 189, 197, 232, 234
$Na_5Zr_2[Si_6O_{18}](Cl, OH) \cdot 2\ H_2O$ 98, 193

Co

$CaCo[Si_2O_6]$ 102
$Co_2[SiO_4]$ 47
$Co_2[Si_2O_6]$ 47, 49
$Co_3[Si_2O_5](OH)_4$ 221, 225
$Co_3Al_2[SiO_4]_3$ 51
$H_2[Co(H_2N \cdot CH_2 \cdot CH_2 \cdot NH_2)_3]_2[Si_8O_{20}] \cdot 16.4\ H_2O$ 99, 194
$KCo_3[Si_3CoO_{10}](OH)_2$ 214
$KFe_3[Si_3CoO_{10}](OH)_2$ 214

Cr

$NaCr[Si_2O_6]$ 2, 50, 102
$NaMg_2Cr[Si_3O_{10}]$ 2

Cs

$Cs[AlSiO_4]$ 256, 258, 261, 263, 275
$Cs[AlSi_2O_6]$ 253, 261
$Cs[AlSi_3O_8]$ 250
$Cs_2[Si_2O_5]$ 199, 200, 205
$Cs_6[Si_{10}O_{23}]$ 126, 189, 205, 244
$Cs_2Cu_2[Si_8O_{19}]$ 232, 233
$(Na, Cs)_3[Al_3Si_9O_{24}] \cdot 11\ H_2O$ 156

Cu

$BaCu[Si_4O_{10}]$ 196, 205
$CaCu[Si_4O_{10}]$ 196, 203, 205
$Ca_2Cu_2[Si_3O_{10}] \cdot 2\ H_2O$ 95
$Ca_2Cu_2Al_2[Si_4O_{12}](OH)_6$ 192
$Cu_5[Si_2O_6]_2(OH)_2$ 80, 102, 174
$Cu_6[Si_6O_{18}] \cdot 6\ H_2O$ 99, 140, 193
$[Cu(H_2N \cdot CH_2 \cdot CH_2 \cdot NH_2)_2]_4[Si_8O_{20}] \cdot 38\ H_2O$ 99, 194
$[Cu(H_2N \cdot CH_2 \cdot CH_2 \cdot NH_2)_2][(C_6H_4O_2)_3Si]$ 91
$K_8Cu_4[Si_{16}O_{40}]$ 109–111, 203, 205

$K_2Mg_3Cu_2[Si_{12}O_{30}]$ 99, 194, 204
$Li_2Cu_5[Si_2O_7]_2$ 18
$Na_2Cu_3[Si_4O_{12}]$ 79, 101, 103, 174
$Na_4Cu_2[Si_8O_{20}]$ 188, 205
$Na_2K_2Cu_2[Si_8O_{20}]$ 112, 140, 188
$NaKMg_3Cu_2[Si_{12}O_{30}]$ 99, 194, 204
$Na_2Mg_3Cu_2[Si_{12}O_{30}]$ 99, 194, 204
$NaRbMg_3Cu_2[Si_{12}O_{30}]$ 99, 194, 204
$SrCu[Si_4O_{10}]$ 196, 203, 205
$Cs_2Cu_2[Si_8O_{19}]$ 232, 233

Er

$Er_2[Si_2O_7]$ 19, 28
$Er_4Pb[Si_2O_7][Si_3O_{10}]$ 270

F

$Al_{13}[Si_5O_{16}](OH, F)_{18}O_4Cl$ 28, 73, 94, 96, 140, 165, 188
$Ca_4[Si_2O_7](OH, F)_2$ 19
$Ca_{13}[Si_5O_{14}]_2F_8(OH)_2 \cdot 6\,H_2O$ 121, 141, 188
$CaB_3Al_6(Al, Li)_3[Si_6O_{18}]O_9(O, OH, F)_4$ 193
$Ca_2Ba_2Pb_2B_2[(Si_{1.5}Be_{0.5})Si_8O_{28}]F$ 100, 188
$H_xNa_{0-2}(Ca, Na)_8(Fe, Mn, Zn, Ti)RE_2[(Si, Be)_{10}(O, F)_{24}]_2$ 123, 189
$KCa_4[Si_4O_{10}]_2(F, OH) \cdot 8\,H_2O$ 73, 115, 141, 196–198, 221, 225–228
$K_2Ca_{10}(Y, RE)_2[Si_2O_7]_2[Si_{12}O_{30}](OH)_2F_2$ 110, 111, 129, 142, 186, 271
$(K, Na)_{2.5}(Ca, Na)_7[Si_8O_{19}]_2(OH, F)_2 \cdot H_2O$ 232
$Li_3KCa_7Ti_2[Si_6O_{18}]_2F_2$ 98
$LiKMg_2[Si_2O_5]_2F_2$ 215
$Li_3(K, Na)Ca_7(Ti, Fe, Mn)_2[Si_6O_{18}]_2(OH, F)_2$ 193
$Na_2Be_2Ca_2[Si_4O_{12}]F_2$ 81, 103, 174
$NaCa_4[Si_4O_{10}]_2F \cdot 8\,H_2O$ 196, 221
$Na_2Ca_4[Si_8O_{20}]F_2$ 187, 188
$(Na, Ca)_4Be_4Ca_4[SiO_4][Si_7O_{20}]F_4$ 121, 129, 188, 271
$(Na, Ca)_3(Ca, Ce)_4(Ti, Nb)[Si_2O_7]_2(O, F)_4$ 19
$(Na, Ca)_2(Ca, Fe, Mn, Ti)(Zr, Nb)[Si_2O_7]OF$ 19
$Na_2Ca_4Zr(Nb, Ti)[Si_2O_7]_2O_2F(O, F)$ 19
$Na_3K_7Ca_5[AlSi_7O_{19}]_2F_4Cl_2$ 124, 141, 189, 197, 232, 234
$Na_4KCa_4[Si_8O_{18}][CO_3]_4(OH, F) \cdot H_2O$ 124, 141, 189, 197, 232, 235
$Na_4K_2Ca_5[Si_{12}O_{30}](OH, F)_4$ 186
$NaK_2Mg_2(Fe, Mn)_5Ti_2[Si_4O_{12}]_2(O, OH, F)_7$ 104, 188, 190
$Na_2(Sr, Ba)_2Ti_3[Si_2O_7]_2(OH, F)_2O_2$ 18

Fe

$Ba_4B_2(Fe, Ti)_4[Si_4O_{12}]_2O_5Cl_x$ 96, 98, 192
$BaFe[Si_4O_{10}]$ 28, 196, 201, 203, 205
$Ba_{12}(Mn, Ti, Fe)_6[Si_4O_{12}]_3(O, OH)_2(OH, H_2O)_7Cl_9$ 98
$Ba_2RE_2FeTi_2[Si_4O_{12}]_2O_2$ 192
$(Ba, Sr)(Mn, Fe)_2(Fe, Ti)[Si_2O_7](O, OH)_2$ 18
$Be_2Ca_2(Ca, Pb)(Fe, Mg)_5[SiO_4]_2[Si_2O_6]_2O_2$ 129, 271
$BeFe_3[Si_3O_9](OH)_2$ 103
$Ca_2Al_2Fe[SiO_4][Si_2O_7]O(OH)$ 140, 270
$Ca_{10}Al_4(Fe, Mg)_2[SiO_4]_5[Si_2O_7]_2(OH)_4$ 270
$Ca_2(Al, Fe, Mn)_3[SiO_4][Si_2O_7]O(OH)$ 270
$Ca_2Ba_4(Fe, Mg)_4[Si_{12}O_{34}]$ 28, 104, 188
$CaFe[Si_2O_6]$ 102
$(Ca, Fe)_3[Si_3O_9]$ 103
$(Ca, Fe)_7[Si_7O_{21}]$ 103
$Ca_3Fe_2[SiO_4]_3$ 34
$Ca_2(Fe, Al)_3[SiO_4][Si_2O_7]O(OH)$ 129
$CaFe_2(Fe, Mn)[Si_2O_7]O(OH)$ 18
$Ca(Mg, Fe)[SiO_4]$ 2
$Ca_2(Mg, Fe)_5[Si_4O_{11}]_2(OH)_2$ 71
$Ca_{10}(Mg, Fe)_2Al_4[SiO_4]_5[Si_2O_7]_2(OH)_4$ 129
$Ca_{0.95}(Mn, Fe, Mg)_{2.05}[Si_3O_9]$ 180
$Ca_{2.35}(Mn, Fe, Mg)_{0.65}[Si_3O_9]$ 180
$(Ca, Sr)Ba_{24}(Mg, Mn, Fe, Al, Ti)_{16}[Si_2O_7]_6[Si_{12}O_{36}](O, OH)_{30}Cl_6 \cdot 14\,H_2O$ 96, 99, 193, 270
$(Fe, Mg)^{[4]}Fe^{[6]}Si^{[6]}O_4$ 2
$Fe_2[SiO_4]$ 47
$Fe[SiO_3]$ (hT, mP) (ferrosilite-III) 101, 140, 173
$Fe_2[Si_2O_6]$ 47, 49, 102, 197
$Fe_3[Si_2O_5](OH)_4$ 221, 225
$Fe_9[Si_9O_{27}]$ 81, 101, 103, 173
$FeAl_2[Si_2O_6](OH)_4$ 80, 174
$Fe_3Al_2[SiO_4]_3$ 51
$(Fe^{+2}, Fe^{+3})_{<6}[Si_4O_{10}](OH)_8$ 2, 229
$Fe_6^{+2}Fe_3^{+3}[Si_6O_{17}]O_3(OH)_5$ 73, 104, 188
$(Fe, Ca)_7[Si_7O_{21}]$ 79, 81, 101, 103
$(Fe, Mg)_6[AlSi_3O_{10}](O, OH)_8$ 2
$(Fe, Mg)_7[Si_4O_{11}]_2(OH)_2$ 50, 80
$(Fe, Mg)_{17}[Si_4O_{11}]_2[Si_6O_{16}]_2(OH)_6$ 129, 271
$(Fe^{+2}, Mg, Mn, Fe^{+3})_{<9}[Si_9(Al, Si)_3O_{30}](OH)_6X_n \cdot 6\,H_2O$ 229
$(Fe, Mn)_7[Si_7O_{21}]$ 103
$(Fe, Mn, Ca)[SiO_3]$ 140
$HCa_2Fe_2[Si_5O_{15}]$ 103, 180
$H_xNa_{0-2}(Ca, Na)_8(Fe, Mn, Zn, Ti)RE_2[(Si, Be)_{10}(O, F)_{24}]_2$ 123, 189
$KFe_3[AlSi_3O_{10}](OH)_2$ 214
$KFe_{13}[(Al, Si)_9O_{21}]_2(OH)_{14}$ 188, 235
$KFe_{24}[(AlSi_{35})O_{84}]O_6(OH)_{24} \cdot n\,H_2O$ 122, 124, 232, 234
$KFe_3[Si_3CoO_{10}](OH)_2$ 214
$K_2Fe_5[Si_{12}O_{30}]$ 205
$K(Fe, Mg, Mn, Al)_{13}[(Al, Si)_9O_{21}]_2(OH)_{14}$ 141
$(K, H_2O)_3(Mn, Fe^{+3}, Mg, Al)_{<9}[Si_{12}O_{30}](OH)_6X_n \cdot 6\,H_2O$ 229

$K_2Mg_3Fe_2[Si_{12}O_{30}]$ 99, 194, 205
$(K, Na, Ca, H_2O)(Mn, Mg, Fe^{+2}, Zn, Al)_{9.6}[Si_{12}O_{30}](OH)_{10} \cdot 4\,H_2O$ 229
$(K, Sr, H_2O)_2(Ca, Na, Mg, Fe)_6[(Al_{3.66}Si_{6.34})O_{22}][SO_4]_{1.4}[CO_3]_{0.5}[O_4H_4]_{0.1}$ 86, 124, 231
$LiFe[Si_2O_6]$ 102, 171
$Li_3(K, Na)Ca_7(Ti, Fe, Mn)_2[Si_6O_{18}]_2(OH, F)_2$ 193
$Li_2Na_4Fe_2[Si_{12}O_{30}]$ 81, 185, 205
$LiNa_2K(Fe, Mg, Mn)_2Ti_2[Si_8O_{22}]O_2$ 128, 144
$(Mg, Fe)_2[SiO_4]$ 2, 94
β-$(Mg, Fe)_2[SiO_4]$ 2
$(Mg, Fe)_5[Si_6O_{16}](OH)_2$ 111, 186
$(Mg, Fe)_7[Si_4O_{11}]_2(OH)_2$ 111, 140
$(Mg, Fe)_{17}[Si_4O_{11}]_2[Si_6O_{16}]_2(OH)_6$ 141, 271
$(Mg, Fe, Al)_4[Si_2O_5]_3(O, OH)_2 \cdot 4\,H_2O$ 196, 216, 217
$(Mg, Fe, Al)_6[(Al, Si)_4O_{10}](OH)_8$ 2
$(Mn, Fe)_8[Si_6O_{15}](OH, Cl)_{10}$ 141, 196, 221, 229
$(Mn, Fe, Ca)_7[Si_7O_{21}]$ 81
$(Na, Ca)_2(Ca, Fe, Mn, Ti)(Zr, Nb)[Si_2O_7]OF$ 19
$Na_{12}Ca_3Fe_2[Si_6O_{18}]_2$ 98, 193
$(Na, Ca)_3(Fe, Al, Mg)_5[(Al, Si)_4O_{11}]_2(OH)_2$ 80
$(Na, Ca)(Mn, Fe, Al, Mg)_{12}[Si_6O_{17}]_2(O, OH)_{10}$ 188
$Na_{12}(Ca, RE)_6(Fe, Mn, Mg)_3(Zr, Nb)_{3+x}[Si_3O_9]_2[Si_9(O, OH)_{27}]_2Cl_y$ 96, 99, 129, 141, 193, 270
$Na_6Fe[(Al_4Si_8)O_{26}]$ 122, 124, 141, 232, 235
$NaFe[Si_2O_6]$ 50, 102
$Na_4Fe_2[Si_8O_{20}]$ 188
$Na(Fe^{+2}, Fe^{+3}, Mn, Al, Mg)_{12}[Si_6O_{17}]_2(O, OH)_{10}$ 188
$Na_2Fe_5Ti[Si_6O_{18}]O_2$ 104, 188
$Na_9KCa_2Ba_6(Mn, Fe)_6(Ti, Nb, Ta)_6B_{12}[Si_{18}O_{54}]_2O_{15}(OH)_2$ 97, 188
$Na_2K_2Fe_2[Si_8O_{20}]$ 71, 188
$(Na, K)_2Fe_2^{+2}Fe_2^{+3}[Si_{12}O_{30}] \cdot H_2O$ 105, 142, 185, 205
$NaK_2Mg_2(Fe, Mn)_5Ti_2[Si_4O_{12}]_2(O, OH, F)_7$ 104, 188, 190
$Na_2Mg_3Fe_2[Si_{12}O_{30}]$ 99, 194, 204
$Na_4(Na, Mn)_2(Sr, Ca)_2(Zn, Mg, Fe, Mn)RE_2[Si_{12}O_{34}]$ 104, 188
$NaRbMg_3Fe_2[Si_{12}O_{30}]$ 99, 194, 204
$Sr_3(Ti, Fe^{+3})[Si_4O_{12}](O, OH) \cdot 2-3\,H_2O$ 81, 103, 174

Gd

$BaGd_4[Si_2O_7][Si_3O_{10}]$ 270
$Gd_2[Si_2O_7]$ 19
$Na_{15}Gd_3[Si_{12}O_{36}]$ 193

H^+

$H_2[Si_2O_5]$ 196, 200
$H_4[SiO_4]$ 47
$H_4[Si_6O_{14}]$ 189
$H_6[Si_2O_7]$ 47
$H_4Ba_2[Si_4O_{12}] \cdot 4\,H_2O$ 79, 81, 103, 174
$H_2(Ba, Ca)_{10}[(Al, Si)_{20}O_{43}][SO_4]_3 \cdot H_2O$ 126, 128, 189
$H_2[Be_4Si_2O_9]$ 142
$H_4Ca_4[(Al_4Si_8)O_{26}](OH)_4 \cdot \sim 6.5\,H_2O$ 126, 189
$HCa[BSiO_5]$ 141
$H_2Ca_3[Si_3O_{10}]$ 73, 95, 140
$H_2Ca_4Ba[Si_8O_{19}]_2 \cdot (8+x)H_2O$ 124, 189, 232
$HCa_2Fe_2[Si_5O_{15}]$ 103, 180
$HCaMn_4[Si_5O_{15}](OH) \cdot H_2O$ 103, 180
$HCaSc[Si_3O_9]$ 103, 180
$H_2[Co(H_2N \cdot CH_2 \cdot CH_2 \cdot NH_2)_3]_2[Si_8O_{20}] \cdot 16.4\,H_2O$ 99, 194
$H_4K_4[Si_4O_{12}]$ 192
$H_2KCa_2[AlSi_7O_{19}] \cdot 6\,H_2O$ 189, 232
$HKCa_2[Si_8O_{19}] \cdot 5\,H_2O$ 124, 189, 232, 234
$H_2K_3Y[Si_3O_{10}]$ 95
$HLi_2Ba_6[SiO_4]_2[Si_2O_7]Cl$ 270
$HLiMn_4[Si_5O_{15}]$ 103, 180
$H(Li, Na)Mn_4[Si_5O_{15}]$ 81, 103, 180
$HMn_4As[Si_3O_{10}]O_3$ 95
$HMn_6V[Si_5O_{16}]O_3$ 94, 95, 165
$H_7[N(n-C_4H_9)_4][Si_8O_{20}] \cdot 5.33\,H_2O$ 99, 194
$HNa_2[(AlSi_3)O_9]$ 126, 142, 144
$H_2Na_6[Be_2Al_2Si_{16}O_{41}] \cdot 1.5\,H_2O$ 142
$H_2Na_2[Si_4O_{10}] \cdot 4\,H_2O$ 196
$HNaCa_2[Si_3O_9]$ 81, 103, 180, 197
$HNa(Ca, Mn)_2[Si_3O_9]$ 180, 198
$HNaCaMn_3[Si_5O_{15}]$ 103, 180
$H_2Na_4(Ca, Mn)[Si_6O_{16}]$ 189
$H_xNa_{0-2}(Ca, Na)_8(Fe, Mn, Zn, Ti)RE_2[(Si, Be)_{10}(O, F)_{24}]_2$ 123, 189
$HNaCd_2[Si_3O_9]$ 180
$HNa_2Ce[Si_6O_{15}] \cdot n\,H_2O$ 118, 196, 219, 223
$HNaK_2Ca_2Ti[Si_7O_{19}]O$ 112, 188
$HNaMg_4[Si_6O_{16}](OH)_2$ 80, 186
$HNaMn_2[Si_3O_9]$ 103
$HNa_{1.15}Mn_5[Si_{5.5}V_{0.5}O_{18}](OH)$ 104, 180, 188
$HNa(Mn, Ca)_2[Si_3O_9]$ 79, 81, 179, 180, 197, 198
$HNaMn_3Ca[Si_5O_{15}]$ 180
$H_2NaNd[Si_6O_{15}] \cdot n\,H_2O$ 115, 118, 196, 219, 223
$H_2Zn_4Si_2O_9 \cdot H_2O$ 142

Ho

$Ho_4[SiO_4][Si_3O_{10}]$ 270

In

$In_2Si_2O_7$(hP) (pyrochlore-type) 31, 91, 92
$NaIn[Si_2O_6]$ 50, 80, 102

K

$(Ca, K)_4[(Al, Si)_5O_{11}][SO_4, CO_3]$ 141
$H_4K_4[Si_4O_{12}]$ 192
$H_2KCa_2[AlSi_7O_{19}] \cdot 6\,H_2O$ 189, 232
$HKCa_2[Si_8O_{19}] \cdot 5\,H_2O$ 124, 189, 232, 234
$H_2K_3Y[Si_3O_{10}]$ 95
$HNaK_2Ca_2Ti[Si_7O_{19}]O$ 112, 188
$KAlSi_3O_8$(hP) (hollandite-type) 90, 91, 92
$K[AlSiO_4]$ 244, 256, 258, 261, 262, 275
$K[AlSi_2O_6]$ 253, 254, 261
$K[(AlSi_3)O_8]$ (feldspar-type) 12, 61, 73, 127, 142, 156, 189, 246, 248, 249, 261, 264
$K_2[Si_2O_5]$ 200
$K_2[Si_4O_9]$ 92, 119, 141, 196
$K_6[Si_2O_7]$ 19, 34
$KAl_2[AlSi_3O_{10}](OH)_2$ 115, 116, 196
$K_2Ba_7[Si_4O_{10}]_4$ 28, 196, 205
$K_2Be[Si_4O_{10}]$ 196, 205
$K_2Be_2[Si_6O_{15}]$ 196, 203, 205
$K_2Be_2Zn_2[SiO_4][Si_2O_7]$ 270
$K_{1.7}Ca_6[(Al_{5.7}Si_{4.3})O_{22}][SO_4]_{1.4}[CO_3]_{0.6}$ 86, 124, 231
$K_5Ca_2[Al_9Si_{23}O_{64}] \cdot 24\,H_2O$ 150, 152
$KCa_4[Si_4O_{10}]_2(F, OH) \cdot 8\,H_2O$ 73, 115, 141, 196–198, 221, 225–228
$K_{16}Ca_4[Si_{12}O_{36}]$ 193
$KCa_2(Be_2Al)[Si_{12}O_{30}] \cdot 0.75\,H_2O$ 97, 99, 194, 204
$(K, Ca, Na)_2[Al_3Si_{11}O_{28}] \cdot 12\,H_2O$ 158
$K_2Ca_{10}(Y, RE)_2[Si_2O_7]_2[Si_{12}O_{30}](OH)_2F_2$ 110, 111, 129, 142, 186, 271
$(K, \square)(Ca, RE, Na)_2(RE, Th)[Si_8O_{20}]$ 99, 194
$K_2Ce[Si_6O_{15}]$ 126, 142, 144, 166, 167, 189, 199, 202, 205, 243
$KCo_3[Si_3CoO_{10}](OH)_2$ 214
$K_8Cu_4[Si_{16}O_{40}]$ 109–111, 203, 205
$KFe_3[AlSi_3O_{10}](OH)_2$ 214
$KFe_{13}[(Al, Si)_9O_{21}]_2(OH)_{14}$ 188, 235
$KFe_{24}[(AlSi_{35})O_{84}]O_6(OH)_{24} \cdot n\,H_2O$ 122, 124, 232, 234
$KFe_3[Si_3CoO_{10}](OH)_2$ 214
$K_2Fe_5[Si_{12}O_{30}]$ 205
$K(Fe, Mg, Mn, Al)_{13}[(Al, Si)_9O_{21}]_2(OH)_{14}$ 141
$(K, H_2O)_3(Mn, Fe^{+3}, Mg, Al)_{<9}[Si_{12}O_{30}](OH)_6X_n \cdot 6\,H_2O$ 229
$K_2Mg_5[Si_{12}O_{30}]$ 99, 194, 204
$KMg_2Al_3[(Al_2Si_{10})O_{30}]$ 99, 194
$K_2Mg_3Cu_2[Si_{12}O_{30}]$ 99, 194, 204
$K_2Mg_3Fe_2[Si_{12}O_{30}]$ 99, 194, 205
$K_2Mg_3Zn_2[Si_{12}O_{30}]$ 99, 194, 204
$K_2Mn_5[Si_{12}O_{30}]$ 99, 194, 205
$K_2Mn_2Zn_4[SiO_4]_2[Si_2O_7]$ 270
$(K, Na, Ca, H_2O)(Mn, Mg, Fe^{+2}, Zn, Al)_{9.6}[Si_{12}O_{30}](OH)_{10} \cdot 4\,H_2O$ 229
$(K, Na)_{2.5}(Ca, Na)_7[Si_8O_{19}]_2(OH, F)_2 \cdot H_2O$ 232
$K_{1-x}(Na, Ca)_{2-y}Th_{1-z}[Si_8O_{20}]$ 97, 99, 141, 194, 204
$KNa_3Mg_4[Si_{12}O_{30}]$ 194
$K_3Nd[Si_6O_{15}]$ 196, 205
$KNi_3[Si_3NiO_{10}](OH)_2$ 214
$K_4Sc_2[Si_4O_{12}](OH)_2$ 98, 192
$K_2Si[Si_3O_9]$(hP) 31, 47, 91, 92, 192
$K_{16}Sr_4[Si_{12}O_{36}]$ 99, 193
$(K, Sr, H_2O)_2(Ca, Na, Mg, Fe)_6[(Al_{3.66}Si_{6.34})O_{22}][SO_4]_{1.4}[CO_3]_{0.5}[O_4H_4]_{0.1}$ 86, 124, 231
$K_2Ti[Si_3O_9]$ 192
$K_2Ti[Si_6O_{15}]$ 196
$K_8Yb_3[Si_6O_{16}]_2OH$ 121, 189
$K_2Zr[Si_3O_9]$ 98, 192
$K_2Zr[Si_3O_9] \cdot H_2O$ 103, 176
$K_2Zr[Si_6O_{15}]$ 115, 118, 141, 196, 201, 205
$Li_3KCa_7Ti_2[Si_6O_{18}]_2F_2$ 98
$LiKMg_2[Si_2O_5]_2F_2$ 215
$LiK_3Mg_4[Si_{12}O_{30}]$ 99, 194, 204
$Li_3(K, Na)Ca_7(Ti, Fe, Mn)_2[Si_6O_{18}]_2(OH, F)_2$ 193
$Li_3KSn_2[Si_{12}O_{30}]$ 99, 194, 204
$LiNa_2K(Fe, Mg, Mn)_2Ti_2[Si_8O_{22}]O_2$ 128, 144
$Na_{1-x}K_x[AlSiO_4]$ 256, 259
$(Na, K)_5[Al_5Si_{11}O_{32}] \cdot 10\,H_2O$ 152
$NaK_2[Al_3Si_9O_{24}] \cdot 7\,H_2O$ 146, 150
$Na_3K[AlSiO_4]_4$ 244, 275
$Na_3K_7Ca_5[AlSi_7O_{19}]_2F_4Cl_2$ 124, 141, 189, 197, 232, 234
$(Na, K)Ca_7[Si_4O_{10}][(AlSi_7)O_{19}](OH)_4 \cdot 3\,H_2O$ 115, 122, 124, 129, 141, 232, 234, 271
$Na_4KCa_4[Si_8O_{18}][CO_3]_4(OH, F) \cdot H_2O$ 124, 141, 189, 197, 232, 235
$Na_4K_2Ca_5[Si_{12}O_{30}](OH, F)_4$ 186
$Na_9KCa_2Ba_6(Mn, Fe)_6(Ti, Nb, Ta)_6B_{12}[Si_{18}O_{54}]_2O_{15}(OH)_2$ 97, 188
$(Na, K, Ca)_3Mg_2Ca[Al_{10}Si_{26}O_{72}] \cdot 28\,H_2O$ 150
$(Na, K)_2CaZr_2[Si_5O_{13}]_2 \cdot 5-6\,H_2O$ 189
$Na_2K_2Cu_2[Si_8O_{20}]$ 112, 140, 188
$Na_2K_2Fe_2[Si_8O_{20}]$ 71, 188
$(Na, K)_2Fe_2^{+2}Fe_2^{+3}[Si_{12}O_{30}] \cdot H_2O$ 105, 142, 185, 205
$NaKMg_5[Si_{12}O_{30}]$ 99, 168, 194, 204
$NaKMg_3Cu_2[Si_{12}O_{30}]$ 99, 194, 204
$NaK_2Mg_2(Fe, Mn)_5Ti_2[Si_4O_{12}]_2(O, OH, F)_7$ 104, 188, 190
$(Na, K)_3Y[Si_8O_{19}] \cdot 5\,H_2O$ 232
$Na_2K_2Zr_2[Si_6O_{18}] \cdot 4\,H_2O$ 103, 174

La

$La_2[Si_2O_7]$ 18

Li

$CaB_3Al_6(Al, Li)_3[Si_6O_{18}]O_9(O, OH, F)_4$ 193
$Ca_4Li_2[SiO_4][Si_3O_9]$ 271

$Ca_4Li_4[Si_{10}O_{26}]$ 188
$HLi_2Ba_6[SiO_4]_2[Si_2O_7]Cl$ 270
$HLiMn_4[Si_5O_{15}]$ 103, 180
$H(Li, Na)Mn_4[Si_5O_{15}]$ 81, 103, 180
$LiAlSiO_4$ 260, 261, 275, 277
$Li[AlSiO_4]$(hT) (quartz-type) 255, 256, 261, 262
$Li[AlSi_2O_6]$-II (keatite-type) 253, 254, 261, 262
$Li[AlSi_2O_6]$-III (quartz-type) 253, 254, 261, 262
$Li[AlSi_2O_6] \cdot H_2O$ 11, 156
$Li[AlSi_3O_8]$ (quartz-type) 246, 248, 252, 261, 264
$Li[AlSi_3O_8]$ (keatite-type) 246, 248, 252, 261, 264
$Li[AlSi_3O_8]$ (feldspar-type) 250
$Li_2[Si_2O_5]$ 34, 196, 197, 200, 202, 203, 205
$Li_4[Si_2O_6]$ 34, 80, 102, 174, 185, 197
$Li_6[Si_2O_7]$ 34
$Li_4[SiGe_3O_{10}]$ 105, 110, 111, 141, 146, 185, 187
$LiAl_2[AlSiO_5](OH)_4$ 220
$LiAl[SiO_4]$(lT) (phenakite-type) 255–257, 261
$LiAl[Si_2O_5]_2$ 12, 28, 196, 201–205
$LiAl[Si_2O_6]$ (pyroxene-type) 12, 102, 253, 254, 261
$LiBa_9[Si_{10}O_{25}][CO_3]Cl_7$ 115, 196
$Li_4Ca_4[Si_{10}O_{26}]$ 112, 188, 190
$Li_2Cu_5[Si_2O_7]_2$ 18
$LiFe[Si_2O_6]$ 102, 171
$Li_3KCa_7Ti_2[Si_6O_{18}]_2F_2$ 98
$LiKMg_2[Si_2O_5]_2F_2$ 215
$LiK_3Mg_4[Si_{12}O_{30}]$ 99, 194, 204
$Li_3(K, Na)Ca_7(Ti, Fe, Mn)_2[Si_6O_{18}]_2(OH, F)_2$ 193
$Li_3KSn_2[Si_{12}O_{30}]$ 99, 194, 204
$Li_2Mg_2[Si_4O_{11}]$ 104, 188, 190
$Li_2Na_4Fe_2[Si_{12}O_{30}]$ 81, 185, 205
$LiNa_2K(Fe, Mg, Mn)_2Ti_2[Si_8O_{22}]O_2$ 128, 144
$LiNa_3Mg_4[Si_{12}O_{30}]$ 99, 194, 205
$Li_2Na_2Sn_2[Si_{12}O_{30}]$ 185, 205
$Li_2Na_2Ti_2[Si_{12}O_{30}]$ 185, 205
$Li_2Na_4Y_2[Si_{12}O_{30}]$ 185, 205
$Li_2Na_2Zr_2[Si_{12}O_{30}]$ 81, 185, 205
$LiSc[Si_2O_6]$ 102
$Li_4Zr_2[Si_{12}O_{30}]$ 185, 205
$(Mg_{38.46}Sc_{3.11})(Li_{1.16}Si_{0.18}Si_{40})O_{124}$ 85
$NaB_3Al_6(Al, Li)_3[Si_6O_{18}]O_9(OH)_4$ 193

Mg

$BaMgSiO_4$ 244
$Be_2Ca_2(Ca, Pb)(Fe, Mg)_5[SiO_4]_2[Si_2O_6]_2O_2(OH)_2$ 129, 271
$Ca_{10}Al_4(Fe, Mg)_2[SiO_4]_5[Si_2O_7]_2(OH)_4$ 270
$Ca_2Ba_4(Fe, Mg)_4[Si_{12}O_{34}]$ 28, 104, 188
$CaMg[Si_2O_6]$ 80, 102, 111
$Ca_2Mg[Si_2O_7]$ 18
$CaMg_2Al[Al_3SiO_{10}](OH)_2$ 214, 215
$Ca_2(Mg, Al)[(Al, Si)_2O_7]$ 2
$Ca(Mg, Fe)[SiO_4]$ 2
$Ca_2(Mg, Fe)_5[Si_4O_{11}]_2(OH)_2$ 71
$Ca_{10}(Mg, Fe)_2Al_4[SiO_4]_5[Si_2O_7]_2(OH)_4$ 129
$\sim CaMg_2Ti[Al_2SiO_{10}]$ 2
$Ca_{0.95}(Mn, Fe, Mg)_{2.05}[Si_3O_9]$ 180
$Ca_{2.35}(Mn, Fe, Mg)_{0.65}[Si_3O_9]$ 180
$(Ca, Sr)Ba_{24}(Mg, Mn, Fe, Al, Ti)_{16}[Si_2O_7]_6[Si_{12}O_{36}](O, OH)_{30}Cl_6 \cdot 14\ H_2O$ 96, 99, 193, 270
$(Fe, Mg)^{[4]}Fe^{[6]}Si^{[6]}O_4$ 2
$(Fe, Mg)_6[AlSi_3O_{10}](O, OH)_8$ 2
$(Fe, Mg)_7[Si_4O_{11}]_2(OH)_2$ 50, 80
$(Fe, Mg)_{17}[Si_4O_{11}]_2[Si_6O_{16}]_2(OH)_6$ 129, 271
$(Fe^{+2}, Mg, Mn, Fe^{+3})_{<9}[Si_9(Al, Si)_3O_{30}](OH)_6X_n \cdot 6\ H_2O$ 229
$HNaMg_4[Si_6O_{16}](OH)_2$ 80, 186
$K(Fe, Mg, Mn, Al)_{13}[(Al, Si)_9O_{21}]_2(OH)_{14}$ 141
$(K, H_2O)_3(Mn, Fe^{+3}, Mg, Al)_{<9}[Si_{12}O_{30}](OH)_6X_n \cdot 6\ H_2O$ 229
$K_2Mg_5[Si_{12}O_{30}]$ 99, 194, 204
$KMg_2Al_3[(Al_2Si_{10})O_{30}]$ 99, 194
$K_2Mg_3Cu_2[Si_{12}O_{30}]$ 99, 194, 204
$K_2Mg_3Fe_2[Si_{12}O_{30}]$ 99, 194, 204
$K_2Mg_3Zn_2[Si_{12}O_{30}]$ 99, 194, 204
$(K, Na, Ca, H_2O)(Mn, Mg, Fe^{+2}, Zn, Al)_{9.6}[Si_{12}O_{30}](OH)_{10} \cdot 4\ H_2O$ 229
$KNa_3Mg_4[Si_{12}O_{30}]$ 194
$(K, Sr, H_2O)_2(Ca, Na, Mg, Fe)_6[(Al_{3.66}Si_{6.34})O_{22}][SO_4]_{1.4}[CO_3]_{0.5}[O_4H_4]_{0.1}$ 86, 124, 231
$LiKMg_2[Si_2O_5]_2F_2$ 215
$LiK_3Mg_4[Si_{12}O_{30}]$ 99, 194, 204
$Li_2Mg_2[Si_4O_{11}]$ 104, 188, 190
$LiNa_2K(Fe, Mg, Mn)_2Ti_2[Si_8O_{22}]O_2$ 128, 144
$LiNa_3Mg_4[Si_{12}O_{30}]$ 99, 194, 204
$MgSiO_3$ (pyroxene-type) see $Mg_2[Si_2O_6]$
$MgSiO_3$(hP) (ilmenite-type) 31, 91, 93
$MgSiO_3$(hP) (perovskite-type) 31, 91, 92
$Mg[Al_2Si_2O_8]$ (quartz-type) 251, 252, 263, 264
$Mg_2[Al_4Si_5O_{18}]$ 2, 128
$Mg_2[SiO_4]$ 9, 34, 39, 43, 47, 49
$Mg_2[Si_2O_6]$ 34, 37, 47–49, 79, 80, 171–175, 190, 197, 208
$Mg_3[Si_2O_5](OH)_4$ 2, 196, 220, 221, 223–226
$Mg_3[Si_2O_5]_2(OH)_2$ 114, 196, 197, 213, 214, 217, 218, 222
$Mg_4[Si_2O_5]_3(OH)_2 \cdot 4\ H_2O$ 2, 116, 196, 217, 219, 222

$Mg_5[Si_6O_{16}](OH)_2$ 141
$Mg_5[Si_2O_5]_4(OH)_2 \cdot 8\,H_2O$ 2, 116, 196, 216, 217, 219, 222
$Mg_{48}[Si_4O_{10}]_{8.5}(OH)_{62}$ 2, 196, 197, 221, 222, 226
$MgAl_2[AlSiO_5]_2(OH)_2$ 218
$Mg_2Al[AlSiO_5](OH)_4$ 220, 223
$Mg_3Al_2[SiO_4]_3$ 34, 51
$Mg_5Ca_2[Si_4O_{11}]_2(OH)_2$ 50, 80, 105, 111, 186
$(Mg, Fe)_2[SiO_4]$ 2, 94
β-$(Mg, Fe)_2[SiO_4]$ 2
$(Mg, Fe)_5[Si_6O_{16}](OH)_2$ 111, 186
$(Mg, Fe)_7[Si_4O_{11}]_2(OH)_2$ 111, 140
$(Mg, Fe)_{17}[Si_4O_{11}]_2[Si_6O_{16}]_2(OH)_6$ 141, 271
$(Mg, Fe, Al)_4[Si_2O_5]_3(O, OH)_2 \cdot 4\,H_2O$ 196, 216, 217
$(Mg, Fe, Al)_6[(Al, Si)_4O_{10}](OH)_8$ 2
$Mg_3(Mg, Si)_2^{[6]}[SiO_4]_3$ 2
$(Mg_{38.46}Sc_{3.11})(Li_{1.16}Si_{0.18}Si_{40})O_{124}$ 85
$(Mg_{15.61}Sc_{1.37})(Mg_{0.30}Si_{0.02})[Si_8O_{25}]_2$ 94, 95
$(Mg_{17.40}Sc_{1.49})(Mg_{0.15}Si_{0.11})[Si_9O_{28}]_2$ 94, 95
$(Mg_{19.60}Sc_{1.28})(Mg_{0.04}Si_{0.22})[Si_{10}O_{31}]_2$ 94, 95
$MgZn[Si_2O_6]$ 102
$(Mn, Mg)_5[Si_5O_{15}]$ 103
$(Na, Ca)_3(Fe, Al, Mg)_5[(Al, Si)_4O_{11}]_2(OH)_2$ 80
$(Na_2, Ca, Mg)_{29}[Al_{58}Si_{134}O_{384}] \cdot 240\,H_2O$ 156
$(Na, Ca)(Mn, Fe, Al, Mg)_{12}[Si_6O_{17}]_2(O, OH)_{10}$ 188
$Na_{12}(Ca, RE)_6(Fe, Mn, Mg)_3(Zr, Nb)_{3+x}[Si_3O_9]_2[Si_9(O, OH)_{27}]_2Cl_y$ 96, 99, 129, 141, 193, 270
$Na(Fe^{+2}, Fe^{+3}, Mn, Al, Mg)_{12}[Si_6O_{17}]_2(O, OH)_{10}$ 188
$(Na, K, Ca)_3Mg_2Ca[Al_{10}Si_{26}O_{72}] \cdot 28\,H_2O$ 150
$NaKMg_5[Si_{12}O_{30}]$ 99, 168, 194, 204
$NaKMg_3Cu_2[Si_{12}O_{30}]$ 99, 194, 204
$NaK_2Mg_2(Fe, Mn)_5Ti_2[Si_4O_{12}]_2(O, OH, F)_7$ 104, 188, 190
$Na_2Mg_5[Si_{12}O_{30}]$ 99, 194, 204
$Na_4Mg_2[Si_3O_{10}]$ 95
$Na_4Mg_4[Si_{12}O_{30}]$ 81, 185, 205
$Na_2Mg_3Al_2[Si_4O_{11}]_2(OH)_2$ 50, 80
$Na_x(Mg, Al)_2[Si_4O_{10}](OH)_2 \cdot 4\,H_2O$ 2
$NaMg_2Cr[Si_3O_{10}]$ 2
$Na_2Mg_3Cu_2[Si_{12}O_{30}]$ 99, 194, 204
$Na_2Mg_3Fe_2[Si_{12}O_{30}]$ 99, 194, 204
$Na_2Mg_3Zn_2[Si_{12}O_{30}]$ 99, 194, 204
$Na_4(Na, Mn)_2(Sr, Ca)_2(Zn, Mg, Fe, Mn)RE_2[Si_{12}O_{34}]$ 104, 188
$NaRbMg_5[Si_{12}O_{30}]$ 99, 194, 204
$NaRbMg_3Cu_2[Si_{12}O_{30}]$ 99, 194, 204
$NaRbMg_3Fe_2[Si_{12}O_{30}]$ 99, 194, 204

Mn

$Ba_{10}(Ca, Mn, Ti)_4[Si_8O_{24}](Cl, O, OH)_{12} \cdot 4\,H_2O$ 96, 99, 140, 193
$Ba_{12}(Mn, Ti, Fe)_6[Si_4O_{12}]_3(O, OH)_2(OH, H_2O)_7Cl_9$ 98
$(Ba, Sr)(Mn, Fe)_2(Fe, Ti)[Si_2O_7](O, OH)_2$ 18
$Ca_2(Al, Fe, Mn)_3[SiO_4][Si_2O_7]O(OH)$ 270
$CaFe_2(Fe, Mn)[Si_2O_7]O(OH)$ 18
$CaMn[Si_2O_6]$ 80, 102
$(Ca, Mn)_3[Si_3O_9]$ 81, 103, 179
$Ca_2Mn_7[Si_{10}O_{28}](OH)_2 \cdot 5\,H_2O$ 81, 105, 140, 186
$Ca_3Mn_2[Si_4O_{12}]O_2$ 103, 174
$Ca_2(Mn, Al)_3[SiO_4][Si_2O_7](OH)_3$ 270
$Ca_{0.95}(Mn, Fe, Mg)_{2.05}[Si_3O_9]$ 180
$Ca_{2.35}(Mn, Fe, Mg)_{0.65}[Si_3O_9]$ 180
$(Ca, Sr)Ba_{24}(Mg, Mn, Fe, Al, Ti)_{16}[Si_2O_7]_6[Si_{12}O_{36}](O, OH)_{30}Cl_6 \cdot 14\,H_2O$ 96, 99, 193, 270
$(Fe^{+2}, Mg, Mn, Fe^{+3})_{<9}[Si_9(Al, Si)_3O_{30}](OH)_6X_n \cdot 6\,H_2O$ 229
$(Fe, Mn)_7[Si_7O_{21}]$ 103
$(Fe, Mn, Ca)[SiO_3]$ 140
$HCaMn_4[Si_5O_{15}](OH) \cdot H_2O$ 103, 180
$HLiMn_4[Si_5O_{15}]$ 103, 180
$H(Li, Na)Mn_4[Si_5O_{15}]$ 81, 103, 180
$HMn_4As[Si_3O_{10}]O_3$ 95
$HMn_6V[Si_5O_{16}]O_3$ 94, 95, 165
$HNa(Ca, Mn)_2[Si_3O_9]$ 180, 198
$HNaCaMn_3[Si_5O_{15}]$ 103, 180
$H_2Na_4(Ca, Mn)[Si_6O_{16}]$ 189
$H_xNa_{0-2}(Ca, Na)_8(Fe, Mn, Zn, Ti)RE_2[(Si, Be)_{10}(O, F)_{24}]_2$ 123, 189
$HNa_{1.15}Mn_5[Si_{5.5}V_{0.5}O_{18}](OH)$ 104, 180, 188
$HNaMn_3Ca[Si_5O_{15}]$ 180
$HNa(Mn, Ca)_2[Si_3O_9]$ 79, 81, 179, 180, 197, 198
$HNaMn_2[Si_3O_9]$ 103
$K(Fe, Mg, Mn, Al)_{13}[(Al, Si)_9O_{21}]_2(OH)_{14}$ 141
$(K, H_2O)_3(Mn, Fe^{+3}, Mg, Al)_{<9}[Si_{12}O_{30}](OH)_6X_n \cdot 6\,H_2O$ 229
$(K, Na, Ca, H_2O)(Mn, Mg, Fe^{+2}, Zn, Al)_{9.6}[Si_{12}O_{30}](OH)_{10} \cdot 4\,H_2O$ 229
$K_2Mn_5[Si_{12}O_{30}]$ 99, 194, 205
$K_2Mn_2Zn_4[SiO_4]_2[Si_2O_7]$ 270
$Li_3(K, Na)Ca_7(Ti, Fe, Mn)_2[Si_6O_{18}]_2(OH, F)_2$ 193
$LiNa_2K(Fe, Mg, Mn)_2Ti_2[Si_8O_{22}]O_2$ 128, 144
$MnSiO_3$(hT) (pyroxene-type) 102
$MnSiO_3$(hT) (wollastonite-ype) 172
$MnSiO_3$(mP) (pyroxmangite-type) 103, 173
$MnSiO_3$(lP) (rhodonite-type) 103, 172, 173
$Mn_2[SiO_4]$ 47
$Mn_2[Si_2O_6]$ 102

$Mn_5[Si_2O_5]_4(OH)_2 \cdot 8\,H_2O$ 196, 216, 219
$Mn_5[Si_5O_{15}]$ 81, 103, 173, 180
$Mn_6[Si_4O_{10}](OH)_8$ 229
$Mn_7[SiO_4]_2[SiO_4](OH)_2$ 93
$Mn_7[Si_6O_{15}](OH)_8$ 196, 221, 227
$Mn_7[Si_7O_{21}]$ 103, 173, 227
$Mn_8[Si_6O_{15}](OH, Cl)_{10}$ 115, 196, 221, 225, 227
$MnAl_2[Si_2O_6](OH)_4$ 102, 174
$Mn_2Al_3[SiO_4][Si_2O_7](OH)_3$ 270
$Mn_3Al_2[SiO_4]_3$ 51
$Mn_4Al_6[(As, V)O_4][SiO_4]_2[Si_3O_{10}](OH)_6$ 140, 270
$(Mn, Ca)_5[Si_5O_{15}]$ 101, 103, 140, 180, 186
$(Mn, Fe)_8[Si_6O_{15}](OH, Cl)_{10}$ 141, 196, 221, 229
$(Mn, Fe, Ca)_7[Si_7O_{21}]$ 81
$(Mn, Mg)_5[Si_5O_{15}]$ 103
$MnPb_8[Si_2O_7]_3$ 94
$Mn_6[Zn(OH)_4][Zn_2SiAs_2O_{12}(OH)_2](OH)_2$ 142
$(Na, Ca)(Mn, Fe, Al, Mg)_{12}[Si_6O_{17}]_2(O, OH)_{10}$ 188
$Na_{12}(Ca, RE)_6(Fe, Mn, Mg)_3(Zr, Nb)_{3+x}[Si_3O_9]_2[Si_9(O, OH)_{27}]_2Cl_y$ 96, 99, 129, 141, 193, 270
$Na(Fe^{+2}, Fe^{+3}, Mn, Al, Mg)_{12}[Si_6O_{17}]_2(O, OH)_{10}$ 188
$Na_9KCa_2Ba_6(Mn, Fe)_6(Ti, Nb, Ta)_6B_{12}[Si_{18}O_{54}]_2O_{15}(OH)_2$ 97, 188
$NaK_2Mg_2(Fe, Mn)_5Ti_2[Si_4O_{12}]_2(O, OH, F)_7$ 104, 188, 190
$Na_6Mn_3[Si_6O_{18}]$ 193
$Na_4(Na, Mn)_2(Sr, Ca)_2(Zn, Mg, Fe, Mn)RE_2[Si_{12}O_{34}]$ 104, 188
$Na_6MnTi[Si_6O_{18}]$ 98, 193
$(Na, Ca)_2(Ca, Fe, Mn, Ti)(Zr, Nb)[Si_2O_7]OF$ 19

Na

$Ca_{1-x}Na_x[Al_{2-x}Si_{2+x}O_8]$ 88
$H(Li, Na)Mn_4[Si_5O_{15}]$ 81, 103, 180
$HNa_2[(AlSi_3)O_9]$ 126, 142, 144
$H_2Na_6[Be_2Al_2Si_{16}O_{41}] \cdot 1.5\,H_2O$ 142
$H_2Na_2[Si_4O_{10}] \cdot 4\,H_2O$ 196
$HNaCa_2[Si_3O_9]$ 81, 103, 180, 197
$HNaCaMn_3[Si_5O_{15}]$ 103, 180
$HNa(Ca, Mn)_2[Si_3O_9]$ 180, 198
$H_2Na_4(Ca, Mn)[Si_6O_{16}]$ 189
$H_xNa_{0-2}(Ca, Na)_8(Fe, Mn, Zn, Ti)RE_2[(Si, Be)_{10}(O, F)_{24}]_2$ 123, 189
$HNaCd_2[Si_3O_9]$ 180
$HNa_2Ce[Si_6O_{15}] \cdot n\,H_2O$ 118, 196, 219, 223
$HNaK_2Ca_2Ti[Si_7O_{19}]O$ 112, 188
$HNaMg_4[Si_6O_{16}](OH)_2$ 80, 186
$HNaMn_2[Si_3O_9]$ 103
$HNa_{1.15}Mn_5[Si_{5.5}V_{0.5}O_{18}](OH)$ 104, 180, 188
$HNaMn_3Ca[Si_5O_{15}]$ 180
$HNa(Mn, Ca)_2[Si_3O_9]$ 79, 81, 179, 180, 197, 198
$H_2NaNd[Si_6O_{15}] \cdot n\,H_2O$ 115, 118, 196, 219, 223
$(K, Ca, Na)_2[Al_3Si_{11}O_{28}] \cdot 12\,H_2O$ 158
$(K, \square)(Ca, RE, Na)_2(RE, Th)[Si_8O_{20}]$ 99, 194
$(K, Na, Ca, H_2O)(Mn, Mg, Fe^{+2}, Zn, Al)_{9.6}[Si_{12}O_{30}](OH)_{10} \cdot 4\,H_2O$ 229
$(K, Na)_{2.5}(Ca, Na)_7[Si_8O_{19}]_2(OH, F)_2 \cdot H_2O$ 232
$K_{1-x}(Na, Ca)_{2-y}Th_{1-z}[Si_8O_{20}]$ 97, 99, 141, 194, 204
$KNa_3Mg_4[Si_{12}O_{30}]$ 194
$(K, Sr, H_2O)_2(Ca, Na, Mg, Fe)_6[(Al_{3.66}Si_{6.34})O_{22}][SO_4]_{1.4}[CO_3]_{0.5}[O_4H_4]_{0.1}$ 86, 124, 231
$Li_3(K, Na)Ca_7(Ti, Fe, Mn)_2[Si_6O_{18}]_2(OH, F)_2$ 193
$Li_2Na_4Fe_2[Si_{12}O_{30}]$ 81, 185, 205
$LiNa_2K(Fe, Mg, Mn)_2Ti_2[Si_8O_{22}]O_2$ 128, 144
$LiNa_3Mg_4[Si_{12}O_{30}]$ 99, 194, 204
$Li_2Na_2Sn_2[Si_{12}O_{30}]$ 185, 205
$Li_2Na_2Ti_2[Si_{12}O_{30}]$ 185, 205
$Li_2Na_4Y_2[Si_{12}O_{30}]$ 185, 205
$Li_2Na_2Zr_2[Si_{12}O_{30}]$ 81, 185, 205
$NaAlSiO_4$(hP) ($CaFe_2O_4$-type) 90, 91, 92
$Na[AlSiO_4]$ 2, 244, 256–261, 275
$Na[AlSi_2O_6] \cdot H_2O$ 156, 158, 253, 254
$Na[AlSi_3O_8]$ 34, 189, 246–249, 261, 264
$Na[AlSi_5O_{12}] \cdot 3\,H_2O$ 149, 156, 158, 245
$Na_2[Al_2Si_3O_{10}] \cdot 2\,H_2O$ 142
$Na_4[Al_3Si_3O_{12}]Cl$ 2, 128, 142, 146, 156, 189
$Na_5[Al_5Si_{19}O_{48}] \cdot 12\,H_2O$ 149
$Na_6[Al_6Si_6O_{24}] \cdot 8\,H_2O$ 245
$Na_8[Al_8Si_8O_{32}] \cdot 16\,H_2O$ 152
$Na_{12}[(Al_{12}Si_{12})O_{48}] \cdot 27\,H_2O$ 128, 156, 245
$Na_2[Si_2O_5]$ 196, 197, 200–202, 205
$Na_4[Si_2O_6]$ 80, 102, 171, 174, 197
$Na_4[Si_6O_{14}]$ 189
$NaAl[Si_2O_6]$ (pyroxene-type) 50, 102, 253, 254, 261
$NaB_3Al_6(Al, Li)_3[Si_6O_{18}]O_9(OH)_4$ 193
$Na_2Ba[Al_4Si_4O_{16}]$ 153
$Na_2Ba[Si_2O_6]$ 102, 171, 174
$NaBaNd[Si_3O_9]$ 98
$NaBa_3Nd_3[Si_2O_7][Si_4O_{13}]$ 94, 96, 164, 188
$Na_2BaNd_2[Si_4O_{12}][CO_3]$ 98
$Na_2BaTi_2[Si_4O_{12}]O_2$ 81, 103, 140, 174, 185
$Na_4BaTi_2[B_2Si_{10}O_{28}]O_2$ 189, 232
$Na_6Be_2[(Al_2Si_{16})O_{39}](OH)_2 \cdot 1.5\,H_2O$ 28, 126, 128, 144, 189
$Na_2Be_2[Si_3O_9]$ 98, 192
$Na_2Be_2[Si_6O_{15}] \cdot H_2O$ 105, 111, 146, 186, 189, 205

$Na_6Be_3[Si_6O_{18}]$ 81, 103, 174, 185
$Na_2Be_2Ca_2[Si_4O_{12}]F_2$ 81, 103, 174
$Na_4Be_2Sn[Si_3O_9]_2 \cdot 2\,H_2O$ 79, 81, 103, 174, 176
$NaCa_2[(Al, Si)_{10}O_{20}] \cdot 6\,H_2O$ 128, 189
$(Na_2, Ca)[Al_2Si_7O_{18}] \cdot 7\,H_2O$ 146
$(Na_2, Ca)_2[Al_4Si_8O_{24}] \cdot 12\,H_2O$ 146
$Na_2Ca_5[Al_{12}Si_{20}O_{64}] \cdot 28\,H_2O$ 152
$Na_6Ca[AlSiO_4]_6[CO_3] \cdot 2\,H_2O$ 146
$NaCa_4[Si_4O_{10}]_2F \cdot 8\,H_2O$ 196, 221
$Na_2Ca_3[Si_3O_{10}]$ 95
$Na_2Ca_4[Si_8O_{20}]F_2$ 187, 188
$Na_4Ca_4[Si_6O_{18}]$ 98, 192
$Na_4Ca_6[Si_{12}O_{32}]$ 186
$Na_{16}Ca_4[Si_{12}O_{36}]$ 193
$NaCaAl[Si_2O_7]$ 18
$(Na, Ca)_4Be_4Ca_4[SiO_4][Si_7O_{20}]F_4$ 121, 129, 188, 271
$(Na, Ca)_3(Ca, Ce)_4(Ti, Nb)[Si_2O_7]_2(O, F)_4$ 19
$(Na, Ca)_2(Ca, Fe, Mn, Ti)(Zr, Nb)[Si_2O_7]OF$ 19
$Na_{12}Ca_3Fe_2[Si_6O_{18}]_2$ 98, 193
$(Na, Ca)_3(Fe, Al, Mg)_5[(Al, Si)_4O_{11}]_2(OH)_2$ 80
$(Na_2, Ca, Mg)_{29}[Al_{58}Si_{134}O_{384}] \cdot 240\,H_2O$ 156
$(Na, Ca)(Mn, Fe, Al, Mg)_{12}[Si_6O_{17}]_2(O, OH)_{10}$ 188
$Na_{12}(Ca, RE)_6(Fe, Mn, Mg)_3(Zr, Nb)_{3+x}[Si_3O_9]_2[Si_9(O, OH)_{27}]_2Cl_y$ 96, 99, 129, 141, 193, 270
$Na_6CaZr[Si_6O_{18}]$ 98, 192
$Na_2Ca_4Zr(Nb, Ti)[Si_2O_7]_2O_2F(O, F)$ 19
$Na_2Cd_3[Si_3O_{10}]$ 95
$Na_4Cd_2[Si_3O_{10}]$ 95
$Na_6Cd_3[Si_6O_{18}]$ 193
$NaCr[Si_2O_6]$ 2, 50, 102
$(Na, Cs)_3[Al_3Si_9O_{24}] \cdot 11\,H_2O$ 156
$Na_2Cu_3[Si_4O_{12}]$ 79, 101, 103, 174
$Na_4Cu_2[Si_8O_{20}]$ 188, 205
$Na_6Fe[(Al_4Si_8)O_{26}]$ 122, 124, 141, 232, 235
$NaFe[Si_2O_6]$ 50, 102
$Na_4Fe_2[Si_8O_{20}]$ 188
$Na(Fe^{+2}, Fe^{+3}, Mn, Al, Mg)_{12}[Si_6O_{17}]_2(O, OH)_{10}$ 188
$Na_2Fe_5Ti[Si_6O_{18}]O_2$ 104, 188
$Na_{15}Gd_3[Si_{12}O_{36}]$ 193
$NaIn[Si_2O_6]$ 50, 80, 102
$Na_{1-x}K_x[AlSiO_4]$ 256, 259
$(Na, K)_5[Al_5Si_{11}O_{32}] \cdot 10\,H_2O$ 152
$NaK_2[Al_3Si_9O_{24}] \cdot 7\,H_2O$ 146, 150
$Na_3K[AlSiO_4]_4$ 244, 275
$Na_3K_7Ca_5[AlSi_7O_{19}]_2F_4Cl_2$ 124, 141, 189, 197, 232, 234
$(Na, K)Ca_7[Si_4O_{10}][(AlSi_7)O_{19}](OH)_4$ $3\,H_2O$ 115, 122, 124, 129, 141, 232, 234, 271
$Na_4KCa_4[Si_8O_{18}][CO_3]_4(OH, F) \cdot H_2O$ 124, 141, 189, 197, 232, 235
$Na_4K_2Ca_5[Si_{12}O_{30}](OH, F)_4$ 186
$Na_9KCa_2Ba_6(Mn, Fe)_6(Ti, Nd, Ta)_6B_{12}[Si_{18}O_{54}]_2O_{15}(OH)_2$ 97, 188
$(Na, K, Ca)_3Mg_2Ca[Al_{10}Si_{26}O_{72}] \cdot 28\,H_2O$ 150
$(Na, K)_2CaZr_2[Si_5O_{13}]_2 \cdot 5-6\,H_2O$ 189
$Na_2K_2Cu_2[Si_8O_{20}]$ 112, 140, 188
$Na_2K_2Fe_2[Si_8O_{20}]$ 71, 188
$(Na, K)_2Fe_2^{+2}Fe_2^{+3}[Si_{12}O_{30}] \cdot H_2O$ 105, 142, 185, 205
$NaKMg_5[Si_{12}O_{30}]$ 99, 168, 194, 204
$NaKMg_3Cu_2[Si_{12}O_{30}]$ 99, 194, 204
$NaK_2Mg_2(Fe, Mn)_5Ti_2[Si_4O_{12}]_2(O, OH, F)_7$ 104, 188, 190
$(Na, K)_3Y[Si_8O_{19}] \cdot 5\,H_2O$ 232
$Na_2K_2Zr_2[Si_6O_{18}] \cdot 4\,H_2O$ 103, 174
$Na_2Mg_5[Si_{12}O_{30}]$ 99, 194, 204
$Na_4Mg_2[Si_3O_{10}]$ 95
$Na_4Mg_4[Si_{12}O_{30}]$ 81, 185, 205
$Na_2Mg_3Al_2[Si_4O_{11}]_2(OH)_2$ 50, 80
$Na_x(Mg, Al)_2[Si_4O_{10}](OH)_2 \cdot 4\,H_2O$ 2
$NaMg_2Cr[Si_3O_{10}]$ 2
$Na_2Mg_3Cu_2[Si_{12}O_{30}]$ 99, 194, 204
$Na_2Mg_3Fe_2[Si_{12}O_{30}]$ 99, 194, 204
$Na_2Mg_3Zn_2[Si_{12}O_{30}]$ 99, 194, 204
$Na_6Mn_3[Si_6O_{18}]$ 193
$Na_4(Na, Mn)_2(Sr, Ca)_2(Zn, Mg, Fe, Mn)RE_2[Si_{12}O_{34}]$ 104, 188
$Na_6MnTi[Si_6O_{18}]$ 98, 193
$NaNd[Si_6O_{14}]$ 189
$NaPr[Si_6O_{14}]$ 119, 121, 189
$Na_{15}RE_3[Si_{12}O_{36}]$ 99, 193
$NaSc[Si_2O_6]$ 50, 102
$Na_4Sc_2[Si_4O_{13}]$ 94, 95
$Na_4Sn_2[Si_5O_{16}] \cdot H_2O$ 94, 95
$Na_8Sn[Si_6O_{18}]$ 98, 193
$Na_2(Sr, Ba)_2Ti_3[Si_2O_7]_2(OH, F)_2O_2$ 18
$Na_2Ti_2[Si_2O_6]O_3$ 102, 174
$Na_4Ti_2[Si_8O_{20}]O_2$ 81, 105, 111, 151, 185, 205
$NaV[Si_2O_6]$ 50
$Na_3Y[SiO_3]_3$ 140
$Na_{15}Y_3[Si_{12}O_{36}]$ 140
$Na_{24}Y_8[Si_{24}O_{72}]$ 79, 81, 101, 103, 140, 171, 174
$Na_2Zn[Si_2O_6]$ 102, 174
$Na_2Zn[Si_3O_8]$ 119, 152, 166, 196
$Na_2Zr[Si_3O_9] \cdot 2\,H_2O$ 98, 192
$Na_2Zr[Si_3O_9] \cdot 3\,H_2O$ 103, 176, 192
$Na_2Zr[Si_6O_{15}] \cdot 3\,H_2O$ 81, 141, 186
$Na_4Zr_2[Si_6O_{18}] \cdot 4\,H_2O$ 81, 103, 174
$Na_4Zr_2[Si_8O_{22}]$ 104, 140, 188, 190
$Na_5Zr_2[Si_6O_{18}](Cl, OH) \cdot 2\,H_2O$ 98, 193
$NaRbMg_5[Si_{12}O_{30}]$ 99, 194, 204
$NaRbMg_3Cu_2[Si_{12}O_{30}]$ 99, 194, 204
$NaRbMg_3Fe_2[Si_{12}O_{30}]$ 99, 194, 204

Nb

$Ba_3Nb_6[Si_2O_7]_2O_{12}$ 19, 28
$Ba_4(Ti, Nb)_8[Si_4O_{12}]O_{16}Cl$ 98, 140, 192
$(Na, Ca)_3(Ca, Ce)_4(Ti, Nb)[Si_2O_7]_2(O, F)_4$ 19
$(Na, Ca)_2(Ca, Fe, Mn, Ti)(Zr, Nb)[Si_2O_7]OF$ 19
$Na_{12}(Ca, RE)_6(Fe, Mn, Mg)_3(Zr, Nb)_{3+x}[Si_3O_9]_2[Si_9(O, OH)_{27}]_2Cl_y$ 96, 99, 129, 141, 193, 270
$Na_2Ca_4Zr(Nb, Ti)[Si_2O_7]_2O_2F(O, F)$ 19
$Na_9KCa_2Ba_6(Mn, Fe)_6(Ti, Nb, Ta)_6B_{12}[Si_{18}O_{54}]_2O_{15}(OH)_2$ 97, 188

Nd

$H_2NaNd[Si_6O_{15}] \cdot n\, H_2O$ 115, 118, 196, 219, 223
$K_3Nd[Si_6O_{15}]$ 196, 205
$NaBaNd[Si_3O_9]$ 98
$NaBa_3Nd_3[Si_2O_7][Si_4O_{13}]$ 94, 96, 164, 188
$Na_2BaNd_2[Si_4O_{12}][CO_3]$ 98
$NaNd[Si_6O_{14}]$ 189
$Nd_2[Si_2O_7]$ 18

Ni

$CaNi[Si_2O_6]$ 102
$KNi_3[Si_3NiO_{10}](OH)_2$ 214
$Ni_3[Si_2O_5](OH)_4$ 221, 225
$[Ni(H_2N \cdot CH_2 \cdot CH_2 \cdot NH_2)_3]_3[Si_6O_{15}] \cdot 26\, H_2O$ 97, 99, 141, 194

O^{-2}, $(OH)^{-1}$

$Al_2[Si_2O_5](OH)_4$ 141, 214, 220, 222, 223, 225, 227
$Al_2[Si_2O_5]_2(OH)_2$ 213, 214, 218
$Al_{13}[Si_5O_{16}](OH, F)_{18}O_4Cl$ 28, 73, 94, 96, 140, 165, 188
$Ba[AlSi_2O_6](OH, Cl)$ 152
$Ba_4B_2(Fe, Ti)_4[Si_4O_{12}]_2O_5Cl_X$ 96, 98, 192
$Ba_4B_2(V, Ti)_4[Si_4O_{12}]_2O_3(O, OH)_2Cl$ 98, 192
$Ba_{10}(Ca, Mn, Ti)_4[Si_8O_{24}](Cl, O, OH)_{12} \cdot 4\, H_2O$ 96, 99, 140, 193
$Ba_{12}(Mn, Ti, Fe)_6[Si_4O_{12}]_3(O, OH)_2(OH, H_2O)_7Cl_9$ 98
$Ba_3Nb_6[Si_2O_7]_2O_{12}$ 19, 28
$Ba_2RE_2FeTi_2[Si_4O_{12}]_2O_2$ 192
$(Ba, Sr)(Mn, Fe)_2(Fe, Ti)[Si_2O_7](O, OH)_2$ 18
$Ba_4(Ti, Nb)_8[Si_4O_{12}]O_{16}Cl$ 98, 140, 192
$Ba_2V_2[Si_4O_{12}]O_2$ 103, 174
$Be_2Ca_3[Si_3O_{10}](OH)_2$ 94, 95
$Be_2Ca_4Al_2[Si_3O_{10}][Si_6O_{16}](OH)_2$ 112, 188, 271
$Be_2Ca_2(Ca, Pb)(Fe, Mg)_5[SiO_4]_2[Si_2O_6]_2O_2(OH)_2$ 129, 271
$BeFe_3[Si_3O_9](OH)_2$ 103
$\sim Ca_4Al_4Si_4O_6(OH)_{24} \cdot 3\, H_2O$ 232
$Ca_4[Si_3O_9](OH)_2$ 81, 103
$Ca_4[Si_2O_7](OH, F)_2$ 19
$Ca_6[SiO_4][Si_2O_7](OH)_2$ 71, 270
$Ca_6[Si_6O_{17}](OH)_2$ 28, 81, 105, 140, 186
$Ca_7[Si_8O_{19}]_2(OH)_2$ 232
$Ca_{10}[SiO_4][Si_2O_7]_2(OH)_2Cl_2$ 71, 140, 270
$Ca_{13}[Si_5O_{14}]_2F_8(OH)_2 \cdot 6\, H_2O$ 121, 141, 188
$CaAl_2[AlSiO_5]_2(OH)_2$ 216, 218, 223
$Ca_2Al[AlSi_3O_{10}](OH)_2$ 85, 121, 188
$CaAl_2[Si_2O_7](OH)_2 \cdot H_2O$ 18
$Ca_2Al_3[SiO_4][Si_2O_7]O(OH)$ 129
$Ca_2Al_2Fe[SiO_4][Si_2O_7]O(OH)$ 140, 270
$Ca_{10}Al_4(Fe, Mg)_2[SiO_4]_5[Si_2O_7]_2(OH)_4$ 270
$Ca_2(Al, Fe, Mn)_3[SiO_4][Si_2O_7]O(OH)$ 270
$Ca_2Al_2Sn[Si_6O_{18}](OH)_2 \cdot 2\, H_2O$ 73, 97, 188
$CaB_2[Si_2O_7]O$ 18, 87
$CaB_3Al_6(Al, Li)_3[Si_6O_{18}]O_9(O, OH, F)_4$ 193
$Ca_2Cu_2Al_2[Si_4O_{12}](OH)_6$ 192
$Ca_2(Fe, Al)_3[SiO_4][Si_2O_7]O(OH)$ 129
$CaFe_2(Fe, Mn)[Si_2O_7]O(OH)$ 18
$CaMg_2Al[Al_3SiO_{10}](OH)_2$ 214, 215
$Ca_2(Mg, Fe)_5[Si_4O_{11}]_2(OH)_2$ 71
$Ca_{10}(Mg, Fe)_2Al_4[SiO_4]_5[Si_2O_7]_2(OH)_4$ 129
$Ca_2Mn_7[Si_{10}O_{28}](OH)_2 \cdot 5\, H_2O$ 81, 105, 140, 186
$Ca_3Mn_2[Si_4O_{12}]O_2$ 103, 174
$Ca_2(Mn, Al)_3[SiO_4][Si_2O_7](OH)_3$ 270
$(Ca_3RE)Y_4[Si_8O_{20}][CO_3]_6(OH) \cdot 7\, H_2O$ 141, 185
$(Ca, Sr)Ba_{24}(Mg, Mn, Fe, Al, Ti)_{16}[Si_2O_7]_6[Si_{12}O_{36}](O, OH)_{30}Cl_6 \cdot 14\, H_2O$ 96, 99, 193, 270
$CaV[Si_4O_{10}]O \cdot 4\, H_2O$ 28, 196, 217, 219
$Co_3[Si_2O_5](OH)_4$ 221, 225
$Cu_5[Si_2O_6]_2(OH)_2$ 80, 102, 174
$Fe_3[Si_2O_5](OH)_4$ 221, 225
$FeAl_2[Si_2O_6](OH)_4$ 80, 174
$(Fe^{+2}, Fe^{+3})_{<6}[Si_4O_{10}](OH)_8$ 2, 229
$Fe^{+2}_6Fe^{+3}_3[Si_6O_{17}]O_3(OH)_5$ 73, 104, 188
$(Fe, Mg)_6[AlSi_3O_{10}](O, OH)_8$ 2
$(Fe, Mg)_7[Si_4O_{11}]_2(OH)_2$ 50, 80
$(Fe, Mg)_{17}[Si_4O_{11}]_2[Si_6O_{16}]_2(OH)_6$ 129, 271
$(Fe^{+2}, Mg, Mn, Fe^{+3})_{<9}[Si_9(Al, Si)_3O_{30}](OH)_6X_n \cdot 6\, H_2O$ 229
$H_4Ca_4[(Al_4Si_8)O_{26}](OH)_4 \cdot \sim 6.5\, H_2O$ 126, 189
$HCaMn_4[Si_5O_{15}](OH) \cdot H_2O$ 103, 180
$HMn_4As[Si_3O_{10}]O_3$ 95
$HMn_6V[Si_5O_{16}]O_3$ 94, 95, 165
$HNaK_2Ca_2Ti[Si_7O_{19}]O$ 112, 188
$HNaMg_4[Si_6O_{16}](OH)_2$ 80, 186
$HNa_{1.15}Mn_5[Si_{5.5}V_{0.5}O_{18}](OH)$ 104, 180, 188
$KAl_2[AlSi_3O_{10}](OH)_2$ 115, 116, 196
$KCa_4[Si_4O_{10}]_2(F, OH) \cdot 8\, H_2O$ 73, 115, 141, 196–198, 221, 225–228
$K_2Ca_{10}(Y, RE)_2[Si_2O_7]_2[Si_{12}O_{30}](OH)_2F_2$ 110, 111, 129, 142, 186, 271

$KCo_3[Si_3CoO_{10}](OH)_2$ 214
$KFe_3[AlSi_3O_{10}](OH)_2$ 214
$KFe_{13}[(Al, Si)_9O_{21}]_2(OH)_{14}$ 188, 235
$KFe_{24}[(AlSi_{35})O_{84}]O_6(OH)_{24} \cdot n\,H_2O$ 122, 124, 232, 234
$KFe_3[Si_3CoO_{10}](OH)_2$ 214
$K(Fe, Mg, Mn, Al)_{13}[(Al, Si)_9O_{21}]_2(OH)_{14}$ 141
$(K, H_2O)_3(Mn, Fe^{+3}, Mg, Al)_{<9}[Si_{12}O_{30}](OH)_6X_n \cdot 6\,H_2O$ 229
$(K, Na, Ca, H_2O)(Mn, Mg, Fe^{+2}, Zn, Al)_{9.6}[Si_{12}O_{30}](OH)_{10} \cdot 4\,H_2O$ 229
$(K, Na)_{2.5}(Ca, Na)_7[Si_8O_{19}]_2(OH, F)_2 \cdot H_2O$ 232
$KNi_3[Si_3NiO_{10}](OH)_2$ 214
$K_4Sc_2[Si_4O_{12}](OH)_2$ 98, 192
$(K, Sr, H_2O)_2(Ca, Na, Mg, Fe)_6[(Al_{3.66}Si_{6.34})O_{22}][SO_4]_{1.4}[CO_3]_{0.5}[O_4H_4]_{0.1}$ 86, 124, 231
$K_8Yb_3[Si_6O_{16}]_2OH$ 121, 189
$LiAl_2[AlSiO_5](OH)_4$ 220
$Li_3(K, Na)Ca_7(Ti, Fe, Mn)_2[Si_6O_{18}]_2(OH, F)_2$ 193
$LiNa_2K(Fe, Mg, Mn)_2Ti_2[Si_8O_{22}]O_2$ 128, 144
$Mg_3[Si_2O_5](OH)_4$ 2, 196, 220–226
$Mg_3[Si_2O_5]_2(OH)_2$ 114, 196, 197, 213, 214, 218, 222
$Mg_4[Si_2O_5]_3(OH)_2 \cdot 4\,H_2O$ 2, 116, 219, 222
$Mg_5[Si_2O_5]_4(OH)_2 \cdot 8\,H_2O$ 2, 116, 196, 216, 217, 219, 222
$Mg_5[Si_6O_{16}](OH)_2$ 141
$Mg_{48}[Si_4O_{10}]_{8.5}(OH)_{62}$ 2, 196, 197, 221, 222, 226
$MgAl_2[AlSiO_5]_2(OH)_2$ 218
$Mg_2Al[AlSiO_5](OH)_4$ 220, 223
$Mg_5Ca_2[Si_4O_{11}]_2(OH)_2$ 50, 80, 105, 111, 186
$(Mg, Fe)_5[Si_6O_{16}](OH)_2$ 111, 186
$(Mg, Fe)_7[Si_4O_{11}]_2(OH)_2$ 111, 140
$(Mg, Fe)_{17}[Si_4O_{11}]_2[Si_6O_{16}]_2(OH)_6$ 141, 271
$(Mg, Fe, Al)_4[Si_2O_5]_3(O, OH)_2 \cdot 4\,H_2O$ 196, 216, 217
$(Mg, Fe, Al)_6[(Al, Si)_4O_{10}](OH)_8$ 2
$Mn_5[Si_2O_5]_4(OH)_2 \cdot 8\,H_2O$ 196, 216, 219
$Mn_6[Si_4O_{10}](OH)_8$ 229
$Mn_7[SiO_4]_2[SiO_4](OH)_2$ 93
$Mn_7[Si_6O_{15}](OH)_8$ 196, 221, 227
$Mn_8[Si_6O_{15}](OH, Cl)_{10}$ 115, 196, 221, 225, 227
$MnAl_2[Si_2O_6](OH)_4$ 102, 174
$Mn_2Al_3[SiO_4][Si_2O_7](OH)_3$ 270
$Mn_4Al_6[(As, V)O_4][SiO_4]_2[Si_3O_{10}](OH)_6$ 140, 270
$(Mn, Fe)_8[Si_6O_{15}](OH, Cl)_{10}$ 141, 196, 221, 229
$Mn_6[Zn(OH)_4][Zn_2SiAs_2O_{12}(OH)_2](OH)_2$ 142
$NaB_3Al_6(Al, Li)_3[Si_6O_{18}]O_9(OH)_4$ 193
$Na_2BaTi_2[Si_4O_{12}]O_2$ 81, 103, 140, 174, 185
$Na_4BaTi_2[B_2Si_{10}O_{28}]O_2$ 189, 232
$Na_6Be_2[(Al_2Si_{16})O_{39}](OH)_2 \cdot 1.5\,H_2O$ 28, 126, 128, 144, 189
$(Na, Ca)_3(Ca, Ce)_4(Ti, Nb)[Si_2O_7]_2(O, F)_4$ 19
$(Na, Ca)_2(Ca, Fe, Mn, Ti)(Zr, Nb)[Si_2O_7]OF$ 19
$(Na, Ca)_3(Fe, Al, Mg)_5[(Al, Si)_4O_{11}]_2(OH)_2$ 80
$(Na, Ca)(Mn, Fe, Al, Mg)_{12}[Si_6O_{17}]_2(O, OH)_{10}$ 188
$Na_{12}(Ca, RE)_6(Fe, Mn, Mg)_3(Zr, Nb)_{3+x}[Si_3O_9]_2[Si_9(O, OH)_{27}]_2Cl_y$ 96, 99, 129, 141, 193, 270
$Na_2Ca_4Zr(Nb, Ti)[Si_2O_7]_2O_2F(O, F)$ 19
$Na(Fe^{+2}, Fe^{+3}, Mn, Al, Mg)_{12}[Si_6O_{17}]_2(O, OH)_{10}$ 188
$Na_2Fe_5Ti[Si_6O_{18}]O_2$ 104, 188
$(Na, K)Ca_7[Si_4O_{10}][(AlSi_7)O_{19}](OH)_4 \cdot 3\,H_2O$ 115, 122, 124, 129, 141, 232, 234, 271
$Na_4KCa_4[Si_8O_{18}][CO_3]_4(OH, F) \cdot H_2O$ 124, 141, 189, 197, 232, 235
$Na_4K_2Ca_5[Si_{12}O_{30}](OH, F)_4$ 186
$Na_9KCa_2Ba_6(Mn, Fe)_6(Ti, Nb, Ta)_6B_{12}[Si_{18}O_{54}]_2O_{15}(OH)_2$ 97, 188
$NaK_2Mg_2(Fe, Mn)_5Ti_2[Si_4O_{12}]_2(O, OH, F)_7$ 104, 188, 190
$Na_2Mg_3Al_2[Si_4O_{11}]_2(OH)_2$ 50, 80
$Na_x(Mg, Al)_2[Si_4O_{10}](OH)_2 \cdot 4\,H_2O$ 2
$Na_2(Sr, Ba)_2Ti_3[Si_2O_7]_2(OH, F)_2O_2$ 18
$Na_2Ti_2[Si_2O_6]O_3$ 102, 174
$Na_4Ti_2[Si_8O_{20}]O_2$ 81, 105, 111, 151, 185, 205
$Na_5Zr_2[Si_6O_{18}](Cl, OH) \cdot 2\,H_2O$ 98, 193
$Ni_3[Si_2O_5](OH)_4$ 221, 225
$Pb_8[Si_4O_{12}]O_4$ 98, 192
$Pb_8[Si_4O_{13}][SO_4]O_2$ 94, 95
$Pb_{11}[SiO_4][Si_2O_7]O_6$ 270
$Sr_3(Ti, Fe^{+3})[Si_4O_{12}](O, OH) \cdot 2-3\,H_2O$ 81, 103, 174
$Sr_2V_2[Si_4O_{12}]O_2$ 81, 103, 174, 190
$Y_3[Si_3O_{10}](OH)$ 95
$Zn_4[Si_2O_7](OH)_2 \cdot H_2O$ 18
$Zn_6[Si_{8-x}Zn_xO_{20}](OH)_4$ 214

Pb

$Be_2Ca_2(Ca, Pb)(Fe, Mg)_5[SiO_4]_2[Si_2O_6]_2O_2(OH)_2$ 129, 271
$Ca_2Ba_2Pb_2B_2[(Si_{1.5}Be_{0.5})Si_8O_{28}]F$ 100, 188
$Ca_2Pb[Si_3O_9]$ 98, 192
$Ca_2Pb_3[SiO_4][Si_2O_7]$ 270
$Er_4Pb[Si_2O_7][Si_3O_{10}]$ 270
$MnPb_8[Si_2O_7]_3$ 94
Pb_2SiO_4 98
$Pb[Al_2Si_2O_8]$ (hexacelsian-type) 232, 233, 238, 239

$Pb[SiO_3]$ 140
$Pb_8[Si_4O_{12}]O_4$ 98, 192
$Pb_8[Si_4O_{13}][SO_4]O_2$ 94, 95
$Pb_{11}[SiO_4][Si_2O_7]O_6$ 270
$Pb_{12}[Si_{12}O_{36}]$ 81, 101, 103, 174
$Pb_3Cd_2[SiO_4][Si_2O_7]$ 270
$Pb_4Zn_2[SiO_4][Si_2O_7][SO_4]$ 270

Pr

$NaPr[Si_6O_{14}]$ 119, 121, 189

Rb

$NaRbMg_5[Si_{12}O_{30}]$ 99, 194, 204
$NaRbMg_3Cu_2[Si_{12}O_{30}]$ 99, 194, 204
$NaRbMg_3Fe_2[Si_{12}O_{30}]$ 99, 194, 204
$Rb[AlSiO_4]$ 256, 259, 261, 263, 275
$Rb[AlSi_2O_6]$ 254, 261
$Rb[AlSi_3O_8]$ (feldspar-type) 246, 248, 249, 250, 261, 264
$Rb[AlSi_3O_8]$(hT) 232, 233, 238
$Rb_2[Si_2O_5]$ 200
$Rb_6[Si_{10}O_{23}]$ 126, 142, 144, 189, 205, 243
$Rb_2Be_2[Si_2O_7]$ 28

Rare earths (RE)

$Ba_2RE_2FeTi_2[Si_4O_{12}]_2O_2$ 192
$(Ca_3RE)Y_4[Si_8O_{20}][CO_3]_6(OH) \cdot 7\,H_2O$ 141, 185
$H_xNa_{0-2}(Ca, Na)_8(Fe, Mn, Zn, Ti)RE_2[(Si, Be)_{10}(O, F)_{24}]_2$ 123, 189
$K_2Ca_{10}(Y, RE)_2[Si_2O_7]_2[Si_{12}O_{30}](OH)_2F_2$ 110, 111, 129, 142, 186, 271
$(K, \square)(Ca, RE, Na)_2(RE, Th)[Si_8O_{20}]$ 99, 194
$Na_{12}(Ca, RE)_6(Fe, Mn, Mg)_3(Zr, Nb)_{3+x}[Si_3O_9]_2[Si_9(O, OH)_{27}]_2Cl_y$ 96, 99, 129, 141, 193, 270
$Na_4(Na, Mn)_2(Sr, Ca)_2(Zn, Mg, Fe, Mn)RE_2[Si_{12}O_{34}]$ 104, 188
$Na_{15}RE_3[Si_{12}O_{36}]$ 99, 193

S^{-2}, $[SO_4]^{-2}$

$Ca_3[Si(OH)_6][SO_4][CO_3] \cdot 12\,H_2O$ 31, 69
$(Ca, K)_4[(Al, Si)_5O_{11}][SO_4, CO_3]$ 141
$H_2(Ba, Ca)_{10}[(Al, Si)_{20}O_{43}][SO_4]_3 \cdot H_2O$ 126, 128, 189
$K_{1.7}Ca_6[(Al_{5.7}Si_{4.3})O_{22}][SO_4]_{1.4}[CO_3]_{0.6}$ 86, 124, 231
$(K, Sr, H_2O)_2(Ca, Na, Mg, Fe)_6[(Al_{3.66}Si_{6.34})O_{22}][SO_4]_{1.4}[CO_3]_{0.5}[O_4H_4]_{0.1}$ 86, 124, 231
$Pb_8[Si_4O_{13}][SO_4]O_2$ 94, 95
$Pb_4Zn_2[SiO_4][Si_2O_7][SO_4]$ 270
$Sm_4[Si_2O_7]S_3$ 18

Sc

$HCaSc[Si_3O_9]$ 103, 180
$K_4Sc_2[Si_4O_{12}](OH)_2$ 98, 192
$LiSc[Si_2O_6]$ 102
$(Mg_{38.46}Sc_{3.11})(Li_{1.16}Si_{0.18}Si_{40})O_{124}$ 85
$(Mg_{15.61}Sc_{1.37})(Mg_{0.30}Si_{0.02})[Si_8O_{25}]_2$ 94, 95
$(Mg_{17.40}Sc_{1.49})(Mg_{0.15}Si_{0.11})[Si_9O_{28}]_2$ 94, 95
$(Mg_{19.60}Sc_{1.28})(Mg_{0.04}Si_{0.22})[Si_{10}O_{31}]_2$ 94, 95
$NaSc[Si_2O_6]$ 50, 102
$Na_4Sc_2[Si_4O_{13}]$ 94, 95
$Sc_2Si_2O_7$(hP) (pyrochlore-type) 31, 91, 92
$Sc_2[Si_2O_7]$ 19, 28, 34, 140

Sm

$Sm_2[Si_2O_7]$(lT) 18
$Sm_4[Si_2O_7]S_3$ 18

Sn

$BaSn[Si_3O_9]$ 192
$Ca_2Al_2Sn[Si_6O_{18}](OH)_2 \cdot 2\,H_2O$ 73, 97, 188
$Ca_2Sn_2[Si_6O_{18}] \cdot 4\,H_2O$ 81, 101, 103, 140, 174
$Li_3KSn_2[Si_{12}O_{30}]$ 99, 194, 204
$Li_2Na_2Sn_2[Si_{12}O_{30}]$ 185, 205
$Na_4Be_2Sn[Si_3O_9]_2 \cdot 2\,H_2O$ 79, 81, 103, 174, 176
$Na_4Sn_2[Si_5O_{16}] \cdot H_2O$ 94, 95
$Na_8Sn[Si_6O_{18}]$ 98, 193

Sr

$(Ba, Sr)(Mn, Fe)_2(Fe, Ti)[Si_2O_7](O, OH)_2$ 18
$(Ca, Sr)Ba_{24}(Mg, Mn, Fe, Al, Ti)_{16}[Si_2O_7]_6[Si_{12}O_{36}](O, OH)_{30}Cl_6 \cdot 14\,H_2O$ 96, 99, 193, 270
$K_{16}Sr_4[Si_{12}O_{36}]$ 99, 193
$(K, Sr, H_2O)_2(Ca, Na, Mg, Fe)_6[(Al_{3.66}Si_{6.34})O_{22}][SO_4]_{1.4}[CO_3]_{0.5}[O_4H_4]_{0.1}$ 86, 124, 231
$Na_4(Na, Mn)_2(Sr, Ca)_2(Zn, Mg, Fe, Mn)RE_2[Si_{12}O_{34}]$ 104, 188
$Na_2(Sr, Ba)_2Ti_3[Si_2O_7]_2(OH, F)_2O_2$ 18
$Sr[Al_2Si_2O_8]$ (feldspar-type) 251, 252, 264
$Sr[Al_2Si_2O_8]$(hT) (hexacelsian-type) 124, 232, 239, 251, 252, 264
$Sr[Al_2Si_2O_8]$ (paracelsian-type) 251, 252, 264
$Sr[SiO_3]$ 98, 172, 175
$Sr[Si_2O_5]$ 202
$Sr_3[Si_3O_9]$ (α) 98, 192
$SrCu[Si_4O_{10}]$ 196, 203, 205
$Sr_3(Ti, Fe^{+3})[Si_4O_{12}](O, OH) \cdot 2-3\,H_2O$ 81, 103, 174
$Sr_2V_2[Si_4O_{12}]O_2$ 81, 103, 174, 190

Ta

$Na_9KCa_2Ba_6(Mn, Fe)_6(Ti, Nb, Ta)_6B_{12}[Si_{18}O_{54}]_2O_{15}(OH)_2$ 97, 188

Th

$Ca_2Th[Si_4O_{10}]_2$ 196, 203, 205
(K, □)(Ca, RE, Na)$_2$(RE, Th)$[Si_8O_{20}]$ 99, 194
$K_{1-x}(Na, Ca)_{2-y}Th_{1-z}[Si_8O_{20}]$ 97, 99, 141, 194, 204

Ti

$Ba_4B_2(Fe, Ti)_4[Si_4O_{12}]_2O_5Cl_x$ 96, 98, 192
$Ba_4B_2(V, Ti)_4[Si_4O_{12}]_2O_3(O, OH)_2Cl$ 98, 192
$Ba_{10}(Ca, Mn, Ti)_4[Si_8O_{24}](Cl, O, OH)_{12} \cdot 4\,H_2O$ 96, 99, 140, 193
$Ba_{12}(Mn, Ti, Fe)_6[Si_4O_{12}]_3(O, OH)_2(OH, H_2O)_7Cl_9$ 98
$Ba_2RE_2FeTi_2[Si_4O_{12}]_2O_2$ 192
$(Ba, Sr)(Mn, Fe)_2(Fe, Ti)[Si_2O_7](O, OH)_2$ 18
$BaTi[Si_3O_9]$ 96, 98, 140, 192
$Ba_4(Ti, Nb)_8[Si_4O_{12}]O_{16}Cl$ 98, 140, 192
$\sim CaMg_2Ti[Al_2SiO_{10}]$ 2
$(Ca, Sr)Ba_{24}(Mg, Mn, Fe, Al, Ti)_{16}[Si_2O_7]_6[Si_{12}O_{36}](O, OH)_{30}Cl_6 \cdot 14\,H_2O$ 96, 99, 193, 270
$H_xNa_{0-2}(Ca, Na)_8(Fe, Mn, Zn, Ti)RE_2[(Si, Be)_{10}(O, F)_{24}]_2$ 123, 189
$HNaK_2Ca_2Ti[Si_7O_{19}]O$ 112, 188
$K_2Ti[Si_3O_9]$ 192
$K_2Ti[Si_6O_{15}]$ 196
$Li_3KCa_7Ti_2[Si_6O_{18}]_2F_2$ 98
$Li_3(K, Na)Ca_7(Ti, Fe, Mn)_2[Si_6O_{18}]_2(OH, F)_2$ 193
$LiNa_2K(Fe, Mg, Mn)_2Ti_2[Si_8O_{22}]O_2$ 128, 144
$Li_2Na_2Ti_2[Si_{12}O_{30}]$ 185, 205
$Na_2BaTi_2[Si_4O_{12}]O_2$ 81, 103, 140, 174, 185
$Na_4BaTi_2[B_2Si_{10}O_{28}]O_2$ 189, 232
$(Na, Ca)_3(Ca, Ce)_4(Ti, Nb)[Si_2O_7]_2(O, F)_4$ 19
$(Na, Ca)_2(Ca, Fe, Mn, Ti)(Zr, Nb)[Si_2O_7]OF$ 19
$Na_2Ca_4Zr(Nb, Ti)[Si_2O_7]_2O_2F(O, F)$ 19
$Na_2Fe_5Ti[Si_6O_{18}]O_2$ 104, 188
$Na_9KCa_2Ba_6(Mn, Fe)_6(Ti, Nb, Ta)_6B_{12}[Si_{18}O_{54}]_2O_{15}(OH)_2$ 97, 188
$NaK_2Mg_2(Fe, Mn)_5Ti_2[Si_4O_{12}]_2(O, OH, F)_7$ 104, 188, 190
$Na_6MnTi[Si_6O_{18}]$ 98, 193
$Na_2(Sr, Ba)_2Ti_3[Si_2O_7]_2(OH, F)_2O_2$ 18
$Na_2Ti_2[Si_2O_6]O_3$ 102, 174
$Na_4Ti_2[Si_8O_{20}]O_2$ 81, 105, 111, 151, 185, 205
$Sr_3(Ti, Fe^{+3})[Si_4O_{12}](O, OH) \cdot 2-3\,H_2O$ 81, 103, 174

V

$Ba_4B_2(V, Ti)_4[Si_4O_{12}]_2O_3(O, OH)_2Cl$ 98, 192
$Ba_2V_2[Si_4O_{12}]O_2$ 103, 174
$CaV[Si_4O_{10}]O \cdot 4\,H_2O$ 28, 196, 217, 219
$HMn_6V[Si_5O_{16}]O_3$ 94, 95, 165
$HNa_{1.15}Mn_5[Si_{5.5}V_{0.5}O_{18}](OH)$ 104, 180, 188
$Mn_4Al_6[(As, V)O_4][SiO_4]_2[Si_3O_{10}](OH)_6$ 140, 270
$NaV[Si_2O_6]$ 50
$Sr_2V_2[Si_4O_{12}]O_2$ 81, 103, 174, 190

Y

$BaY_4[Si_2O_7][Si_3O_{10}]$ 270
$(Ca_3RE)Y_4[Si_8O_{20}][CO_3]_6(OH) \cdot 7\,H_2O$ 141, 185
$Ca_2Y_2[Si_4O_{12}][CO_3] \cdot H_2O$ 192
$H_2K_3Y[Si_3O_{10}]$ 95
$K_2Ca_{10}(Y, RE)_2[Si_2O_7]_2[Si_{12}O_{30}](OH)_2F_2$ 110, 111, 129, 142, 196, 271
$Li_2Na_4Y_2[Si_{12}O_{30}]$ 185, 205
$(Na, K)_3Y[Si_8O_{19}] \cdot 5\,H_2O$ 232
$Na_3Y[SiO_3]_3$ 140
$Na_{15}Y_3[Si_{12}O_{36}]$ 140
$Na_{24}Y_8[Si_{24}O_{72}]$ 79, 81, 101, 103, 140, 171, 174
$Y_3[Si_3O_{10}](OH)$ 95

Yb

$K_8Yb_3[Si_6O_{16}]_2OH$ 121, 189
$Yb_2[Si_2O_7]$ 19, 28

Zn

$H_xNa_{0-2}(Ca, Na)_8(Fe, Mn, Zn, Ti)RE_2[(Si, Be)_{10}(O, F)_{24}]_2$ 123, 189
$H_2[Zn_4Si_2O_9] \cdot H_2O$ 142
$K_2Be_2Zn_2[SiO_4][Si_2O_7]$ 270
$K_2Mg_3Zn_2[Si_{12}O_{30}]$ 99, 194, 204
$K_2Mn_2Zn_4[SiO_4]_2[Si_2O_7]$ 270
$(K, Na, Ca, H_2O)(Mn, Mg, Fe^{+2}, Zn, Al)_{9.6}[Si_{12}O_{30}](OH)_{10} \cdot 4\,H_2O$ 229
$MgZn[Si_2O_6]$ 102
$Mn_6[Zn(OH)_4][Zn_2SiAs_2O_{12}(OH)_2](OH)_2$ 142
$Na_2Mg_3Zn_2[Si_{12}O_{30}]$ 99, 194, 204
$Na_4(Na, Mn)_2(Sr, Ca)_2(Zn, Mg, Fe, Mn)RE_2[Si_{12}O_{34}]$ 104, 188
$Na_2Zn[Si_2O_6]$ 102, 174
$Na_2Zn[Si_3O_8]$ 119, 152, 166, 196
$Pb_4Zn_2[SiO_4][Si_2O_7][SO_4]$ 270
$ZnSiO_3$(hP) (ilmenite-type) 69, 91, 93
$Zn_2[SiO_4]$ 255
$Zn_2[Si_2O_6]$ 102
$Zn_4[Si_2O_7](OH)_2 \cdot H_2O$ 18
$Zn_6[Si_{8-x}Zn_xO_{20}](OH)_4$ 214

Zr

$BaZr[Si_3O_9]$ 192
$CaZr[Si_6O_{15}] \cdot 2.5\,H_2O$ 118, 196, 217, 219
$K_2Zr[Si_3O_9]$ 98, 192
$K_2Zr[Si_3O_9] \cdot H_2O$ 103, 176
$K_2Zr[Si_6O_{15}]$ 115, 118, 141, 196, 201, 205

$Li_2Na_2Zr_2[Si_{12}O_{30}]$ 81, 185, 205
$Li_4Zr_2[Si_{12}O_{30}]$ 185, 205
$Na_{12}(Ca, RE)_6(Fe, Mn, Mg)_3(Zr, Nb)_{3+x}[Si_3O_9]_2[Si_9(O, OH)_{27}]_2Cl_y$ 96, 99, 129, 141, 193, 270
$Na_6CaZr[Si_6O_{18}]$ 98, 192
$(Na, Ca)_2(Ca, Fe, Mn, Ti)(Zr, Nb)[Si_2O_7]OF$ 19
$Na_2Ca_4Zr(Nb, Ti)[Si_2O_7]_2O_2F(O, F)$ 19
$(Na, K)_2CaZr_2[Si_5O_{13}]_2 \cdot 5-6\,H_2O$ 189
$Na_2K_2Zr_2[Si_6O_{18}] \cdot 4\,H_2O$ 103, 174
$Na_2Zr[Si_3O_9] \cdot 2\,H_2O$ 98, 192
$Na_2Zr[Si_3O_9] \cdot 3\,H_2O$ 103, 176, 192
$Na_2Zr[Si_6O_{15}] \cdot 3\,H_2O$ 81, 141, 186
$Na_4Zr_2[Si_8O_{22}]$ 104, 140, 188, 190
$Na_4Zr_2[Si_6O_{18}] \cdot 4\,H_2O$ 81, 103, 174
$Na_5Zr_2[Si_6O_{18}](Cl, OH) \cdot 2\,H_2O$ 98, 193
$Zr[SiO_4]$ 2, 34, 140

Organic

$C_8H_6O_2BrF_3Si$ 40
$C_8H_{17}NO_2Si$ 40
$C_{12}H_8O_4Si$ 38
$C_{12}H_{16}N_2O_5Si$ 40
$C_{12}H_{17}NO_3Si$ 40
$C_{18}H_{22}N_2O_2Si_2$ 40
$C_{22}H_{28}N_4O_6CuSi$ 91
$C_{22}H_{25}NO_4Si$ 40
$C_{24}H_{17}NO_3Si$ 40
$C_{28}H_{24}N_2O_6Si$ 31, 91
$C_{30}H_{90}O_{17}Si_{16}$ 188
$C_{30}H_{90}O_{25}Si_{20}$ 97, 99, 194
$C_{32}H_{96}N_8[Si_8O_{20}] \cdot 64.8\,H_2O$ 99
$C_{36}H_{30}OSi_2$ 28
$[Cu(H_2N \cdot CH_2 \cdot CH_2 \cdot NH_2)_2]_4[Si_8O_{20}] \cdot 38\,H_2O$ 99, 194
$[Cu(H_2N \cdot CH_2 \cdot CH_2 \cdot NH_2)_2][(C_6H_4O_2)_3Si]$ 91
$H_2[Co(H_2N \cdot CH_2 \cdot CH_2 \cdot NH_2)_3]_2[Si_8O_{20}] \cdot 16.4\,H_2O$ 99, 194
$H_7[N(n-C_4H_9)_4][Si_8O_{20}] \cdot 5.33\,H_2O$ 99, 194
$[N(CH_3)_4]_8[Si_8O_{20}] \cdot 64.8\,H_2O$ 168, 194
$[Ni(H_2N \cdot CH_2 \cdot CH_2 \cdot NH_2)_3]_3[Si_6O_{15}] \cdot 26\,H_2O$ 97, 99, 141, 194
$X_3Si-O-SiX_3$ 51
$96\,SiO_2 \cdot 4[N(C_3H_7)_4]F$ 156

SiO_2 frameworks

SiO_2 2, 21–29, 34, 36, 47, 93, 128, 129, 142, 156, 161, 225, 240, 260–264, 274, 275
$SiO_2(hP)$ 28–36, 47, 91, 92, 156
$34\,SiO_2 \cdot 3\,M^{12} \cdot 2\,M^{12\prime} \cdot M^{20}$ 156, 241
$46\,SiO_2 \cdot 2\,M^{12} \cdot 6\,M^{14}$ 128, 142, 156, 241
$96\,SiO_2 \cdot 4[N(C_3H_7)_4]F$ 156
$136\,SiO_2 \cdot 16\,M^{12} \cdot 8\,M^{16}$ 241

General formula

$A_xM_2M'_3[Si_{12}O_{30}]$ (milarite-type) 99, 194
$Ba_{M+1}[Si_{2M}O_{5M+1}]$ 206
$M[AlSiO_4]$ 235, 236, 255–256, 274, 275
$M[AlSi_2O_6]$ 253–255, 260–265
$M[AlSi_3O_8]$ 246, 248–252, 260–265
$M[Al_2Si_2O_8]$ 251–253, 263, 264
$M[(Al, Si)_4O_8]$ 146, 148, 152, 246–253
$(M^+, M^{+2})[(Al, Si)_4O_8]$ 2
$M_2^+(M^{+2}, M^{+3})_5[(Al, Si)_{12}O_{30}]$ 2
$(M^{+2}, M^{+3})_2[(Al, Si)_2O_6]$ 2
$M_2[SiO_4]$ 49
$M^{+2}[Si_2O_5]$ 200–202
$M_2^+[Si_2O_5]$ 200
$M_2[Si_2O_6]$ 2
$M_2^{+3}[Si_2O_5]_3$ 202
$M_r\{\boldsymbol{B}, \boldsymbol{M}_\infty^D\}[Si_xO_y]$ 73
$MM'[Si_2O_6]$ (pyroxenes) 49
$M_2M'_5[Si_4O_{11}]_2(OH)_2$ (amphiboles) 49
$M_3^{+2}M_2^{+3}[SiO_4]_3$ 2
$M_3^{[12]}M_2^{[6]}[SiO_4]_3$ (garnets) 49
$M[T_4O_8]$ 12
$M_{16}Ca_4[Si_{12}O_{36}]$ 99
$34\,SiO_2 \cdot 3\,M^{12} \cdot 2\,M^{12\prime} \cdot M^{20}$ 156, 241
$46\,SiO_2 \cdot 2\,M^{12} \cdot 6\,M^{14}$ 128, 142, 156, 241
$136\,SiO_2 \cdot 16\,M^{12} \cdot 8\,M^{16}$ 241
$XY_3Z_6B_3[Si_6O_{18}]O_9(O, OH, F)_4$ 98

Non-silicates

$Cd(CN)_2$ 129
$Cs[Be_2F_5]$ 199, 201
$CuGeO_3$ 104
H_2O 129
$HLiCd_4[Ge_5O_{15}]$ 180
$HNaCd_4[Ge_5O_{15}]$ 180
$HNa_2[P_3O_9]$ 180
$K_2Ba_4[P_{10}O_{30}]$ 104
$M_m[PO_3]_n$ 181
$Na_3[P_3O_9]$ 180
$NaBePO_4$ 257
$(NH_4)_2Si[P_4O_{13}]$ 31, 91
$Rb[Be_2F_5]$ 201
Si_2N_2O 2
SiO 1, 2
$Si[P_2O_7]$ 31, 34, 51, 69, 91
$Si_5[PO_4]_6O$ 28, 31, 91, 92, 129
SiS 1, 2
$Sr[ZnO_2]$ 118
$Tl[Be_2F_5]$ 201
$Zn(CN)_2$ 129

O. P. Mchedlov-Petrossyan, V. I. Babushkin, G. M. Matveyev

Thermodynamics of Silicates

Editor in Chief: **O. P. Mchedlov-Petrossyan**
Translated from the Russian by B. N. Frenkel, V. A. Terentyev

1984. 93 figures, 135 tables. Approx. 475 pages
ISBN 3-540-12750-X

Contents: The Basic Concepts and Laws of Thermodynamics. - Pyrosilicate Reactions. - Hydration Reactions. - Corrosion Reactions. - References. - Appendix 1: Standard Enthalpies, Gibbs Free Energies and Heat Capacity Equations of Some Elements, Simple Substances and Compounds. - Appendix 2: Thermodynamic Properties of Ions and Molecules in Aqueous Solutions. - References to Appendices 1 and 2. - Appendix 3: Heats of Polymorphic Transitions and Melting. - References to Appendix 3. - Subject Index.

The thermodynamic method is of great importance for studying chemical reactions of silicate technology. Together with a study of the rate and mechanism of substance transfer, it yields data necessary for the efficient operation of technological processes.
In this book, a translation for the fourth edition, all sections of the book dealing with the practical application of thermodynamics to silicate and concrete technology have been substantially revised. All the calculations have been carried out with refined interrelated thermodynamic data. The book comprises two parts: the theoretical, in which general thermodynamic ideas are stated, experimental and approximate methods of obtaining the necessary initial thermodynamic data are discussed and the expediency and limits of the application of thermodynamics to the study of silicates are grounded; and the applied, in which thermodynamic calculation of reactions of theoretical and practical importance in silicate systems are carried out and their theoretical interpretation is given.

Springer-Verlag
Berlin
Heidelberg
New York
Tokyo